AUTOPHAGY IN CURRENT TRENDS IN CELLULAR PHYSIOLOGY AND PATHOLOGY

Edited by **Nikolai V. Gorbunov**
and **Marion Schneider**

Autophagy in Current Trends in Cellular Physiology and Pathology

http://dx.doi.org/10.5772/61911

Edited by Nikolai V. Gorbunov and Marion Schneider

Contributors

Gary Warnes, Benjamin Loos, Claudia Ntsapi, Chrisna Swart, Dumisile Lumkwana, Daolin Tang, Rui Kang, Paul Sylvester, Roshan Tiwari, Mauricio S. Baptista, Waleska K. Martins, Dexian Zheng, Zhenyu Yao, Blanca-Estela García-Pérez, Kahiry Leyva-Paredes, Nayeli Shantal Castrejon-Jimenez, Hugo Ivan Arrieta-Oliva, Shantal Lizbeth Baltierra-Uribe, Ligen Li, Mengjing Xiao, Amer Amal, Estelle Cormet-Boyaka, Vincent Kam Wai Wong, Betty Yuen Kwan Law, Damla Arisan, Pinar Obakan-Yerlikaya, Ajda Coker-Gurkan, Narcin Palavan-Unsal, Shilpa Buch, Palsamy Periyasamy, Ming-Lei Guo, Elite Possik, Arnim Pause, Fu-Cheng Lin, Wilhelmus Kwanten, Wim Martinet, Sven Francque, Guillermo Blanco, Laura Kornblihtt, María Carreras, Guang-Chao Chen, Yi-Ting Wang, Cornelia Wilson, Michael P. A. Davies, Michael Boulton, Sayak Mitter, Kuang-Den Chen, Chih-Che Lin, Chih-Wen Lin, Ming-Chao Tsai, Kuang-Tzu Huang, Chao-Long Chen, Heesun Cheong, Akitoshi Nakashima, Aiko Aoki, Shigeru Saito, Nikolai Gorbunov

Published by InTech

Janeza Trdine 9, 51000 Rijeka, Croatia

Publishing Process Manager Andrea Koric

Technical Editor SPi Global

Cover InTech Design team

First published October, 2016

Printed in Croatia

Additional hard copies can be obtained from orders@intechopen.com

Autophagy in Current Trends in Cellular Physiology and Pathology, Edited by Nikolai V. Gorbunov and Marion Schneider

p. cm.

Print ISBN 978-953-51-2726-0

Online ISBN 978-953-51-2727-7

[33] Guo, J.Y., et al., Autophagy suppresses progression of K-ras-induced lung tumors to oncocytomas and maintains lipid homeostasis. Genes & Devel, 2013. 27(13): pp. 1447–1461.

[34] Rao, S., et al., A dual role for autophagy in a murine model of lung cancer. Nat Commun, 2014. 5: p. 3056.

[35] Viale, A., et al., Oncogene ablation-resistant pancreatic cancer cells depend on mitochondrial function. Nature, 2014.514(7524):pp.628–32.

[36] Yang, Z., J.J. Goronzy, and C.M. Weyand, The glycolytic enzyme PFKFB3/phosphofructokinase regulates autophagy. Autophagy, 2014. 10(2): pp. 382–3.

[37] Roberts, D.J., et al., Hexokinase-II positively regulates glucose starvation-induced autophagy through TORC1 inhibition. Mol Cell, 2014. 53(4): pp. 521–33.

[38] Klarer, A.C., et al., Inhibition of 6-phosphofructo-2-kinase (PFKFB3) induces autophagy as a survival mechanism. Cancer Metab, 2014. 2(1): p. 2.

[39] Roberts, D.J., et al., Akt phosphorylates HK-II at Thr-473 and increases mitochondrial HK-II association to protect cardiomyocytes. J Biol Chem, 2013. 288(33): pp. 23798–806.

[40] Strohecker, A.M., et al., Identification of 6-phosphofructo-2-kinase/fructose-2,6-bisphosphatase as a novel autophagy regulator by high content shRNA screening. Oncogene, 2015. 34(45): pp. 5662–76.

[41] Mason, E.F., et al., Aerobic glycolysis suppresses p53 activity to provide selective protection from apoptosis upon loss of growth signals or inhibition of BCR-Abl. Cancer Res, 2010. 70(20): pp. 8066–76.

[42] Xia, H.G., et al., Degradation of HK2 by chaperone-mediated autophagy promotes metabolic catastrophe and cell death. J Cell Biol, 2015. 210(5): pp. 705–16.

[43] Macintyre, A.N. and J.C. Rathmell, PKM2 and the tricky balance of growth and energy in cancer. Mol Cell, 2011. 42(6): pp. 713–4.

[44] Lv, L., et al., Acetylation targets the M2 isoform of pyruvate kinase for degradation through chaperone-mediated autophagy and promotes tumor growth. Mol Cell, 2011. 42(6): pp. 719–30.

[45] Anastasiou, D., et al., Pyruvate kinase M2 activators promote tetramer formation and suppress tumorigenesis. Nat Chem Biol, 2012. 8(10): pp. 839–47.

[46] Marino, G., et al., Regulation of autophagy by cytosolic acetyl-coenzyme A. Mol Cell, 2014. 53(5): pp. 710–25.

[47] Vander Heiden, M.G., L.C. Cantley, and C.B. Thompson, Understanding the Warburg effect: the metabolic requirements of cell proliferation. Science, 2009. 324(5930): pp. 1029–33.

[48] Wise, D.R. and C.B. Thompson, Glutamine addiction: a new therapeutic target in cancer. Trends Biochem Sci, 2010. 35(8): pp. 427–33.

[49] Rajagopalan, K.N. and R.J. DeBerardinis, Role of glutamine in cancer: therapeutic and imaging implications. J Nucl Med, 2011. 52(7): pp. 1005–8.

[50] Son, J., et al., Glutamine supports pancreatic cancer growth through a KRAS-regulated metabolic pathway. Nature, 2013. 496(7443): pp. 101–5.

[51] Nicklin, P., et al., Bidirectional transport of amino acids regulates mTOR and autophagy. Cell, 2009. 136(3): p. 521.

[52] Duran, R.V., et al., Glutaminolysis activates Rag-mTORC1 signaling. Mol Cell, 2012. 47(3): pp. 349–58.

[53] Lorin, S., et al., Glutamate dehydrogenase contributes to leucine sensing in the regulation of autophagy. Autophagy, 2013. 9(6): pp. 850–60.

[54] Eng, C.H., et al., Ammonia derived from glutaminolysis is a diffusible regulator of autophagy. Sci Signal, 2010. 3(119): p. ra31.

[55] Cheong, H., et al., Ammonia-induced autophagy is independent of ULK1/ULK2 kinases. Proc Natl Acad Sci U S A, 2011. 108(27): pp. 11121–6.

[56] Lin, T.C., et al., Autophagy: resetting glutamine-dependent metabolism and oxygen consumption. Autophagy, 2012. 8(10): pp. 1477–93.

[57] Chen, R., et al., The general amino acid control pathway regulates mTOR and autophagy during serum/glutamine starvation. J Cell Biol, 2014. 206(2): pp. 173–82.

[58] Tanaka, K., et al., Compensatory glutamine metabolism promotes glioblastoma resistance to mTOR inhibitor treatment. J Clin Invest, 2015. 125(4): pp. 1591–602.

[59] Bar-Sagi, D. and J.R. Feramisco, Induction of membrane ruffling and fluid-phase pinocytosis in quiescent fibroblasts by ras proteins. Science, 1986. 233(4768): pp. 1061–8.

[60] Mercer, J. and A. Helenius, Virus entry by macropinocytosis. Nat Cell Biol, 2009. 11(5): pp. 510–20.

[61] Commisso, C., et al., Macropinocytosis of protein is an amino acid supply route in Ras-transformed cells. Nature, 2013. 497(7451): pp. 633–7.

[62] Palm, W., et al., The utilization of extracellular proteins as nutrients is suppressed by mTORC1. Cell, 2015. 162(2): pp. 259–70.

[63] Sung, S., J. Choi, and H. Cheong, Catabolic pathways regulated by mTORC1 are pivotal for survival and growth of cancer cells expressing mutant Ras. Oncotarget, 2015. 6(38): pp. 40405–17.

For EU product safety concerns:
IN TECH d.o.o., Prolaz Marije Krucifikse Kozulić 3, 51000 Rijeka, Croatia,
info@intechopen.com or visit our website at intechopen.com.

Meet the editors

Dr. Nikolai V. Gorbunov received his PhD degree in biology from the Russian Academy Sciences and then he was awarded by NRC NAS (http://sites.nationalacademies.org/pga/rap/) and the Department of Energy to pursue postdoctoral training in translational science at the University of Pittsburgh and the Pacific Northwest National Laboratory (https://www.emsl.pnl.gov/emslweb; Washington, USA). His translational research area has encompassed molecular pathology of trauma and countermeasures against acute radiation injury that he explored at the Walter Reed Army Institute of Research (http://wrair-www.army.mil) and the Uniformed Services University of the Health Sciences (https://www.usuhs.edu). *In this* framework, the focus of interest was on interplay of the etiological factors leading to injury and the mechanisms driving resilience, defense, and morphogenic responses in organelles, cells, and tissues arranging biological barriers. The research objectives are (i) to define the key structural components and pathways, which regulate and sustain intrinsic resistance, recovery, or remodeling of cells and systems following harmful exposures and damage and (ii) to employ the acquired knowledge for further development of new *injury-specific treatment* models.

Marion Schneider studied biology and medicine at Bonn University, Germany. Her postdoctoral fellowship was on T-cell deficiencies and stem cell transplantation at the University of Tubingen and HIV1 infection in macrophages and myeloid cells at the Institut Pasteur in Paris (1981–1985), where for the first time she got interested in vacuolization and prolonged viability as well as persistence of macrophages even under conditions of virus infections. Her next topics were hemophagocytic diseases (hemophagocytic lymphohistiocytosis, HLH) and macrophage activation syndromes (MAS) related to immune dysfunction and chronic virus infections as well as severe sepsis and septic shock. When taking the professorship for Experimental Anesthesiology at Ulm University, Ulm, Germany, in 1998, she concentrated on biomarker analysis combining soluble and membrane-bound characteristics of major inflammatory diseases related to inflammasome activation. Inflammation appears to be a major risk factor for sepsis (systemic inflammation) and also for tumor manifestation. In collaboration with the Department of Neurosurgery in the Ulm University, an intense analysis of autophagy has been initiated to overcome the upregulation of autophagy in type IV glioblastoma with novel chemotherapeutic agents.

Contents

Preface

What this book is about?

This book is a multinational effort of dozens of contributors with expertise in translational research and thus reflects growing global research interests in autophagy biology and medicine. The intent of the book is to present an overview of diverse aspects of biogenesis of autophagy and autophagy-mediated barriers in cell and tissue development and aging, its pro-survival effects in injury and degenerative diseases, and its arrangement of intrinsic resistance to stress and harmful impacts. Along with this, a particular attention is paid to controversial yet detrimental implication of autophagy in cancerogenesis and a divergent role of autophagy in cell death, chronic inflammation and infectious diseases.

Focusing on autophagy function in both tissue's normal development and pathology, a large volume of the book discussions is dealing with investigation of (i) a variety of active intrinsic barrier functions mediated by autophagy; (ii) autophagy-controlled cellular biogenesis, metabolome, and proteostasis; and (iii) interplay of key pathways, hub proteins, and organelles featuring autophagy interactome in order to execute selective degradation of cell constituents and xenobiotics that is termed as "autophagic flux." In this account, investigators who are working on development of new remedies for therapy of stress, trauma-related injury, degenerative diseases, and cancer could avail themselves an in-depth understanding of the autophagy biology, the mechanisms of autophagy biogenesis, and the regulation of autophagic flux. Giving a wide-angle perspective on biomedical aspects of autophagy, we addressed this book to a broad audience of readers from students to practicing clinicians and experts in autophagy research.

Autophagy and the necessity for barrier formation and metabolic control

Autophagy is a catabolic pathway of lysosomal recycling of cell constituents and xenobiotics by the mechanisms of sequestration of targeted biomolecules and compromised organelles within isolating membranes, that is, *autophagosomes*, or protein complexes and processing them in lysosomal machinery. By this means autophagy mediates cell and organelle biogenesis, provides energy supplies, and sustains system integrity and homeostasis. Autophagy is evolutionarily conserved and ubiquitously present in eukaryotes, for example, multicellular organisms such as fungi, *plants* and *animals*. Therefore, many details related to autophagy signaling, target selection, autophagic flux, as well as autophagy-targeted modulation of cell and tissue biogenesis and morphogenesis have been recently investigated using in vitro models, vertebrates, and lower vertebrate animals. Note that these aspects of autophagy translational research are elegantly presented in several chapters of this book and discussed in conjunction with developments of new models for therapy of cancer and degenerative diseases.

Although autolysosomal process per se is still considered to be a bulk hydrolytic degradation, a growing number of evidence indicate that autophagy biogenesis is strictly regulated by autophagy-related genes (i.e., *ATG* genes) and their protein products, numerous signaling cascades, adaptors, chaperones, modifiers, cell energetic conditions, and intimate interactions in the organelle networks (e.g., the endoplasmic reticulum-mitochondrial interplay). Overall, that determines cargo-selectivity to proteins and organelles as well as the pathway specificity (e.g., macroautophagy vs. chaperone-mediated autophagy). This network governs a crucial autophagy feature, which is the execution of barrier functions by targeting, sequestration and compartmentalization for recycling of damaged cytotoxic constituents, acquired xenobiotics and invading pathogens. These barrier functions "arm" organisms with the ability to specifically respond to starvation and oxidative, electrophilic, and hypoxic stress related to acute injury and hyperinflammation as well as to mediate innate and adaptive immunity and to control aging, infections, degenerative disease, cancer development, etc. Remarkably, the interplay between autophagy biogenesis and the endoplasmic reticulum-mitochondrial axis, cell metabolome, proteostasis, and energetic machinery defines the capacity of cell intrinsic resistance to stress impacts and impairments. These evidences can imply novel concepts for therapy of numerous illnesses such as cystic fibrosis, preeclampsia, drug addiction-related disease, and dysfunctions of the heart, lung, and nervous tissue.

Autophagy subjects addressed

In conjunction with the above, the scientific reviews and research data presented in the chapters of this book are focused on several interconnected autophagy areas such as (i) autophagy signaling pathways, mediators, and autophagy interactome; (ii) autophagy and biological barriers such as defense, offense, survival, or death; (iii) autophagy in tissue morphogenesis, remodeling, and regeneration; (iv) selective autophagy in health, disease, and therapy; and (v) autophagy in metabolic homeostasis.

The selected topics encompass personal experience and visions of the chapter contributors and editors as well as a broad analysis of literature on biology of autophagy.

Dr. Nikolai V. Gorbunov
Department of Anaesthesiology,
Uniformed Services University of the Health Sciences,
Bethesda, Maryland, USA

Prof. Dr. Marion Schneider
Head of Division of Experimental Anesthesiology
Department of Anaesthesiology, University Hospital Ulm
Ulm, Germany

Section 1

Introduction

Chapter 1

Introductory Chapter: Overview on Autophagy in Burden of Functions

Nikolai V. Gorbunov and E. Marion Schneider

Additional information is available at the end of the chapter

http://dx.doi.org/10.5772/65435

> "The Nobel Assembly at Karolinska Institutet has today decided to award the 2016 Nobel Prize in Physiology or Medicine to Yoshinori Ohsumi for his discoveries of mechanisms for autophagy. 2016-10-03".
> [q]:www.nobelprize.org/nobel_prizes/medicine/laureates/2016/press.pdf
> This Nobel Assembly announcement culminates enormous efforts devoted over past decades to elucidate the molecular network of autophagy – fundamental "self-eating" machinery for decomposition and remodeling of cellular components - and thus, closes another chapter in the History of Cell Physiology.

1. Introduction to Autophagy: "Self-Eating" in the Pursuit of Cell Health and "Happiness"

Multicellular organisms evolved, adopted and conserved numerous mechanisms and pathways, which allow them to sustain essential metabolism and morphogenesis over life period and to proceed through aging. Remarkably, often the same mechanisms orchestrate response to impacts of environmental and oxidative stressogens, toxins and adverse conditions caused by a variety of infections, and degenerative and malignant transformations. Among these mechanisms, a crucial role is borne by autophagy-lysosomal catabolicmachinery or autophagy, which is constituted by a large interactome of autophagy-related genes and organelle networks, that is integrated within a distinct singularity along with newly introduced "omes" such as "exposome" and "stressome" [1–5]. The normal "housekeeping task burden" of the autophagy system (i.e., chaperone-mediated autophagy, microautophagy and the nonselective and selective macroautophagy) is to sustain, adjust, reconstitute and to remodel the cellular contents in order to maintain cell metabolism, phenotypes and structural integrity, especially in the case of starvation when energy molecules are depleted, nutrition is limited, and thus death risk is high [2, 5]. In these events, autophagy biogenesis is regulated with precision by the autophagy-response receptors and signaling systems at metabolic and immune checkpoints that operate in conjunction with other cellular elements and machineries, such as cellular energy sensors, the proteotoxic stress sensors the pattern recognition receptors, the ubiquitin system, galectins, danger and redox signaling cascades, the endoplasmic reticulum-mitochondrial system, *etc.* [2, 5, 6, current book Chpt. 4]. Overall, that phenomena make the autophagy-lysosomal pathway to be a universal mechanism for decomposition of biological targets regardless their biochemical nature yet vital for response to stress and damage [1, 2]. The functional role of the autophagy machinery in immunochemical and structural cellular homeostasis can be further elaborated within a framework of "entity component system architecture" with a perspective on "the autophagy-mediated intrinsic barrier network".

Historically, invention of the term "autophagy" is credited to Nobel Laureate Christian de Duve. Since his mid-1950s pioneering work on compartmentalization of hydrolytic enzymes in subcellular structures (defined thereafter as "lysosomes"), and then his postulation that the end-point of all organelles subjected to destruction is sequestration in lysosomal apparatus, it took Christian de Duve another decade to evolve a concept of biodegradation of cell constituents via ubiquitous lysosomal pathway, which he named as "autophagy" (from the Greek for self-eating) [7]. Over that time period, the crucial evidence supporting de Duve's concept was provided by several other scientists (Novikoff AB, Essner E, Clark SL, Holt SJ, Hruban Z, Spargo B, Swift H, Ashford TP, Porter KR), who developed and implemented new cell fractional analysis, organelle contrasting techniques and ultrastructural analyses of lysosomal sequestration of organelles and proteins with electron microscopy—the only advanced cell research technique available at that time [8–9]. The research conducted globally by thousands scientists over the following fifty years lead to development of new autophagy techniques and brought enormous progress in clarification of many fundamental aspects of autophagy biology and the pathway details [7, 10, 11]. This research includes remarkable work of Drs, Mortimore GE, Dice JF, Ohsumi Y, Mizushima N, Klionsky DJ, Levine B, Johansen T, Yoshimori T, and others who built up milestones toward development of the autophagy paradigm. That progress was culminated by the 2016 Nobel Prize Award to Ohsumi Yoshinori for his achievements in the molecular dissection of key autophagy pathways leading to discovery of two ubiquitin-like conjugation system (see below) essential for development of autophagic vesicles (i,e., autophagosomes) [7, 10]. But still a lot needs to be done, especially when it comes to understanding autophagy signaling, the biogenesis of phagophores (another name - "isolation membranes"), mechanisms of selective autophagy and autophagosome trafficking.

2. Autophagy Molecular Paradigm: the Universal Mechanisms for Defining "to Be or not to Be"

According to the modern paradigm, autophagy represents a system of evolutionary conserved and strictly regulated "multitasking" mechanisms which is employed by eukaryotic cells (i) to control quality of cellular constituents and to maintain AMP and ATP balance and cell metabolism—all by refurbishing unnecessary or compromised organelles and biomolecules; (ii) to compensate stress and danger and, thus to sustain the intrinsic resistance to harmful exposures; (iii) to decompose acquired xenobiotics (e.g., microorganisms and viruses) and, by these means, to mediate innate and adaptive immunity [1, 2, 5, 7, 10–16, current book Chpt. 18]. Detailed analyses of the autophagy interactive system revealed many aspects of the stress/damage-activated "eat-me" signaling mechanisms. Thus referring to macroautophagy the phagophore initiation, nucleation and elongation, the autophagic flux, the autophagy-regulated proteostasis, organelle biogenesis, pathogen sequestration, etc. are all borne by the highly conserved *ATG*-genes, Atg-proteins as well as by several signaling modules (e.g., mTOR, AMPK, the PI3K complexes, CREB transcriptional factor), sequestosome 1/p62-like receptors—adaptor proteins (e.g., p62/SQSTM1, NBR1, NDP52, T6BP, optineurin) and quality control modifiers (e.g., ubiquitins, galectins, STING) [1, 2, 5–7, 10, 11, 13, 16 and current book Chpts. 2, 3, 16, 17, 21]. In conjunction with this, interestingly that "membrane structural modules" arranged by mitochondria and endoplasmic reticulum (or the ER-mitochondrial axes), which are other key players in cell bioenergetics and proteostasis, are also involved in the emerging macroautophagy signaling mechanisms [5, 6, 12, 13]. Thus, numerous reports indicate that the ER-mitochondrial membrane modules along with their contact membranes (i.e., mitochondria-associated membranes – MAMs) can operate as a "fine stress-sensing interface", which either triggers prosurvival reconstitution of the damaged organelles or diverges the pathway to cell death [5, 12, 13]. Moreover, referring to the macroautophagy biogenesis, the ER-MAM-mitochondrial structures can originate omegasomes yet essentially contribute to formation, nucleation and elongation of the isolation membranes (phagophores), which further became sources of autophagosomes [5, 6, 10, 16]. Note the phagophore formation is mediated by the Atg12-Atg5-Atg16L and LC3/GABARAP/Atg8-phosphatidylethanolamine autophagy conjugates produced via activation of two ubiquitin-like enzymes E1 and E2 (i.e., Atg7 and Atg10—for Atg12 conjugation system and Atg7 and Atg3—for Atg8 conjugation system), and is assisted by an autophagosome-specific pool of phosphatidylinositol-3-phosphate and syntaxin-17 [1, 5, 13, current book Chpts. 2, 6, 12, 16, 21]. These observations are crucial for understanding efficacy of the macroautophagy target-sequestration from topological perspective. Indeed, according to the above model, potential sources of phagophores (i.e., the ER-MAM-mitochondrial structures) are ubiquitously present across cell volume (except other organelles), and thus the phagophore and autophagosome biogenesis can be specifically activated at the "target site". Further accomplishment of autophagic flux occurs with fusion of spatially separated autophagosomes, endosomes and lysosomes [5]. It should be denoted that spatiotemporal dynamics of autophagosome-to-lysosome trafficking through the cytoplasm still remains obscure.

Corroborating importance of crosstalk between autophagy and the ER-mitochondrial membrane modules in control of cell homeostasis, we can refer to the fact that suppression of *ATG*-genes and dysregulation of autophagic flux results in accumulation of damaged and therefore cytotoxic mitochondria and misfolded and oxidized proteins [11–14, 17, current book Chpts. 3, 14, 22]. These observations link impairment of autophagy machinery with pathogenesis of severe degenerative diseases and with promotion of aging, chronic viral infections and tumorigenesis, but also tumor cell death [11–14, 17], current book Chpts. 3,

14, 22]. Taking all the above into consideration, autophagy appears to be an efficient prosurvival adaptive and protective cellular mechanism [2, 5, 6, 12–16].

Evidently, autophagy-lysosomal pathway is adapted to recycle diverse intracellular components regardless of their size and biological nature, i.e., from polypeptide molecules (10^{-8} m) through microorganisms (10^{-6} m). That makes autophagy extremely efficient in barrier functions. From the evolutionary perspective it is interesting that host cells can eliminate bacteria and the "alleged" bacteria-derived endosymbionts, i.e., mitochondria, via selective autophagy, i.e., xenophagy and mitophagy, respectively [5, 6, 12–14, current book Chpt. 19]. Both xenophagy and mitophagy require implication of the "core" autophagy proteins [e.g., ULKs (Atg1), LC3/GABARAP, Atg5-Atg12, Atg9, Atg16L1], autophagy adaptors ("cargo" receptors) (e.g., p62, NBR1, NDP52), factor FIP200 and the poly-ubiquitin-modifiers—for the sequence of phagophore formation, cargo-selection and autophagosome enclosure and for autophagic flux. However, activation of these pathways and their spatial arrangements are regulated by distinct signaling mechanisms triggered by either invaded bacteria or damaged and depolarized mitochondria. Thus, in the first case the signaling cascade is comprised of pathogen recognition sensors (e.g., Toll and NOD-like receptors, antimicrobial GTPase proteins, STING, galectin-8), which can respond to entire microorganism or to the bacterium-containing vacuoles and their fragments [6, 15, 18]. While in the second case the signaling mechanism is initiated by mitochondrial expression of either Nix/Bnip3L and Bnip3 receptors of LC3/GABARAP or—PINK1 kinase followed by recruitment of cytosolic Parkin (E3 ubiquitin ligase), Mfn2, ubiquitin and p62 linking the mitochondrial cargo with LC3/GABARAP on isolation membrane [5, 6]. Seemingly, host cells endowed with these signaling mechanisms are capable of pursuing selective "multitargeting" autophagy, and thus sustaining resistance to invading pathogens, pathogenic factors as well as to the associated damage to mitochondria and ER [19].

Paradoxically, as many other crucial pathways, autophagy plays a dual role under normal and pathological conditions. In addition to the well-known role of autophagy in cell survival, autophagy-mediated type II programmed cell death has long been proposed [5, 12–14 and current book Chpt. 12]. The autophagic cell death was originally reported in tissues subjected to extensive development and remodeling. That effect seems to be analogous to apoptosis in similar metamorphoses [8, 14, current book Chpt. 13]. However, the autophagy implication in cell death is not restricted to the morphogenetic events; it can also drive cell death under various pathological conditions such as acute inflammation, ageing, malignancies and intracellular pathogens such as tuberculosis, borreliosis, etc., [5, 6, 8, 12–14, current book Chpt. 21]. Thus, many highly virulent intracellular pathogens can subvert and adapt different components of autophagy machinery to establish replicative niches eventually leading to host pathology and death [14, 16, 18, 20]. In conjunction with this, interestingly that upon invagination of hepatocytes, malaria sporozoites are able to adapt autophagy LC3-II protein to form host cell-derived parasitophorous vacuoles; and then to subvert amphisomes—originated from another autophagy pathway—as a nutritional source arranging "own feeding mechanisms", while avoiding degradation by lysosomes [20].

In addition, regulation of autophagic flux with pro- and anti-proteolytic enzymes can be a leading key for development of either "excessive autophagy" or "defective autophagy", such as described in the context of persistent chronic infections during sepsis [5, 11, 13–15, current book Chpt. 10]. These both autophagy conditions can affect the ER-mitochondrial network resulting in alterations in oxidative phosphorylation, energetic collapse, impairment of proteostasis, inflammation and the detrimental course of immune dysfunction [5, 12–15].

3. Autophagy and Intrinsic Biological Barriers

In the light of the above considerations, we can say that autophagy mechanism is indispensable contributor to the interactive system of intrinsic biological barriers in living organisms, i. e., the barriers which are essential to maintain nonequilibrium dynamics of organic and inorganic metabolites, to control bioenergetics, antigen and redox status, to protect against thermal impacts and electromagnetic radiation, to interact with microbiota, etc. Based on biological nature of the elements which conduct sequestration, spatial isolation, shielding and target-processing, the barrier functions can be carried out at: (i) molecular level by e.g., ligands, carriers/transporters, proteasomes and redox "converters"; (ii) membrane subcellular level by e.g., mitochondria, endoplasmic reticulum, phagophores, the plasma membrane; (iii) extracellular level by e.g., mucus and other extracellular matrices. Moreover, at cellular level these barriers represent numerous interfaces of tissues and organs with ambient and internal environment to sustain structural, immune and metabolic integrity. From this perspective, while macroautophagy, microphagy and chaperone-mediated autophagy execute barrier functions at molecular and membrane levels [2, 5, 6, 13–17], it would be reasonable to assume that dynamics and efficacy of autophagy function can determine performance of the barrier-forming cells. Note, in the vertebrates the infection cellular barriers are constituted by multidimensional interactive networks of mesenchymal, epithelial, reticuloendothelial, endothelial and hematopoietic cells, where along with monocytes and polymorphonuclear granulocytes, a particular role in xenobiotic control and "cleaning function" is attributed to nonprofessional phagocytes, e.g., skin fibroblasts, bone marrow stromal cells, endothelial and epithelial cells [18, 19]. Evidently, nonprofessional phagocytes are very efficient in phagocytosis with "autophagy-to-pathogen" response mechanism and therefore, can compensate professional phagocyte function, when the last one declines [19]. Thus, a lack of the "canonical" phagocytic features (such as phagosome biogenesis, the oxidative burst, etc.) in nonprofessional phagocytes is presumably compensated by empowering xenophagy to control and execute all events from pathogen sequestration through degradation [5, 6, 18–21, current book Chpt. 15]. That infers increasing burden of autophagy function in nonprofessional phagocytes when professional phagocytes are depleted due to pathological conditions. Furthermore, considering that nonprofessional phagocytes can also orchestrate response to acute stress or trauma by expression and massive release of paracrine and endocrine factors, such as damage-associated molecular patterns (DAMPs), inflammatory cytokines, proteases, chemokines, defensins, nitric oxide, ROS, fragmented DNA, exosomes and microvesicles, which in turn can trigger and propagate autophagy stress response [6, 14–19].

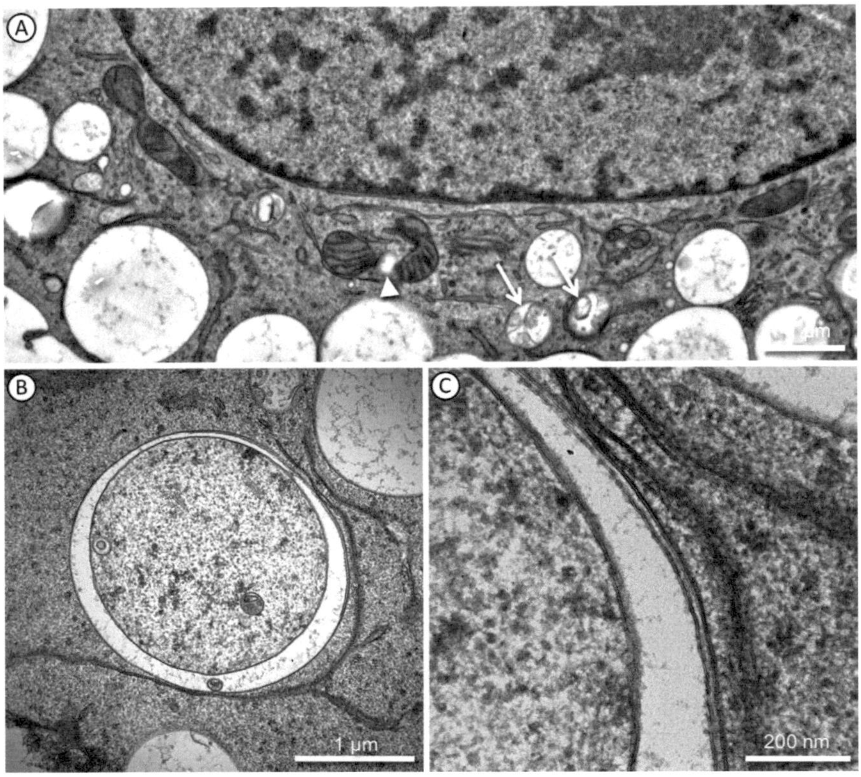

Figure 1. Various states of mitophagy in a glioblastoma cell and a clear contribution by endoplasmic reticulum forming the phagophore can be deduced; lysosomes are found adjacent to autophagosomes (arrow head) as well as following fusion (arrows) (A). Two lipid bilayers of a completed phagophore engulfing cytoplasm are surrounded by endoplasmic reticulum membranes (B). Lipid bilayers of the phagophore double membrane can be identified by higher magnification. The high resolution also shows the typical asymmetry of phagophore (and autophagosome membranes). Lower staining intensity is seen in the vesicle-faced bilayers whereas higher contrast is seen in the outer lipid bilayers (C) (trans electron microscopy performed with a JEOL 1400 at 120 keV).

These effects may suggest a presence of cross-talk in the "barrier network" assisted by autophagy mechanism.

4. Conclusion

Overall, it is hard to overestimate the vital role of autophagy in function of the intrinsic cellular barriers. Thus, autophagy machinery bears specific types of physical barriers emerging from activation and interaction of autophagy scaffolds, membrane-assembling proteins, ubiquitin-like modifiers and autophagy adaptors, which sustain autophagic flux. In this event, the hallmark of macroautophagy is formation of new organelles, i.e., double-membrane phagophores and then sequestration and compartmentalization of cellular constituents within autophagosomes (exemplified in **Figure 1** and **Supplement 1**).

Supplement 1

A tomogram depicting the glioblastoma cells (line #12537 GBM) with active autophagy and mitophagy. Different stages of mitophagy progression [e,g., advanced mitophagy (upper left) and just initiated mitophagy (lower left)] are detectable within numerous autophagosomes and newly formed intravesicular membranes. The residual mitochondrial constituents are observable at higher magnification (*see in the middle*). The autophagy tunneling system, shielded from the rest of the cell's cytoplasm can be best envisaged in a tomogram. Note numerous ribosomes and structural elements comprised of actin, intermediate filaments and microtubules.

The tomogram was taken by Paul Walther in the Central Facility for Electron Microscopy at Ulm University using Jeol 2100F TEM in STEM mode at 200 keV.

Acknowledgements

We thank Paul Walther (Central Facility for Electron Microscopy at Ulm University) for specimen processing and imaging. NVG thanks the Department of Anesthesiology at USUHS for the encouragement and support in this effort.

Author details

Nikolai V. Gorbunov[1]* and E. Marion Schneider[2]

*Address all correspondence to: gorbunov.nikolaiv@gmail.com

1 Department of Anaesthesiology, Uniformed Services University of the Health Sciences, Bethesda, Maryland, USA

2 Division of Experimental Anesthesiology, Department of Anaesthesiology, University Hospital Ulm, Ulm, Germany

References

[1] Behrends C, Sowa ME, Gygi SP, Harper JW. Network organization of the human autophagy system. Nature. 2010;466(7302):68–76. doi: 10.1038/nature09204.

[2] Jegga AG, Schneider L, Ouyang X, Zhang J. Systems biology of the autophagy-lysosomal pathway. Autophagy. 2011;7(5):477–489.

[3] Wu CW, Schmoller SK, Shin SJ, Talaat AM. Defining the stressome of *Mycobacterium avium* subsp. paratuberculosis in vitro and in naturally infected cows. J Bacteriol. 2007;189(21):7877–7886.

[4] Dennis KK, Auerbach SS, Balshaw DM, Cui Y, Fallin MD, Smith MT, Spira A, Sumner S, Miller GW. The importance of the biological impact of exposure to the concept of the exposome. Environ Health Perspect. 2016. doi: 10.1289/EHP140. http://ehp.niehs.nih.gov/wp-content/uploads/advpub/2016/6/EHP140.acco.pdf.

[5] Kawabata T, Yoshimori T. Beyond starvation: an update on the autophagic machinery and its functions. J Mol Cell Cardiol 2016;95:2–10.

[6] Randow F, Youle RJ. Self and nonself: how autophagy targets mitochondria and bacteria. Cell Host Microbe. 2014;15(4):403–411. doi: 10.1016/j.chom.2014.03.012.

[7] Ohsumi Y. Historical landmarks of autophagy research. Cell Res. 2014 Jan;24(1):9-23. doi: 10.1038/cr.2013.169.

[8] Novikoff AB, Essner E (1962) Cytolysomes and mitochondrial degeneration. J Cell Biol 15:140–146.

[9] Hruban Z, Spargo B, and Swift H. (1963). Focal cytoplasmic degradation. The American Journal of Pathology and Zoology, 42(6), 657–683.

[10] Ichimura, Y., Kirisako T., Takao, T., Satomi, Y., Shimonishi, Y., Ishihara, N., Mizushima, N., Tanida, I., Kominami, E., Ohsumi, M., Noda, T. and Ohsumi, Y. A ubiquitin-like system mediates protein lipidation. Nature, 2000; 408, 488-492.

[11] Klionsky DJ, et al. Guidelines for the use and interpretation of assays for monitoring autophagy (3rd edition). Autophagy. 2016;12(1):1–222. doi: 10.1080/15548627.2015.1100356.

[12] Dorn GW 2nd, Kitsis RN. The mitochondrial dynamism-mitophagy-cell death interactome: multiple roles performed by members of a mitochondrial molecular ensemble. Circ Res. 2015;116(1):167–182. doi: 10.1161/CIRCRESAHA.116.303554.

[13] Khaminets A, Heinrich T, Mari M, Grumati P, Huebner AK, Akutsu M, Liebmann L, Stolz A, Nietzsch S, Koch N, Mauthe M, Katona I, Qualmann B, Weis J, Reggiori F, Kurth I, Hübner CA, Dikic I. Regulation of endoplasmic reticulum turnover by selective autophagy. Nature. 2015;522(7556):354–358. doi: 10.1038/nature14498.

[14] Levine B, Kroemer G. Autophagy in the pathogenesis of disease. Cell 2008;132:27–42.

[15] Schneider EM, Flacke S, Liu F, Lorenz MR, Schilling P, Nass ME, Foehr KJ, Huber-Lang M, Weiss ME. Autophagy and ATP-induced anti-apoptosis in antigen presenting cells (APC) follows the cytokine storm in patients after major trauma. J Cell Commun Signal 2011;5:145–156.

[16] Deretic V, Saitoh T, Akira S. Autophagy in infection, inflammation and immunity. Nat Rev Immunol. 2013;13(10):722–737.

[17] Qin W, et al. Inhibition of autophagy promotes metastasis and glycolysis by inducing ROS in gastric cancer cells. Oncotarget. 2015;6(37):39839–39854. doi: 10.18632/oncotarget.5674.

[18] Huang J, Brumell JH. Bacteria-autophagy interplay: a battle for survival. Nat Rev Microbiol. 2014;12(2):101–114.

[19] Gorbunov NV, McDaniel DP, Zhai M, Liao PJ, Garrison BR, Kiang JG. Autophagy and mitochondrial remodelling in mouse mesenchymal stromal cells challenged with Staphylococcus epidermidis. J Cell Mol Med. 2015;19(5):1133–1150.

[20] Thieleke-Matos C, Lopes da Silva M, Cabrita-Santos L, Portal MD, Rodrigues IP, Zuzarte-Luis V, Ramalho JS, Futter CE, Mota MM, Barral DC, Seabra MC. Host cell autophagy contributes to Plasmodium liver development. Cell Microbiol. 2016;18(3):437–450. doi: 10.1111/cmi.12524.

[21] Sorbara MT, Girardin SE. Emerging themes in bacterial autophagy. Curr Opin Microbiol 2015;23:163–170.

Section 2

Autophagy mediators and signaling mechanisms

Chapter 2

Aging-Related Diseases and Autophagy

Elif Damla Arisan, Pinar Obakan-Yerlikaya,
Ajda Coker-Gurkan and Narçin Palavan Unsal

Additional information is available at the end of the chapter

http://dx.doi.org/10.5772/64540

Abstract

Autophagy is fundamental, evolutionary conserved physiological process at molecular level which targets long-lived cytosolic proteins and organelles to be recycled through lysosomal degradation. Diminished autophagic activity caused cellular stress in many organisms following aging, and inhibition of autophagy in model organisms causes degenerative changes and pathologic diseases observed with high incidence ratio generally in older ages. Consequently the delayed senescence or increased longevity in model organisms often stimulate autophagy, and autophagy inhibition compromises anti-aging effects. The cytoprotective function of autophagy is presented in various human diseases such as lung, liver, cardiovascular diseases, neurodegeneration, myopathies, cancer, stroke, infections and metabolic diseases which are found associated with autophagic targets. These pathologies are defined with their age-dependent characteristics, is not fully understood that how autophagy network regulates metabolism and may cause diseases in age-related manner. In this book chapter, we are going to discuss the autophagy and aging relationship in three different parts. In the first section autophagy and aging relationship is going to be presented through explaining responsible signalling network. The autophagy and age-related neurological disorders, genetic basis of age-dependent diseases and the functional role of autophagy is going to be discussed in the second and third part of the chapter.

Keywords: autophagy, aging, metabolism, diseases, sirtuins, AMPKa

Definitions [1]:

Macroautophagy: Macroautophagy is a complex process that involves the formation of subcellular and typically double membrane vesicles. These subcellular compartments are called as autophagosomes, which is used for sequestration of cytoplasmic materials and cargo them into lysosomes to be recycled. The process of macroautophagy starts with the initiation of the formation of the phagophore. The growth of the phagophore terminates in completion of the autophagosome.

Chaperone-mediated autophagy (CMA): Chaperone-mediated autophagy is the only type of autophagy in mammalian cells that able to selectively degrade cytosolic proteins in lysosomes. All CMA substrates contain in their amino acid sequence a motif biochemically related to the pentapeptide KFERQ required for their selective recognition by the CMA cytosolic chaperone complex. These proteins are recognized by a chaperone complex and are translocated into the lysosome through a specific receptor called the lysosome-associated membrane protein (LAMP)-2A. Reduced CMA activity has been observed in many cell types and tissues of old rodents, as well as in cells derived from aged individuals. This mechanism is also linked to "lipophagy."

Lipophagy: Cellular lipid stores are also targeted for lysosomal degradation through a process termed "lipophagy." Therefore during lipophagy, autophagy and lipases can act together to mobilize lipids stored in lipid droplets. According to *Caenorhabditis elegans* studies, less fat content may promote longevity through inactivating mTOR downstream targets.

Microautophagy: During microautophagy, cytoplasmic content is sequestered into lysosomes through direct invagination of lysosomal membranes. This process can be observed in different organisms: yeast and mammalians. However, there are less information about microautophagy in mammalians compared to other distinguished autophagic processes.

Mitophagy: The process of removal of damaged mitochondria through autophagy is called mitophagy. Accumulating evidence points that the maintenance of mitochondrial homeostasis is strongly associated with the onset and the progression of several age-associated neurodegenerative diseases, such as Parkinson's disease, Alzheimer's disease, and Huntington's disease.

Moreover, the selective degradation of endoplasmic reticulum, mitochondria, ribosomes, and peroxisomes are referred to as ERphagy, mitophagy, ribophagy, and pexophagy, respectively.

1. Introduction

Autophagy is an evolutionary conserved process, characterized by massive degradation of cytosolic contents [2]. The typical autophagy process is finalized by fusion of autophagosome to endosomes and lysosomes, which engulf cytoplasmic contents within a double-membrane vacuole [3]. Autophagy process has important physiological functions including the degradation of misfolded proteins and organelle turnover [4]. Recent studies also showed that there is a functional role between autophagy and apoptosis, which has been introduced as an important regulation of cell death in response to chemotherapeutic drugs [5]. The process is regulated by several proteins such as Atg protein family, essential for the initial building of the autophagosome, and phosphoinositide-3 kinase (PI3K) important in the early stages of autophagic vesicle formation. This complex is called autophagosomes, which are specific cytoplasmic compartments to degrade useless cellular components to reutilize in cellular processes. It is well established that a number of molecular targets are critical in autophagosome complexes [2]. One of the most remarkable autophagic marker, Beclin-1, has Bcl-2 homology (BH) 3 domain and it is a linker protein between apoptosis and autophagy due to its interaction with antiapoptotic Bcl-2 family members. Therefore, forced expression of Bcl-2 renders autophagy-related processes and thereby prevented autophagy as well as apoptosis [6].

In addition, LC3, a cytosolic soluble protein, is cleaved during autophagic induction and involved in the autophagic vacuole membrane formation. When autophagic process starts, LC3-I (16 kDa) is converted to LC3-II (14 kDa). Recent studies showed that another autophagic key molecule, p62, is integrated in the autophagosome complexes during autophagy and reduced level of cytoplasmic-free p62 level could be accepted as an autophagic marker in the cells. Autophagy is also classified as the second type of cell death. A number of reports showed that drug-induced apoptosis mechanism could be postponed in cancer cells by activating autophagy [5]. Recent reports showed that inhibition of autophagy by the treatment of specific inhibitors for autophagic regulators, 3-MA, or suppression of autophagy regulatory pathways [7, 8] may provoke apoptotic efficiency of chemotherapeutic agents in prostate, breast, colon, lung, and HeLa cancer cells [9–11]. Mammalian target of rapamycin (mTOR) signaling pathway is one of the leading pathways that orchestrates autophagy in the cells [12]. Normally, mTOR is activated and autophagy is suppressed in the presence of insulin. Insulin binds to its specific receptor and caused autophosphorylation by the recruitment and phosphorylation of its major substrates insulin receptor substrate 1 and 2 (IRS1 and IRS2). Phosphorylated partners then recruit class I PI3K. Rapamycin is a lipophilic, macrolide antibiotic which has been shown to induce autophagy by inactivating mTOR [13, 14]. Therefore, rapamycin-mimicking agents, rapalogs, are natural autophagy inducers through inhibiting mTOR downstream signaling cascade.

Autophagy is also referred as a catabolic process, which involves the formation of a double membrane structure around damaged organelles and cellular compartments which lead to growth arrest [4]. It has been shown that mTOR negatively regulates autophagy in response to cellular conditions and environmental stress [15]. mTOR consists of two complexes: mTOR complex I (mTORC1) and mTOR complex II (mTORC2). mTORC1 has specific protein, raptor, which is sensitive to rapamycin. mTORC2 associates with Rictor, which is considered to be insensitive to rapamycin (**Figure 1**).

The PI3K/Akt pathway is an important intracellular signaling pathway in the regulation of cell survival through activating mTOR. Its downstream targets are translational regulators: p70S6K and eukaryotic initiation factor 4E (eIF4E) binding protein-1 (4EBP1) [12]. Raptor binds to mTOR substrates, including 4E-BP1 and p70 S6 kinase, through their TOR signaling (TOS) motifs and is required for mTOR-mediated phosphorylation of these substrates. Furthermore, the Rictor-mTOR complex has been identified as the previously elusive PDK2 responsible for the phosphorylation of Akt/PKB on Ser473, facilitating phosphorylation of Akt/PKB on Thr308 by PDK1 and required for the full activation of Akt/PKB [16]. The PI3K/Akt signaling proteins also interfere with apoptotic regulators, Forkhead box O (FoxO) and glycogen synthase kinase 3 (GSK3β), to inhibit apoptosis and autophagy in cancer cells [17, 18].

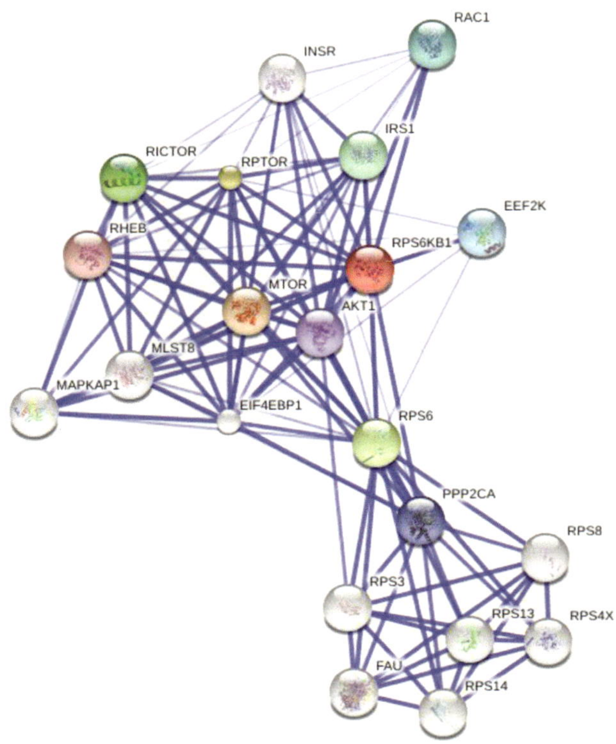

Figure 1. STRING mTOR interacting partner analysis.

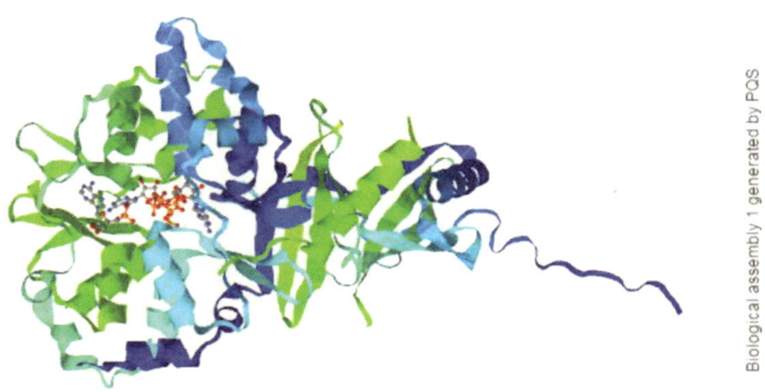

Figure 2. Crystal structure of regulatory fragment of mammalian AMPK in complexes with ATP-AMP through PDB database ribbon presentation.

All the mentioned molecular targets are well documented with their autophagy-related characteristics in different cells or organisms. Beside these highlighted targets, there are other important regulators, which also act as stress sensors in the cells [19]. One of the leading targets is AMP-activated protein kinase (AMPK), energy-sensing kinase, which is a downstream target of mTOR (**Figure 2**). AMPK is a heterotrimeric protein complex that is regulated by different modulators in the cells. This serine/threonine kinase is a heterotrimer composed of a catalytic (AMPKα) subunit and two regulatory (AMPKβ and AMPKγ) subunits. The phosphorylation of a conserved threonine residue (T172) in the activation domain of catalytic α-subunit by a number of kinases is crucial for the activity of AMPK [20]. Activated AMPK typically phosphorylates TSC2 tumor suppressor and leads to inactivation of Rheb, which is an interacting partner of mTORC1. Alternatively, it is shown that AMPK can regulate mTOR signaling by phosphorylating Raptor at Ser722 and Ser792, which leads to 14-3-3 binding to Raptor, and induces cell-cycle arrest triggered by impaired energy balance [21].

Since the ratio of AMP to ATP exerts the intracellular energy measurement, these substrates determine the AMPK activity in the cells [22, 23]. AMPK can be also activated by metabolic stress factors, hypoxia or ATP consuming catabolic processes in the cells [17, 24, 25]. For this purpose, it can be emphasized that there is strong relationship between energy balance and autophagy regulation in the cells. To point this relationship, it is critical to put forward AMPK activation status in different conditions. In a brief presentation, AMPK is referred as a central metabolic sensor found in a variety of organisms that regulates glucose and lipid metabolism in response to alterations in nutrients and intracellular energy levels [23, 24]. However, the functional role of autophagy in energy balance conditions is not fully understood.

As an example for this issue, although glucose starvation can activate AMPK-mediated signaling route and trigger autophagy through phosphorylating Ulk1 at Ser 317 and Ser 777, nutrient deficiency in high mTOR activity can also prevent Ulk1 activation by phosphorylating Ulk1 Ser 757 and disrupting the interaction between Ulk1 and AMPK. In addition, an established marker for autophagy, p62, can accumulate in AMPK-deficient livers [26]. Since p62 is involved in mitochondria clearance, the defects in selective degradation of mitochondria by autophagy (mitophagy) and a corresponding mitochondria accumulation was also shown in the same study with presence of severe abnormalities in AMPK- or ULK1-deficient hepatocytes.

Therefore, there is still need to evaluate the AMPK activation status in autophagy-related issues. The supporting data for this manner are also observed in the treatment of type 2 diabetes (T2D). The well-known T2D treating drugs, metformin, thiazolidinediones, etc., can activate AMPK and improve insulin sensitivity and metabolic health [27–29]. It can be concluded that AMPK as a critical autophagy regulator has a great impact on human metabolic diseases. Thereby, diminished cellular energy capacity can stimulate glucose uptake in skeletal muscles or fatty acid (FA) oxidation in tissues through modulating AMPK.

AMPK is a central molecular target that orchestrates metabolic stress and energy balance in the cells. One of the critical mechanisms regulated by AMPK is fatty acid synthesis, which is generally age-related problem due to nutritional habits or genetic background. When activated AMPK acetyl-CoA carboxylase (ACC) is phosphorylated at Ser79 (an inhibitory site), it prevents the conversion of acetyl-CoA to malonyl CoA. This action allows long-chain FAs to enter the mitochondria for oxidation. Concomitantly, HMG-CoA reductase leads to the inhibition of cholesterol synthesis, peroxisome proliferator-activated receptor-gamma coactivator (PPARα) 1α, which stimulates mitochondrial biogenesis and many others [19, 23]. The inhibition of FA synthase (FAS) expression due to AMPK was previously reported in primary cultured hepatocytes [30, 31]. Supporting this finding, it was shown that AMPK can suppress FAS gene expression either by AMPK activating AICAR or an antidiabetic drug metformin treatment in liver cells [29]. Indeed, activation of AMPK by either AICAR or rosiglitazone reduces expression of FAS and ACC resulting in the suppression of proliferation of prostate cancer cells [32]. Of note, physical exercise and calorie restriction (CR) may exert similar beneficial effects on metabolic health and reduce risk of several diseases, including T2D and cardiovascular diseases via targeting previously mentioned pathways [33]. Both exercise and CR are shown as the frequently observed metabolic stresses that increase the AMP: ATP ratio in an organism's cells, which led to activation of AMPK. Similar to AMPK, silent information regulator 1 (SIRT1) signaling pathways are evolutionarily conserved energy sensors in cells responding to the increase in cellular AMP and NAD^+ concentrations, respectively [34–36]. SIRT1 is a member of sirtuins, which is discussed later in detail to evaluate the autophagy and aging relationship.

2. Aging: energy balance and stress management

It is well studied that over 30 proteins orchestrated autophagy-related processes in the cells, which differ due to stress stimuli or depends on intrinsic molecular mechanism. Autophagy is a complex process in development, metabolism, and aging. In order to evaluate the potential characteristics of autophagy in aging, researchers pointed out that energy balance and stress factors should be discussed. Both factors are critical initiators of autophagy and play a role in cell decision signaling routes. For this reason, in this part, we will discuss the energy metabolism-related signaling cascades and stress-related cellular responses.

Aging is strongly correlated with autophagy in different organisms from fungi to humans. It is well documented that protein degradation ratio is decreased due to aging [35, 36], which presents similar observations of diminished levels of age-related autophagic/proteolytic activity [37]. Therefore, it can be emphasized that there is strong relationship between autophagy, aging, and lifespan. The genetic basis of this connection was established in *C. elegans* daf-2 mutants, which have diminished insulin-signaling cascade and extended lifespan. Similar to this finding, mTOR or p53 mutants show lifespan extension [38–40].

It is well documented that CR or late findings also showed the potential effect of resveratrol or spermidine treatment causing upregulation of sirtuins and led to increased lifespan in the cells. Similar findings were also shown for *Caenorhabditis elegans* and *Drosophila melanogaster* species [33, 41–44]. For this reason, sirtuins (mammalian protein family members 1–7) are also termed as antiaging proteins, class III histone deacetylases (HDACs), exerting function as protein deacetylases/ADP ribosyltransferases that target a wide range of cellular proteins in the nucleus, cytoplasm, and mitochondria for posttranslational modification by acetylation (SIRT1, -2, -3, and -5) or ADP ribosylation (SIRT4 and 6). Sirtuins have conserved NAD+-dependent deacetylase domain, which is known to regulate cellular senescence and lifespan. SIRT1 is generally found in nucleus, but there are remarkable data about its presence in cytoplasm. SIRT2 is the dominant member found in cytoplasm. SIRT3, SIRT4, and SIRT5 are localized to the mitochondria with different enzymatic activities. SIRT6 is a chromatin-associated nuclear protein and SIRT7 is found in nucleoli. Early data about involvement of sirtuins in autophagy-related longevity was shown with CR experiments [43, 44]. The reduced food intake without malnutrition caused increased autophagy via upregulation of AMPK and SIRT1 and inhibition of insulin/insulin-like growth factor (IGF) signaling. mTOR inhibition also has a remarkable data with these alterations [45]. Rapamycin is an mTOR complex I inhibitor that altered sirtuins and caused autophagy responses in the cells [46]. In a similar way, researchers highlighted that increased levels of acetate, acetyl-CoA, could inactivate autophagy in yeast models [47, 48]. These substances are generated through mitochondrial energy regulator networks such as acetyl-CoA hydrolase-1 (ACH1) and mitochondrial pyruvate carrier-1 (MPC1)-dependent pathway and the acetyl-CoA synthetase-2 (ACS2)-dependent nucleo-cytoplasmic pathway. The hyperactivation of these targets led to hyperacetylation of histones and ATG genes [48]. Similar findings were also shown for a number of pharmacological drugs, such as rapamycin, spermidine, or resveratrol, in different organisms.

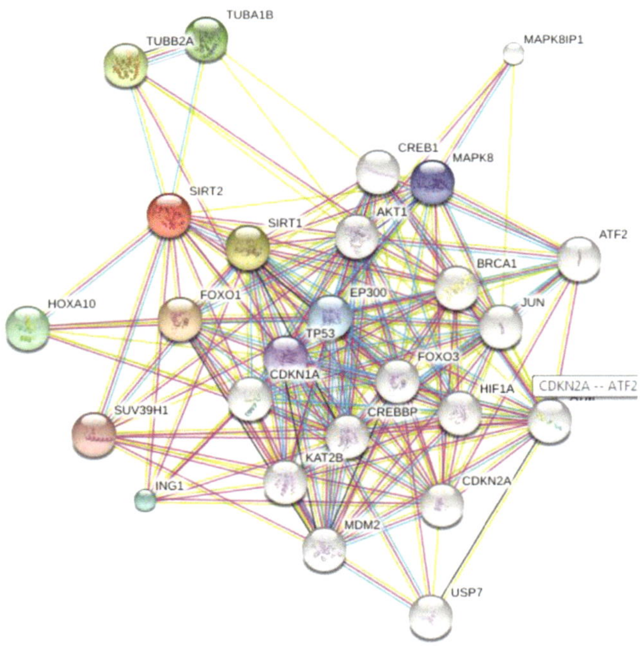

Figure 3. STRING analysis of *Homo sapiens* SIRT2.

The supporting data are observed within 25 years of experiment with primates. CR reduced mortality rates and age-related diseases in Rhesus monkeys [49]. Therefore, the identification of CR-altered molecular mechanisms has gained importance to evaluate the main reason of human diseases arisen during aging. *In vivo* and *in vitro* evidences highlighted that CR or fasting without malnutrition upregulated SIRT1, which regulates several transcription factors that regulate stress responses, energy metabolism, and endocrine signaling, including peroxisome proliferator-activated receptor (PPAR)γ, PPARγ coactivator 1 (PGC1)-α, forkhead box transcription factors (FOXOs), liver X receptor (LXR), and p53 [50]. In addition to these observations, we search SIRT2 on STRING (**Figure 3**). The analysis results showed that SIRT2 and SIRT1 have strong interactions with cellular dynamic proteins tubulins, cell survival, and death decision maker proteins: p53, MDM2, FoXO1, FoXO3, DNA repair proteins (BRCA1), and PI3K/AKT/MAPKs signaling axis proteins [51]. For this reason, it can be suggested that longevity, which is a final destination of sirtuins, is a complex cellular decision.

When we checked protein atlas database for cancer-related SIRT1 and 2 expression profiles, we observed that SIRT1 is the most critical target in a number of cancer cases (**Figure 4A** and **B**). Overexpression of SIRT1 is regulated at the transcriptional level through p53 binding sites, as SIRT1 promoter normally repress SIRT1 expression. However, in the absence of nutrients, FoXO3a translocates to the nucleus, interacts with p53, inhibits its suppressive activity, and leads to increased SIRT1 expression [52]. Moreover, double knockout p53 mice show increased basal expression of SIRT1 in selective tissues, including adipose tissue, but SIRT1 levels were not further elevated upon nutrient withdrawal [52]. The loss of functional p53 in carcinogenesis might increase SIRT1 levels [53].

It is noteworthy that the clarification of several indicators is required to determine critical molecular factors in disease progression related to autophagy in age-dependent manner. The well-established models in this concept are nutrient deficiency with CR or physical exercise, a metabolic stress inducer. According to previous results both CR and physical exercise exert beneficial effects on metabolic health and reduce risk of several diseases, including T2D and cardiovascular diseases through targeting previously mentioned pathways. These factors are also accepted as metabolic stressors that increase the AMP: ATP ratio in an organism's cells, which led to activation of AMPK [54].

A number of studies showed that AMPK activation may slow aging [55, 56]. In contrary, the decline in AMPK activation with aging causes diminished autophagic regulation, increased oxidative stress, endoplasmic stress, apoptotic resistance, inflammation, fat deposition, hyperglycemia, and finally metabolic disorders. The key molecule AMPK gains more importance in age-related disease progression. While AMPK stimulates energy production from glucose and FA during metabolic stress and depress energy consumption for macromolecule synthesis [57, 58], it is not a new paradigm that nutritional overload breaks the functional AMPK status and induces insulin resistance which trigger metabolic syndromes such as obesity, diabetes, and cardiovascular diseases [59]. According to the findings obtained from model organisms, metformin treatment increases lifespan of *C. elegans* model organism [60]. The AMPK ortholog, AAK-2, can be activated through metformin treatment. Similar findings were also observed in *Drosophila* model organism [61]. However, all findings indicate that there is a clear deficiency in the sensitivity of AMPK activation in aged tissues. This might be a reason of systemic alterations such as function of protein phosphatases, which could be involved in the suppression of AMPK activation with aging.

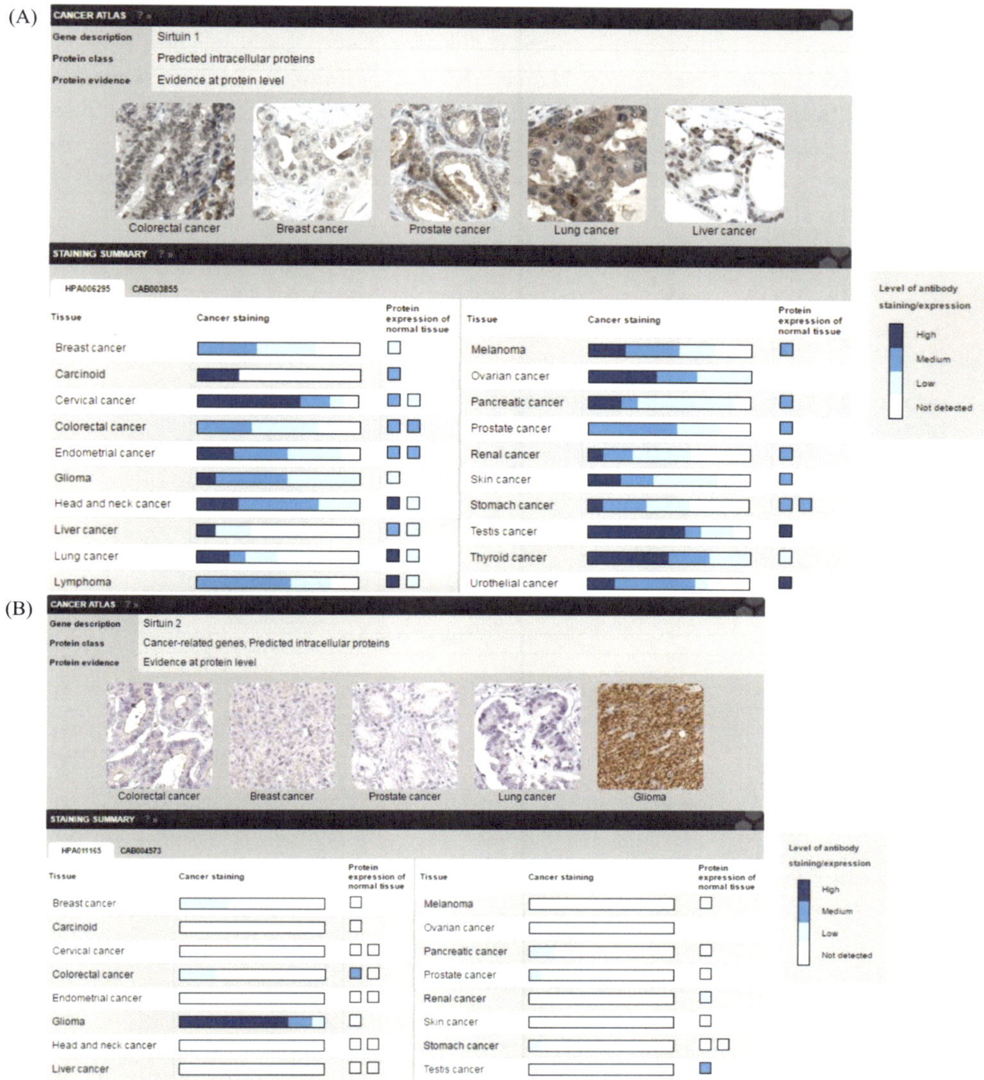

Figure 4. Human protein atlas database query results for SIRT1 (A) and SIRT2 (B). http://www.proteinatlas.org/.

Since it is well established that CR might reduce the tumor weight, the energy balance mechanism is investigated by cancer researchers. Supporting this observation, it is highlighted that cancer is a disease of aging, and the incidence of most of the cancers are increased with age due to genomic stability problems in genome [62–64]. DNA replication errors, reactive oxygen species (ROS) generation due to intrinsic cellular stress factors, or extrinsic stress inducers increased genomic instability. In correlation, in a number organisms which have lower reactive oxygen species are shown with increased lifespan. Therefore, increased lifespan may be causative factor for cancer development.

In contrary to CR, high fat diet or increased calorie intake leads to obesity with a number of comorbidities, including cancer, cardiovascular diseases, and diabetes [65–68]. The main idea is to understand energy balance and human diseases relationship through investigating molecular targets in the cells (**Figure 5**). High calorie intake or fat oxidative stress cause metabolic dysfunction of critical pathways. During aging, slow rate of autophagy decision mediates a number of pathogenesis related to functional status of mTOR, AMPKa, and sirtuins, which are cellular stress and nutrient sensors.

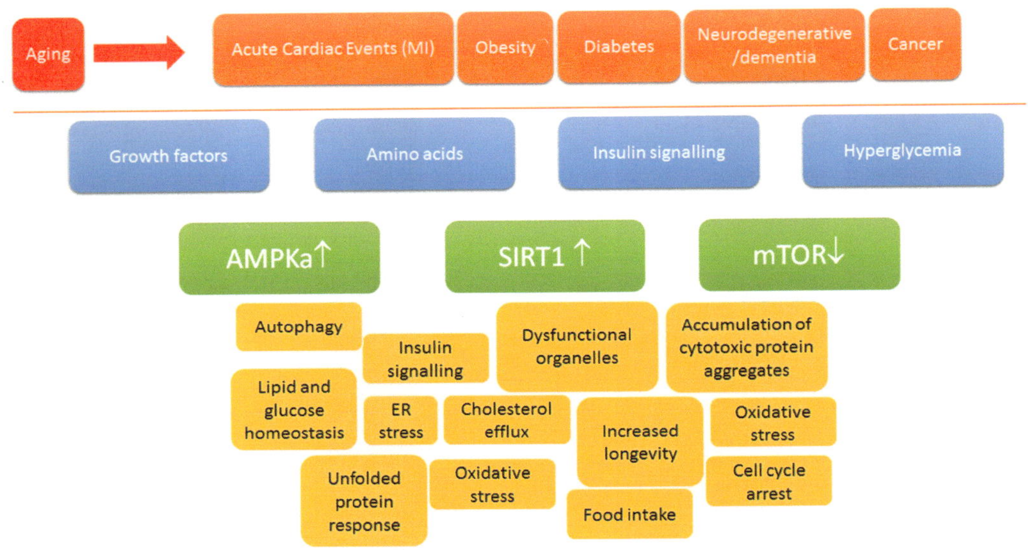

Figure 5. Aging-related diseases and autophagy-related molecular signaling cascades.

3. Aging and related neurological disorders

Human aging, the gradual harmful effect of time on an organism, is comprised of physiological changes leading to senescence and inability to adapt to metabolic stress. Although aging has been considered as a natural process, age-associated diseases are found as leading causes of death. The dramatic increase in average life expectancy in the last century has been accompanied by an equivalent increase in age-related disease diagnosis, such as cancer, neurodegenerative diseases, type II diabetes, and cardiovascular diseases.

The basis of aging process is examined under several hypothesis: the free radical, the immunologic, the inflammation, and the mitochondrial theories [69–72]. However, aging seems more likely a multifactorial process rather than a single cause [73, 74]. In this context, studies on longevity focus more than one target at a time, and animal models showed that aging rates and life expectancy can be modified by multitarget modulators. The consensus among researchers in the field indicates that aging can be retarded by dietary and pharmaceutical interventions, which let to delays in age-related diseases.

Aging is an occasional process in different individuals in contrast to the programmed events in early development. Although recent studies indicate that *klotho* gene mutation could cause premature aging or telomeres are tightly linked to senescence, the variation of aging initiation pose an obstacle to interfere to the progress.

The promising strategies to slow aging have suggested suppressing glucose production by the liver, inhibition of inflammation, and protein restriction [75]. The prevention of high glucose production, hyperglycemia, the condition in which excessive amount of glucose is found in the blood plasma, by metformin is under investigation for its potential effect on slowing aging in different organisms such as *C. elegans* [76]. Studies indicate that worms treated with metformin have stable body volume with reduced deformation of cuticle [76, 77].

One of the possible mechanisms of metformin to decrease hyperglycemia is the activation of AMPK. AMPK is the primary activator of cellular response to lowered ATP levels [78]. As mentioned in previous section, AMPK targets mTOR signaling pathway, which affects transcription and translation through effector proteins 4E-BP1 and p70S6. The involvement of AMPK/mTOR axis in the suppressing glucose production strategy also overlaps with the inhibition of inflammation and protein restriction targets against aging. mTOR signaling regulates inflammatory responses after bacterial stimulation in monocytes, macrophages, and primary dendritic cells [79]. mTOR following tuberous sclerosis complex 2 activation (TSC2) has been shown to diminish proinflammatory cytokines production through nuclear factor (NF)-κB [79]. On the other hand, mTOR inhibition by AMPK is also a critical step in the control of translation attenuation [80]. In normal cellular conditions, nutrients induce mTOR and its downstream target S6K to promote growth and proliferation. However, in nonproliferating cells, this signaling axis has been shown to initiate cellular senescence, the phenomenon by which cells cease to divide. mTOR inhibitor rapamycin has been proposed for decelerating aging and age-related pathologies in *D. melanogaster, C. elegans*, and yeast [81–83].

Therefore, the inhibition of mTOR is a critical phenomenon to balance cell survival and death signaling in eukaryotic organisms. Nutrient starvation can directly cause mTOR inhibition and induction of autophagy, a process that optimize the usage of limited energy supply. Autophagy is generally referred as a catabolic process during which autophagosomic-lysosomal degradation of cytoplasmic proteins, macromolecules, and damaged or aged organelles occur.

The hallmarks of neurogedenerative diseases are generally described with the accumulation of abnormal proteins forming aggregates. These aggregates usually cause toxic effects, such as defective axonal transport, inactivation of transcription factors, reactive oxygen species generation, and consequently neuronal death [84]. Since differentiated neural cells lose their ability to divide, except the granule cell layer of the olfactory bulb, and the dentate gyrus of the hippocampus, a well-organized protein quality-control complex is needed in neural cells.

Autophagy, as a process of cellular recycling for aggregated proteins, might be a critical target in the treatment of neurodegenerative diseases; however, altered autophagic activity has also been implicated in their pathogenesis [85].

3.1. Alzheimer's disease

The main manifestations of Alzheimer's disease (AD) are selective memory impairment and degenerative dementia in the elderly people. AD is characterized by the formation of neuro-fibrillary tangles and extracellular senile plaques. Tau protein and amyloid beta-peptide (Aβ) are involved in these two processes, respectively. Tau is a soluble microtubule-associated protein playing a role in microtubule stabilization and vesicle transport along the axon. Tau proteins have six isoforms with different size of amino acid chain. All isoforms are present usually in central nervous system, and upon hyperphosphorylated, they paired as helical filaments, a characteristic feature of AD. The hyperphosphorylation might be due to mutations in tau isoforms that alter their function and expression or in tau-kinases capable of phosphorylating tau such as glycogen synthase kinase 3 β (GSK3β), cyclin-dependent kinase 5 (CDK5), the mitogen-activated protein kinase (MAPK) extracellular-regulated kinases 1 and 2 (ERK1/2), p38, and the c-Jun NH2-terminal kinases (JNKs) [86–88]. Hyperphosphorylated tau disassembles microtubules and aggregates with MAP 1 (microtubule-associated protein1), MAP 2, and ubiquitin forming tangles. These aggregations are insoluble and cause neuronal dysfunctions in axonal transport resulting in cell death [89].

Apart from Tau protein, amyloid plaques, which consist of aggregates of Aβ peptide, are responsible for AD development. Neurotoxic Aβ42 peptide is generated by the irregular proteolytic cleavages of transmembrane amyloid precursor protein (APP) extracellular domain by β- and γ-secretases. The cleaved intracellular parts form fibrils due to protein misfolding and can induce tau hyperphosphorylation, disruption of proteasome, mitochondria, and synapses as well [90, 91].

Both tau and Aβ neurotoxicity exhibit altered protein aggregate formation, therefore the clearance of these structures is extremely important in neurons which are unable to eliminate them by dilution through cell division [92]. Thus, a protein quality-control system is needed in neural cells. Autophagy, due to its role of degrading nonfunctional proteins, is one of the candidates to process against neurodegenerative disorders. The increasing autophagosome formation augmented Aβ42 in autophagy-deficient conditions and reduced Beclin1 expression, which provided evidence for the importance of autophagy in AD [93–95]. However, autophagy might not always be the answer. The effect of autophagy is divided into two stages during neurodegenerative disorder: the acute and the chronic condition. Although acute autophagy helps neurons to eliminate neurotoxic aggregates, studies showed that chronic autophagy may be implicated in AD pathogenesis [96]. When autophagy is induced by rapamycin, due to mTOR inhibition γ-secretase activity and Aβ production was found increased by two fold compared to autophagy suppressed mouse fibroblasts [97]. Similar results were also found by serum starvation, where threefold increase in Aβ levels was also observed in human neurons [98].

Consequently, the role of the autophagy is elusive for AD at the initial steps; however, later stages of the same pathway might affect the prognosis negatively. Therapies based on autophagy will require attentive targeting of specific steps of the process for efficient digestion of the aggregates without worsening the disease stage.

3.2. Parkinson's disease

The loss of dopaminergic neurons in the substantia nigra of the central nervous system is the primary cause of Parkinson's disease (PD). The pathology of the disease requires characteristic Lewy bodies in the nuclei of neurons [99]. Lewy bodies contain insoluble α-synuclein aggregates. α-synuclein, in nonpathological conditions, has the ability to bind membrane phospholipids and involved in presynaptic membrane procedures during neurotransmitter release, especially dopamine [100, 101]. The accumulation of α-synuclein occurs due to two missense mutations during PD: A53T and A30P [102]. A small percentage of the aggregates carrying these two mutations have been shown to recycle by the proteasomal degradation or CMA in dopaminergic neurons. During CMA, pathologic α-synucleins are directly targeted to lysosomes by HSC70 due to their Lys-Phe-Glu-Arg-Gln (KFERQ) amino acid sequence without involvement of vesicle formation apart from macroautophagy [103]. Mutated α-synucleins can accumulate with extra phosphate groups which led to the loss of the recognition sequence for CMA. In this case, the accumulation cannot be tolerated by CMA and dopaminergic neurons die via apoptosis [104]. Other than α-synuclein, parkin and PINK1 (PTEN-induced putative kinase 1) are PD promoting molecules.

Parkin is an ubiquitin E3 ligase, located in mitochondria, which regulates variety of cellular processes in neural cells. Loss of parkin has been suggested as the second most common cause of PD. On the other hand, PINK1, with parkin, manage mitochondrial quality control. Recent studies indicate that upon mitochondria membrane potential loss, PINK1 cannot be imported to mitochondria and it accumulates in the cytoplasm where it recruits parkin to induce mitophagy [105]. Therefore, dysfunctional mitochondria are degraded under normal cellular conditions. However, when PINK1 is mutated, altered parkin activity leads to autophagy impairment and mitochondria imbalance which has been reported for animal models of PD [106, 107]. In addition, PINK1 also interplays with Beclin1. PINK1 mutation during PD cause defective or loss of PINK1-Beclin1 interaction and thus resulted in insufficient autophagic activity [108]. Taken together, PD promoting proteins having a role in either cell membrane or mitochondria integrity are also involved in the induction of autophagy. Therefore, mutations in their genes cause defective autophagy process and leads to the accumulation of both α-synuclein and unhealthy mitochondria leading to the apoptotic cell death of dopaminergic neurons in the substantia nigra.

3.3. Amyotrophic lateral sclerosis

Amyotrophic lateral sclerosis (ALS) is a neurodegenerative disease characterized by the degeneration of both upper and lower motor neurons resulting in paralysis. ALS patients suffer from muscle weakness, atrophy, and spasticity. Denervation of the respiratory muscles and diaphragm is the fatal event of ALS. Although most incidences of ALS are sporadic, 20% of all cases are hereditary and caused by mutations in the superoxide dismutase 1 gene (SOD1) [109]. SOD1 is responsible to convert the natural byproduct of respiration, superoxide, to water and hydrogen peroxide. Therefore, defective SOD1 is not able to work properly causing loss of detoxification in motor neurons [110]. The alanine-to-valine substitution at position 4 of SOD1 is responsible for most of the cases, and patients carrying the mutation have a mean survival of 1 year after onset. Mice having the mutant SOD1 gene have been shown also to develop progressive motor neuron degeneration [111, 112]. ALS mice expressing mutant SOD1 have defective protein folding, mitochondrial dysfunction, oxidative stress, inflammation, and toxicity. More importantly, these mice exhibited aberrant neuronal aggregates composed by insoluble forms of SOD1 in their motor neurons, which suggested a pathological hallmark of ALS. These aggregates, also detected in sporadic ALS patients, were shown to carry not only SOD1 but also neurofilaments, peripherin, an intermediate filament subunit, and ubiquitin [110, 113, 114]. Therefore, it is concluded that although cells are willing to eliminate aggregates following ubiquitination through proteasomal degradation, misfolding due to mutations provides an obstacle for this process, which prevents them from degrading. In addition, proteasome malfunction has been implicated in motor neuron death during ALS [115]. The experimental models of ALS suggested that the above aggregates are cleared by autophagy. When autophagy was inhibited by 3MA or bafilomycin, cell viability was found further decreased in *in vitro* ALS models [116]. In contrary, autophagy inducer lithium increased the number of Renshaw cells, interneurons found in the spinal cord, which are affected early during experimental ALS [117]. Therefore, it is concluded that a proper autophagy mechanism is needed for the elimination of the aggregates for ALS treatment.

3.4. Multiple sclerosis

Studies on age-related diseases revealed that there is a relationship between age and the rate of disability progression of multiple sclerosis (MS). Although MS patients are usually diagnosed between the ages of 20 and 50, the relapse-remitting form of MS exhibits active symptomatic period by the age, indicating a faster rate of disease progression in older patients. MS is the most common autoimmune inflammatory demyelinating disease of the central nervous system. Demyelination occurs due to T cells and activated microglia attack to myelin proteins resulting in axonal injury and loss of oligodendrocytes. Findings also indicate that MS patients have increased T and B lymphocyte levels in demyelinated areas due to blood-brain barrier disruption [118]. In addition, dysfunction of mitochondria is one of the important factors in the pathogenesis of MS [119]. The decreased expression of cytochrome *c* oxidase impairs the function of mitochondria [120]. Dysfunction of mitochondria induces reactive oxygen species generation, contributing demyelination, and axonal loss [121]. Recent studies revealed that autophagy plays a role in the progress of MS and experimental autoimmune encephalomyelitis (EAE), which is accepted as the mouse model of MS. Studies indicated that depolarized mitochondria is engulfed and degraded in autophagic vacuoles to reduce the excessive production of ROS, which was supported by the increase of Beclin1 and Atg4 expression in MS brains [122]. On the other hand, exposure of rapamycin, mTOR inhibitor, prevented relapsing-remitting EAE.

It was also shown that Atg5 was increased, whereas Atg16L2 is reduced in T cells in EAE and relapsing-remitting MS brains [123]. Atg-5-deficient mice were reported to have impaired T-cell function and survival [124]. All these data suggest that autophagy relates to both prevention of MS by degrading defective mitochondria and inducer of MS through Atg5 to extend T-cell survival [125].

3.5. Huntington disease

Huntington disease (HD) is an autosomal-dominant neurodegenerative disorder with a distinct phenotype, including cognitive decline, muscle incoordination. The HD symptoms are noticeable between the ages 35–45; however, the case gradually worsens at the old age with dementia, pneumonia, and heart diseases. HD develops due to a mutation in huntingtin protein, an expanded CAG repeat leading to a toxic polyglutamine strand of variable length at the *N*-terminus. Normally associated with vesicle and microtubule function, mutated huntingtin accumulates in tissues causing undegradable molecules by proteosomal degradation [126]. In this step, macroautophagy acts as a compensatory mechanism for the elimination of huntingtin [127]. Studies indicated that HD is associated with impaired degradation process of autophagosomes resulting in the accumulation of highly ubiquitinated aggregates of huntingtin in the endosomal-lysosomal organelles. Moreover, mutations in Atg genes, especially in Atg7 (V471A), have been linked to disease onset [128].

4. Genetic basis of autophagy-related genes and diseases

Since Human Genome Project is completed in 2003 and HapMap Project in 2005, valuable bioinformatics data were gained and published for research interested tools. Instead of classical Sanger-type DNA sequencing, next-generation DNA sequencing equipment accelerated the human gene-related alteration and human disease among ethnic population all around the world [129, 130]. While Mendelian-mediated monogenic and multifactorial-induced polygenic genes responsible for diseases were determined and localized within the human genome, association mediated regression analysis of human population genetics revealed some specific genes related with various cellular process involved in diseases has been assumed. One of the essential cellular processes, autophagy, is the process of long-lived proteins and organelles that are nonfunctional or damaged, maintaining the cellular homeostasis mediated by autophagosome and autolysosome formation [131]. Autophagy is demostrated as a protective event against oxidizable substrates, various pathological processes such as aging, neurodegeneration, cancer, diabetes, obesity, cardiac disease infection, and immunity [132–136]. These diseases are linked with various autophagy process key elements expressing genes via etiology of them. All the mutated autophagy-related genes and linked disease are presented in **Table 1**.

Aging slows autophagy and prevents cellular defense mechanism against metabolic stress factor. However, all findings indicate that the functional role of autophagy differs in conditional manner.

Genes	Mutation type	Associated human disease	Ref.
Gene products required for autophagosome formation			
ATG5	Polymorphism	Asthma and risk of systemic lupus erythematosus	[137]
ATG16L	T300A	Crohn's disease	[138]
BECN1	Monoallelic deletion	Risk and prognosis of human breast, ovarian, prostate and colorectal cancer	[139–142]
EI24/PIG8	Mutations and deletions	Risk of early breast cancer	[143]
TECPR2	Frameshift mutation	Hereditary spastic paraparesis	[144]
UVRAG	Deletion mutation	Static encephalopathy of childhood with neurodegeneration in adulthood	[145]
Gene products required for autophagosome maturation/degradation			
EPG5	Recessive mutations	Vici Syndrome	[146]
IRGM	SNPs, deletions	Risk of Crohn's disease	[147]
ZFYVE26/SPG15	Mutations	Hereditary spastic paraparesis type 15	[148]
Gene products required for induction of mitophagy			
PARK2/Parkin	Mutation	Autosomal recessive or sporadic early-onset Parkinson's disease	[149]
PARK6/PINK1	Mutations	Autosomal recessive or sporadic early-onset Parkinson's disease	[150]
Gene products involved in autophagosomal sequestration, movement or maturation			
SQSTM1/P62	Mutations	Paget disease of bone and amyotrophic lateral sclerosis	[151]
CLN3	Mutations	Batten disease	[152]
LAMP-2	Mutations	Danon disease	[153]
Dynactin subunit p150	Mutations	Spinal or bulbar muscular atrophy	[154]

Table 1. Germline and somatic mutations in human diseases-related with autophagy.

4.1. Static encephalopathy of childhood with neurodegeneration in adulthood (SENDA)

SENDA begins with early childhood intellectual impairment. Unlike the other forms of NBIA, however, the cognitive dysfunction remains nonprogressive, sometimes for decades, after first being recognized. Then, in adulthood, affected patients develop severe dystonia-parkinsonism and later exhibit signs of a progressive dementia. Although no etiology has yet been identified for SENDA, autophagy is focused on the pathogenesis of the disease [155, 156]. The neuroimaging of SENDA is distinct. In addition to iron deposition in the globus pallidus and substantia nigra, SENDA features T1 hyperintensity of the substantia nigra with a central band of T1 hypointensity. Significant cerebral and milder cerebellar atrophy also occur in elder age [157].

Recent studies about autophagy-related genes has been identified, *de novo* mutations within WD Repeat-Containing Protein 45 (WDR45) gene [158]. WDR45 is a member of WD repeat protein family, encodes WD repeats which has minimally 40 amino acid conserved region leads to heterotrimeric or multiprotein complex generation [159]. WDR family member proteins are involved in cell-cycle progression, signal transduction, apoptosis, and gene regulation [160]. WDR45 gene is located at Xp11.23 band, and 25.9 kb length gene composed of 12 exons and 11 introns [161]. WDR45 gene expressed one of the four mammalian homologs of yeast Atg18 protein. Besides, Atg18 is the major autophagosome formation-related protein, with WIPIs proteins Atg18 belonging to PROPPIN family proteins [162]. Atg18/WIPIs protein complexes interact directly with Atg2 and this complex cross-talk with class III PtIns 3-kinase during autophagosome formation [163]. Homologs of Atg18 are ATG-18 and EPG-6 in *C. elegans*. Although *C. elegans* needs each homolog proteins at the same time to form autophagosome, homology of WDR45/WIPI4 in human shows powerful correlation with EPG-6 than ATG-18 [164]. Next-generation whole-exome sequencing results revealed that static encephalopathy of childhood with neurodegeneration in adulthood is classified as a subtype of neurodegenerative disease category. The etiology of this disease is mainly related to iron accumulation in brain leading to paraplegia and mental retardation at early onset. Other symptoms of this disease are aggressive behavior, abnormality of eye movement, absent speech, cerebellar atrophy, cerebral atrophy, dementia, dystonia, neurodegeneration, parkinsonism, and spastic paraparesis [165]. Lymphoblastoid cell lines derived from SENDA patients highlighted the reduced level of WIP14 expression compared to healthy control cases. In affected patients lymphoblastoid cell lines demonstrated abnormal accumulation of ATG9A and LC3-double positive components leading to autophagy blockage [159]. As both WDR45 and WIPI4 genes are localized within the same locus at chromosome X. Although the gender-dependent gene expression through X inactivation has not been determined yet, female-type mosaics and male-type hemizyotic lethally confused the molecular processes [166].

4.2. Vici syndrome

Vici syndrome is a very rare and severe congenital multisystem disorder characterized by the principal features of agenesis of the corpus callosum, cataracts, oculocutaneous hypopigmentation, cardiomyopathy, and combined immunodeficiency. The pathogenesis of Vici syndrome is related with autophagy because of the putative role of autolysosome formation gene, EPG-5 [Ectopic P-Granules Autophagy Protein 5 Homolog (*C. elegans*)] [167]. EPG-5 is a metazoan-specific autophagy gene that encodes a large coiled coil domain-containing protein that functions in autophagy during starvation conditions. Note that 5.9 kb long EPG-5 gene is localized in the Xp11.23 band and it is composed of 16 exons and 15 introns. Mutations within EPG-5 reported to be autophagy defective profile in *C. elegans* [146]. Moreover, lack of *C. elegans* EPG-5 demonstrated an accumulation of nondigested autolysosome in mammalian cells. Accumulation of SSTM/P62 and NBR1 are the leading cause of autophagy flux blockage in *in vitro* culture of fibroblasts from Vici syndrome patients [168]. In addition, the role of EPG-5 in

Vici syndrome also indicated by reporting of dysregulation in endocytic pathway. In addition, EPG5$^{-/-}$ mice displayed some of the symptoms of Vici syndrome such as facial dysmorphism and cataracts [167].

4.3. Danon disease

Danon disease is a X-linked recessive disease characterized by weakening of the heart muscle (cardiomyopathy); weakening of the muscles used for movement, called skeletal muscles (myopathy); and intellectual disability [169]. Age and gender are the main risk factors for Danon disease as males develop Danon disease earlier than female and the symptoms come up in childhood or adolescence in most affected males and in early adulthood in most affected females. Heart-related signs and symptoms, including a sensation of fluttering or pounding in the chest (palpitations), an abnormal heartbeat (arrhythmia), or chest pain are the symptoms of Danon disease [170]. The association between autophagy and Danon disease is dependent on the gene that is responsible for Danon disease formation, LAMP2. LAMP2 gene encodes integral lysosomal membrane proteins that is an essential protein involved in the autophagosome vesicle formation *via* interaction with LAMP-2A [171]. The gene responsible for Danon disease is lysosomal-associated membrane protein 2 (LAMP2) that is localized at Xq24 band [172]. LAMP2 protein is a member of a family of membrane glycoproteins, which provides selectins with carbohydrate ligands and it may also function in the protection, maintenance, and adhesion of the lysosome. Alternative splicing of this gene results in multiple transcript variants encoding distinct proteins [173]. LAMP2 gene is composed of 9 exons and 8 introns, 43.2 kb in length. GLN174TER, VAL310ILE, and TRP321ARG mutations within the LAMP2 gene lead to Danon disease [174].

4.4. Liver disease

Both liver disease and lung disease may be developed with the deficiency of alpha-1 antitrypsin gene [175]. Alpha trypsin deficiency prevalence is 1/1500–3500 individuals among European ancestry population. The disease onset varies at age range among individuals. First signs of lung disease in alpha-1 antitrypsin deficiency are generally observed between ages 20 and 50 [176]. Among alpha-1 antitrypsin deficiency patients, 10% of them develop liver disease, which can be diagnosed by yellowing of the skin and whites of the eyes (jaundice). Approximately 15% of adults with alpha-1 antitrypsin deficiency develop liver damage (cirrhosis) due to the formation of scar tissue in the liver [177]. Since liver disease is a multifactorial polygenic disease and alcohol and hepatotoxic agents are the major environmental risk factors for liver disease causing cirrhosis include a swollen abdomen, swollen feet or legs, and jaundice. The most common genetic risk factor for liver disease is the alpha-antitrypsin gene deficiency [178]. The protein encoded by SERPINA1 [serpin peptidase inhibitor, clade A (alpha-1 antiproteinase, antitrypsin)] is secreted as a serine protease inhibitor. It has a number of targets including elastase, plasmin, thrombin, trypsin, chymotrypsin, and plasminogen activator. The mutations on SERPINA1 can cause emphysema or liver disease. Several transcript variants encoding the same protein have been found for this gene. SERPINA1 is composed of 7 exons and 6 introns with a length of 13.9 kb that is located in the 14q32.1 band [179]. SERPINA1 gene product plays an essential role in the hepato-detoxification process of ZZ genotype of alpha antitrypsin deficiency syndrome diagnosed by PCR amplification and RFLP analysis [180]. By using 19-mer synthetic oligonucleotide probes, SZ phenotype is reported to be associated with M/S difference in exon 3 and M/Z difference in exon 5, whereas phenotype of MZ heterozygotes showed a low Z expression [181]. By routine isoelectric focusing of affected Z type and MZ (her husband genotype) of an obligate carrier mother of PI(M)/PI(null), heterozygote showed atypically low concentrations of circulating Z peptides, which were demonstrated by Harrison et al. [182]. Accumulation of ZZ peptides as intracellular inclusion bodies was reported by Lomas [183] in the ZZ homozygote. Moreover, it was shown that only about 15% of the AAT protein is secreted in the plasma in ZZ homozygotes and the rest of 85% of the protein is not secreted and accumulates in the endoplasmic reticulum (ER) of the hepatocyte. Thus, about 10% of newborn ZZ homozygotes develop liver disease that often leads to fatal childhood cirrhosis. Antitrypsin is an acute phase protein and undergoes a manifold increase in association with temperature elevations during triggered inflammation. Regulation of triggered inflammation and pyrexia symptoms in ZZ homozygote infants is found critical [183]. Wild-type protein primarily degraded by proteosomal activity, mutant alpha-ATZ protein, is reported to be digested autophagy-mediated degradation. According to Yorimitsu and Klionsky et al [1], depletion of Atg-5 in hepatocytes leads to the formation of insoluble aggregates of ATZ proteins and increased production of inclusion bodies. Although the protective or tumor suppressor effect of ATZ protein via autophagy regulation has not demonstrated yet, general evidences support the role of ATZ as a protection against alcohol and hepatotoxic agents [184] (**Figure 6**).

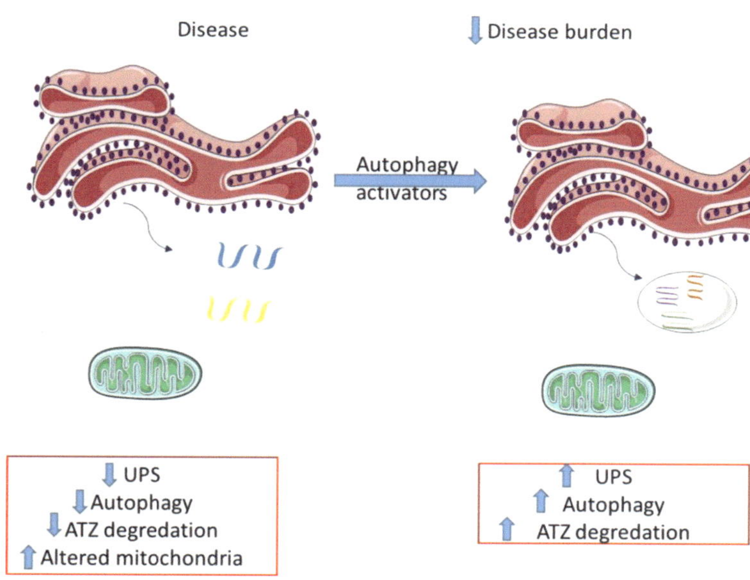

Figure 6. The role of autophagy in alpha antitrypsin deficiency syndrome etiology.

4.5. Myopathy

Myopathies are neuromuscular disorders in which the primary symptom is muscle weakness due to dysfunction of muscle fiber. As different genes are responsible for various types of myopathy, the symptoms of myopathy can include muscle cramps, stiffness, and spasm [185]. Moreover, myopathies can be inherited as the muscular dystrophies or acquired as common muscle cramps [186]. Congenita (developmental delays in motor skills; skeletal and facial abnormalities are occasionally evident at birth), muscular dystrophies (progressive weakness in voluntary muscles; sometimes evident at birth), mitochondrial myopathies (such as in Kearns-Sayre syndrome, MELAS, and MERRF), glycogen storage diseases of muscle (Pompe's, Andersen's, and Cori's diseases), dermatomyositis (inflammatory myopathy of skin and muscle), myositis ossificans (bone growing in muscle tissue), polymyositis, inclusion body myositis, and related myopathies (inflammatory myopathies of skeletal muscle), neuromyotonia (alternation episodes of twitching and stiffness), tetany (characterized by prolonged spasms of the arms and legs is defined as myopathy), and major symptoms of Danon disease is linked with myopathy. There are also various autophagy-related myopathic disorders such as X-linked congenital autophagic vascular myopathy and adult onset vacuolar myopathy with multiorgan involvement that the etiology machinery has not been highlighted [187–190]. All these emphases disorders are predicted to be associated with autophagosome-lysosome fusion. Among muscle diseases sporadic inclusion body myositis, limb girdle muscular dystrophy type 2B, and miyoshi myopathy are shown to be associated with autophagy *via* clearance of the disease causing proteins during molecular pathogenesis [191].

4.6. Cardiac disease

X-linked Danon disease or lysosomal storage disorder and Pompe diseases are rare hereditary diseases of heart, and they are associated with imperfect autophagy processes due to impaired autophagosome lysosome fusion. Patients with coronary artery disease, hypertension, aortic valvular disease, and congestive heart failure are associated with autophagy [135, 192]. The cardiomyocytes isolation from cardiac disease rodent models showed that an obvious accumulation of autophagosomes is distinguishable [193]. Although it is not well clarified that autophagy might exert cytoprotective effects in these models via regulating ATP production, protein, and organelle quality control, or other mechanisms [136]. Atg5 knockout heart tissue models in adult mice results cardiac hypertrophy and contractile dysfunction. The heart consumes more energy per gram than any other tissue in the body. Therefore, energy turnover mechanisms are strictly orchestrated in normal heart tissue. In contrary, cell homeostasis is not properly regulated in a number of cardiac disorders such as cardiac ischemia and heart failure, which are characterized by a reduction in the availability of energy substrates [194]. Furthermore, long-term cardiac stress may remodel in myocytes through inducing elongation and hypertrophy to adapt to stress factors [195]. According to previous data, it is well established that heart tissue required more energy substrates under stress conditions. Therefore, active autophagy may increase the survival of heart cells when they were exposed to stress. Cardiac-specific Atg-5 knockout models did not exert any physiological change under normal conditions. However, stress induction caused more severe pathophysiological processes. Therefore, these data suggest that upregulation of autophagy in failing hearts is an adaptive response that protects against hemodynamic or neurohormonal stresses. Furthermore, it was shown that Beclin-1 protects contractile functions in the myocytes after stress overload [196]. While heterozygous disruption of Beclin-1 mediated decrease in the size of the myocardial infarction after ischemia/reperfusion, Beclin-1 overexpression decreased cell injury in an *in vitro* model of cardiac ischemia/reperfusion. Dominant-negative Atg5 overexpression increased cell injury, suggesting a protective effect for both ATG genes in ischemia/reperfusion [196].

5. Conclusion

Under physiological conditions, autophagy-related processes are important to provide unique cell homeostasis. In addition, when the cells are exposed to a number of environmental or cellular stress factors, full functional autophagy may protect cells against stress factors. However, as we discussed in previous parts, aging is a multifactorial process, which renders functional regulatory pathways and cause a pathophysiological problems in the cells. Aging in the presence of metabolic diseases causes the impairment of a number of critical genes, which orchestrates autophagy. For this reason, the determination of potential role of autophagic processes in aging-related diseases has potential to provide better therapeutic strategies in the treatment of diseases.

Briefly all mentioned signalling cascades related to aging are altered during autophagic processes. As shown in **Figure 5**, the incidence of orange-colored diseases is increased in an age-dependent manner. Under normal conditions physiological energy balance is orchestrated by several factors placed in blue boxes. Intracellular stress and nutrient sensors labeled in green exert their functional roles through modulating different signaling cascades, which produce a variety of cellular responses as placed in yellow boxes.

The functional role of autophagy in diseases is controversial. Also nutritional status, oxidative stress, and genetic basis of autophagy-related molecular targets determine the disease progression. In this section, we discuss the functional role of autophagy in genetic manner in common diseases and rare diseases (Vici syndrome, SENDA, and Danon disease).

Author details

Elif Damla Arisan*, Pinar Obakan-Yerlikaya, Ajda Coker-Gurkan and Narçin Palavan Unsal

*Address all correspondence to: d.arisan@iku.edu.tr

Department of Molecular Biology and Genetics, Istanbul Kultur University, Istanbul, Turkey

References

[1] Klionsky, D.J., et al., Guidelines for the use and interpretation of assays for monitoring autophagy (3rd edition). Autophagy, 2016. 12(1): 1–222.

[2] Klionsky, D.J. and P. Codogno, The mechanism and physiological function of macroautophagy. J Innate Immun, 2013. 5(5): 427–433.

[3] Klionsky, D.J., Why just eat in, when you can also eat out? Autophagy, 2013. 9(2): 119.

[4] Lee, W.S., W.H. Yoo, and H.J. Chae, ER stress and autophagy. Curr Mol Med, 2015. 15(8): 735–745.

[5] Jordan, K., F. Jahn, and M. Aapro, Recent developments in the prevention of chemotherapy-induced nausea and vomiting (CINV): a comprehensive review. Ann Oncol, 2015. 26(6): 1081–1090.

[6] Sahni, S., et al., Gene of the month: BECN1. J Clin Pathol, 2014. 67(8): 656–660.

[7] Coker-Gurkan, A., et al., Purvalanol induces endoplasmic reticulum stress-mediated apoptosis and autophagy in a time-dependent manner in HCT116 colon cancer cells. Oncol Rep, 2015. 33(6): 2761–2770.

[8] Coker-Gurkan, A., et al., Inhibition of autophagy by 3-MA potentiates purvalanol-induced apoptosis in Bax deficient HCT 116 colon cancer cells. Exp Cell Res, 2014. 328(1): 87–98.

[9] Tacar, O. and C.R. Dass, Doxorubicin-induced death in tumour cells and cardiomyocytes: is autophagy the key to improving future clinical outcomes? J Pharm Pharmacol, 2013. 65(11): 1577–1589.

[10] Tachibana, K. and C. Kitanaka, Autophagy and autophagic cell death. Tanpakushitsu Kakusan Koso, 2006. 51(10 Suppl): 1519–1524.

[11] Yip, K.W. and J.C. Reed, Bcl-2 family proteins and cancer. Oncogene, 2008. 27(50): 6398–6406.

[12] Kim, Y.C. and K.L. Guan, mTOR: a pharmacologic target for autophagy regulation. J Clin Invest, 2015. 125(1): 25–32.

[13] Huber, K.M., et al., Dysregulation of mammalian target of rapamycin signaling in mouse models of autism. J Neurosci, 2015. 35(41): 13836–13842.

[14] Tsun, Z.Y. and R. Possemato, Amino acid management in cancer. Semin Cell Dev Biol, 2015. 43: 22–32.

[15] Cai, Y. and Y.H. Wei, Stress resistance and lifespan are increased in C. elegans but decreased in S. cerevisiae by mafr-1/maf1 deletion. Oncotarget, 2016. 7(10):10812–26.

[16] Strozyk, E. and D. Kulms, The role of AKT/mTOR pathway in stress response to UV-irradiation: implication in skin carcinogenesis by regulation of apoptosis, autophagy and senescence. Int J Mol Sci, 2013. 14(8): 15260–15285.

[17] Maiese, K., Targeting molecules to medicine with mTOR, autophagy, and neurodegenerative disorders. Br J Clin Pharmacol, 2015 (no information)

[18] Cai, Z., et al., Activation of mTOR: a culprit of Alzheimer's disease? Neuropsychiatr Dis Treat, 2015. 11: 1015–1030.

[19] Cetrullo, S., et al., mTOR, AMPK, and Sirt1: key players in metabolic stress management. Crit Rev Eukaryot Gene Expr, 2015. 25(1): 59–75.

[20] Krishan, S., D.R. Richardson, and S. Sahni, Gene of the month. AMP kinase (PRKAA1). J Clin Pathol, 2014. 67(9): 758–763.

[21] Alers, S., et al., Role of AMPK-mTOR-Ulk1/2 in the regulation of autophagy: cross talk, shortcuts, and feedbacks. Mol Cell Biol, 2012. 32(1): 2–11.

[22] Qi, D. and L.H. Young, AMPK: energy sensor and survival mechanism in the ischemic heart. Trends Endocrinol Metab, 2015. 26(8): 422–429.

[23] Sanli, T., et al., AMP-activated protein kinase (AMPK) beyond metabolism: a novel genomic stress sensor participating in the DNA damage response pathway. Cancer Biol Ther, 2014. 15(2): 156–169.

[24] Ha, J., K.L. Guan, and J. Kim, AMPK and autophagy in glucose/glycogen metabolism. Mol Aspects Med, 2015. 46: 46–62.

[25] Salminen, A. and K. Kaarniranta, AMP-activated protein kinase (AMPK) controls the aging process via an integrated signaling network. Ageing Res Rev, 2012. 11(2): 230–241.

[26] Egan, D.F., et al., Phosphorylation of ULK1 (hATG1) by AMP-activated protein kinase connects energy sensing to mitophagy. Science, 2011. 331(6016): 456–461.

[27] Bourron, O., et al., Biguanides and thiazolidinediones inhibit stimulated lipolysis in human adipocytes through activation of AMP-activated protein kinase. Diabetologia, 2010. 53(4): 768–778.

[28] Kefas, B.A., et al., Metformin-induced stimulation of AMP-activated protein kinase in beta-cells impairs their glucose responsiveness and can lead to apoptosis. Biochem Pharmacol, 2004. 68(3): 409–416.

[29] Zhou, G., et al., Role of AMP-activated protein kinase in mechanism of metformin action. J Clin Invest, 2001. 108(8): 1167–1174.

[30] Foretz, M., et al., AMP-activated protein kinase inhibits the glucose-activated expression of fatty acid synthase gene in rat hepatocytes. J Biol Chem, 1998. 273(24): 14767–14771.

[31] Leclerc, I., A. Kahn, and B. Doiron, The 5'-AMP-activated protein kinase inhibits the transcriptional stimulation by glucose in liver cells, acting through the glucose response complex. FEBS Lett, 1998. 431(2): 180–184.

[32] Xiang, X., et al., AMP-activated protein kinase activators can inhibit the growth of prostate cancer cells by multiple mechanisms. Biochem Biophys Res Commun, 2004. 321(1): 161–167.

[33] Speakman, J.R. and S.E. Mitchell, Caloric restriction. Mol Aspects Med, 2011. 32(3): 159–221.

[34] Ruderman, N.B., et al., AMPK and SIRT1: a long-standing partnership? Am J Physiol Endocrinol Metab, 2010. 298(4): E751–E760.

[35] Dice, J.F., Altered degradation of proteins microinjected into senescent human fibroblasts. J Biol Chem, 1982. 257(24): 14624–14627.

[36] Makrides, S.C., Protein synthesis and degradation during aging and senescence. Biol Rev Camb Philos Soc, 1983. 58(3): 343–422.

[37] Cuervo, A.M., Calorie restriction and aging: the ultimate "cleansing diet." J Gerontol A Biol Sci Med Sci, 2008. 63(6): 547–549.

[38] Hansen, M., et al., A role for autophagy in the extension of lifespan by dietary restriction in C. elegans. PLoS Genet, 2008. 4(2): e24.

[39] Partridge, L., Some highlights of research on aging with invertebrates, 2008. Aging Cell, 2008. 7(5): 605–608.

[40] Tavernarakis, N., et al., The effects of p53 on whole organism longevity are mediated by autophagy. Autophagy, 2008. 4(7): 870–873.

[41] Guarente, L., Sirtuins in aging and disease. Cold Spring Harb Symp Quant Biol, 2007. 72: 483–488.

[42] Guarente, L. and F. Picard, Calorie restriction--the SIR2 connection. Cell, 2005. 120(4): 473–482.

[43] Allard, J.S., et al., Dietary activators of Sirt1. Mol Cell Endocrinol, 2009. 299(1): 58–63.

[44] Blagosklonny, M.V., Linking calorie restriction to longevity through sirtuins and autophagy: any role for TOR. Cell Death Dis, 2010. 1: e12.

[45] Dali-Youcef, N., et al., Sirtuins: the 'magnificent seven', function, metabolism and longevity. Ann Med, 2007. 39(5): 335–345.

[46] Yu, Z., et al., Rapamycin and dietary restriction induce metabolically distinctive changes in mouse liver. J Gerontol A Biol Sci Med Sci, 2015. 70(4): 410–420.

[47] Mirzaei, H. and V.D. Longo, Acetyl-CoA synthetase is a conserved regulator of autophagy and life span. Cell Metab, 2014. 19(4): 555–557.

[48] Schroeder, S., et al., Acetyl-coenzyme A: a metabolic master regulator of autophagy and longevity. Autophagy, 2014. 10(7): 1335–1337.

[49] Mattison, J.A., et al., Calorie restriction in rhesus monkeys. Exp Gerontol, 2003. 38(1-2): 35–46.

[50] Camici, G.G., et al., Anti-aging medicine: molecular basis for endothelial cell-targeted strategies - a mini-review. Gerontology, 2011. 57(2): 101–108.

[51] Matsui, K., et al., NAD-dependent histone deacetylase, SIRT1, plays essential roles in the maintenance of hematopoietic stem cells. Biochem Biophys Res Commun, 2012. 418(4): 811–817.

[52] Nemoto, S., M.M. Fergusson, and T. Finkel, Nutrient availability regulates SIRT1 through a forkhead-dependent pathway. Science, 2004. 306(5704): 2105–2108.

[53] Gonfloni, S., et al., P53 and Sirt1: routes of metabolism and genome stability. Biochem Pharmacol, 2014. 92(1): 149–156.

[54] Wohlgemuth, S.E., et al., Skeletal muscle autophagy and apoptosis during aging: effects of calorie restriction and life-long exercise. Exp Gerontol, 2010. 45(2): 138–148.

[55] Bergamini, E., et al., The role of macroautophagy in the ageing process, anti-ageing intervention and age-associated diseases. Int J Biochem Cell Biol, 2004. 36(12): 2392–3404.

[56] Menendez, J.A., et al., mTOR-regulated senescence and autophagy during reprogramming of somatic cells to pluripotency: a roadmap from energy metabolism to stem cell renewal and aging. Cell Cycle, 2011. 10(21): 3658–3677.

[57] Steinberger, J., et al., Progress and challenges in metabolic syndrome in children and adolescents: a scientific statement from the American Heart Association Atherosclerosis, Hypertension, and Obesity in the Young Committee of the Council on Cardiovascular Disease in the Young; Council on Cardiovascular Nursing; and Council on Nutrition, Physical Activity, and Metabolism. Circulation, 2009. 119(4): 628–647.

[58] Hardie, D.G., AMPK and autophagy get connected. EMBO J, 2011. 30(4): 634–635.

[59] Lage, R., et al., AMPK: a metabolic gauge regulating whole-body energy homeostasis. Trends Mol Med, 2008. 14(12): 539–549.

[60] Narbonne, P. and R. Roy, Caenorhabditis elegans dauers need LKB1/AMPK to ration lipid reserves and ensure long-term survival. Nature, 2009. 457(7226): 210–214.

[61] Ulgherait, M., et al., AMPK modulates tissue and organismal aging in a non-cell-autonomous manner. Cell Rep, 2014. 8(6): 1767–1780.

[62] Anisimov, V.N., Life span extension and cancer risk: myths and reality. Exp Gerontol, 2001. 36(7): 1101–1136.

[63] Hursting, S.D., et al., Calorie restriction, aging, and cancer prevention: mechanisms of action and applicability to humans. Annu Rev Med, 2003. 54: 131–152.

[64] Mai, V., et al., Calorie restriction and diet composition modulate spontaneous intestinal tumorigenesis in Apc(Min) mice through different mechanisms. Cancer Res, 2003. 63(8): 1752–1755.

[65] Ghanayem, B.I., et al., Diet-induced obesity in male mice is associated with reduced fertility and potentiation of acrylamide-induced reproductive toxicity. Biol Reprod, 2010. 82(1): 96–104.

[66] Hill-Baskin, A.E., et al., Diet-induced hepatocellular carcinoma in genetically predisposed mice. Hum Mol Genet, 2009. 18(16): 2975–2988.

[67] Paik, J., et al., High-fat diet-induced obesity exacerbates inflammatory bowel disease in genetically susceptible Mdr1a-/- male mice. J Nutr, 2013. 143(8): 1240–1247.

[68] Tajima, K., et al., Metformin prevents liver tumorigenesis induced by high-fat diet in C57Bl/6 mice. Am J Physiol Endocrinol Metab, 2013. 305(8): E987–E998.

[69] Harman, D., The free radical theory of aging. Antioxid Redox Signal, 2003. 5(5): 557–561.

[70] Franceschi, C., et al., Inflamm-aging. An evolutionary perspective on immunosenescence. Ann N Y Acad Sci, 2000. 908: 244–254.

[71] Chung, H.Y., et al., Molecular inflammation: underpinnings of aging and age-related diseases. Ageing Res Rev, 2009. 8(1): 18–30.

[72] Cadenas, E. and K.J. Davies, Mitochondrial free radical generation, oxidative stress, and aging. Free Radic Biol Med, 2000. 29(3–4): 222–230.

[73] Kowald, A. and T.B. Kirkwood, A network theory of ageing: the interactions of defective mitochondria, aberrant proteins, free radicals and scavengers in the ageing process. Mutat Res, 1996. 316(5–6): 209–236.

[74] Weinert, B.T. and P.S. Timiras, Invited review: Theories of aging. J Appl Physiol (1985), 2003. 95(4): 1706–1716.

[75] Longo, V.D., et al., Interventions to slow aging in humans: are we ready? Aging Cell, 2015. 14(4): 497–510.

[76] De Haes, W., et al., Metformin promotes lifespan through mitohormesis via the peroxiredoxin PRDX-2. Proc Natl Acad Sci U S A, 2014. 111(24): E2501–E2509.

[77] Cabreiro, F., et al., Metformin retards aging in C. elegans by altering microbial folate and methionine metabolism. Cell, 2013. 153(1): 228–239.

[78] Mihaylova, M.M. and R.J. Shaw, The AMPK signalling pathway coordinates cell growth, autophagy and metabolism. Nat Cell Biol, 2011. 13(9): 1016–1023.

[79] Weichhart, T., et al., The TSC-mTOR signaling pathway regulates the innate inflammatory response. Immunity, 2008. 29(4): 565–577.

[80] Choo, A.Y., et al., Rapamycin differentially inhibits S6Ks and 4E-BP1 to mediate cell-type-specific repression of mRNA translation. Proc Natl Acad Sci U S A, 2008. 105(45): 17414–17419.

[81] Demidenko, Z.N. and M.V. Blagosklonny, Growth stimulation leads to cellular senescence when the cell cycle is blocked. Cell Cycle, 2008. 7(21): 3355–3361.

[82] Demidenko, Z.N., et al., Rapamycin decelerates cellular senescence. Cell Cycle, 2009. 8(12): 1888–1895.

[83] Johnson, S.C., P.S. Rabinovitch, and M. Kaeberlein, mTOR is a key modulator of ageing and age-related disease. Nature, 2013. 493(7432): 338–345.

[84] Ravikumar, B. and D.C. Rubinsztein, Can autophagy protect against neurodegeneration caused by aggregate-prone proteins? Neuroreport, 2004. 15(16): 2443–2445.

[85] Rami, A., Review: autophagy in neurodegeneration: firefighter and/or incendiarist? Neuropathol Appl Neurobiol, 2009. 35(5): 449–461.

[86] Flaherty, D.B., et al., Phosphorylation of human tau protein by microtubule-associated kinases: GSK3beta and cdk5 are key participants. J Neurosci Res, 2000. 62(3): 463–472.

[87] Ferrer, I., et al., Phosphorylated map kinase (ERK1, ERK2) expression is associated with early tau deposition in neurones and glial cells, but not with increased nuclear DNA vulnerability and cell death, in Alzheimer disease, Pick's disease, progressive supranuclear palsy and corticobasal degeneration. Brain Pathol, 2001. 11(2): 144–158.

[88] Lee, M.S., et al., Neurotoxicity induces cleavage of p35 to p25 by calpain. Nature, 2000. 405(6784): 360–364.

[89] Mudher, A. and S. Lovestone, Alzheimer's disease-do tauists and baptists finally shake hands? Trends Neurosci, 2002. 25(1): 22–26.

[90] Stein, T.D., et al., Neutralization of transthyretin reverses the neuroprotective effects of secreted amyloid precursor protein (APP) in APPSW mice resulting in tau phosphorylation and loss of hippocampal neurons: support for the amyloid hypothesis. J Neurosci, 2004. 24(35): 7707–7717.

[91] Pakaski, M. and J. Kalman, Interactions between the amyloid and cholinergic mechanisms in Alzheimer's disease. Neurochem Int, 2008. 53(5): 103–111.

[92] Komatsu, M., et al., Loss of autophagy in the central nervous system causes neurodegeneration in mice. Nature, 2006. 441(7095): 880–884.

[93] Nixon, R.A., et al., Extensive involvement of autophagy in Alzheimer disease: an immuno-electron microscopy study. J Neuropathol Exp Neurol, 2005. 64(2): 113–122.

[94] Ling, D. and P.M. Salvaterra, A central role for autophagy in Alzheimer-type neurodegeneration. Autophagy, 2009. 5(5): 738–740.

[95] Pickford, F., et al., The autophagy-related protein beclin 1 shows reduced expression in early Alzheimer disease and regulates amyloid beta accumulation in mice. J Clin Invest, 2008. 118(6): 2190–2199.

[96] Nah, J., J. Yuan, and Y.K. Jung, Autophagy in neurodegenerative diseases: from mechanism to therapeutic approach. Mol Cells, 2015. 38(5): 381–389.

[97] Yu, W.H., et al., Macroautophagy-a novel Beta-amyloid peptide-generating pathway activated in Alzheimer's disease. J Cell Biol, 2005. 171(1): 87–98.

[98] LeBlanc, A.C., R. Xue, and P. Gambetti, Amyloid precursor protein metabolism in primary cell cultures of neurons, astrocytes, and microglia. J Neurochem, 1996. 66(6): 2300–2310.

[99] Schulz-Schaeffer, W.J., Neurodegeneration in Parkinson disease: moving Lewy bodies out of focus. Neurology, 2012. 79(24): 2298–2299.

[100] Ulmer, T.S., et al., Structure and dynamics of micelle-bound human alpha-synuclein. J Biol Chem, 2005. 280(10): 9595–9603.

[101] Abeliovich, A., et al., Mice lacking alpha-synuclein display functional deficits in the nigrostriatal dopamine system. Neuron, 2000. 25(1): 239–252.

[102] Polymeropoulos, M.H., et al., Mutation in the alpha-synuclein gene identified in families with Parkinson's disease. Science, 1997. 276(5321): 2045–2047.

[103] Cuervo, A.M., Chaperone-mediated autophagy: selectivity pays off. Trends Endocrinol Metab, 2010. 21(3): 142–150.

[104] Stefanis, L., Caspase-dependent and -independent neuronal death: two distinct pathways to neuronal injury. Neuroscientist, 2005. 11(1): 50–62.

[105] Vincow, E.S., et al., The PINK1-Parkin pathway promotes both mitophagy and selective respiratory chain turnover in vivo. Proc Natl Acad Sci U S A, 2013. 110(16): 6400–6405.

[106] Tait, S.W. and D.R. Green, Mitochondria and cell signalling. J Cell Sci, 2012. 125(Pt 4): 807–815.

[107] Dawson, T.M. and V.L. Dawson, The role of parkin in familial and sporadic Parkinson's disease. Mov Disord, 2010. 25 Suppl 1: S32–S39.

[108] Michiorri, S., et al., The Parkinson-associated protein PINK1 interacts with Beclin1 and promotes autophagy. Cell Death Differ, 2010. 17(6): 962–974.

[109] Rosen, D.R., et al., Mutations in Cu/Zn superoxide dismutase gene are associated with familial amyotrophic lateral sclerosis. Nature, 1993. 362(6415): 59–62.

[110] Wong, P.C., et al., An adverse property of a familial ALS-linked SOD1 mutation causes motor neuron disease characterized by vacuolar degeneration of mitochondria. Neuron, 1995. 14(6): 1105–1116.

[111] Jonsson, P.A., et al., Minute quantities of misfolded mutant superoxide dismutase-1 cause amyotrophic lateral sclerosis. Brain, 2004. 127(Pt 1): 73–88.

[112] Wang, J., et al., Coincident thresholds of mutant protein for paralytic disease and protein aggregation caused by restrictively expressed superoxide dismutase cDNA. Neurobiol Dis, 2005. 20(3): 943–952.

[113] Gros-Louis, F., et al., A frameshift deletion in peripherin gene associated with amyotrophic lateral sclerosis. J Biol Chem, 2004. 279(44): 45951–45956.

[114] Lim, K.L. and J.M. Tan, Role of the ubiquitin proteasome system in Parkinson's disease. BMC Biochem, 2007. 8 Suppl 1: S13.

[115] Tummala, H., et al., Inhibition of chaperone activity is a shared property of several Cu,Zn-superoxide dismutase mutants that cause amyotrophic lateral sclerosis. J Biol Chem, 2005. 280(18): 17725–17731.

[116] Scotter, E.L., et al., Differential roles of the ubiquitin proteasome system and autophagy in the clearance of soluble and aggregated TDP-43 species. J Cell Sci, 2014. 127(Pt 6): 1263–1278.

[117] Pasquali, L., et al., Autophagy, lithium, and amyotrophic lateral sclerosis. Muscle Nerve, 2009. 40(2): 173–194.

[118] Freedman, M.S., et al., Recommended standard of cerebrospinal fluid analysis in the diagnosis of multiple sclerosis: a consensus statement. Arch Neurol, 2005. 62(6): 865–870.

[119] Liang, P. and W. Le, Role of autophagy in the pathogenesis of multiple sclerosis. Neurosci Bull, 2015. 31(4): 435–444.

[120] Broadwater, L., et al., Analysis of the mitochondrial proteome in multiple sclerosis cortex. Biochim Biophys Acta, 2011. 1812(5): 630–641.

[121] van Horssen, J., et al., Radical changes in multiple sclerosis pathogenesis. Biochim Biophys Acta, 2011. 1812(2): 141–150.

[122] Chen, Y. and S.B. Gibson, Is mitochondrial generation of reactive oxygen species a trigger for autophagy? Autophagy, 2008. 4(2): 246–248.

[123] Alirezaei, M., et al., Elevated ATG5 expression in autoimmune demyelination and multiple sclerosis. Autophagy, 2009. 5(2): 152–158.

[124] Francois, A., et al., Impairment of autophagy in the central nervous system during lipopolysaccharide-induced inflammatory stress in mice. Mol Brain, 2014. 7: 56.

[125] Ghavami, S., et al., Autophagy and apoptosis dysfunction in neurodegenerative disorders. Prog Neurobiol, 2014. 112: 24–49.

[126] Cattaneo, E., C. Zuccato, and M. Tartari, Normal huntingtin function: an alternative approach to Huntington's disease. Nat Rev Neurosci, 2005. 6(12): 919–930.

[127] He, L.Q., J.H. Lu, and Z.Y. Yue, Autophagy in ageing and ageing-associated diseases. Acta Pharmacol Sin, 2013. 34(5): 605–611.

[128] Metzger, S., et al., The V471A polymorphism in autophagy-related gene ATG7 modifies age at onset specifically in Italian Huntington disease patients. PLoS One, 2013. 8(7): e68951.

[129] Conneally, P.M., The human genome project: Its impact on complex disorders. International Seminar on Nuclear War and Planetary Emergencies - 27th Session, 2003: pp. 407–409.

[130] Thorisson, G.A., et al., The International HapMap Project Web site. Genome Res, 2005. 15(11): 1592–1593.

[131] Klionsky, D.J., Autophagy. Curr Biol, 2005. 15(8): R282–R283.

[132] Rubinszfein, D.C., et al., Autophagy and its possible roles in nervous system diseases, damage and repair. Autophagy, 2005. 1(1): 11–22.

[133] Gu, Y.Y., et al., Autophagy-related prognostic signature for breast cancer. Mol Carcinog, 2016. 55(3): 292–299.

[134] Masini, M., et al., Autophagy in human type 2 diabetes pancreatic beta cells. Diabetologia, 2009. 52(6): 1083–1086.

[135] Essick, E.E. and F. Sam, Oxidative stress and autophagy in cardiac disease, neurological disorders, aging and cancer. Oxid Med Cell Longev, 2010. 3(3): 168–177.

[136] Hill, J.A., Autophagy in cardiac plasticity and disease. Pediat Cardiol, 2011. 32(3): 282–289.

[137] Martin, L.J., et al., Functional variant in the autophagy-related 5 gene promotor is associated with childhood asthma. PLoS One, 2012. 7(4): e33454.

[138] Prescott, N.J., et al., A non-synonymous SNP in an autophagy-related gene is associated with Crohn's disease. Gut, 2007. 56: A118–A119.

[139] Saito, H., et al., Detailed deletion mapping of chromosome 17q in ovarian and breast cancers - 2 cm region on 17q21.3 often and commonly deleted in tumors. Am J Hum Genet, 1993. 53(3): 28–28.

[140] Gao, X., et al., Loss of heterozygosity of the Brca1 and other loci on chromosome 17q in human prostate-cancer. Cancer Res, 1995. 55(5): 1002–1005.

[141] Aita, V.M., et al., Cloning and genomic organization of beclin 1, a candidate tumor suppressor gene on chromosome 17q21. Genomics, 1999. 59(1): 59–65.

[142] Liang, X.H., et al., Induction of autophagy and inhibition of tumorigenesis by beclin 1. Nature, 1999. 402(6762): 672–676.

[143] Gentile, M., et al., Candidate tumour suppressor genes at 11q23-q24 in breast cancer: evidence of alterations in PIG8, a gene involved in p53-induced apoptosis. Oncogene, 2001. 20(53): 7753–7760.

[144] Oz-Levi, D., et al., Mutation in TECPR2 reveals a role for autophagy in hereditary spastic paraparesis. Am J Hum Genet, 2012. 91(6): 1065–1072.

[145] Liang, C., et al., Autophagic and tumour suppressor activity of a novel Beclin1-binding protein UVRAG. Nat Cell Biol, 2006. 8(7): 688-694.

[146] Cullup, T., et al., Recessive mutations in EPG5 cause Vici syndrome, a multisystem disorder with defective autophagy. Nat Genet, 2013. 45(1): 83–U122.

[147] Massey, D.C.O. and M. Parkes, Genome-wide association scanning highlights two autophagy genes, ATG16L1 and IRGM, as being significantly associated with Crohn's disease. Autophagy, 2007. 3(6): 649–651.

[148] Vantaggiato, C., et al., Defective autophagy in spastizin mutated patients with hereditary spastic paraparesis type 15. Brain, 2013. 136: 3119–3139.

[149] Trinh, J. and M. Farrer, Advances in the genetics of Parkinson disease. Nat Rev Neurol, 2013. 9(8): 445–454.

[150] Valente, E.M., et al., Hereditary early-onset Parkinson's disease caused by mutations in PINK1. Science, 2004. 304(5674): 1158–1160.

[151] Faccio, D., et al., Cavitation dynamics and directional microbubble ejection induced by intense femtosecond laser pulses in liquids. Phys Rev E, 2012. 86(3): 036304-1-6.

[152] Puls, I., et al., Distal spinal and bulbar muscular atrophy caused by dynactin mutation. Ann Neurol, 2005. 57(5): 687–694.

[153] Nishino, I., et al., Primary LAMP-2 deficiency causes X-linked vacuolar cardiomyopathy and myopathy (Danon disease). Nature, 2000. 406(6798): 906–910.

[154] Vesa, J., et al., Neuronal ceroid lipofuscinoses are connected at molecular level: Interaction of CLN5 protein with CLN2 and CLN3. Mol Biol Cell, 2002. 13(7): 2410–2420.

[155] Schneider, S.A. and K.P. Bhatia, Syndromes of neurodegeneration with brain iron accumulation. Semin Pediatr Neurol, 2012. 19(2): 57–66.

[156] Schneider, S.A., G. Zorzi, and N. Nardocci, Pathophysiology and treatment of neurodegeneration with brain iron accumulation in the pediatric population. Curr Treat Options Neurol, 2013. 15(5): 652–667.

[157] Ebrahimi-Fakhari, D., Autophagy and neurodegeneration - genetic findings in SENDA syndrome, a subtype of neurodegeneration with brain iron accumulation, provide a novel link. Mov Disord, 2013. 28(8): 1050–1050.

[158] Guan, J., et al., Cvt18/Gsa12 is required for cytoplasm-to-vacuole transport, pexophagy, and autophagy in Saccharomyces cerevisiae and Pichia pastoris. Mol Biol Cell, 2001. 12(12): 3821–3838.

[159] Saitsu, H., et al., De novo mutations in the autophagy gene WDR45 cause static encephalopathy of childhood with neurodegeneration in adulthood. Nat Genet, 2013. 45(4): 445–449.

[160] Hoffjan, S., et al., WDR45 mutations in Rett (-like) syndrome and developmental delay: Case report and an appraisal of the literature. Mol Cell Probes, 2016. 30(1): 44–49.

[161] Haack, T.B., et al., Exome sequencing reveals de novo WDR45 mutations causing a phenotypically distinct, X-linked dominant form of NBIA. Am J Hum Genet, 2012. 91(6): 1144–1149.

[162] Proikas-Cezanne, T., et al., WIPI-1 alpha (WIPI49), a member of the novel 7-bladed WIPI protein family, is aberrantly expressed in human cancer and is linked to starvation-induced autophagy. Oncogene, 2004. 23(58): 9314–9325.

[163] Dove, S.K., et al., Svp1p defines a family of phosphatidylinositol 3,5-bisphosphate effectors. EMBO J, 2004. 23(9): 1922–1933.

[164] Lu, Q., et al., The WD40 repeat PtdIns(3)P-binding protein EPG-6 regulates progression of omegasomes to autophagosomes. Develop Cell, 2011. 21(2): 343–357.

[165] Dusek, P. and S.A. Schneider, Neurodegenerative disorder with brain iron accumulation previously known as SENDA syndrome now genetically determined. Movement Disorders, 2013. 28(8): 1051–1052.

[166] Hayflick, S.J., et al., Beta-propeller protein-associated neurodegeneration: a new X-linked dominant disorder with brain iron accumulation. Brain, 2013. 136: 1708–1717.

[167] Zhao, Y.G., et al., Role of Epg5 in selective neurodegeneration and Vici syndrome. Autophagy, 2013. 9(8): 1258–1262.

[168] Tian, Y., et al., C. elegans screen identifies autophagy genes specific to multicellular organisms. Cell, 2010. 141(6): 1042–1055.

[169] Colomer, J., et al., Danon disease: a treacherous illness. Neuromuscul Disord, 2002. 12(7–8): 730–730.

[170] Fidzianska, A., et al., Morphological and clinical aspects of Danon disease. Kardiologia Polska, 2008. 66(3): 302–306.

[171] Tunon, T., et al., Danon disease: A novel Lamp-2 gene mutation in a family with four affected members. Neuromuscul Disord, 2008. 18(2): 167–174.

[172] Di Blasi, C., L. Jarre, and M. Mora, Danon disease: A novel donor site LAMP2 mutation causing alternative transcripts. Neuromuscular Disorders, 2008. 18(9–10): 829–830.

[173] Di Blasi, C., et al., Danon disease: A novel LAMP2 mutation affecting the pre-mRNA splicing and causing aberrant transcripts and partial protein expression. Neuromuscul Disord, 2008. 18(12): 962–966.

[174] Luo, S.S., et al., Novel LAMP2 mutations in Chinese patients with Danon disease cause varying degrees of clinical severity. Clin Neuropathol, 2014. 33(4): 284–291.

[175] Sablayrolles, B., et al., Alpha-1 anti-trypsin. Revue De Medecine De Toulouse, 1981. 17(7): 427–437.

[176] Hutchison, D.C.S., M.J. Tobin, and P.J.L. Cook, Alpha-1 anti-trypsin deficiency - clinical and physiological features in heterozygotes of Pi-type-Sz - a survey by the British-Thoracic-Association. Br J Dis Chest, 1983. 77(1): 28–34.

[177] Hunt, J.M. and R. Tuder, Alpha 1 anti-trypsin: one protein, many functions. Curr Mol Med, 2012. 12(7): 827–835.

[178] Silverthorn, A., Cirrhosis associated with alpha-1 anti-trypsin deficiency. Arizona Med, 1982. 39(8): 519–521.

[179] Greene, C.M., et al., The role of proteases, endoplasmic reticulum stress and SERPINA1 heterozygosity in lung disease and alpha-1 anti-trypsin deficiency. Expert Rev Respir Med, 2011. 5(3): 395–411.

[180] Abbott, C.M., et al., Prenatal-diagnosis of alpha-1-antitrypsin deficiency using polymerase chain-reaction. Lancet, 1988. 1(8588): 763–764.

[181] Nukiwa, T., et al., Identification of a 2nd mutation in the protein-coding sequence of the Z-type alpha-1-antitrypsin gene. J Biol Chem, 1986. 261(34): 5989–5994.

[182] Harrison, H.H., K.L. Miller, and P.F. Whitington, Improved detection, via Iso-Dalt 2-dimensional electrophoresis, of low Z expresser individuals with alpha-1-antitrypsin Mz phenotype. Clin Chem, 1990. 36(10): 1850–1851.

[183] Lomas, D.A., New insights into the structural basis of alpha(1)-antitrypsin deficiency. QMJ, 1996. 89(11): 807–812.

[184] Yorimitsu, T. and D.J. Klionsky, Eating the endoplasmic reticulum: quality control by autophagy. Trends Cell Biol, 2007. 17(6): 279–285.

[185] Hedberg, C., et al., Myopathy in a woman and her daughter associated with a novel splice site MTM1 mutation. Neuromuscul Disord, 2012. 22(3): 244–251.

[186] Sas, A.M.C., et al., Atypical myopathy: a review and description of the outbreak in the Netherlands during autumn 2009 and spring 2010. Tijdschrift Voor Diergeneeskunde, 2012. 137(8): 514–521.

[187] Arzuffi, P., et al., Partial tandem duplication of mtDNA-tRNA(Phe) impairs mtDNA translation in late-onset mitochondrial myopathy. Neuromuscul Disord, 2012. 22(1): 50–55.

[188] Doherty, K.M., et al., Familial camptocormia: from dystonia to myopathy. J Neurol Neurosurg Psychiat, 2012. 83(3): 350–351.

[189] Wagner, M., et al., Using whole exome sequencing to identify the mutation causing oculopharyngodistal myopathy. Neuromuscul Disord, 2012. 22: S31–S32.

[190] Albert, D.A., et al., Mitochondrial myopathy in rheumatology: comment on the article by Albert et al Reply. Arthritis Care Res, 2012. 64(7): 1099–1100.

[191] Mussini, J.M., et al., Structural changes in X-linked myopathy with excessive autophagy: What do they look like? Neuromuscul Disord, 2012. 22(9–10): 820–820.

[192] Fiuza-Luces, C., et al., Exercise training can induce cardiac autophagy at end-stage chronic conditions: Insights from a graft-versus-host-disease mouse model. Brain Behav Immun, 2014. 39: 56–60.

[193] Terman, A. and U.T. Brunk, Autophagy in cardiac myocyte homeostasis, aging, and pathology. Cardiovas Res, 2005. 68(3): 355–365.

[194] Hamacher-Brady, A., et al., Autophagy as a protective response to Bnip3-mediated apoptotic signaling in the heart. Autophagy, 2006. 2(4): 307–309.

[195] Brady, N.R., et al., A wave of reactive oxygen species (ROS)-induced ROS release in a sea of excitable mitochondria. Antioxid Redox Signal, 2006. 8(9–10): 1651–1665.

[196] Hamacher-Brady, A., N.R. Brady, and R.A. Gottlieb, Enhancing macroautophagy protects against ischemia/reperfusion injury in cardiac myocytes. J Biol Chem, 2006. 281(40): 29776–29787.

Chapter 3

Cell Cycle Analysis of ER Stress and Autophagy

Ashik Asvin Patel and Gary Warnes

Additional information is available at the end of the chapter

http://dx.doi.org/10.5772/63653

Abstract

Cell cycle-arresting drugs, thapsigargin (Tg) and chloroquine (CQ), are employed to study endoplasmic reticulum (ER) stress and the autophagic process using cell lines without measuring the cell cycle of such cells. The potential cell cycle-dependent aspect of such processes in cell lines may impact upon the degree of ER stress and autophagy measured. ER stress is known to be caused by a build-up of misfolded proteins within the ER, which may then undergo ER phagy or reticulophagy. The cell cycle-dependent nature of all these processes is not well studied, so we investigated ER stress and autophagy by use of a combination of flow cytometric assays. These included cell cycle-dependent measurement of reticulophagy, misfolded protein levels and autophagic marker LC3-II in K562 and Jurkat cells. ER stress-inducing drug Tg caused significant reticulophagy in both cell types. This was cell cycle dependent in K562 cells only, with proliferating cells undergoing more reticulophagy. In contrast, autophagy-initiating drug CQ caused reticulophagy at higher doses in Jurkat cells, whereas K562 cells showed a cell cycle-dependent elongation of the ER, which was less pronounced in proliferating cells. The level of cellular misfolded protein in response to both drugs was high in K562 cells when either undergoing reticulophagy or elongation in a non-cell cycle-dependent manner, whereas the misfolded protein levels in Jurkat cells in response to both drugs were lower than those observed in K562 cells. Both cell lines employed in this study showed no increase of LC3-II above controls in response to Tg treatment. However, CQ induced a cell cycle-dependent increase of LC3-II in both cell types. Thus, the type of cell employed and the cell cycle dependent modulation of thebiological processes involved in ER stress and autophagy should be considered when designing studies in ER stress and autophagy.

Keywords: reticulophagy, ER stress, misfolded proteins, autophagy, flow cytometry

1. Introduction

Endoplasmic reticulum (ER) stress has been implicated in numerous degenerative neurological diseases and cancer; elucidation of the mechanisms involved may determine drug targets for the treatment of such diseases [1, 2]. ER stress is known to cause autophagy and ultimately cell death via apoptosis, the mechanism of which is only beginning to be understood [3–5]. ER stress can be induced by a range of drugs, including thapsigargin (Tg) which acts by inhibition of ER ATPase located in the ER, resulting in the accumulation of misfolded proteins within the ER [6]. Tg is also known to cause cell arrest in G_1 phase of the cell cycle and has been shown to inhibit autophagic flux by inhibition of the translocation of Rab7, a protein required for the fusion of lysosomes with autophagosomes [6–8]. ER stress resulting from a build-up of misfolded proteins occurs once a threshold of misfolded protein accumulation has been reached. This then initiates the unfolded protein response (UPR) by the ER stress sensor signalling proteins, IRE1, PERK and ATF6 which initiate protein refolding and elongation of the ER until ER homeostasis is returned [3, 9, 10]. This process is a coping mechanism which reduces the mass of misfolded protein per unit volume of ER. If the high level of misfolded protein persists, then reticulophagy (phagy of the ER) occurs [11]. Chloroquine (CQ) is also known to initiate autophagy and block completion of the autophagic process by increasing the pH within the lysosome, with resultant inhibition of the lysosome fusion process with autophagosomes [12]. CQ was investigated for its ER stress-inducing qualities by comparison with Tg upon Jurkat T cells and K562 erythromyeloid cell lines. Thus measurement of changes in ER mass and misfolded protein levels would give a measure of the degree of ER stress within a cell.

The term autophagy or type II cell death is derived from the Greek roots "auto" (self) and "phagy" (eat) and was first observed by Porter in 1962 [13, 14]. Autophagy or macroautophagy is an intracellular degradation system that maintains cell homeostasis and is characterised by the formation of a double membrane around the cytosolic components to be degraded, forming an autophagosome of sequestered malfunctioning components ranging from misfolded proteins to organelles such as stressed ER [15, 16]. An autophagosome then fuses with lysosomes, giving rise to an autolysosome, where the intracellular components are degraded by hydrolases which produces energy, thus promoting cellular haemostasis [2, 15, 17–20]. The main biological autophagy marker is the microtubule-associated protein LC3-II or LC3B. LC3-I is normally located in the cytoplasm but when cleaved and lipidated by phosphatidylethanolamine is then incorporated into the autophagosome inner leaflet of the membrane in the form of LC3-II [21–23].

Methods for monitoring autophagy began with the initial discovery of the process by the use of electron microscopy, which showed the presence of autophagosome and autolysosome or autophagolysosome [13, 14]. Biochemical techniques such as Western blotting can be used to quantitate the degree of autophagy in cells by measuring the autophagy marker proteins, LC3-II and LC3-I [21–23]. Fluorescently tagged LC3-II can also be imaged and flow cytometrically analysed through transfections with GFP-RFP through transfections with GFP-RFP, with the benefit that GFP fluorescence is dissipated by the acidic conditions prevailing in autolysosomes, whereas RFP is not, thus making LC3-II-GFP detection specific for autophagosomes and LC3-II-RFP specific for autolysosomes [22, 24–26]. The number and intensity of fluorescently labelled anti-LC3-II-positive puncta (autophagosomes-autolysosomes) can also be quantitated by time-consuming image analysis, whereas measuring the increase in median fluorescent values of LC3-II antigen level flow cytometrically makes the process significantly less burdensome, especially when combined with cell cycle analysis [25, 27–29]. Here we show how flow cytometry can be used to measure not only the autophagy marker LC3-II but also the cellular end products of ER stress which include reticulophagy and cellular levels of misfolded proteins and their cell cycle distribution. Most studies using cell lines have not generally focussed upon the cell cycle distribution of the autophagy marker even though autophagy-inducing drugs such as rapamycin are known to cause G_1 cell cycle arrest. However a few studies have investigated the cell cycle distribution of LC3-II with variable results which seem to vary depending upon the cell line employed [17, 30, 31]. Thus this chapter investigates whether the autophagic biological marker, LC3-II, and end products of ER stress (ER mass and misfolded protein) showed a cell cycle-dependent nature during ER stress. Drugs such as Tg and CQ, which cause the build-up of misfolded proteins and induce ER stress, were then compared in terms of cell cycle-dependent reticulophagy, misfolded protein levels and LC3-II [32].

Here, we employed flow cytometric cell cycle analysis of live cells to measure reticulophagy combined with an assay employing fixed-cell immunofluorescence analysis of LC3-II developed in this laboratory and the cell cycle analysis of cellular misfolded proteins [31, 33, 34]. The use of ER Tracker was used as previously described but was now employed in a cell cycle-dependent manner to determine the relative change in ER mass compared to untreated control cells [35, 36]. We also determined the degree of misfolded protein aggregate formation in a cell cycle-dependent manner using the Proteostat probe (Enzo Life Sciences) that fluoresces when bound to misfolded proteins in fixed cells [32]. Thus, we were able to investigate whether the observed drug-induced ER stress had a cell cycle-dependent nature by the measurement of changes in ER mass, misfolded proteins and the autophagy marker LC3-II.

2. Materials and methods

2.1. Induction of ER stress

Jurkat and K562 cells were grown in RPMI-1640 with L-glutamine with 10% FBS (Invitrogen, UK) and penicillin and streptomycin or treated with CQ at 25, 50 and 75 μM (Sigma Chemicals, UK) or 0.1, 0.5 and 1 μM Tg (Santa Cruz, US) for 24 h (n = 3). Cells were harvested and processed as described in the sections below.

2.2. Cell cycle and ER Tracker

Jurkat and K562 cells with or without treatment were adjusted to 1×10^6/ml and loaded with Hoechst 33342 (15 μg/ml, Sigma Chemicals, UK) and ER Tracker Red dye (ERTR 100 nM, Invitrogen, UK), by incubating cells with the dyes for 1 and 0.5 h at 37°C, respectively. Dead cells were detected by DRAQ7 (2.5 μM, Biostatus, UK), incubated with cells for 10 min at 37°C. Live cells were analysed for ERTR Area MFI levels from Ho33343 Area and Width dot plots; see **Figure 1**. The percentage change in ER mass test samples was determined by comparing test and control values for each phase of the cell cycle comparing control values for each phase of the cell cycle: 30,000 events collected by flow cytometry, with Ho33342 excited at 350–60 nm (UV laser) and emission collected at 450/50 nm; ERTR Area excited at 488 nm (blue laser) and emission collected at 610/10 nm; and DRAQ7 excited at 633 nm (red laser) and emission collected at 780/60 nm (n = 3).

2.3. Misfolded protein labelling

Jurkat and K562 cells with or without treatment were pelleted and resuspended in 200 μl of Solution A fixative for 15 min at room temperature (RT) (Caltag, UK). Cells were then

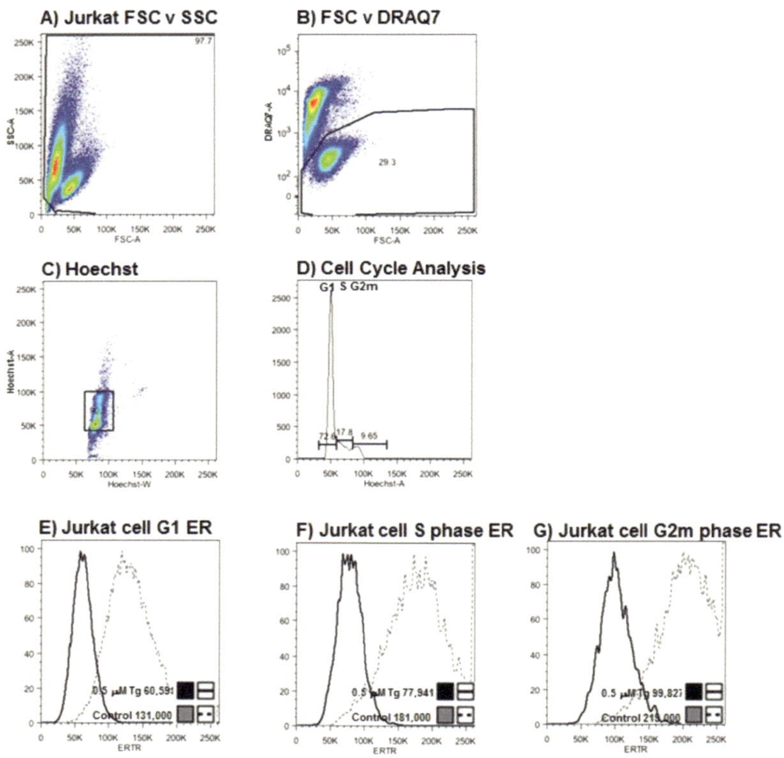

Figure 1. Cells were gated on forward-scatter (FSC) vs. side-scatter (SSC) dot plot (A). Live cells were then gated by their exclusion of cell viability dye DRAQ7 from a FSC vs. DRAQ7 dot plot (B). Live single cells were then gated through Ho33342 Area and Width parameter analysis (C). Cell cycle analysis of these single cells into G_1, S and G_2m phases of the cell cycle by virtue that Ho33342 fluorescence intensity being proportional to DNA content as shown in (D). The ERTR fluorescence signal is proportional to ER mass; the phases of the cell cycle in cells treated with Tg were compared to untreated cells in G_1 (E), S phase (F) and G_2m (G). ERTR MFI were normalised as a percentage change in the test ERTR signal was made to the control ERTR signal.

washed in PBS buffer (Invitrogen, UK) and cell pellets permeabilised with 0.25% Triton X-100 (Sigma Chemicals, UK) for 15 min at RT. Cells were then washed in PBS buffer and a 1:20,000 dilution of Proteostat (Enzo Life Sciences, UK) was made in Proteostat buffer (1 μl of Proteostat reagent in 20 ml of buffer) and 400 μl added to the cell pellet with 1 μg/ml DAPI and incubated for 30 min at RT. Flow cytometric analysis was performed with excitation of Proteostat with the blue 488 nm laser and excitation of DAPI with UV laser line and emissions collected at 610/20 nm and 440/40 nm, respectively (n = 3). Cell cycle analysis was performed by collecting 30,000 events with the DAPI 440/40 Area and Width parameter for doublet discrimination; see **Figure 3**. Then Proteostat Area MFI for cells in G_1, S phase and G_2m cycling cells was determined by comparison of control and test MFI for each phase of the cell cycle by the formula given in the manufacturer's instructions, $(MFI_{test} - MFI_{con})/MFI_{test} \times 100$ = misfolded protein level.

2.4. Indirect immunofluorescence LC3-II labelling and cell cycle analysis

Jurkat and K562 cells with or without treatment were pelleted and resuspended in 200 μl of Solution A fixative for 15 min at RT (Caltag, UK). Cells were then washed in PBS buffer and cell pellets permeabilised with 0.25% Triton X-100 for 15 min at RT. Cells were washed in PBS and anti-LC3-II monoclonal antibody (1:400 dilution) (Cat. No. 3868, Cell Signalling Technology Inc., USA) incubated for 0.5 h at RT. Cells were then washed in PBS and labelled with 0.125 μg of secondary fluorescent conjugate Alexa Fluor-647 goat anti-rabbit IgG (Invitrogen, UK) for 0.5 h at RT. Cells were then washed in PBS buffer and resuspended in 400 μl of PBS with DAPI (1 μg/ml). Analysis of upregulated LC3-II-Alexa Fluor-647 MFI signal in treated samples (above control levels) was determined for each phase of the cell cycle by gating cells on a FSC versus SSC dot plot, with doublet discrimination achieved by use of the DAPI Area and Width parameters with gating for G_1, S and G_2m on the DAPI-A parameter; see **Figure 5**. Thirty thousand events were collected by flow cytometry. DAPI was excited at 350–60 nm (UV laser) and emission collected at 450/50 nm; LC3-II-AF-647 was excited at 633 nm (red laser) and emission collected at 660/20 nm (n = 3).

2.5. Flow cytometry

A Becton Dickinson LSRII with FACSDiva software (ver 6.3.1) fitted with blue (488 nm), red (633 nm), violet (405 nm) and UV (350–360 nm) lasers. The optical filters (fitted in 2005) used were for ERTR (610/10 nm), Ho33342 and DAPI (450/50 nm), DRAQ7 (780/60 nm), LC3B-AF-647 (660/20 nm) and Proteostat (610/10 nm).

2.6. Statistics

Student *t* tests were performed in GraphPad software with $P > 0.05$ not significant (NS), $P < 0.05$*, $P < 0.01$**, n = 3.

3. Results

3.1. Flow cytometric analysis of cell cycle-dependent reticulophagy

For estimating the degree of reticulophagy after treatment with Tg or CQ, known ER stress inducers and interceptor of autophagy, cells were labelled with Ho33342, ERTR and DRAQ7. Live cells were analysed for reticulophagy after flow cytometric analysis on a Becton Dickinson LSRII after UV, blue and red laser excitation and fluorescence collection at 460, 610 and 780 nm, respectively. The gating strategy for the analysis of live cell cycle-dependent reticulophagy of such cells is shown in **Figure 1**.

Tg induced a high degree of K562 cell cycle-dependent reticulophagy (29–76%), which was greater in S, G_2m and G_1 phases with increasing [Tg] ($P < 0.05$, 0.01, n = 3, **Figure 2A**), whereas Jurkat cells displayed a lower level of reticulophagy 35 to >55%, which was not cell cycle dependent ($P < 0.01$, n = 3, **Figure 2B**). The K562 response to CQ was very different to that of Tg with an ER elongation of >10 to >40%, with the highest degree of ER stress (smallest amount of elongation) present in G_2m cells (NS, $P < 0.05$, <0.01, n = 3, **Figure 2C**). CQ treatment of Jurkat

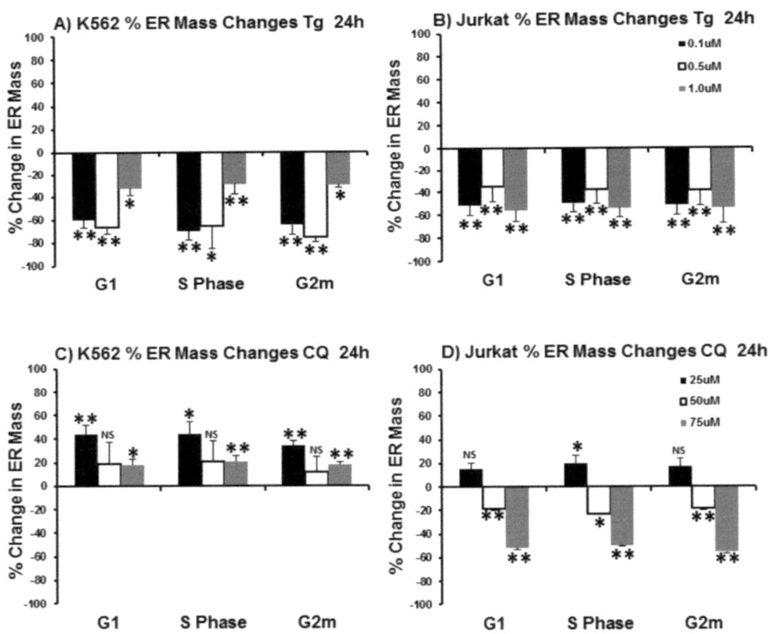

Figure 2. Jurkat and K562 cell lines were untreated or treated with Tg (0.1, 0.5, 1 μM) or CQ (25, 50, 75 μM) for 24 h. Cells were loaded with Hoechst 33342 (15 μg/ml), ERTR (100 nM) and DRAQ7 (2.5 μM) for 1, 0.5 and 0.25 h at 37°C, respectively. Cells were gated as described in Section 2; see **Figure 1**. ERTR test MFI were normalised against the control and a percentage change in the test ERTR signal was calculated. Percentage changes in ER mass of K562 (A) and Jurkat cells (B) treated with Tg were made for G_1, S and G_2m phases of the cell cycle. Percentage changes in ER mass of K562 (C) and Jurkat cells (D) treated with Tg or CQ were made for G_1, S and G_2m phases of the cell cycle. Student *t* test, *P <0.05; **P < 0.01; NS, not significant; error bars indicate SEM, n = 3.

cells displayed a lower level of non-cell cycle-dependent ER elongation than that observed in K562 cells (<20%, $P < 0.05$, NS, n = 3, **Figure 2D**). However the higher dose CQ induced a reticulophagy (<20–>50%) which was also non-cell cycle dependent ($P < 0.05$, $P < 0.01$, n = 3, **Figure 2D**).

3.2. ER stress responses: protein misfolding

For estimating the overall level of misfolded proteins after ER stress induction, Tg or CQ cells were fixed, permeabilised and labelled with DAPI and the misfolded protein detection reagent, Proteostat [32]. Flow cytometric analysis of the cell cycle and misfolded protein levels was performed on a Becton Dickinson LSRII after UV and blue laser excitation and fluorescence collected at 460 and 610, nm respectively. The gating strategy for the cell cycle analysis of misfolded proteins is shown in **Figure 3**.

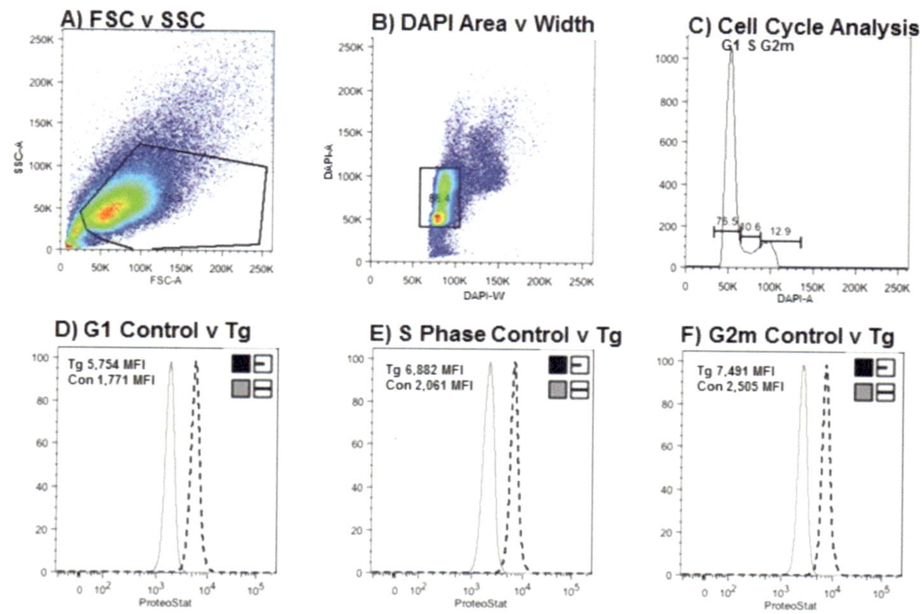

Figure 3. Jurkat and K562 cell lines were untreated or treated with Tg (0.1, 0.5, 1 μM) or CQ (25, 50, 75 μM) for 24 h. Cells were fixed and permeabilised and stained with Proteostat probe (Enzo Life Sciences) according to that described in Section 2. Misfolded protein levels were calculated from the test and untreated Proteostat MFI according to that described in Section 2. The gating strategy employed was to gate on cells through FSC vs. SSC (A) and then gate on single cells using DAPI Area and Width parameters (B). Followed by cell cycle analysis on the DAPI Area parameter, marking off G_1, S and G_2m phases of the cell cycle by virtue that DAPI fluorescence intensity is proportional to DNA content is shown in (C). The Proteostat fluorescence signal is proportional to the level of cellular misfolded proteins, in cells treated with Tg, and was compared to untreated cells in G_1 (D), S (E) and G_2m (F) phases of the cell cycle.

Tg treatment of K562 cells showed a moderately high level of misfolded proteins which were neither dose nor non-cell cycle dependent (>40, $P < 0.05$, $P < 0.01$, n = 3, **Figure 4A**). Jurkat cells showed a low level of misfolded proteins when treated with Tg which was also not dose nor cell cycle dependent (<35, $P < 0.01$, NS, n = 3, **Figure 4B**). CQ treatment of K562 cells showed a variable level of misfolded proteins which was not cell cycle dependent (25–80, $P < 0.01$, n = 3, **Figure 4C**). In contrast Jurkat cells produced a lower degree of misfolded proteins when treated with CQ (25–40 $P < 0.05$, $P < 0.01$, n = 3)) which was cell cycle dependent in that cells in G_2m had lower amounts of misfolded proteins (**Figure 4D**).

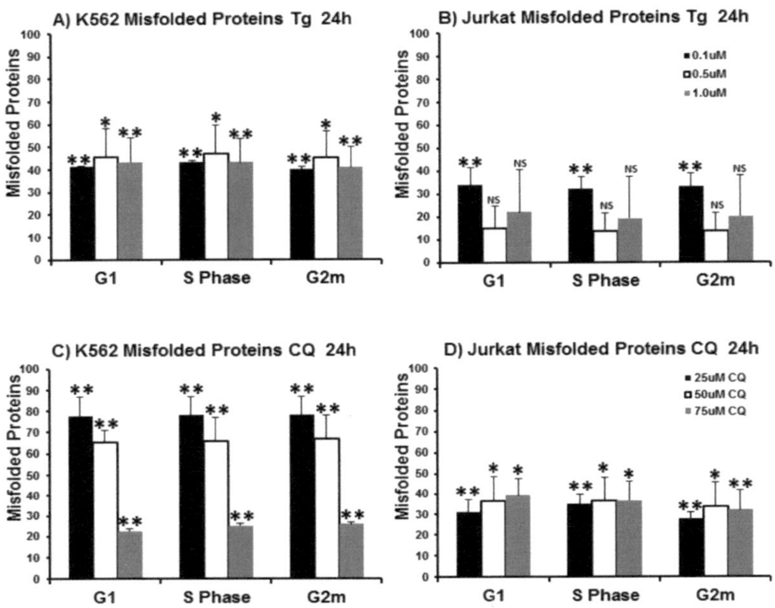

Figure 4. Jurkat and K562 cell lines were untreated or treated with Tg (0.1, 0.5, 1 μM) or CQ (25, 50, 75 μM) for 24 h. Cells were fixed and permeabilised and stained with Proteostat probe (Enzo Life Sciences) according to that described in Section 2; see **Figure 3**. Misfolded protein levels for each phase of the cell cycle (G_1, S and G_2m) were calculated from the test and untreated Proteostat MFI according to that described in Section 2. The level of misfolded proteins in K562 cells is shown in (A) Tg, (C) CQ and Jurkat cells treated with (B) Tg, (D) CQ. Student *t* test, $^*P < 0.05$; $^{**}P < 0.01$; NS, not significant; error bars denote SEM, n = 3.

3.3. ER stress responses: LC3-II-associated autophagy is cell cycle dependent

For estimating the level of autophagy occurring in cells treated with ER stress-inducing drugs, Tg and CQ cells were fixed, permeabilised and labelled with DAPI cells and anti-LC3-II-Alexa Fluor-647 which were collected on a Becton Dickinson LSRII after UV and red laser excitation and fluorescence collected at 460 and 660 nm, respectively. The gating strategy for the cell cycle analysis of LC3-II is shown in **Figure 5**.

Although Tg induced a high degree of ER stress in K562 cells and a lower order in Jurkat cells after 24 h, LC3-II did not increase above control levels (data not shown). However CQ did induce a cell cycle-dependent increase of LC3-II for both cell lines employed in this study

http://dx.doi.org/10.5772/63653

(**Figure 6A** and **B**). Both cell types showed a similar level of LC3-II in that 25 and 75 μM CQ showed a dose and cell cycle-dependent increase above control levels (**Figure 6A** and **B**), whereas 50 μM CQ showed a lower degree of LC3-II upregulation than the other doses of CQ employed in this study (**Figure 6A** and **B**).

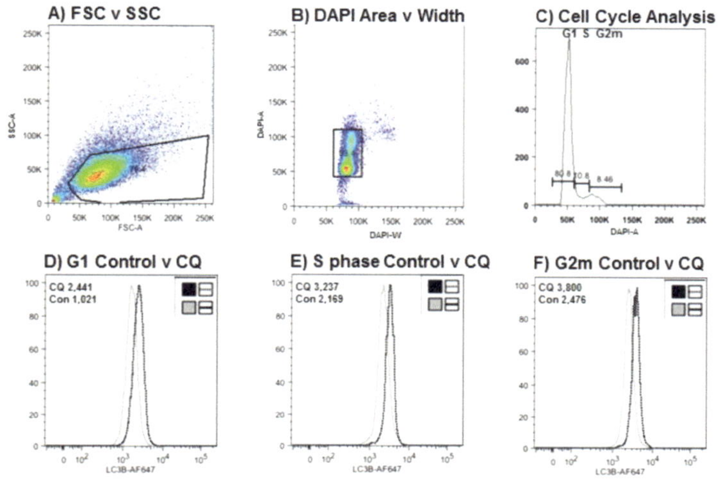

Figure 5. Jurkat and K562 cell lines were untreated or treated with Tg (0.1, 0.5, 1 μM) or CQ (25, 50, 75 μM) for 24 h. Cells were fixed and permeabilised and stained with anti-LC3-II and Alexa Fluor-647 according to that described in Section 2. LC3-II MFI test samples were subtracted from untreated cells according to that described in Section 2. The gating strategy employed was to gate on cells through FSC vs. SSC (A). Single cells were gated using DAPI Area and Width parameters (B). Followed by cell cycle analysis on the DAPI Area parameter, marking off G_1, S and G_2m phases of the cell cycle by virtue that DAPI fluorescence intensity is proportional to DNA content is shown in (C). The LC3-II MFI of CQ treated cells was subtracted from untreated cells in G_1 (D), S (E) and G_2m (F) phases of the cell cycle.

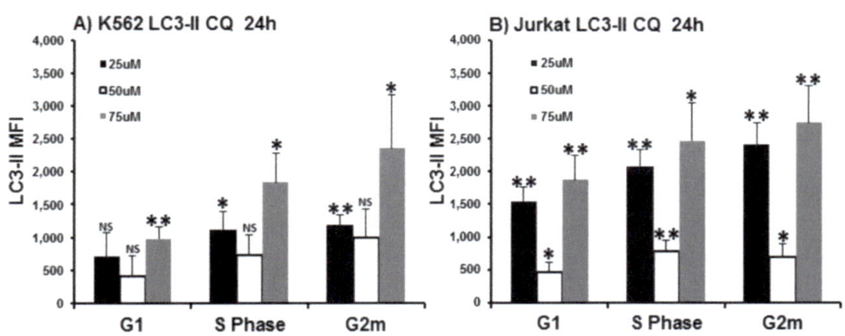

Figure 6. Jurkat and K562 cell lines were untreated or treated with CQ (25, 50, 75 μM) for 24 h. Cells were then fixed, permeabilised and labelled with rabbit anti-LC3-II-AF-647 and DAPI for cell cycle analysis. After gating as described in Section 2 (see **Figure 5**), the LC3-II MFI levels in test samples (test MFI-control MFI) were determined for all cell cycle phases. The levels of LC3-II in K562 (A) and Jurkat cells (B) treated with CQ were calculated for G_1, S and G_2m phases of the cell cycle. Student *t* test, $^*P < 0.05$; $^{**}P < 0.01$; NS, not significant; error bars indicate SEM, n = 3.

4. Discussion

Flow cytometry can be used to investigate ER stress, a biological process not much analysed by this experimental approach to date. Here we measured the end products of ER stress, that is, reticulophagy and misfolded proteins rather second messengers associated with the process, for example, PERK, IRE-1 and ATF6. However the use of live cells and ER Tracker probes to measure reticulophagy has been previously used to estimate the degree of reticulophagy [35, 36], although not in a cell cycle-dependent manner. Analysis of these cells under fixed conditions allowed flow cytometry to estimate the level of misfolded proteins present in the cell (rather than exclusively in the ER) in a cell cycle-dependent manner. The assays employed using just two or three fluorescent probes allowed for a relatively easy flow cytometric analysis, without the need for colour compensation or correction of the bleed through of the different fluorophores into each other, thus avoiding false readings and incorrect conclusions from the acquired data sets. Gating strategies used were as simple as possible, but given the relatively complex nature of the gating needed, they all employed the same approach. This study employed known ER stress-inducing drug Tg, as well as by comparison CQ, and used two cell types to show potential different responses to these drugs. To this end Tg was shown to induce ER stress after 24 h in both cell lines with the ER in such cells undergoing a significant degree of reticulophagy. K562 cells were more affected than Jurkat T cells, whereas only K562 cells showed a reticulophagy, which was more pronounced in specific phases of the cell cycle with the different concentrations of the drug.

The drug CQ, a known initiator of autophagy and apoptosis (although the drug also blocks the process at the lysosome-autophagosome fusion step), was also tested for its ER stress-inducing qualities [12, 32, 33]. CQ appeared to induce ER stress in both cell lines, with Jurkat cells treated with high concentration of CQ displaying a high level of reticulophagy like that observed with Tg. However lower doses of CQ induced ER stress which was typified by an elongation of the ER in Jurkat cells which was again not cell cycle dependent. This mode of action of CQ causing an elongation of the ER was repeated in K562 cells which was cell cycle dependent. Here the degree of elongation reduced as the cell moved through the cell cycle. Thus G_2m cells showed the least amount of ER elongation and hence were more stressed than cells in the G_1 phase of the cell cycle.

Induction of ER stress by both Tg and CQ was further confirmed by the detection of misfolded proteins in a non-cell cycle-dependent manner above that found in untreated cells (except G_2m phase Jurkat cells treated with CQ). K562 cells had a high level of misfolded proteins in response to both Tg and CQ even though Tg induced a reticulophagy and CQ an ER elongation. Similarly Jurkat cells responded to Tg and CQ with a moderate level of misfolded proteins, with most cells displaying a reticulophagy in this instance. Thus, although cells were displaying ER elongation as well as reticulophagy with the different drugs, misfolded proteins were detectable and the highest level found in cells undergoing ER elongation rather than reticulophagy, this perhaps being a reflection that cells with less ER have less space for misfolded proteins and thus a lower level of misfolded proteins.

Although Tg induced a high degree of reticulophagy and misfolded proteins in both cell lines, there was no evidence of autophagy as LC3-II levels did not increase above control levels. However CQ shown previously and in this study to induce ER stress did show an increase LC3-II in a cell cycle-dependent manner in both cell lines.

Thus, the degree and mode of action of these two drugs appear to be cell type dependent, with K562 cells displaying a cell cycle-dependent ER stress response to both drugs, whereas Jurkat cells did not. However, the induction of the autophagic response (the process being blocked at the lysosome fusion step) to CQ was cell cycle dependent in both cell lines. While Tg did not induce an autophagic response in either cell line after 24 h. The type of cell employed and the cell cycle-dependent modulation of these biological processes involved in ER stress and autophagy should be considered when designing studies in ER stress and autophagy. Flow cytometry makes the analysis of these cell cycle-dependent events in the ER stress process easily measureable.

Author details

Ashik Asvin Patel and Gary Warnes*

*Address all correspondence to: g.warnes@qmul.ac.uk

Flow Cytometry Core Facility, Blizard Institute, Barts and The London School of Medicine and Dentistry, Queen Mary University London, London, UK

References

[1] Rosello A, Warnes G, Meier U. Cell death pathways and autophagy in the central nervous system and its involvement in neurodegeneration, immunity and CNS infection: to die or not to die—that is the question. Clinical and Experimental Immunology. 2012;168:52–7.

[2] Yang Z, Klionsky D. Eaten alive: a history of macroautophagy. Nature Cell Biology. 2010;12(9):814–22.

[3] Ogata M, Hino S, Saito A, et al. Autophagy is activated for cell survival after endoplasmic reticulum stress. Molecular and Cellular Biology. 2006;26(24):9220–31.

[4] Chen Y, Brandizzi F. IRE1: ER stress sensor and cell fate executor. Trends in Cell Biology. 2013;23(11):547–55.

[5] Moretti L, Cha YI, Niermann KJ, Lu B. Switch between apoptosis and autophagy: radiation-induced endoplasmic reticulum stress? Cell Cycle. 2007;6(7):793–8.

[6] Furuya Y, Lundmo P, Short AD, Gill DL, Isaacs JT. The role of calcium, pH, and cell proliferation in the programmed (apoptotic) death of androgen-independent prostatic cancer cells induced by thapsigargin. Cancer Research. 1994;54:6167–75.

[7] Lin X, Denmeade R, Cisek L, Isaccs J. Mechanism and role of growth arrest in programmed (apoptotic) death of prostatic cancer cells induced by Thapsigargin. The Prostate. 1997;33:201–7.

[8] Ganley I, Wong P, Gammoh N, Jiang X. Distinct autophagosomal–lysosomal fusion mechanism revealed by thapsigargin-induced autophagy arrest. Molecular Cell. 2011;42(6):731–43.

[9] Yorimitsu T, Klionsky DJ. Eating the endoplasmic reticulum: quality control by autophagy. Trends in Cell Biology. 2007;17(6):279–85.

[10] Hoyer-Hansen M, Jaattela M. Connecting endoplasmic reticulum stress to autophagy by unfolded protein response and calcium. Cell Death and Differentiation. 2007;14(9): 1576–82.

[11] Klionsky DJ. Guidelines for the use and interpretation of assays for monitoring autophagy (3rd edition). Autophagy. 2016;12(1):1–222.

[12] Choi JH, Yoon JS, Won YW, Park BB, Lee YY. Chloroquine enhances the chemotherapeutic activity of 5-fluorouracil in a colon cancer cell line via cell cycle alteration. APMIS: Acta Pathologica, Microbiologica, et Immunologica Scandinavica. 2012;120(7): 597–604.

[13] Ashford TP, Porter K. Cytoplasmic components in hepatic cell lysosomes. The Journal of Cell Biology. 1962;12:198–202.

[14] Deter RL, Duve CD. Influence of glucagon, an inducer of cellular autophagy, on some physical properties of rat liver lysosomes. The Journal of Cell Biology. 1967;33:437–49.

[15] Tooze SA, Yoshimori T. The origin of the autophagosomal membrane. Nature Cell Biology. 2010;12(9):831–5.

[16] Mehrpour M, Esclatine A, Beau I, Codogno P. Overview of macroautophagy regulation in mammalian cells. Cell Research. 2010;20(7):748–62.

[17] Ge JN, Huang D, Xiao T, et al. Effect of starvation-induced autophagy on cell cycle of tumor cells. Chinese Journal of Cancer. 2008;27(8):102–8.

[18] Komatsu M, Ichimura Y. MBSJ MCC Young Scientist Award 2009: selective autophagy regulates various cellular functions. Genes to Cells. 2010;15(9):923–33.

[19] Hailey DW, Rambold AS, Satpute-Krishnan P, et al. Mitochondria supply membranes for autophagosome biogenesis during starvation. Cell. 2010;141(4): 656–67.

[20] Kim I, Rodriguezenriquez S, Lemasters J. Selective degradation of mitochondria by mitophagy. Archives of Biochemistry and Biophysics. 2007;462(2):245–53.

[21] Kabeya Y, Mizushima N, Ueno T, et al. LC3, a mammalian homologue of yeast Apg8p, is localized in autophagosome membranes after processing. EMBO. 2000;19(21):5720–8.

[22] Barth S, Glick D, Macleod K. Autophagy: assays and artifacts. The Journal of Pathology. 2010;221(2):117–24.

[23] Hansen TE, Johansen T. Following autophagy step by step. BMC Biology. 2011;9(39):1–4.

[24] Shvets E, Fass E, Elazar Z. Utilizing flow cytometry to monitor autophagy in living mammalian cells. Autophagy. 2008;4(5):621–8.

[25] Kimura S, Noda T, Yoshimori T. Dissection of the autophagosome maturation process by a novel reporter protein, tandem fluorescent-tagged LC3. Autophagy. 2007;3(5):452–60.

[26] Wu YT, Tan HL, Huang Q, et al. Autophagy plays a protective role during zVAD-induced necrotic cell death. Autophagy. 2008;4(4):457–66.

[27] Thomas S, Thurn KT, Biçaku E, Marchion DC, Münster N. Addition of a histone deacetylase inhibitor redirects tamoxifen-treated breast cancer cells into apoptosis, which is opposed by the induction of autophagy. Breast Cancer Research and Treatment. 2011;130(2):437–47.

[28] Geng Y, Kohil L, Klocke BJ, Roth KA. Chloroquine-induced autophagic vacuole accumulation and cell death in glioma cells is p53 independent. Neuro-Oncology. 2010;12(5):473–81.

[29] Chen Y, McMillan-Ward E, Kong J, Israels SJ, Gibson SB. Oxidative stress induces autophagic cell death independent of apoptosis in transformed and cancer cells. Cell Death and Differentiation. 2007;15(1):171–82.

[30] Kaminskyy V, Abdi A, Zhivotovosky B. A quantitative assay for the monitoring of autophagosome accumulation in different phases of the cell cycle. Autophagy. 2011;7(1):83–90.

[31] Warnes G. Flow cytometric assays for the study of autophagy. Methods. 2015;82:21–8.

[32] Shen D, Coleman J, Chan E, et al. Novel cell- and tissue-based assays for detecting misfolded and aggregated protein accumulation within aggresomes and inclusion bodies. Cell Biochemistry and Biophysics. 2011;60(3):173–85.

[33] Chikte S, Panchal N, Warnes G. Use of lysotracker dyes: a flow cytometric study of autophagy. Cytometry A. 2014;85A(2):169–78.

[34] Warnes G. Measurement of autophagy by flow cytometry. Current Protocols in Cytometry. 2014:9.45.1–10.

[35] Panchal N, Chikte S, Wilbourn BR, Meier UC, Warnes G. Autophagy—a double-edged sword—cell survival or death? Flow cytometric measurement of cell organelle autophagy. Intech Open Access Publisher. 2013.

[36] Jia W, Pua HH, Li QJ, He Y. Autophagy regulates endoplasmic reticulum homeostasis and calcium mobilization in T lymphocytes. The Journal of Immunology. 2010;186(3): 1564–74.

Chapter 4

High-Mobility Group Box 1 and Autophagy

Daolin Tang and Rui Kang

Additional information is available at the end of the chapter

http://dx.doi.org/10.5772/63476

Abstract

High-mobility group box 1 (HMGB1) is an architectural chromosomal protein and stress sensor that plays a critical role in various physiological and pathological processes, including cell death and survival. Autophagy is the major pathway involved in the degradation of proteins and organelles, maintenance of cellular homeostasis, and promotion of survival during environmental stress. HMGB1 plays an important location-dependent role in the regulation of autophagy. Nuclear HMGB1 contributes to mitophagy by regulation of heat shock protein beta-1 (HSPB1) expression and cytoskeleton dynamics. Cytoplasmic HMGB1 is a novel coiled-coil myosin-like BCL2-interacting protein (BECN1)-binding protein in the induction of autophagosome formation. Extracellular-reduced HMGB1 triggers autophagy in an advanced glycosylation end product-specific receptor (AGER)-dependent manner. HMGB1-dependent autophagy promotes chemotherapy resistance, sustains the tumor metabolism requirement and T-cell survival, prevents polyglutamine aggregates and excitotoxicity, and protects against endotoxemia, bacterial infection, and ischemia-reperfusion injury in vitro or in animal studies. Targeting the HMGB1-mediated autophagy pathway may be required to address whether or not this approach is therapeutically advantageous in human disease.

Keywords: HMGB1, autophagy, ATG, disease, pathway

1. Introduction

The autophagic network is complex and requires a core regulator: autophagy-related genes/proteins (ATGs) [1]. The study of the molecular basis of autophagy started with the discovery of Aut1 (now Atg3) [2], Apg13 (now Atg13) [3], and Apg1 (now Atg1; the mammalian homolog is ULK1 [unc-51 like autophagy-activating kinase 1]) [4] in *Saccharomyces cerevisiae* in 1997. Currently, over 38 ATGs that control membrane dynamics during autophagy have been identified in the cells of different organisms, ranging from yeasts to mammals [5–7].

In addition to ATGs, several non-ATG proteins are involved in the regulation of autophagy through direct or indirect interplay with ATGs. High-mobility group protein 1 (HMGB1) belongs to the family of the high-mobility group (HMG) nuclear proteins [8]. Here, we highlight the emerging role of HMGB1 as an important non-ATG protein in the regulation of autophagy.

2. HMGB1 structure

HMGB1 is a highly conserved protein and present in almost all cell types [9]. The genetic sequence of human HMGB1 is located on chromosome 13q12–13 and the protein sequence of human HMGB1 is composed of 215 amino acids (AAs). Human HMGB1 is 99% AA identical to mouse, rat, bovine, and porcine HMGB1. HMGB1 structurally consists of three different domains: two DNA-binding domains (HMG boxes A and B) and a negatively charged 30 AA C-terminal region that contains only Asp and Glu. Both the HMGB1 A and B boxes are about 75–80 AAs long and are formed by two short and one long α-helixes that upon folding, produce an L- or V-shaped three-dimensional domain structure [10–12]. The cytokine activity of HBMG1 is restricted to the HMG B box because AA89–108 of HMGB1 can be recognized by Toll-like receptor (TLR)-4 to induce the release of proinflammatory cytokines [13]. In contrast, the purified recombinant A box has been identified as an antagonistic and anti-inflammatory factor [14]. AA150–183 of HMGB1 is responsible for binding to the receptor for advanced glycation end products (AGER/RAGE) to induce cell migration [15]. In addition, to mediate cell migration, AGER is also important for HMGB1-induced autophagy [16], metabolism [17], and inflammation [18] in a context-dependent manner. AA 27–43 and 178–184 of HMGB1 contain two nuclear localization signals, respectively. Acetylation of nuclear localization signals triggers HMGB1 translocation from the nucleus to the cytosol. HMGB1 is a redox protein and contains three cysteines (C23, C45, and C106). C23 and C45 form an intramolecular disulfide bond within the A-domain, while C106 is redox inactive and remains reduced [19, 20]. In general, reduced HMGB1 exhibits immune activity, whereas oxidized HMGB1 displays immune tolerance [21, 22]. In addition to regulating activity, redox also affects HMGB1 translocation and release [23, 24]. For example, mutant C106 can cause HMGB1 translocation from the nucleus to the cytosol. Indeed, oxidative stress plays a central role in mediating active HMGB1 secretion as well as passive release [25]. Inhibition of HMGB1 release by antioxidant compounds such as *N*-acetyl-cysteine [26], quercetin [27], edaravone [28], epigallocatechin gallate [29], and resveratrol [30] improves animal survival and limits the inflammatory response in infection and tissue damage.

3. HMGB1 function

Under normal conditions, over 95% of intracellular HMGB1 is located in the nucleus and functions as a deoxyribonucleic acid (DNA) chaperone. Under stress conditions, HMGB1 can be released from the intracellular to the extracellular space [31]. Extracellular HMGB1 acts as a damage-associated molecular pattern molecule (DAMP) to mediate the inflammation, immunity, and metabolism responses in human disease [9, 32]. Thus, the function of HMGB1 depends on its location. Below, we summarize the major functions of nuclear, cytosolic, and extracellular HMGB1, respectively.

3.1. Nuclear HMGB1

Like the histones, HMGB1 is among the most important chromatin proteins. In particular, the DNA-binding domains confers HMGB1 the ability to recognize, bind, and bend different DNA structures such as DNA mini-circles, four-way junctions, looped structures, hemicatenated DNA, and triplex DNA [33]. This DNA chaperone activity is critical for HMGB1-mediated nuclear homeostasis and genome stability.

3.1.1. Nucleosome dynamics and quantity

The nucleosome is the fundamental subunit of chromatin. Each nucleosome is composed of a core particle, DNA, and a linker protein. The proteins in the core particle and linker proteins are called histones. HMGB1 can bind to histones and DNA to promote nucleosome sliding, relax nucleosome structure, and make chromatin more accessible [34, 35]. Loss of HMGB1 leads to the loss and release of nucleosomes [36, 37]. Extracellular nucleosomes, including histones and DNA, are inflammatory mediators in cancer, sepsis, and pancreatitis [38].

3.1.2. Gene transcription

HMGB1 knockout mice die shortly after birth due to hypoglycemia and exhibit a defect in the transcriptional enhancement of the glucocorticoid receptor [39]. In addition to the glucocorticoid receptor, HMGB1 interacts with a number of transcription factors (e.g., p53, p73, the retinoblastoma protein, nuclear factor kappa B [NF-κB], and estrogen receptor) to either activate or repress the transcription of specific genes.

3.1.3. DNA repair

The major forms of DNA damage include single-strand breaks, double-strand breaks, alteration of bases, hydrolytic depurination, hydrolytic deamination of cytosine and 5-methylcytosine bases, formation of covalent adducts with DNA, and oxidative damage to bases and to the phosphodiester backbone of DNA. Loss of HMGB1 increases these lesions [40, 41] and nuclear HMGB1 contributes to base excision repair, nucleotide excision repair, and mismatch repair [33].

3.1.4. V(D)J recombination

VDJ recombination is the process by which T cells and B cells randomly assemble different gene segments—known as variable (V), diversity (D), and joining (J) genes—in order to generate unique receptors (known as antigen receptors) that can collectively recognize many different types of molecules. HMG proteins, including HMGB1, are important components of the V(D)J recombinase complex [42].

3.1.5. Telomere homeostasis

Telomeres are caps with a region of repetitive nucleotide sequences at the end of chromosomes. Telomerase is an enzyme made of protein and ribonucleic acid (RNA) subunits that elongates chromosomes by adding TTAGGG sequences to the end of existing chromosomes. Telomere shortening is involved in the aging process. Loss of HMGB1 reduces telomerase activity, decreases telomere length, and increases chromosomal stability on a cellular level [43, 44].

3.2. Cytosolic HMGB1

Early studies have shown that the expression of HMGB1 in hepatic and brain tissues is high; it has been suggested a functional role of HMGB1 in both the nucleus and the cytoplasm [45, 46]. Recent studies have demonstrated that various tissues have a near-universal high expression of HMGB1. Cytoplasmic localization of HMGB1 has been observed in living fibroblasts [47], thymocytes [48], and several different tissues (e.g., liver, kidney, heart, and lung) [49]. Normally, the nuclear-to-cytoplasmic HMGB1 ratio is about 30:1 and this ratio is significantly reduced in cellular stress [49]. HMGB1 translocates from the nucleus to the cytoplasm, including the mitochondria and lysosomes, following various types of stressors such as inflammatory cytokines/chemokines and thermal and hypoxic stress. Although the study of the function of cytosolic HMGB1 remains poor, our research indicates that the main function of HMGB1 in the cytoplasm is to function as a positive regulator of autophagy and mitophagy (discussed later at section 4.2).

3.3. Extracellular HMGB1

HMGB1 can be actively secreted by immune cells or passively released by dead, dying, or injured cells [50]. Extracellular HMGB1 has multiple functions and is involved in several processes.

3.3.1. Cell differentiation

The first reported activity of extracellular HMGB1 is that HMGB1 promotes murine erythroleukemia cell differentiation [51, 52]. Structurally, the N-terminal region of HMGB1 is responsible for promoting murine erythroleukemia cell differentiation [53]. In addition to murine erythroleukemia cells, extracellular HMGB1 triggers the differentiation of chronic lymphocytic leukemia, stem cells, dendritic cells (DCs), and T cells [54, 55].

http://dx.doi.org/10.5772/63476

3.3.2. Inflammation and immunity

HMGB1 is an important late mediator released by macrophages in sepsis [31]. In addition to macrophages, many immune cells (e.g., neutrophils, mast cells, eosinophils, DCs, T cells, and natural killer cells) can release HMGB1 in response to infection [56]. HMGB1 cannot be actively secreted via the classical endoplasmic reticulum -Golgi secretory pathway due to lacking a leader signal sequence [57]. In turn, secretory lysosome contributes to HMGB1 secretion [58]. This process is regulated by metabolism [59]. Once released, HMGB1 can activate immune cells to sustain the inflammatory response. This process is regulated by redox status, receptor, and partner of HMGB1. For example, HMGB1 can bind and activate different signaling transduction cell receptors such as AGER, TLRs (e.g., TLR-2, -4, and -9), CD24, and TIM3 [60–62]. HMGB1 is very "sticky" and can bind to various extracellular pathogen-associated molecular patterns (PAMPs) (e.g., lipopolysaccharide) and DAMPs (e.g., DNA and histones) to amply inflammatory and immune responses [60, 63].

3.3.3. Cell migration

HMGB1 has chemokine activity to induce cell invasion and migration, a key process during the development of most organisms [64]. The potential mechanism includes HMGB1-mediated signaling transduction (e.g., ERK [65, 66] and Cdc42 [67]), transcriptional factor activation (e.g., NF-κB), and chemokine production.

3.3.4. Tissue regeneration

Tissue regeneration is the body's autohealing reaction once it gets injured or damaged. HMGB1 can stimulate myocardial regeneration, which may facilitate cardiac repair [68–71], cardiomyocyte hypertrophy [72], or cardiac fibrosis [73].

3.3.5. Angiogenesis

Angiogenesis is the growth of blood vessels from the existing vasculature. Treatment with HMGB1 protein increases angiogenesis by the secretion of vascular endothelial growth factor, an important inducer of angiogenesis [74].

4. HMGB1 and autophagy

4.1. Nuclear HMGB1 in autophagy

Mitophagy is an important mitochondrial quality control mechanism to sustain mitochondrial structure and function. We recently demonstrated that the nuclear protein HMGB1 modulates mitochondrial respiration and morphology by sustaining mitophagy through the regulation of heat shock protein β-1 (HSPB1) gene expression [75]. Metabolic activities in normal cells rely primarily on mitochondrial oxidative phosphorylation (OXPHOS) to generate adenosine triphosphate (ATP) for energy. In contrast, cancer cells mainly use glycolysis to generate ATP for energy. This type of energy reprogramming is called the Warburg effect. Interestingly, knockout or knockdown of HMGB1 or HSPB1 significantly inhibits OXPHOS and glycolysis in cancer cells or fibroblasts [75]. As expected, ATP production is decreased in HMGB1- or HSPB1-deficient cells. HSPB1 is a member of the small heat shock proteins (HSPs), which are important for protein folding [75]. HSPB1, but not other HSPs, is significantly inhibited in HMGB1-/-cells. Transfection of HMGB1 complementary DNA (cDNA) into HMGB1 cells restores HSPB1 expression at messenger RNA (mRNA) and protein levels. This process is not dependent on heat shock factor 1 (HSF1), the major transcription factor for HSP expression. Importantly, forced expression of HSPB1 by gene transfection corrects the deficiency in mitochondrial respiration, ATP production, and mitochondrial fragmentation, which is observed in HMGB1-deficient cells [75]. Thus, HSPB1 is the primary downstream mediator of HMGB1's effect on the regulation of mitochondrial homeostasis.

Alterations to the cytoskeleton during cell death and autophagy have been described in a variety of different cells. Previous studies have suggested that HSPB1 has a direct influence on the dynamics of cytoskeletal elements by HSPB1 phosphorylation [76, 77]. Similarly, by using cytoskeleton inhibitor cytochalasin D, loss of HSPB1 or mutation of its phosphorylation sites at serines 15 and 86 decreases starvation and rotenone-induced autophagy and mitophagy and impairs autophagosome and lysosome fusion [75]. These findings suggest that impaired cytoskeleton is involved in HMGB1-HSPB1 pathway-mediated mitophagy.

PTEN-induced putative kinase-1 (PINK1) is a kinase of the outer mitochondrial membrane, and PARK2 is a protein implicated in autosomal recessive juvenile Parkinsonism. The PINK1-PARK2 pathway has been largely implicated in the removal of damaged mitochondria with depolarized membranes in mammalian cells [78]. Upon mitochondrial membrane depolarization, PINK1 mediates the stress-induced mitochondrial translocation of PARK2. Subsequently, mitochondrial PARK2 drives the formation of Lys27-linked ubiquitin chains on the outer membrane of voltage-dependent anion channel 1 (VDAC1) [78]. These chains are then recognized by the autophagic adapter protein sequestosome 1 (SQSTM1/p62). SQSTM1 binds directly to LC3 to facilitate the formation of autophagosomes engulfing damaged mitochondria. HMGB1 and HSPB1 regulate PARK2 translocation and VDAC1 ubiquitination during mitochondrial depolarization. Knockdown of PINK1 or PARK2 abolishes the HSPB1-induced restoration in ATP production and reduction in mitochondrial fragmentation in HMGB1-deficient cells. Collectively, activation of the PINK1-PARK2 pathway is required for the HMBG1-HSPB1-dependent autophagic clearance of mitochondria. HMGB1 and HSPB1 translocate into the mitochondria during cellular stress. Whether these proteins interact directly with PINK1 or PARK2 remains unknown.

4.2. Cytosolic HMGB1 in autophagy

Release of HMGB1 has been observed in different types of cell death such as apoptosis, necrosis, and necroptosis [50, 79–84]. Similarly, classical autophagic stimuli such as rapamycin or starvation trigger HMGB1 translocation and release [81]. This process is not associated with lactate dehydrogenase release in the early stage, suggesting that translocation and release of HMGB1 in autophagy is an active process. Reactive oxygen species (ROS) generated in cell stress induce cell death, survival, or senescence, depending on the concentration of ROS. ROS quencher (e.g. *N*-acetyl cysteine) inhibits starvation- and rapamycin-induced HMGB1 translocation and subsequent autophagy [81]. Knockdown of antioxidant enzyme superoxide dismutase 1 also promotes HMGB1 cytosolic translocation and release in autophagy [85]. These findings suggest that oxidative stress is required for the translocation and release of HMGB1 in autophagy.

BECN1 was originally discovered as a Bcl-2-interacting protein. Bcl-2 binds to BECN1, leading to repression of autophagy [86]. We now know that BECN1 participates in autophagosome formation and plays an important role in the regulation of interplay between autophagy and apoptosis [87]. The levels of HMGB1 affect the interaction between Bcl-2 and BECN1 in autophagy. On one hand, HMGB1 is involved in the regulation of Bcl-2 phosphorylation by activation of the ERK pathway. Ablation of HMGB1 diminishes starvation-induced phosphorylation of both ERK1/2 and Bcl-2 [87]. Phosphorylation of Bcl-2 inhibits interaction between Bcl-2 and BECN1. On the other hand, cytosolic HMGB1 has been identified as a direct BECN1-binding protein in tumor and nontumor cells. HMGB1 competes with Bcl-2 for interaction with BECN1 and orients BECN1 to autophagosomes in response to starvation. Structurally, C23 and C45 are required for HMGB1 to bind to BECN1 [87]. Mutation of C23 and C45 in HMGB1 results in the loss of their ability to mediate autophagy. Moreover, C106S mutation of HMGB1 results in much higher cytoplasmic levels of HMGB1 and demonstrates enhanced binding to BECN1, leading to the subsequent dissociation of Bcl-2 from BECN1. Knockdown of HMGB1 finally inhibits the formation of the BECN1- PIK3C3 complex in autophagy.

In addition to the redox state of HMGB1, several proteins such as ULK1, FIP200, nuclear accumbens-1 (NAC1), p53, SNCA/α-synuclein, and gamma-interferon inducible lysosomal thiol reductase (GILT) have been demonstrated to positively or negatively regulate HMGB1-BECN1 complex formation in several cells.

Different from other ATGs, ULK1 is a serine/threonine-protein kinase. FIP200 (FAK family kinase-interacting protein of 200 kDa) was identified in a two-hybrid screen with the tyrosine kinase Pyk2. Both ULK1 and FIP200 are involved in the formation of ULK1-ATG13-FIP200 complex in triggering vesicle nucleation during autophagy [88–91]. The formation of the ULK1-ATG13-FIP200 complex is not affected by HMGB1. However, knockdown of ULK1 or FIP200 inhibits HMGB1-BECN1 complex formation. This increases cell death in osteosarcoma cells following anticancer agent treatment [92]. Thus, the HMGB1-BECN1 complex functions as a downstream signal from ULK1-mATG13-FIP200 complex formation in the induction of autophagy.

NAC1 is a nuclear protein that belongs to the POZ/BTB (Pox virus and zinc finger/bric-a-brac tramtrack broad complex) domain family. NAC1 can bind and increase HMGB1 translocation from the nucleus to the cytosol and subsequent HMGB1-BECN1 complex formation in response to cisplatin [93]. Suppression of NAC1 expression limits HMGB1-BECN1 complex formation and impairs the autophagic response and enhanced anticancer activity of cisplatin in tumor cells [93].

p53, the most common tumor suppressor, plays both transcription-dependent and -independent roles in the regulation of apoptosis, autophagy, metabolism, cell cycle progression, and many other processes. Cytosolic p53 is a negative regulator of autophagy through a transcription-independent mechanism [94] whereas nuclear p53 is a positive regulator of autophagy by a transcription-dependent mechanism [95, 96]. A number of studies have demonstrated a nuclear interaction between HMGB1 and p53 in the regulation of gene expression [97–100]. The interaction between HMGB1 and p53 in the nucleus and cytosol is increased in colon cancer cells following starvation-induced autophagy [101]. Importantly, p53-HMGB1 complexes regulate cytosolic translocation of the reciprocal protein and levels of autophagy. Loss of p53 increases HMGB1 cytosolic translocation and HMGB1- BECN1 complex formation, which results in autophagy induction [101]. In contrast, loss of HMGB1 increases p53 cytosolic translocation, which leads to autophagy inhibition [101]. This dynamic location change between p53 and HMGB1 affects the levels of autophagy and anticancer activity of chemotherapy in colon cancer cells [102].

SNCA is expressed predominantly in the brain, where it is concentrated in presynaptic nerve terminals. The deposition of the abundant presynaptic brain protein SNCA as an aggregating fibrillary in neurons or glial cells is a hallmark lesion in a subset of neurodegenerative disorders. Autophagy contributes to SNCA clearance. Interestingly, aggregated SNCA may inhibit autophagy by blocking the cytosolic translocation of HMGB1 and subsequent HMGB1-BECN1 binding in PC12 cells [103]. Thus, HMGB1 may be a new target for drug intervention to restore the deficient autophagy caused by SNCA in neurodegenerative disorders.

GILT is a lysosomal thiol reductase, which can reduce protein disulfide bonds at a low pH. The enzyme is expressed constitutively in antigen-presenting cells (e.g., B cells, DCs, monocytes, and macrophages) and is induced by γ interferon in endothelial cells and tumor cells. GILT may negatively regulate HMGB1-BECN1 complex formation in response to oxidative stress [104]. Loss of GILT increases the cytosolic translocation of HMGB1 and subsequent autophagy, which contributes to diminished superoxide dismutase 2 expression and elevated superoxide production.

4.3. Extracellular HMGB1 in autophagy

HMGB1 release is a critical regulator of apoptosis and autophagy in response to metabolic and therapeutic stress [105]. Treatment with reduced but not oxidized HMGB1 protein increases the accumulation of LC3 puncta associated with induced LC3-II formation, reduced expression of SQSTM1, and suppressed BECN1-Bcl-2 complex formation [16]. However, the HMGB1 C106A mutant protein significantly decreases autophagy compared with wild-type reduced HMGB1 protein [16]. Moreover, knockout of BECN1 inhibits reduced HMGB1-induced autophagy, suggesting that BECN1 is required for reduced HMGB1-induced autophagy. Interestingly, oxidized HMGB1 may trigger mitochondria-mediated apoptosis by activation of CASP-3 and -9 [16].

Multiple surface receptors, including TLR2, TLR4, and AGER have been demonstrated to mediate HMGB1 activity. C106 is required for HMGB1 binding to TLR4 and activation of cytokine release in macrophages. AGER is a transmembrane receptor of the immunoglobulin gene superfamily encoded within the class III region of the major histocompatibility locus. RAGE activation has been implicated in infection and sterile inflammation, as well as in cancer, diabetes, and Alzheimer's disease. The interaction between AGER and its ligands, including HMGB1, promotes proinflammatory signal pathway activation and the formation of neutrophil extracellular traps partly through upregulation of autophagy [17, 106, 107]. AGER promotes anticancer agent-induced autophagy by regulating MTOR activation and BECN1-PIK3C3 complex formation [108, 109]. Knockdown of AGER, but not TLR4 in cancer cells diminishes HMGB1-induced autophagy [16]. Moreover, AGER contributes to HMGB1-induced autophagy in a BECN1-dependent manner in cancer cells. However, it is unclear which receptor is required for oxidized HMGB1-induced apoptosis. In addition to HMGB1, AGER is required for IL6- and hypoxia-induced autophagy in pancreatic cancer cells, suggesting an important role of RAGE-mediated autophagy in the pancreatic tumor microenvironment [106, 110].

5. Transcriptional regulation of HMGB1 in autophagy

Transcription factors such as p53 [111], c-Myc [112], and Kruppel-like factor-4 [113] have been reported to regulate mRNA expression of HMGB1 in several cells. These transcription factors are also important for autophagy. In addition to transcription factors, microRNAs (miRNAs) play an important role in the regulation of HMGB1 expression. miRNAs are a class of post-transcriptional regulators of gene expression. They are short (about 22 nucleotide) RNA sequences that bind to complementary sequences in the 3' untranslated region (3'UTR) of multiple target messenger RNAs (mRNAs). At the molecular level, miRNAs restrain the production of proteins by affecting the stability of their target mRNA and/or by downregulating their translation. We recently demonstrated that MIR34A is a potent inhibitor of autophagy by suppression of HMGB1 (but not sirtuin 1) expression in the retinoblastoma cell [114]. MIR34A directly targets HMGB1 mRNA and inhibits HMGB1 protein levels, thereby preventing autophagosome activation [114]. Targeting the MIR34A-HMGB1 pathway inhibits autophagy and increases apoptosis in response to chemotherapy.

Another study suggests that MIR22 controls autophagy by regulating HMGB1 protein levels [115]. MIR22 is an evolutionally conserved miRNA that is highly expressed in various tissues and cancer cells. MIR22-mediated transcriptional regulation of HMGB1 inhibits autophagy and chemotherapy resistance in osteosarcoma cells [115]. The human let-7 family of microRNAs contains 13 members that are major players in the regulation of gene expression. HMGB1 is another important direct target of MIR-let-7f-1 in medulloblastoma cells [116]. Overexpression of MIR-let-7f-1 inhibits HMGB1 expression and subsequent autophagy in medulloblastoma cells following treatment with cisplatin [116]. The complex interactions of HMGB1 expression by miRNAs and transcription factors in autophagy must be further investigated and will likely impact tumor treatment in the future.

6. Post-translational modification of HMGB1 in autophagy

Autophagy is mainly regulated by post-translational and lipid modifications of ATG proteins. HMGB1 also undergoes extensive post-translational modifications, including reversible and terminal acetylation [117], poly-ADP-ribosylation [118, 119], phosphorylation [120], and oxidation [121]. These post-translational modifications have been demonstrated to influence HMGB1's DNA chaperone activity, subcellular localization, and extracellular DAMP activity. We have discussed above that the redox status of HMGB1 affects autophagy. Unlike other members of the tumor necrosis factor (TNF) superfamily, the TNF (ligand) superfamily, member 10 (TNFSF10/TRAIL) selectively activates CASP8 and induces apoptosis in cancer cells (but not normal cells) *in vitro* and *in vivo*. HMGB1 is specifically poly-ADP-ribosylated (PAR) by PAR polymerase-1 (PARP1) in pancreatic cancer cells. This HMGB1 modification contributes to TNFSF10 resistance through upregulation of autophagy and suppression of apoptosis [122]. PARP1 is an ADP-ribosylating enzyme critical for initiating various forms of DNA repair in nucleus. Activation of PARP1 mediates TNFSF10-induced poly-ADP-ribosylation and subsequent translocation of HMGB1 from the nucleus to the cytosol. Inhibition of PARP1 expression or activity via shRNA knockdown or pharmacologic inhibitor PJ-34 significantly limits TNFSF10-induced poly-ADP-ribosylation and the subsequent cytoplasmic translocation of HMGB1 in human pancreatic cancer cells [122]. Importantly, activation of PARP1 promotes HMGB1-BECN1 complex formation, which leads to autophagy following TNFSF10 treatment. Transfection of HMGB1 C106S mutant cDNA into PARP1-knockdown cancer cells increases cytosolic HMGB1 level, LC3-II expression, and TNFSF10 resistance. These findings suggest that cytoplasmic HMGB1 is sufficient to trigger autophagy and TNFSF10 resistance in PARP1-deficient cancer cells [122]. Compared with C106S mutation, the C23S and C45S mutations fail to restore TNFSF10-induced HMGB1-BECN1 complex formation, LC3 turnover, and resistance to apoptosis in HMGB1-knockdown pancreatic cancer cells. Thus, PARP1-HMGB1-BECN1-mediated autophagy inhibits TNFSF10-induced apoptosis by suppression of CASP8 activity [122]. It will be interesting to test whether other post-translational modifications of HMGB1 directly activate autophagy under stress.

7. HMGB1-mediated autophagy in disease

The role of autophagy in cancer is complex and is likely dependent on tumor type and stage [123]. On one hand, autophagy plays a tumor suppressor role by preventing genome instability, limiting oxidative injury, reducing the inflammatory response, and inhibiting angiogenesis. On the other hand, autophagy functions as a survival mechanism in tumor development. Upregulation of autophagy promotes the growth of established tumors by sustaining energy metabolism and cell proliferation. In addition, increased autophagy leads to therapy resistance by diminishing regulated cell death. A number of studies indicate that HMGB1-mediated autophagy can enable tumor cell survival by inhibition of apoptosis, which can lead to therapeutic resistance [92, 93, 102, 122, 124–130]. It remains to be determined whether HMGB1-mediated autophagy contributes to the suppression of tumorigenesis [131, 132].

Autophagy regulates inflammation through interfering with innate immune signaling pathways, including inflammasome activation and proinflammatory cytokine release [133, 134]. We and others have demonstrated that activation of autophagy contributes to HMGB1 release in immune and nonimmune cells [84]. Inflammasomes are protein complexes in the innate immune system that regulate the activation of CASP1 or CASP11 and induce IL-1β and IL-18 release in response to infection or tissue injury. Conditional depletion of HMGB1 in myeloid cells renders mice more sensitive to *Listeria monocytogenes* infection and endotoxic shock [135] partly through downregulation of autophagy. This in turn promotes inflammasome activation and IL-1β release in macrophages [135]. Cytosolic HMGB1 in intestinal epithelial cells suppresses inflammation-associated cellular injury by controlling the switch between the proautophagic and proapoptotic functions of BECN1 and ATG5 during inflammation [136]. Moreover, conditional knockout of HMGB1 in the pancreas and liver promotes pancreatitis [137] and liver ischemic reperfusion [138], which are sterile inflammatory diseases without infection. Additionally, HMGB1 and BECN1 are co-expressed in the invading T cells in the muscle tissue of myositis patients, which is required for T-cell survival and function [139]. The underlying molecular mechanism of HMGB1-mediated autophagy in inflammation and immunity remains to be further explored.

Most neurodegenerative diseases that afflict humans are associated with the intracytoplasmic deposition of aggregate-prone proteins in neurons and with autophagy dysfunction. Impairment of HMGB1-mediated autophagy has been implicated in the increased protein misfolding and aggregation in neurodegenerative disease [140, 141]. In addition, HMGB1 has recently been indicated to be involved in the autophagy inhibition caused by SNCA overexpression, implying a direct role in modulating the autophagic degradation of SNCA [103].

8. Conclusion

HMGB1 is a nuclear protein and stress sensor that plays a critical role in various physiological and pathological processes, including autophagy. Autophagy is the major pathway involved in the degradation of proteins and organelles, cellular remodeling, and survival during stress. HMGB1 plays important intranuclear, cytosolic, and extracellular roles in the regulation of autophagy [142]. Cytoplasmic HMGB1 is a novel BECN1-binding protein active in autophagy. Extracellular HMGB1 induces autophagy in an AGER-dependent manner. Nuclear HMGB1 contributes to mitophagy by regulation of HSPB1 expression. HMGB1-dependent autophagy promotes chemotherapy resistance, [92, 93, 101, 108, 114, 127, 128, 143–145], sustains the tumor metabolism requirement [16, 146] and T-cell survival, [139], prevents polyglutamine aggregates [140] and excitotoxicity [141], and protects against endotoxemia, bacterial infection, and ischemia-reperfusion injury [135, 147–149]. The role of HMGB1 in autophagy is clearly complex and tissue dependent [142]. HMGB1 is not required for starvation-induced autophagy in mice with hepatocyte-specific HMGB1 deletion, suggesting that an HMGB1-independent autophagy pathway exists in different organs [150]. Indeed, mice with hepatocyte-specific HMGB1 deletion have a different phenotype following different stressors [151, 152]. Targeting the HMGB1-mediated autophagy pathway may be required to address whether or not this approach is therapeutically advantageous in human disease.

Acknowledgements

We apologize to the researchers who were not referenced due to space limitations. We thank Christine Heiner (Department of Surgery, University of Pittsburgh) for her critical reading of the manuscript. This work was supported by the USA National Institutes of Health (R01GM115366 and R01CA160417 to D.T.) and the Natural Science Foundation of Guangdong Province (2016A030308011 to D.T.).

Author details

Daolin Tang[1,2*] and Rui Kang[2*]

*Address all correspondence to: tangd2@upmc.edu and kangr@upmc.edu

1 The Center for DAMP Biology, The Third Affiliated Hospital of Guangzhou Medical University, Guangzhou, Guangdong, China

2 Department of Surgery, University of Pittsburgh, Pittsburgh, Pennsylvania, USA

References

[1] Behrends, C., Sowa, M.E., Gygi, S.P., Harper, J.W. (2010) Network organization of the human autophagy system. Nature 466, 68–76.

[2] Schlumpberger, M., Schaeffeler, E., Straub, M., Bredschneider, M., Wolf, D.H., Thumm, M. (1997) AUT1, a gene essential for autophagocytosis in the yeast Saccharomyces cerevisiae. J Bacteriol 179, 1068–1076.

[3] Funakoshi, T., Matsuura, A., Noda, T., Ohsumi, Y. (1997) Analyses of APG13 gene involved in autophagy in yeast, Saccharomyces cerevisiae. Gene 192, 207–213.

[4] Matsuura, A., Tsukada, M., Wada, Y., Ohsumi, Y. (1997) Apg1p, a novel protein kinase required for the autophagic process in Saccharomyces cerevisiae. Gene 192, 245–250.

[5] Boya, P., Reggiori, F., Codogno, P. (2013) Emerging regulation and functions of autophagy. Nat Cell Biol 15, 713–720.

[6] Subramani, S., Malhotra, V. (2013) Non-autophagic roles of autophagy-related proteins. EMBO Rep 14, 143–151.

[7] Xie, Y., Kang, R., Sun, X., Zhong, M., Huang, J., Klionsky, D.J., Tang, D. (2015) Post-translational modification of autophagy-related proteins in macroautophagy. Autophagy 11, 28–45.

[8] Goodwin, G.H., Sanders, C., Johns, E.W. (1973) A new group of chromatin-associated proteins with a high content of acidic and basic amino acids. Eur J Biochem 38, 14–19.

[9] Kang, R., Chen, R., Zhang, Q., Hou, W., Wu, S., Cao, L., Huang, J., Yu, Y., Fan, X.G., Yan, Z., Sun, X., Wang, H., Wang, Q., Tsung, A., Billiar, T.R., Zeh, H.J. III, Lotze, M.T., Tang, D. (2014) HMGB1 in health and disease. Mol Aspects Med 40, 1–116.

[10] Weir, H.M., Kraulis, P.J., Hill, C.S., Raine, A.R., Laue, E.D., Thomas, J.O. (1993) Structure of the HMG box motif in the B-domain of HMG1. Embo J 12, 1311–1319.

[11] Hardman, C.H., Broadhurst, R.W., Raine, A.R., Grasser, K.D., Thomas, J.O., Laue, E.D. (1995) Structure of the A-domain of HMG1 and its interaction with DNA as studied by heteronuclear three- and four-dimensional NMR spectroscopy. Biochemistry 34, 16596–16607.

[12] Read, C.M., Cary, P.D., Crane-Robinson, C., Driscoll, P.C., Norman, D.G. (1993) Solution structure of a DNA-binding domain from HMG1. Nucleic Acids Res 21, 3427–3436.

[13] Li, J., Kokkola, R., Tabibzadeh, S., Yang, R., Ochani, M., Qiang, X., Harris, H.E., Czura, C.J., Wang, H., Ulloa, L., Wang, H., Warren, H.S., Moldawer, L.L., Fink, M.P., Andersson, U., Tracey, K.J., Yang, H. (2003) Structural basis for the proinflammatory cytokine activity of high mobility group box 1. Mol Med 9, 37–45.

[14] Yang, H., Ochani, M., Li, J., Qiang, X., Tanovic, M., Harris, H.E., Susarla, S.M., Ulloa, L., Wang, H., DiRaimo, R., Czura, C.J., Wang, H., Roth, J., Warren, H.S., Fink, M.P., Fenton, M.J., Andersson, U., Tracey, K.J. (2004) Reversing established sepsis with antagonists of endogenous high-mobility group box 1. Proc Natl Acad Sci U S A 101, 296–301.

[15] Huttunen, H.J., Fages, C., Kuja-Panula, J., Ridley, A.J., Rauvala, H. (2002) Receptor for advanced glycation end products-binding COOH-terminal motif of amphoterin inhibits invasive migration and metastasis. Cancer Res 62, 4805–4811.

[16] Tang, D., Kang, R., Cheh, C.W., Livesey, K.M., Liang, X., Schapiro, N.E., Benschop, R., Sparvero, L.J., Amoscato, A.A., Tracey, K.J., Zeh, H.J., Lotze, M.T. (2010) HMGB1 release and redox regulates autophagy and apoptosis in cancer cells. Oncogene 29, 5299–5310.

[17] Kang, R., Tang, D., Schapiro, N.E., Loux, T., Livesey, K.M., Billiar, T.R., Wang, H., Van Houten, B., Lotze, M.T., Zeh, H.J. (2014) The HMGB1/RAGE inflammatory pathway

promotes pancreatic tumor growth by regulating mitochondrial bioenergetics. Oncogene 33, 567–577.

[18] Andersson, U., Tracey, K.J. (2011) HMGB1 is a therapeutic target for sterile inflammation and infection. Annu Rev Immunol 29, 139–162.

[19] Yang, H.A., Hreggvidsdottir, H.S., Palmblad, K., Wang, H.C., Ochani, M., Li, J.H., Lu, B., Chavan, S., Rosas-Ballina, M., Al-Abed, Y., Akira, S., Bierhaus, A., Erlandsson-Harris, H., Andersson, U., Tracey, K.J. (2010) A critical cysteine is required for HMGB1 binding to Toll-like receptor 4 and activation of macrophage cytokine release. Proc Natl Acad Sci U S A 107, 11942–11947.

[20] Venereau, E., Casalgrandi, M., Schiraldi, M., Antoine, D.J., Cattaneo, A., De Marchis, F., Liu, J., Antonelli, A., Preti, A., Raeli, L., Shams, S.S., Yang, H., Varani, L., Andersson, U., Tracey, K.J., Bachi, A., Uguccioni, M., Bianchi, M.E. (2012) Mutually exclusive redox forms of HMGB1 promote cell recruitment or proinflammatory cytokine release. J Exp Med209, 1519–1528.

[21] Tang, D., Billiar, T.R., Lotze, M.T. (2012) A Janus tale of two active high mobility group box 1 (HMGB1) redox states. Mol Med 18, 1360–1362.

[22] Venereau, E., Casalgrandi, M., Schiraldi, M., Antoine, D.J., Cattaneo, A., De Marchis, F., Liu, J., Antonelli, A., Preti, A., Raeli, L., Shams, S.S., Yang, H., Varani, L., Andersson, U., Tracey, K.J., Bachi, A., Uguccioni, M., Bianchi, M.E. (2012) Mutually exclusive redox forms of HMGB1 promote cell recruitment or proinflammatory cytokine release. J Exp Med 209, 1519–1528.

[23] Tang, D., Shi, Y., Kang, R., Li, T., Xiao, W., Wang, H., Xiao, X. (2007) Hydrogen peroxide stimulates macrophages and monocytes to actively release HMGB1. J Leukoc Biol 81, 741–747.

[24] Tang, D., Kang, R., Zeh, H.J. III, Lotze, M.T. (2011) High-mobility group box 1, oxidative stress, and disease. Antioxid Redox Signal 14, 1315–1335.

[25] Yu, Y., Tang, D., Kang, R. (2015) Oxidative stress-mediated HMGB1 biology. Front Physiol 6, 93.

[26] Tsung, A., Klune, J.R., Zhang, X., Jeyabalan, G., Cao, Z., Peng, X., Stolz, D.B., Geller, D.A., Rosengart, M.R., Billiar, T.R. (2007) HMGB1 release induced by liver ischemia involves Toll-like receptor 4 dependent reactive oxygen species production and calcium-mediated signaling. J Exp Med 204, 2913–2923.

[27] Tang, D., Kang, R., Xiao, W., Zhang, H., Lotze, M.T., Wang, H., Xiao, X. (2009) Quercetin prevents LPS-induced high-mobility group box 1 release and proinflammatory function. Am J Respir Cell Mol Biol 41, 651–660.

[28] Kato, S., Hussein, M.H., Kakita, H., Goto, T., Daoud, G.A., Kato, T., Sugiura, T., Nobata, M., Nakajima, Y., Endo, T., Mizuno, K., Ito, T., Kato, I., Suzuki, S., Togari, H. (2009)

Edaravone, a novel free radical scavenger, reduces high-mobility group box 1 and prolongs survival in a neonatal sepsis model. Shock 32, 586–592.

[29] Li, W., Ashok, M., Li, J., Yang, H., Sama, A.E., Wang, H. (2007) A major ingredient of green tea rescues mice from lethal sepsis partly by inhibiting HMGB1. PLoS ONE One 2, e1153.

[30] Xu, W., Lu, Y., Yao, J., Li, Z., Chen, Z., Wang, G., Jing, H., Zhang, X., Li, M., Peng, J., Tian, X. (2014) Novel role of resveratrol: suppression of high-mobility group protein box 1 nucleocytoplasmic translocation by the upregulation of sirtuin 1 in sepsis-induced liver injury. Shock 42, 440–447.

[31] Wang, H., Bloom, O., Zhang, M., Vishnubhakat, J.M., Ombrellino, M., Che, J., Frazier, A., Yang, H., Ivanova, S., Borovikova, L., Manogue, K.R., Faist, E., Abraham, E., Andersson, J., Andersson, U., Molina, P.E., Abumrad, N.N., Sama, A., Tracey, K.J. (1999) HMG-1 as a late mediator of endotoxin lethality in mice. Science 285, 248–251.

[32] Huang, J., Xie, Y., Sun, X., Zeh, H.J. III, Kang, R., Lotze, M.T., Tang, D. (2015) DAMPs, ageing, and cancer: the "DAMP Hypothesis". Ageing Res Rev 24, 3–16.

[33] Lange, S.S., Vasquez, K.M. (2009) HMGB1: the jack-of-all-trades protein is a master DNA repair mechanic. Mol Carcinog 48, 571–580.

[34] Bonaldi, T., Langst, G., Strohner, R., Becker, P.B., Bianchi, M.E. (2002) The DNA chaperone HMGB1 facilitates ACF/CHRAC-dependent nucleosome sliding. Embo J 21, 6865–6873.

[35] Gerlitz, G., Hock, R., Ueda, T., Bustin, M. (2009) The dynamics of HMG protein-chromatin interactions in living cells. Biochem Cell Biol 87, 127–137.

[36] Celona, B., Weiner, A., Di Felice, F., Mancuso, F.M., Cesarini, E., Rossi, R.L., Gregory, L., Baban, D., Rossetti, G., Grianti, P., Pagani, M., Bonaldi, T., Ragoussis, J., Friedman, N., Camilloni, G., Bianchi, M.E., Agresti, A. (2011) Substantial histone reduction modulates genomewide nucleosomal occupancy and global transcriptional output. PLoS Biol 9, e1001086.

[37] Kang, R., Zhang, Q., Hou, W., Yan, Z., Chen, R., Bonaroti, J., Bansal, P., Billiar, T.R., Tsung, A., Wang, Q., Bartlett, D.L., Whitcomb, D.C., Chang, E.B., Zhu, X., Wang, H., Lu, B., Tracey, K.J., Cao, L., Fan, X.G., Lotze, M.T., Zeh, H.J. III, Tang, D. (2014) Intracellular Hmgb1 inhibits inflammatory nucleosome release and limits acute pancreatitis in mice. Gastroenterology 146, 1097–1107.

[38] Chen, R., Kang, R., Fan, X.G., Tang, D. (2014) Release and activity of histone in diseases. Cell Death Dis 5, e1370.

[39] Calogero, S., Grassi, F., Aguzzi, A., Voigtlander, T., Ferrier, P., Ferrari, S., Bianchi, M.E. (1999) The lack of chromosomal protein Hmg1 does not disrupt cell growth but causes lethal hypoglycaemia in newborn mice. Nat Genet 22, 276–280.

[40] Lange, S.S., Mitchell, D.L., Vasquez, K.M. (2008) High mobility group protein B1 enhances DNA repair and chromatin modification after DNA damage. Proc Natl Acad Sci U S A 105, 10320–10325.

[41] Ohndorf, U.M., Rould, M.A., He, Q., Pabo, C.O., Lippard, S.J. (1999) Basis for recognition of cisplatin-modified DNA by high-mobility-group proteins. Nature 399, 708–712.

[42] van Gent, D.C., Hiom, K., Paull, T.T., Gellert, M. (1997) Stimulation of V(D)J cleavage by high mobility group proteins. Embo J 16, 2665–2670.

[43] Giavara, S., Kosmidou, E., Hande, M.P., Bianchi, M.E., Morgan, A., d'Adda di Fagagna, F., Jackson, S.P. (2005) Yeast Nhp6A/B and mammalian Hmgb1 facilitate the maintenance of genome stability. Curr Biol 15, 68–72.

[44] Polanska, E., Dobsakova, Z., Dvorackova, M., Fajkus, J., Stros, M. (2012) HMGB1 gene knockout in mouse embryonic fibroblasts results in reduced telomerase activity and telomere dysfunction. Chromosoma 121, 419–431.

[45] Bustin, M., Neihart, N.K. (1979) Antibodies against chromosomal HMG proteins stain the cytoplasm of mammalian cells. Cell 16, 181–189.

[46] Mosevitsky, M.I., Novitskaya, V.A., Iogannsen, M.G., Zabezhinsky, M.A. (1989) Tissue specificity of nucleo-cytoplasmic distribution of HMG1 and HMG2 proteins and their probable functions. Eur J Biochem 185, 303–310.

[47] Einck, L., Soares, N., Bustin, M. (1984) Localization of HMG chromosomal proteins in the nucleus and cytoplasm by microinjection of functional antibody fragments into living fibroblasts. Exp Cell Res 152, 287–301.

[48] Guillet, F., Tournefier, A., Denoulet, P., Capony, J.P., Kerfourn, F., Charlemagne, J. (1990) High levels of HMG1-2 protein expression in the cytoplasm and nucleus of hydrocortisone sensitive amphibian thymocytes. Biol Cell 69, 153–160.

[49] Kuehl, L., Salmond, B., Tran, L. (1984) Concentrations of high-mobility-group proteins in the nucleus and cytoplasm of several rat tissues. J Cell Biol 99, 648–654.

[50] Scaffidi, P., Misteli, T., Bianchi, M.E. (2002) Release of chromatin protein HMGB1 by necrotic cells triggers inflammation. Nature 418, 191–195.

[51] Melloni, E., Sparatore, B., Patrone, M., Pessino, A., Passalacqua, M., Pontremoli, S. (1995) Identity in molecular structure between "differentiation enhancing factor" of murine erythroleukemia cells and the 30 kD heparin-binding protein of developing rat brain. Biochem Biophys Res Commun 210, 82–89.

[52] Melloni, E., Sparatore, B., Patrone, M., Pessino, A., Passalacqua, M., Pontremoli, S. (1995) Extracellular release of the "differentiation enhancing factor," a HMG1 protein type, is an early step in murine erythroleukemia cell differentiation. FEBS Lett 368, 466–470.

[53] Sparatore, B., Melloni, E., Patrone, M., Passalacqua, M., Pontremoli, S. (1996) A 6 kDa protein homologous to the N-terminus of the HMG1 protein promoting stimulation of murine erythroleukemia cell differentiation. FEBS Lett 386, 95–98.

[54] Jia, L., Clear, A., Liu, F.T., Matthews, J., Uddin, N., McCarthy, A., Hoxha, E., Durance, C., Iqbal, S., Gribben, J.G. (2014) Extracellular HMGB1 promotes differentiation of nurse-like cells in chronic lymphocytic leukemia. Blood123, 1709–1719.

[55] Pistoia, V., Raffaghello, L. (2011) Damage-associated molecular patterns (DAMPs) and mesenchymal stem cells: a matter of attraction and excitement. Eur J Immunol 41, 1828–1831.

[56] Lotze, M.T., Tracey, K.J. (2005) High-mobility group box 1 protein (HMGB1): nuclear weapon in the immune arsenal. Nat Rev Immunol 5, 331–342.

[57] Bonaldi, T., Talamo, F., Scaffidi, P., Ferrera, D., Porto, A., Bachi, A., Rubartelli, A., Agresti, A., Bianchi, M.E. (2003) Monocytic cells hyperacetylate chromatin protein HMGB1 to redirect it towards secretion. Embo J 22, 5551–5560.

[58] Gardella, S., Andrei, C., Ferrera, D., Lotti, L.V., Torrisi, M.R., Bianchi, M.E., Rubartelli, A. (2002) The nuclear protein HMGB1 is secreted by monocytes via a non-classical, vesicle-mediated secretory pathway. EMBO Rep 3, 995–1001.

[59] Yang, L., Xie, M., Yang, M., Yu, Y., Zhu, S., Hou, W., Kang, R., Lotze, M.T., Billiar, T.R., Wang, H., Cao, L., Tang, D. (2014) PKM2 regulates the Warburg effect and promotes HMGB1 release in sepsis. Nat Commun 5, 4436.

[60] Urbonaviciute, V., Furnrohr, B.G., Weber, C., Haslbeck, M., Wilhelm, S., Herrmann, M., Voll, R.E. (2007) Factors masking HMGB1 in human serum and plasma. J Leukoc Biol 81, 67–74.

[61] Tang, D., Lotze, M.T. (2012) Tumor immunity times out: TIM-3 and HMGB1. Nat Immunol 13, 808–810.

[62] Chiba, S., Baghdadi, M., Akiba, H., Yoshiyama, H., Kinoshita, I., Dosaka-Akita, H., Fujioka, Y., Ohba, Y., Gorman, J.V., Colgan, J.D., Hirashima, M., Uede, T., Takaoka, A., Yagita, H., Jinushi, M. (2012) Tumor-infiltrating DCs suppress nucleic acid-mediated innate immune responses through interactions between the receptor TIM-3 and the alarmin HMGB1. Nat Immunol 13, 832–842.

[63] Chen, R., Fu, S., Fan, X.G., Lotze, M.T., Zeh, H.J. III, Tang, D., Kang, R. (2015) Nuclear DAMP complex-mediated RAGE-dependent macrophage cell death. Biochem Biophys Res Commun 458, 650–655.

[64] Degryse, B., de Virgilio, M. (2003) The nuclear protein HMGB1, a new kind of chemokine? FEBS Lett 553, 11–17.

[65] Degryse, B., Bonaldi, T., Scaffidi, P., Muller, S., Resnati, M., Sanvito, F., Arrigoni, G., Bianchi, M.E. (2001) The high mobility group (HMG) boxes of the nuclear protein

HMG1 induce chemotaxis and cytoskeleton reorganization in rat smooth muscle cells. J Cell Biol 152, 1197–1206.

[66] Ranzato, E., Patrone, M., Pedrazzi, M., Burlando, B. (2009) HMGb1 promotes scratch wound closure of HaCaT keratinocytes via ERK1/2 activation. Mol Cell Biochem 332, 199–205.

[67] Fages, C., Nolo, R., Huttunen, H.J., Eskelinen, E., Rauvala, H. (2000) Regulation of cell migration by amphoterin. J Cell Sci 113 (Pt 4), 611–620.

[68] Germani, A., Limana, F., Capogrossi, M.C. (2007) Pivotal advances: high-mobility group box 1 protein—a cytokine with a role in cardiac repair. J Leukoc Biol 81, 41–45.

[69] Limana, F., Germani, A., Zacheo, A., Kajstura, J., Di Carlo, A., Borsellino, G., Leoni, O., Palumbo, R., Battistini, L., Rastaldo, R., Muller, S., Pompilio, G., Anversa, P., Bianchi, M.E., Capogrossi, M.C. (2005) Exogenous high-mobility group box 1 protein induces myocardial regeneration after infarction via enhanced cardiac C-kit+ cell proliferation and differentiation. Circ Res 97, e73–83.

[70] Abarbanell, A.M., Hartley, J.A., Herrmann, J.L., Weil, B.R., Wang, Y., Manukyan, M.C., Poynter, J.A., Meldrum, D.R. (2011) Exogenous high-mobility group box 1 improves myocardial recovery after acute global ischemia/reperfusion injury. Surgery 149, 329–335.

[71] Limana, F., Esposito, G., Fasanaro, P., Foglio, E., Arcelli, D., Voellenkle, C., Di Carlo, A., Avitabile, D., Martelli, F., Russo, M.A., Pompilio, G., Germani, A., M, C.C. (2013) Transcriptional profiling of HMGB1-induced myocardial repair identifies a key role for Notch signaling. Mol Ther 21, 1841–1851.

[72] Su, F.F., Shi, M.Q., Guo, W.G., Liu, X.T., Wang, H.T., Lu, Z.F., Zheng, Q.S. (2012) High-mobility group box 1 induces calcineurin-mediated cell hypertrophy in neonatal rat ventricular myocytes. Mediators Inflamm 2012, 805149.

[73] Wang, W.K., Wang, B., Lu, Q.H., Zhang, W., Qin, W.D., Liu, X.J., Liu, X.Q., An, F.S., Zhang, Y., Zhang, M.X. (2014) Inhibition of high-mobility group box 1 improves myocardial fibrosis and dysfunction in diabetic cardiomyopathy. Int J Cardiol.172, 202–212.

[74] van Beijnum, J.R., Dings, R.P., van der Linden, E., Zwaans, B.M., Ramaekers, F.C., Mayo, K.H., Griffioen, A.W. (2006) Gene expression of tumor angiogenesis dissected: specific targeting of colon cancer angiogenic vasculature. Blood 108, 2339–2348.

[75] Tang, D., Kang, R., Livesey, K.M., Kroemer, G., Billiar, T.R., Van Houten, B., Zeh, H.J. III, Lotze, M.T. (2011) High-mobility group box 1 is essential for mitochondrial quality control. Cell Metab 13, 701–711.

[76] Pichon, S., Bryckaert, M., Berrou, E. (2004) Control of actin dynamics by p38 MAP kinase - Hsp27 distribution in the lamellipodium of smooth muscle cells. J Cell Sci 117, 2569–2577.

[77] Robitaille, H., Simard-Bisson, C., Larouche, D., Tanguay, R.M., Blouin, R., Germain, L. (2010) The small heat-shock protein Hsp27 undergoes ERK-dependent phosphorylation and redistribution to the cytoskeleton in response to dual leucine zipper-bearing kinase expression. J Invest Dermatol 130, 74–85.

[78] Geisler, S., Holmstrom, K.M., Skujat, D., Fiesel, F.C., Rothfuss, O.C., Kahle, P.J., Springer, W. (2010) PINK1/Parkin-mediated mitophagy is dependent on VDAC1 and p62/SQSTM1. Nat Cell Biol 12, 119–131.

[79] Bell, C.W., Jiang, W., Reich, C.F. III, Pisetsky, D.S. (2006) The extracellular release of HMGB1 during apoptotic cell death. Am J Physiol Cell Physiol 291, C1318–1325.

[80] Jiang, W., Bell, C.W., Pisetsky, D.S. (2007) The relationship between apoptosis and high-mobility group protein 1 release from murine macrophages stimulated with lipopolysaccharide or polyinosinic-polycytidylic acid. J Immunol 178, 6495–6503.

[81] Tang, D., Kang, R., Livesey, K.M., Cheh, C.W., Farkas, A., Loughran, P., Hoppe, G., Bianchi, M.E., Tracey, K.J., Zeh, H.J. III, Lotze, M.T. (2010) Endogenous HMGB1 regulates autophagy. J Cell Biol 190, 881–892.

[82] Thorburn, J., Horita, H., Redzic, J., Hansen, K., Frankel, A.E., Thorburn, A. (2009) Autophagy regulates selective HMGB1 release in tumor cells that are destined to die. Cell Death Differ 16, 175–183.

[83] Dupont, N., Jiang, S., Pilli, M., Ornatowski, W., Bhattacharya, D., Deretic, V. (2011) Autophagy-based unconventional secretory pathway for extracellular delivery of IL-1beta. Embo J 30, 4701–4711.

[84] Zhang, Q., Kang, R., Zeh, H.J. III, Lotze, M.T., Tang, D. (2013) DAMPs and autophagy: cellular adaptation to injury and unscheduled cell death. Autophagy 9, 451–458.

[85] Tang, D., Kang, R., Livesey, K.M., Zeh, H.J. III, Lotze, M.T. (2011) High mobility group box 1 (HMGB1) activates an autophagic response to oxidative stress. Antioxid Redox Signal 15, 2185–2195.

[86] Pattingre, S., Tassa, A., Qu, X., Garuti, R., Liang, X.H., Mizushima, N., Packer, M., Schneider, M.D., Levine, B. (2005) Bcl-2 antiapoptotic proteins inhibit Beclin 1-dependent autophagy. Cell 122, 927–939.

[87] Kang, R., Zeh, H.J., Lotze, M.T., Tang, D. (2011) The Beclin 1 network regulates autophagy and apoptosis. Cell Death Differ 18, 571–580.

[88] Kim, J., Kundu, M., Viollet, B., Guan, K.L. (2011) AMPK and mTOR regulate autophagy through direct phosphorylation of Ulk1. Nat Cell Biol 13, 132–141.

[89] Shang, L., Chen, S., Du, F., Li, S., Zhao, L., Wang, X. (2011) Nutrient starvation elicits an acute autophagic response mediated by Ulk1 dephosphorylation and its subsequent dissociation from AMPK. Proc Natl Acad Sci U S A 108, 4788–4793.

[90] Bach, M., Larance, M., James, D.E., Ramm, G. (2011) The serine/threonine kinase ULK1 is a target of multiple phosphorylation events. Biochem J 440, 283–291.

[91] Dorsey, F.C., Rose, K.L., Coenen, S., Prater, S.M., Cavett, V., Cleveland, J.L., Caldwell-Busby, J. (2009) Mapping the phosphorylation sites of Ulk1. J Proteome Res 8, 5253–5263.

[92] Huang, J., Ni, J., Liu, K., Yu, Y., Xie, M., Kang, R., Vernon, P., Cao, L., Tang, D. (2012) HMGB1 promotes drug resistance in osteosarcoma. Cancer Res 72, 230–238.

[93] Zhang, Y., Cheng, Y., Ren, X., Zhang, L., Yap, K.L., Wu, H., Patel, R., Liu, D., Qin, Z.H., Shih, I.M., Yang, J.M. (2012) NAC1 modulates sensitivity of ovarian cancer cells to cisplatin by altering the HMGB1-mediated autophagic response. Oncogene 31, 1055–1064.

[94] Tasdemir, E., Maiuri, M.C., Galluzzi, L., Vitale, I., Djavaheri-Mergny, M., D'amelio, M., Criollo, A., Morselli, E., Zhu, C., Harper, F., Nannmark, U., Samara, C., Pinton, P., Vicencio, J., Carnuccio, R., Moll, U., Madeo, F., Paterlini-Brechot, P., Rizzuto, R., Szabadkai, G., Pierron, G., Blomgren, K., Tavernarakis, N., Codogno, P., Cecconi, F., Kroemer, G. (2008) Regulation of autophagy by cytoplasmic p53. Nat Cell Biol 10, 676–687.

[95] Crighton, D., Wilkinson, S., O'Prey, J., Syed, N., Smith, P., Harrison, P.R., Gasco, M., Garrone, O., Crook, T., Ryan, K.M. (2006) DRAM, a p53-induced modulator of autophagy, is critical for apoptosis. Cell 126, 121–134.

[96] Gao, W., Shen, Z., Shang, L., Wang, X. (2011) Upregulation of human autophagy-initiation kinase ULK1 by tumor suppressor p53 contributes to DNA-damage-induced cell death. Cell Death Differ 18, 1598–1607.

[97] Rowell, J.P., Simpson, K.L., Stott, K., Watson, M., Thomas, J.O. (2012) HMGB1-facilitated p53 DNA binding occurs via HMG-Box/p53 transactivation domain interaction, regulated by the acidic tail. Structure 20, 2014–2024.

[98] Stros, M., Muselikova-Polanska, E., Pospisilova, S., Strauss, F. (2004) High-affinity binding of tumor-suppressor protein p53 and HMGB1 to hemicatenated DNA loops. Biochemistry 43, 7215–7225.

[99] Zhang, G., Kobayashi, T., Kamitani, W., Komoto, S., Yamashita, M., Baba, S., Yanai, H., Ikuta, K., Tomonaga, K. (2003) Borna disease virus phosphoprotein represses p53-mediated transcriptional activity by interference with HMGB1. J Virol 77, 12243–12251.

[100] Stros, M., Ozaki, T., Bacikova, A., Kageyama, H., Nakagawara, A. (2002) HMGB1 and HMGB2 cell-specifically down-regulate the p53- and p73-dependent sequence-specific transactivation from the human Bax gene promoter. J Biol Chem 277, 7157–7164.

[101] Livesey, K.M., Kang, R., Vernon, P., Buchser, W., Loughran, P., Watkins, S.C., Zhang, L., Manfredi, J.J., Zeh, H.J. III, Li, L., Lotze, M.T., Tang, D. (2012) p53/HMGB1 complexes regulate autophagy and apoptosis. Cancer Res 72, 1996–2005.

[102] Livesey, K.M., Kang, R., Zeh, H.J. III, Lotze, M.T., Tang, D. (2012) Direct molecular interactions between HMGB1 and TP53 in colorectal cancer. Autophagy 8, 846–848.

[103] Song, J.X., Lu, J.H., Liu, L.F., Chen, L.L., Durairajan, S.S., Yue, Z., Zhang, H.Q., Li, M. (2014) HMGB1 is involved in autophagy inhibition caused by SNCA/alpha-synuclein overexpression: a process modulated by the natural autophagy inducer corynoxine B. Autophagy 10, 144–154.

[104] Chiang, H.S., Maric, M. (2011) Lysosomal thiol reductase negatively regulates autophagy by altering glutathione synthesis and oxidation. Free Radic Biol Med 51, 688–699.

[105] Tang, D., Loze, M.T., Zeh, H.J., Kang, R. (2010) The redox protein HMGB1 regulates cell death and survival in cancer treatment. Autophagy 6, 1181–1183.

[106] Kang, R., Loux, T., Tang, D., Schapiro, N.E., Vernon, P., Livesey, K.M., Krasinskas, A., Lotze, M.T., Zeh, H.J.III (2012) The expression of the receptor for advanced glycation endproducts (RAGE) is permissive for early pancreatic neoplasia. Proc Natl Acad Sci U S A 109, 7031–7036.

[107] Maugeri, N., Campana, L., Gavina, M., Covino, C., De Metrio, M., Panciroli, C., Maiuri, L., Maseri, A., D'Angelo, A., Bianchi, M.E., Rovere-Querini, P., Manfredi, A.A. (2014) Activated platelets present high mobility group box 1 to neutrophils, inducing autophagy and promoting the extrusion of neutrophil extracellular traps. J Thromb Haemost 12, 2074–2088.

[108] Kang, R., Tang, D., Schapiro, N.E., Livesey, K.M., Farkas, A., Loughran, P., Bierhaus, A., Lotze, M.T., Zeh, H.J. (2010) The receptor for advanced glycation end products (RAGE) sustains autophagy and limits apoptosis, promoting pancreatic tumor cell survival. Cell Death Differ 17, 666–676.

[109] Kang, R., Tang, D., Loze, M.T., Zeh, H.J. (2011) Apoptosis to autophagy switch triggered by the MHC class III-encoded receptor for advanced glycation endproducts (RAGE). Autophagy 7, 91–93.

[110] Kang, R., Hou, W., Zhang, Q., Chen, R., Lee, Y.J., Bartlett, D.L., Lotze, M.T., Tang, D., Zeh, H.J. (2014) RAGE is essential for oncogenic KRAS-mediated hypoxic signaling in pancreatic cancer. Cell Death Dis 5, e1480.

[111] Uramoto, H., Izumi, H., Nagatani, G., Ohmori, H., Nagasue, N., Ise, T., Yoshida, T., Yasumoto, K., Kohno, K. (2003) Physical interaction of tumour suppressor p53/p73 with

CCAAT-binding transcription factor 2 (CTF2) and differential regulation of human high-mobility group 1 (HMG1) gene expression. Biochem J 371, 301–310.

[112] Rothermund, K., Rogulski, K., Fernandes, E., Whiting, A., Sedivy, J., Pu, L., Prochownik, E.V. (2005) C-Myc-independent restoration of multiple phenotypes by two C-Myc target genes with overlapping functions. Cancer Res 65, 2097–2107.

[113] Liu, J., Liu, Y., Zhang, H., Chen, G., Wang, K., Xiao, X. (2008) KLF4 promotes the expression, translocation, and releas eof HMGB1 in RAW264.7 macrophages in response to LPS. Shock 30, 260–266.

[114] Liu, K., Huang, J., Xie, M., Yu, Y., Zhu, S., Kang, R., Cao, L., Tang, D., Duan, X. (2014) MIR34A regulates autophagy and apoptosis by targeting HMGB1 in the retinoblastoma cell. Autophagy 10, 442–452.

[115] Li, X., Wang, S., Chen, Y., Liu, G., Yang, X. (2014) miR-22 targets the 3′ UTR of HMGB1 and inhibits the HMGB1-associated autophagy in osteosarcoma cells during chemotherapy. Tumour Biol. 35, 6021–6028.

[116] Pannuru, P., Dontula, R., Khan, A.A., Herbert, E., Ozer, H., Chetty, C., Lakka, S.S. (2014) miR-let-7f-1 regulates SPARC mediated cisplatin resistance in medulloblastoma cells. Cell Signal 26, 2193–2201.

[117] El Gazzar, M. (2007) HMGB1 modulates inflammatory responses in LPS-activated macrophages. Inflamm Res 56, 162–167.

[118] Ditsworth, D., Zong, W.X., Thompson, C.B. (2007) Activation of poly(ADP)-ribose polymerase (PARP-1) induces release of the pro-inflammatory mediator HMGB1 from the nucleus. J Biol Chem 282, 17845–17854.

[119] Zong, W.X., Ditsworth, D., Bauer, D.E., Wang, Z.Q., Thompson, C.B. (2004) Alkylating DNA damage stimulates a regulated form of necrotic cell death. Genes Dev 18, 1272–1282.

[120] Lee, H., Park, M., Shin, N., Kim, G., Kim, Y.G., Shin, J.S., Kim, H. (2012) High mobility group box-1 is phosphorylated by protein kinase C zeta and secreted in colon cancer cells. Biochem Biophys Res Commun 424, 321–326.

[121] Kohlstaedt, L.A., King, D.S., Cole, R.D. (1986) Native state of high mobility group chromosomal proteins 1 and 2 is rapidly lost by oxidation of sulfhydryl groups during storage. Biochemistry 25, 4562–4565.

[122] Yang, M., Liu, L., Xie, M., Sun, X., Yu, Y., Kang, R., Yang, L., Zhu, S., Cao, L., Tang, D. (2015) Poly-ADP-ribosylation of HMGB1 regulates TNFSF10/TRAIL resistance through autophagy. Autophagy. 11, 214–224.

[123] White, E. (2012) Deconvoluting the context-dependent role for autophagy in cancer. Nat Rev Cancer 12, 401–410.

[124] Amornsupak, K., Insawang, T., Thuwajit, P., P, O.C., Eccles, S.A., Thuwajit, C. (2014) Cancer-associated fibroblasts induce high mobility group box 1 and contribute to resistance to doxorubicin in breast cancer cells. BMC Cancer 14, 955.

[125] Pan, B., Chen, D., Huang, J., Wang, R., Feng, B., Song, H., Chen, L. (2014) HMGB1-mediated autophagy promotes docetaxel resistance in human lung adenocarcinoma. Mol Cancer 13, 165.

[126] Liu, K., Huang, J., Xie, M., Yu, Y., Zhu, S., Kang, R., Cao, L., Tang, D., Duan, X. (2014) MIR34A regulates autophagy and apoptosis by targeting HMGB1 in the retinoblastoma cell. Autophagy 10, 442–452.

[127] Zhao, M., Yang, M., Yang, L., Yu, Y., Xie, M., Zhu, S., Kang, R., Tang, D., Jiang, Z., Yuan, W., Wu, X., Cao, L. (2011) HMGB1 regulates autophagy through increasing transcriptional activities of JNK and ERK in human myeloid leukemia cells. BMB Rep 44, 601–606.

[128] Yang, L., Yu, Y., Kang, R., Yang, M., Xie, M., Wang, Z., Tang, D., Zhao, M., Liu, L., Zhang, H., Cao, L. (2012) Up-regulated autophagy by endogenous high mobility group box-1 promotes chemoresistance in leukemia cells. Leuk Lymphoma 53, 315–322.

[129] Su, Z., Wang, T., Zhu, H., Zhang, P., Han, R., Liu, Y., Ni, P., Shen, H., Xu, W., Xu, H. (2014) HMGB1 modulates Lewis cell autophagy and promotes cell survival via RAGE-HMGB1-Erk1/2 positive feedback during nutrient depletion. Immunobiology.220, 539–544.

[130] Liu, Y., Zhao, L., Ju, Y., Li, W., Zhang, M., Jiao, Y., Zhang, J., Wang, S., Wang, Y., Zhao, M., Zhang, B., Zhao, Y. (2014) A novel androstenedione derivative induces ROS-mediated autophagy and attenuates drug resistance in osteosarcoma by inhibiting macrophage migration inhibitory factor (MIF). Cell death & disease 5, e1361.

[131] Kang, R., Zhang, Q., Zeh, H.J. III, Lotze, M.T., Tang, D. (2013) HMGB1 in cancer: good, bad, or both? Clin Cancer Res 19, 4046–4057.

[132] Tang, D., Kang, R., Zeh, H.J. III, Lotze, M.T. (2010) High-mobility group box 1 and cancer. Biochim Biophys Acta 1799, 131–140.

[133] Tang, D., Kang, R., Coyne, C.B., Zeh, H.J., Lotze, M.T. (2012) PAMPs and DAMPs: signal 0s that spur autophagy and immunity. Immunol Rev 249, 158–175.

[134] Deretic, V., Saitoh, T., Akira, S. (2013) Autophagy in infection, inflammation and immunity. Nat Rev Immunol 13, 722–737.

[135] Yanai, H., Matsuda, A., An, J., Koshiba, R., Nishio, J., Negishi, H., Ikushima, H., Onoe, T., Ohdan, H., Yoshida, N., Taniguchi, T. (2013) Conditional ablation of HMGB1 in mice reveals its protective function against endotoxemia and bacterial infection. Proc Natl Acad Sci U S A 110, 20699–20704.

[136] Zhu, X., Messer, J.S., Wang, Y., Lin, F., Cham, C.M., Chang, J., Billiar, T.R., Lotze, M.T., Boone, D.L., Chang, E.B. (2015) Cytosolic HMGB1 controls the cellular autophagy/apoptosis checkpoint during inflammation. J Clin Invest 125, 1098–1110.

[137] Kang, R., Zhang, Q., Hou, W., Yan, Z., Chen, R., Bonaroti, J., Bansal, P., Billiar, T.R., Tsung, A., Wang, Q., Bartlett, D.L., Whitcomb, D.C., Chang, E.B., Zhu, X., Wang, H., Lu, B., Tracey, K.J., Cao, L., Fan, X.G., Lotze, M.T., Zeh, H.J. III, Tang, D. (2013) Intracellular hmgb1 inhibits inflammatory nucleosome release and limits acute pancreatitis in mice. Gastroenterology. 146, 1097–1107.

[138] Huang, H., Nace, G.W., McDonald, K.A., Tai, S., Klune, J.R., Rosborough, B.R., Ding, Q., Loughran, P., Zhu, X., Beer-Stolz, D., Chang, E.B., Billiar, T., Tsung, A. (2013) Hepatocyte specific HMGB1 deletion worsens the injury in liver ischemia/reperfusion: a role for intracellular HMGB1 in cellular protection. Hepatology. 59,1984–1997.

[139] Zong, M., Jorholt, J., Winter, J., Lindroos, E., Harris, H.E., Lundberg, I.E. (2014) A8.24 autophagy may contribute to glucocorticoid resistance in myositis patients by maintaining muscle T cells homeostasis. Ann Rheum Dis 73(Suppl 1), A85–86.

[140] Min, H.J., Ko, E.A., Wu, J., Kim, E.S., Kwon, M.K., Kwak, M.S., Choi, J.E., Lee, J.E., Shin, J.S. (2013) Chaperone-like activity of high-mobility group box 1 protein and its role in reducing the formation of polyglutamine aggregates. J Immunol 190, 1797–1806.

[141] Perez-Carrion, M.D., Cena, V. (2013) Knocking down HMGB1 using dendrimer-delivered siRNA unveils its key role in NMDA-induced autophagy in rat cortical neurons. Pharm Res 30, 2584–2595.

[142] Sun, X., Tang, D. (2014) HMGB1-dependent and -independent autophagy. Autophagy 10, 1873–1876.

[143] Liu, L., Yang, M., Kang, R., Wang, Z., Zhao, Y., Yu, Y., Xie, M., Yin, X., Livesey, K.M., Lotze, M.T., Tang, D., Cao, L. (2011) HMGB1-induced autophagy promotes chemotherapy resistance in leukemia cells. Leukemia 25, 23–31.

[144] Zhan, Z., Li, Q., Wu, P., Ye, Y., Tseng, H.Y., Zhang, L., Zhang, X.D. (2012) Autophagy-mediated HMGB1 release antagonizes apoptosis of gastric cancer cells induced by vincristine via transcriptional regulation of Mcl-1. Autophagy 8, 109–121.

[145] Ni, Z., Dai, X., Wang, B., Ding, W., Cheng, P., Xu, L., Lian, J., He, F. (2013) Natural Bcl-2 inhibitor (-)- gossypol induces protective autophagy via reactive oxygen species-high mobility group box 1 pathway in Burkitt lymphoma. Leuk Lymphoma 54, 2263–2268.

[146] Luo, Y., Yoneda, J., Ohmori, H., Sasaki, T., Shimbo, K., Eto, S., Kato, Y., Miyano, H., Kobayashi, T., Sasahira, T., Chihara, Y., Kuniyasu, H. (2014) Cancer usurps skeletal muscle as an energy repository. Cancer Res 74, 330–340.

[147] Hagiwara, S., Iwasaka, H., Hasegawa, A., Kudo, K., Kusaka, J., Oyama, Y., Noguchi, T. (2012) Infusion of a glucose solution reduces autophagy in the liver after LPS-induced systemic inflammation. Inflammation 35, 249–258.

[148] Shen, M., Lu, J., Dai, W., Wang, F., Xu, L., Chen, K., He, L., Cheng, P., Zhang, Y., Wang, C., Wu, D., Yang, J., Zhu, R., Zhang, H., Zhou, Y., Guo, C. (2013) Ethyl pyruvate ameliorates hepatic ischemia-reperfusion injury by inhibiting intrinsic pathway of apoptosis and autophagy. Mediators Inflamm 2013, 461536.

[149] Fang, H., Liu, A., Dahmen, U., Dirsch, O. (2013) Dual role of chloroquine in liver ischemia reperfusion injury: reduction of liver damage in early phase, but aggravation in late phase. Cell Death Dis 4, e694.

[150] Huebener, P., Gwak, G.Y., Pradere, J.P., Quinzii, C.M., Friedman, R., Lin, C.S., Trent, C.M., Mederacke, I., Zhao, E., Dapito, D.H., Lin, Y., Goldberg, I.J., Czaja, M.J., Schwabe, R.F. (2014) High-mobility group box 1 is dispensable for autophagy, mitochondrial quality control, and organ function in vivo. Cell Metab 19, 539–547.

[151] Sun, X., Tang, D. (2015) Hepatocyte-specific Hmgb1 Deletion. Autophagy 11, 1189–1191.

[152] Tang, D., Kang, R., Van Houten, B., Zeh, H.J., Billiar, T.R., Lotze, M.T. (2014) High mobility group box 1 (HMGB1) phenotypic role revealed with stress. Mol Med 20, 359–362.

Chapter 5

Role of Autophagy in Mediating the Anticancer Effects of Tocotrienols

Paul W. Sylvester and Roshan V. Tiwari

Additional information is available at the end of the chapter

http://dx.doi.org/10.5772/64131

Abstract

γ-Tocotrienol, a natural isoform of vitamin E, is a potent anticancer agent. Autophagy is a highly regulated process by which debris is eliminated from a cell, but can also play a role in cellular survival or death. The role of autophagy in mediating the anticancer effects of γ-tocotrienol is not clearly understood. This chapter reviews the mechanism(s) involved in γ-tocotrienol-induced autophagy in breast cancer cells. Treatment with γ-tocotrienol increased conversion of microtubule-associated protein, 1A/1B-light chain 3, from its cytosolic form (LC3B-I) to its lipidated form (LC3B-II), and the accumulation of autophagy-related proteins Beclin-1 (Atg6) and Atg5-Atg12. Additional studies confirmed that transfection with Beclin-1 siRNA or pretreated with 3-methyladenine (3-MA), an inhibitor of autophagy, blocked these effects. γ-Tocotrienol treatment also induced a time-responsive increase in autolysosome markers LAMP-1 and cathepsin-D, and pretreatment with bafilomycin A1 (Baf1), an inhibitor of late phase autophagy, blocked these effects and caused a significant reduction in γ-tocotrienol-induced cytotoxicity. γ-Tocotrienol also induced a decrease in ERK, an increase in p-38 and JNK activation, and endoplasmic reticulum (ER) stress apoptotic markers including phospho-PERK, phospho-elf2α, Bip, IRE1α, ATF-4, CHOP, and TRB3. In summary, γ-tocotrienol-induced autophagy is intimately involved in promoting ER-stress-mediated apoptosis in human breast cancer cells.

Keywords: γ-tocotrienol, autophagy, breast cancer, endoplasmic reticulum stress, oridonin

1. Introduction to autophagy

Autophagy (Greek for "self-eating") is hallmarked by the formation of double-membrane-bound organelles known as autophagosomes and is a lysosome-dependent pathway for the degradation of damaged cytoplasmic organelles and proteins [1]. There are multiple subtypes of autophagy that are classified according to mechanism and function [2, 3]. The present review will focus on macroautophagy, which is characterized by the engulfment and degradation of cytoplasmic materials in bulk in a selective/nonselective manner. Autophagy is the only mechanism that is involved in the degradation of large structures such as organelles and protein aggregates. In the absence of stress, autophagy serves a housekeeping function [2, 3]. However, during starvation autophagy provides material that can be used as a source of nutrition to promote survival. Autophagy can be induced by a broad range of other stressors to aid in the degradation of protein aggregates, oxidized lipids, damaged organelles, and even intracellular pathogens [2, 3]. Defects in autophagy are linked to liver disease, neurodegeneration, Crohn's disease, aging, cancer, and metabolic syndrome [4].

Autophagy is a self-catabolic cellular mechanism in which damaged cellular organelles in the cytoplasmic compartment are sequestered into the double-membrane vesicles known as autophagosomes that further fuse with lysosomes to form autophagolysosomes. Excessive formation of autophagic vesicles interferes with normal membrane functioning and can lead to autophagy-associated programmed cell death [3–5]. Autophagy is regulated by class I and class III phosphatidylinositol 3-kinase (PI3K)-signaling pathways. Activation of class I PI3K leads to the phosphorylation of plasma membrane lipids that play a role in the recruitment and activation of Akt, a downstream negative regulator of autophagy. The tumor suppressor, phosphatase and tensin homolog (PTEN), dephosphorylates lipids in the plasma membrane and thereby acts to prevent the activation of Akt [3–5]. Activated Akt inhibits the tuberous sclerosis complex 1 (TSC1) and TSC2 proteins that act as positive regulators of autophagy by repressing the activity of the small G protein Rheb, which modulates mammalian target of rapamycin (mTOR) [3–5]. mTOR functions to inhibit autophagy [3–5]. Although the exact mechanism by which mTOR inhibits autophagy is not completely understood, it appears to be involved in suppressing Atg autophagy-related genes [6].

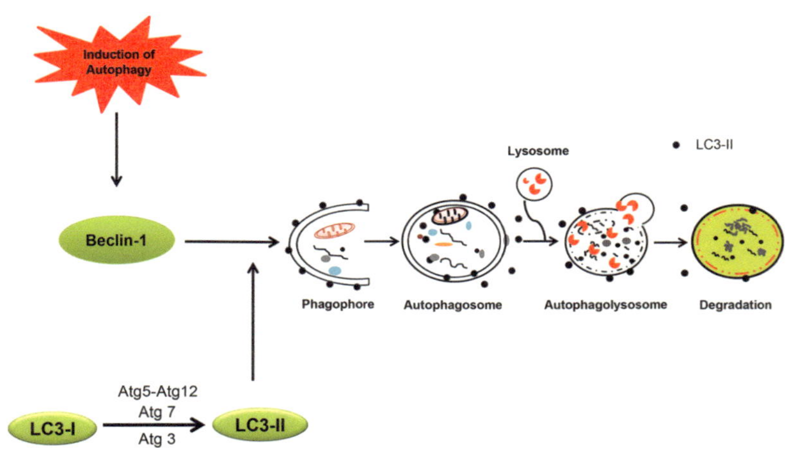

Figure 1. Schematic representation of autophagy.

The execution and regulation of the autophagy pathway is governed by Atg [6]. The initial nucleation and assembly of the primary autophagosomal membrane forms a complex of Beclin-1, the mammalian homolog of Atg6, with class III phosphatidylinositol 3-kinase (PI3K) that mediates the localization of autophagy-targeted proteins into the autophagic vesicles [4, 6]. Elongation of this isolated membrane is governed by two ubiquitin-like conjugation systems, Atg5-Atg12 complex, and microtubule-associated protein 1A/1B-light chain (LC3). Upon the activation of Atg12 by Atg7, Atg12 is transferred to Atg10 and is eventually conjugated to Atg5, which subsequently forms a complex with Atg16. Under basal conditions, LC3B exists in its cytosolic form LC3B-I. However, during autophagy LC3B-I is converted to its lipid-conjugated membrane-bound form LC3B-II, a process that is dependent on the Atg5-Atg12 complex throughout the course of membrane elongation. LC3B-II is associated with autophagosomes and promotes the formation of autophagic vacuoles [4, 6]. A brief summary of the autophagic process is shown in **Figure 1**.

2. Autophagy and cancer

The autophagic process of recycling damaged cytoplasmic organelles and proteins can serve as an alternative energy source during the period of metabolic stress and can play a role in the maintenance of homeostasis and viability. However, in cancer cells autophagy appears to play a dual role that can either allow prolonged cell survival or promote cell death [7, 8]. This dual role of autophagy in cancer, as both tumor suppressor and a protector of cancer cell survival, is not yet clearly understood. Autophagy dysregulation is observed in a wide spectrum of human cancers. For example, the altered expression of several autophagy proteins such as LC3B and Beclin-1 has been observed in brain, esophageal, breast, colon, gastric, liver, and pancreatic cancer as well as osteosarcoma and melanoma [9, 10]. Mutations of various autophagy-related genes have also been reported in gastrointestinal cancers [1]. It is clearly evident that greater understanding of the specific role of autophagy in the etiology of various types of cancer and at various stages of cancer progression will provide useful insights for the development of novel and more effective strategies for the prevention and treatment of cancer.

3. γ-Tocotrienol and cancer

Vitamin E is a generic term that represents eight chemically similar natural products that is subdivided into two subgroups classified as tocopherols and tocotrienols. The four tocopherol isoforms of vitamin E are more common and found in abundance in animal and vegetable oils. By contrast, the remaining four tocotrienol isoforms are quite rare, but found in high concentrations in palm oil [11, 12]. Both tocopherols and tocotrienols are characterized by a chromanol ring structure methylated to varying degrees at the 5, 7, and 8 positions to form eight different isoforms classified as α-, β-, γ-, and δ-tocopherols and α-, β-, γ-, and δ-tocotrienols. Attached to the chromanol ring is a long phytyl that is saturated on tocopherols and unsaturated on tocotrienols (**Figure 2**).

Isomer	R_1	R_2
α	CH_3	CH_3
β	CH_3	H
γ	H	CH_3
δ	H	H

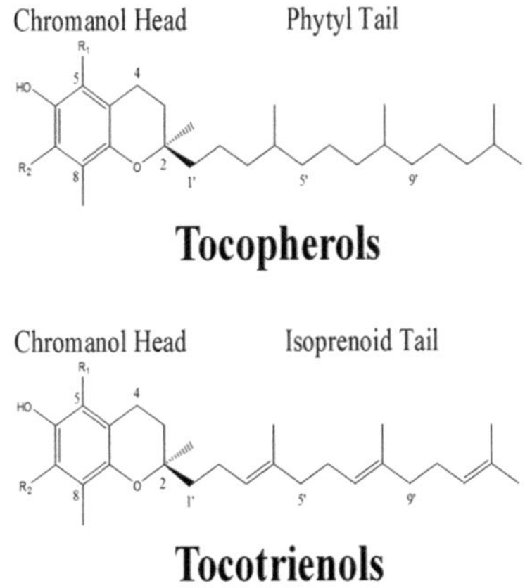

Figure 2. Chemical structure of tocopherols and tocotrienols.

Numerous in vitro and in vivo investigations have demonstrated the anticancer effects of tocotrienols [13–15]. Further studies showed that tocotrienols are preferentially and selectively taken up into mammary tumor cells as compared to tocopherols [16, 17]. Thus, tocotrienol displays potent anticancer activity at treatment doses that have little or no effect on the growth and viability of primary epithelial cells isolated from the mammary gland or immortalized mouse (CL-S1) and human (MCF-10A) normal mammary epithelial cell lines [16–18]. Additional reports have shown that the combined use of tocotrienols with other chemotherapeutic agents results in a synergistic anticancer response [19–23].

4. γ-Tocotrienol effects on mitogenic and apoptotic signaling

Various downstream signaling pathways are initiated by ligand-induced ErbB/HER receptor activation, including the mitogen-activated kinase (MAPK) and PI3K/Akt/mTOR, and activation of these cascades is associated with cellular growth, survival, and motility [24–26]. Specifically, studies have shown that tocotrienol inhibits PI3K/Akt-signaling pathways through the inhibition of EGF-dependent Akt phosphorylation (activation) in mammary cancer cells [27]. Moreover, inhibition of MAPK such as ERK, p38 MAPK, and activation of JNK is critical to the antiproliferative effects of tocotrienols [23, 28, 29].

Studies conducted by Wali et al. [30] were the first to demonstrate that treatment with 15–40 μM of γ-tocotrienol induced mouse +SA mammary tumor cell death in a dose-dependent manner. Specifically, γ-tocotrienol induced cytotoxicity in these cells as associated with an increase in poly (ADP-ribose) polymerase (PARP) cleavage, an established cellular marker for apoptosis, as well as increased signaling of the protein kinase-like endoplasmic reticulum (ER) kinase/eukaryotic translational initiation factor/activating transcription factor 4 (PERK/eIF2α/ATF-4) pathway, a marker of endoplasmic reticulum stress. These studies also showed

http://dx.doi.org/10.5772/64131

that γ-tocotrienol treatment also induced a large increase in C/EBP homologous protein (CHOP) levels and tribbles 3 (TRB3) expression [30]. ER-stress response also cleaved caspase-12 (activated), which is responsible for the disruption of ER calcium homeostasis and the accumulation of excess proteins in ER, and thus initiating the apoptosis signaling was observed following cytotoxic treatment of γ-tocotrienol for 24 h [30]. Studies have also shown that γ-tocotrienol treatment led to apoptosis, necrosis, and autophagy in human prostate PC-3 and LNCap cancer cells, as it causes an increase in the accumulation of dihydrosphingosine and dihydroceramide, important sphingolipids in de novo biosynthesis pathway, but have no effects on ceramide or sphingosine [31]. Other studies have shown that the γ-tocotrienol-induced autophagy is associated with the inhibition of mTOR activation [32].

5. Oridonin and cancer

Oridonin (7,20-epoxy-ent-kauranes), a diterpenoid isolated from the Chinese medicinal herb *Rabdosia rubescens,* is shown in **Figure 3**. Oridonin has also been shown to display potent anticancer activity [33]. Oridonin treatment was found to significantly inhibit tumor growth and induced cancer cell death in vivo [34], and was effective in suppressing tumor development, growth, and progression [33]. The antitumor effects of oridonin appear to include the suppression of cell cycle progression and/or the initiation of apoptosis [35]. Experimental investigations showed that following oridonin treatment, murine melanoma K1735M2 cells [35], DU-145 cells [36], and L929 cells [37] showed cell cycle arrest in the G2/M phase. In addition, other investigators have found that oridonin induced cell cycle arrest in the G1/S phase and this was associated with a corresponding inhibition in cdc2 and cyclin B activation in MCF-7 breast cancer cells [38]. Furthermore, oridonin-induced autophagy has been shown to be a prerequisite for the initiation of apoptosis in breast cancer cells [39].

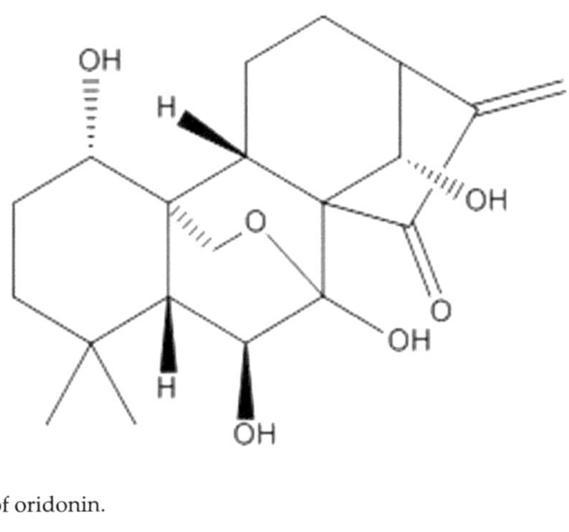

Figure 3. Chemical structure of oridonin.

6. Combination treatment of γ-tocotrienol with other chemotherapeuticagents

Most traditional cancer chemotherapies are not very selective and can cause damage to normal cells. It has become evident that a better approach is to use combination therapy that is more effective and produces less adverse side effects. Furthermore, the use of phytochemicals in the prevention and treatment of cancer has recently gained much interest, and combination therapies are attractive because an additive or synergistic therapeutic response can result. The rationale for using tocotrienols in combination therapy is based on the findings that a form of vitamin E has a broad range of anticancer actions and the principle that resistance to any single agent can be overcome by using multiple agents with complimentary mechanisms of action [11, 12, 19–23, 40]. Previous studies have shown that combined low dose of γ-tocotrienol with other chemotherapeutic or phytochemical agents displays significantly enhanced anticancer effects, as compared to that of individual treatment alone [11]. It has also been shown that γ-tocotrienol synergizes with other phytochemical agents such as resveratrol to induce autophagy accompanied by the activation of Beclin and LC3-II and by decreasing mTOR signaling [32].

7. γ-Tocotrienol-induced autophagy in breast cancer cells

Previous investigations have shown that tocotrienol treatment induces autophagy in various cell types [31, 32, 41, 42]. Treatment with γ-tocotrienol was found to induce autophagy and apoptosis in rat pancreatic stellate [41] and prostate cancer cells [31], whereas other studies have shown that γ-tocotrienol treatment was cardioprotective and prevented apoptosis in ischemic cardiomyocytes [32]. However, the exact role of autophagy in mediating γ-tocotrienol-induced cytotoxicity has only recently been investigated [18]. In these studies, experiments were conducted to characterize γ-tocotrienol-induced autophagy in highly malignant mouse (+SA), and human estrogen-dependent (MCF-7) and estrogen-independent (MDA-MB-231) malignant mammary cancer cell lines. Results showed that γ-tocotrienol treatment significantly reduced cell viability in these breast cancer cell lines in a dose-responsive manner [18]. These same treatments also induced a corresponding increase in autophagy markers as determined by an increase in monodansylcadaverine (MDA) autofluorescence and flow cytometric analysis of positive acridine orange staining [18]. In addition, parallel studies determined that treatment with these same doses of γ-tocotrienol induced an increased conversion of microtubule-associated protein, 1A/1B-light chain 3, from its cytosolic form (LC3B-I) to its lipidated form (LC3B-II), the phosphatidylethanolamine-conjugated form associated with autophagosomes, and a corresponding increase in Beclin-1 and ATG6 in +SA, MCF-7, and MDA-MB-231 breast cancer cells.

These findings confirm and extend previous findings that showed that tocotrienol treatment promotes the conversion of LC3B-I to LC3B-II in other cell types [32, 41, 42]. By contrast, similar treatment with γ-tocotrienol was not found to increase autophagy marker expression in immortalized mouse (CL-S1) and human (MCF10A) normal mammary epithelial cell lines, indicating that γ-tocotrienol displays selective action against cancer cells. Additional studies showed that γ-tocotrienol treatment also caused a reduction in PI3K/Akt/mTOR signaling and a corresponding increase in the Bax/Bcl-2 ratio, cleaved caspase-3, and cleaved PARP levels in these cancer cell lines, suggesting that γ-tocotrienol-induced autophagy may be involved in the initiation of apoptosis [18]. Since mTOR activity is directly associated with a suppression of autophagy [6] and Bcl-2 acts to suppress Beclin-1 levels [43], these findings indicate possible intracellular-signaling mechanisms that may be involved in mediating tocotrienol-induced autophagy and promote cytotoxicity in breast cancer cells. Selective effects of γ-tocotrienol on autophagy cellular markers in MDA-MB-231 human breast cancer cells are shown in **Figure 4**.

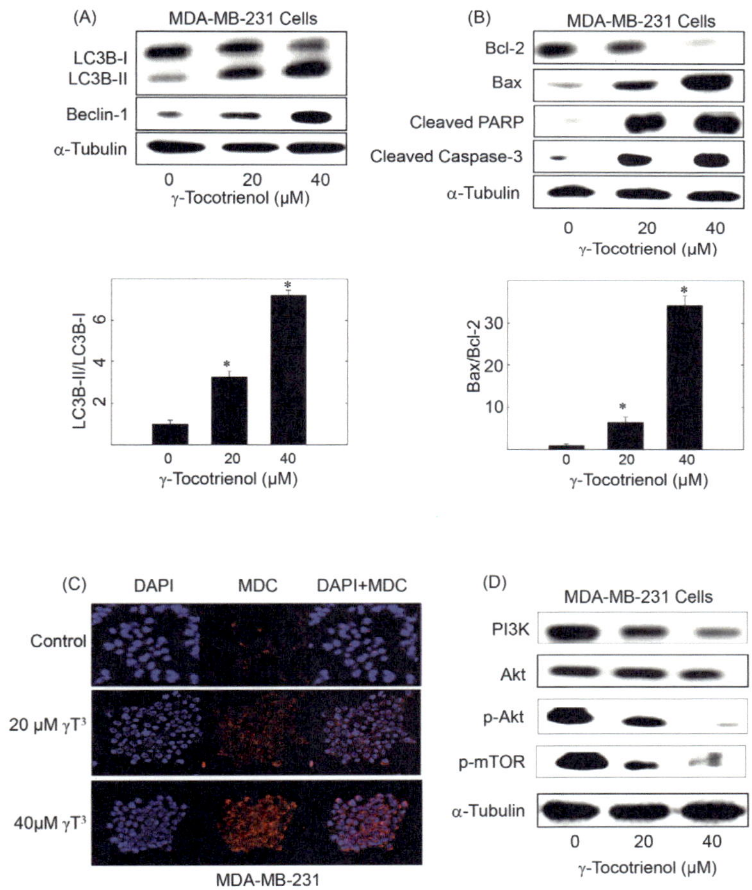

Figure 4. (A) Western blot analysis of γ-tocotrienol effects on the relative protein levels that serve as markers for autophagy in mammary cancer cells and scanning densitometric analysis of Western blots shown above. (B) Western blot analysis of γ-tocotrienol effects on the relative levels of apoptotic protein markers and scanning densitometric analysis of Western blots shown above. (C) Effects of γ-tocotrienol treatment on autophagic vacuoles MDC fluorescence intensity. (D) Western blot analysis of γ-tocotrienol effects on the relative levels of PI3K/Akt/mTOR signaling proteins. *$P < 0.05$ as compared to the vehicle-treated control group.

In summary, these initial studies provided evidence to support the suggestion that the cytotoxic effects of γ-tocotrienol are associated with the induction of autophagy in mouse and human breast cancer cells.

8. γ-Tocotrienol simultaneously induces autophagy and endoplasmic reticulum stress-mediated apoptosis in breast cancer cells

The endoplasmic reticulum is an intracellular organelle that is involved in protein synthesis, but during times of stress the ER plays an important role in programmed cell death [30]. ER-stress-mediated apoptosis is associated with an increased expression of several proteins including protein kinase-like endoplasmic reticulum kinase (PERK), activating transcription factor 6 (ATF-6), and inositol-requiring kinase 1 (IRE1) [44–46]. In addition, the increased expression of phosphorylated eukaryotic translational initiation factor 2 (eIF2α), C/EBP homology protein (CHOP), tribbles 3, and ATF-4 also occurs during the initial phases of ER-stress-mediated apoptosis [30, 47–52].

Previous studies have shown that the anticancer effects of tocotrienols are associated with the induction of autophagy and endoplasmic reticulum-stress-mediated apoptosis [18, 30, 31, 53]. However, a direct causal relationship between tocotrienol-induced autophagy and ER stress had not yet been established. Recently, studies were conducted to characterize the interrelationship between γ-tocotrienol-induced cytotoxicity, autophagy, and ER-stress-mediated apoptosis in human breast cancer cells [54]. In these studies, γ-tocotrienol treatment causes an increase in the appearance of damaged and/or dying MCF-7 and MDA-MB-231 cancer cells together with a corresponding increase in the appearance of large autophagic vacuoles as visualized by Giemsa staining, a large increase in positive MDC fluorescent staining, a positive marker for autophagic vacuole formation [2, 55], and a large increase in positive LC3B fluorescent staining, a marker for autophagosomes [2, 55] in these same cells [54]. In addition, transfection with small interfering RNA (siRNA) targeting Beclin-1 prior to γ-tocotrienol treatment resulted in a modest but significant reduction in γ-tocotrienol-induced cytotoxicity as compared to cells transfected with scrambled siRNA (negative control) and then treated with γ-tocotrienol, and these effects were correlated with a corresponding large increase in Beclin-1 levels and LC3B-II/LC3B-I ratio [54].

Subsequent studies investigated the effects of γ-tocotrienol on the activation of stress-signaling pathways in these same MCF-7 and MDA-MB-231 breast cancer cells. Results showed that γ-tocotrienol induced an increase in the activation of p38 and JNK1/2, and simultaneous decrease in Erk1/2 signaling. This same treatment also induced a reduction in Bcl-2 (antiapoptotic), and an increase in Bax (proapoptotic), cleaved caspase-3 (activated), and cleaved-PARP (activated) protein levels in these breast cancer cells [54]. In addition, γ-tocotrienol treatment significantly increases the ratio of Bax/Bcl-2, and induced a time-dependent increase in the relative levels of ER-stress markers including Bip, IRE1, phosphorylated PERK (activated), phosphorylated eIF2α (inactivated), ATF-4, CHOP, and TRB3 in MCF-7 and MDA-MB-231 breast cancer cells [54]. Combined treatment with the pan caspase inhibitor, zVADfmk, with γ-tocotrienol resulted in a complete blockade of γ-tocotrienol-induced cytotoxicity in these cells [54]. During this same time period, γ-tocotrienol caused a time-dependent decrease in mitogenic Erk and Akt signaling, a corresponding increase in stress-dependent p38 and JNK activation [54].

Since autophagy may have a dual function in cancer that can either promote or suppress tumor cell survival [3, 4, 9, 10, 56], it is unclear whether autophagic activity in dying cancer cells is reflective of a compensatory mechanism that is trying to prevent death or is directly involved in promoting cell destruction. Previous studies have clearly demonstrated that γ-tocotrienol-induced cytotoxicity is not dependent on estrogen receptor status of breast cancer cells [57, 58]. Although it remains unclear why the γ-tocotrienol appears to selectively target cancer cells and not normal cells [14, 18], the possibility exists that noncancerous cells possess specific compensatory mechanisms that provide protection against γ-tocotrienol-induced ER-stress-mediated apoptosis and autophagy, and these self-survival mechanisms may be dysfunctional in malignant cells.

In summary, results obtained in the studies described above strongly suggest that γ-tocotrienol-induced breast cancer cell death is intimately related to the simultaneous initiation of autophagy and ER-stress-mediated apoptosis. This suggestion is further evidenced by the finding that pretreatment with agents that block the induction of autophagy resulted in a suppression in γ-tocotrienol-induced apoptosis. Taken together, these data indicate that γ-tocotrienol-induced autophagy and ER stress act concurrently to enhance the self-destruction of human breast cancer cells. The intracellular mechanism by which γ-tocotrienol-induced autophagy and ER stress initiate breast cancer-programmed cell death is shown as a schematic representation in **Figure 5**.

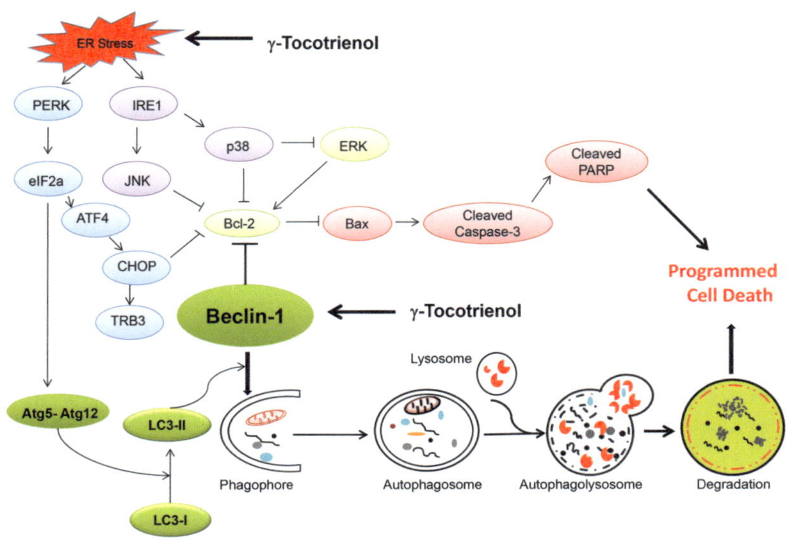

Figure 5. Schematic representation of the proposed molecular mechanism mediating γ-tocotrienol concurrent induction of autophagy and ER-stress-mediated apoptosis in breast cancer cells.

9. Synergistic anticancer effects of combined γ-tocotrienol and oridonin treatment

A great deal of interest has recently been generated in the development of novel therapies that target specific signaling pathways associated with neoplastic transformation, growth, and progression, because not only do these therapies provide enhanced anticancer efficacy, they also display significant less adverse or toxic effects on normal tissue. Phytochemicals appear to have great potential in this area not only for use as anticancer agents alone but also for use as potent supplemental agents that when combined with traditional anticancer therapies provide a synergistic therapeutic responsive [19, 59, 60]. γ-Tocotrienol displays potent antiproliferative, autophagic, and apoptotic activity against cancer cells at treatment doses that have little or no effect on normal cell growth and viability [12, 54]. Oridonin is another such phytochemical isolated from the herb *R. rubescens*, which displays potent anticancer activity against a wide range of cancer cell types [61–63]. Furthermore, oridonin-induced autophagy has been shown to be a prerequisite for the initiation of apoptosis in breast cancer cells [39]. Based on findings in the current literature, γ-tocotrienol-induced autophagy in breast cancer cells is associated with a reduction in mitogen-dependent PI3K/Akt/mTOR signaling and Bcl-2 expression [18]. These effects were also found to be directly related to an increase in Beclin-1 levels and the conversion of LC3B-I to LC3B-II in mouse and human breast cancer cell lines [18]. By contrast, oridonin-induced autophagy was shown to be directly related to a decrease in MAPK signaling and an increase in JNK- and p38-stress pathway signaling [39]. Combination therapy has also been shown to provide therapeutic advantages over monotherapy because a synergistic response is often observed using very low treatment doses and thereby reduce the emergence of toxic side effects [11].

Recent studies have also showed that combined treatment of subeffective doses of γ-tocotrienol and oridonin resulted in a significantly greater reduction in mammary tumor cell viability as compared to cells treated with either drug along [64]. Isobologram analysis of combination treatment with γ-tocotrienol and oridonin determined that these effects were synergistic. By contrast, similar combination treatment had no effect on the viability of normal mammary epithelial cells [64]. In addition, combined therapy significantly increased the conversion of LC3B-I to LC3B-II, as well as the expression of Beclin-1, Atg3, Atg7, Atg5-Atg12, LAMP-1, and cathepsin-D, established cellular markers of autophagy [64]. Furthermore, pretreatment with 3-methyladenine or bafilomycin A1, agents that prevent the induction of autophagy, blocked these effects induced by combined treatment with γ-tocotrienol and oridonin [64]. Additional studies showed that combination treatment with these phytochemicals also induced a large suppression in Akt/mTOR mitogenic signaling and corresponding increase in the levels of apoptotic cellular marker including cleaved caspase-3 and PARP, and Bax/Bcl-2 ratio in these same mammary tumor cells [64].

In summary, findings from these studies demonstrate that combined low-dose treatment of γ-tocotrienol and oridonin acts synergistically to induce autophagy and apoptosis in mammary tumor cells. Since these effects were associated with a large reduction in PI3K/Akt/mTOR signaling, these findings suggest that the combined use of the phytochemicals γ-tocotrienol and oridonin may provide some benefit as supplemental or adjuvant therapy in the treatment of breast cancer.

10. Conclusion

Experimental data summarized above in this review provide convincing evidence that the anticancer effects of γ-tocotrienol are directly associated with the simultaneous initiation of autophagy and ER-stress-mediated apoptosis in breast cancer cells. This suggestion is further supported by the finding that chemical-induced blockade of γ-tocotrienol-induced autophagy significantly reduces γ-tocotrienol-induced apoptosis and cell death. These findings also show that γ-tocotrienol-induced autophagy is directly associated with a significant reduction in PI3K/Akt/mTOR signaling and corresponding increase in intracellular levels of Beclin-1 and conversion of LC3B-I to LC3B-II in these breast cancer cells. Furthermore, combination treatment with subeffective doses of γ-tocotrienol and oridonin acts synergistically in promoting the initiation of autophagy and apoptosis, indicating that the combined use of these natural phytochemicals may have value in the treatment of breast cancer in women.

Author details

Paul W. Sylvester* and Roshan V. Tiwari

*Address all correspondence to: sylvester@ulm.edu

School of Pharmacy, University of Louisiana at Monroe, Monroe, LA, USA

References

[1] Wu WK, Coffelt SB, Cho CH, Wang XJ, Lee CW, Chan FK, Yu J, Sung JJ (2012) The autophagic paradox in cancer therapy. *Oncogene*. 31, 939–953.

[2] Chen Y, Azad MB, Gibson SB (2010) Methods for detecting autophagy and determining autophagy-induced cell death. *Can J Physiol Pharmacol*. 88, 285–295.

[3] Mizushima N, Levine B, Cuervo AM, Klionsky DJ (2008) Autophagy fights disease through cellular self-digestion. *Nature*. 451, 1069–1075.

[4] Rabinowitz JD, White E (2010) Autophagy and metabolism. *Science*. 330, 1344–1348.

[5] Mizushima N, Yoshimori T, Levine B (2010) Methods in mammalian autophagy research. *Cell*. 140, 313–326.

[6] Baehrecke EH (2005) Autophagy: dual roles in life and death? *Nat Rev Mol Cell Biol*. 6, 505–510.

[7] Shintani T, Klionsky DJ (2004) Autophagy in health and disease: a double-edged sword. *Science*. 306, 990–995.

[8] White E, DiPaola RS (2009) The double-edged sword of autophagy modulation in cancer. *Clin Cancer Res*. 15, 5308–5316.

[9] Chen N, Karantza V (2011) Autophagy as a therapeutic target in cancer. *Cancer Biol Ther*. 11, 157–168.

[10] Mathew R, Karantza-Wadsworth V, White E (2007) Role of autophagy in cancer. *Nat Rev Cancer*. 7, 961–967.

[11] Sylvester PW (2012) Synergistic anticancer effects of combined gamma-tocotrienol with statin or receptor tyrosine kinase inhibitor treatment. *Genes Nutr*. 7, 63–74.

[12] Sylvester PW, Akl MR, Malaviya A, Parajuli P, Ananthula S, Tiwari RV, Ayoub NM (2014) Potential role of tocotrienols in the treatment and prevention of breast cancer. *Biofactors*. 40, 49–58.

[13] Aggarwal B, Nesaretnam K (2012) Vitamin E tocotrienols: life beyond tocopherols. *Genes Nutr*. 7, 1.

[14] Sylvester PW (2007) Vitamin E and apoptosis. *Vitam Horm*. 76, 329–356.

[15] Sylvester PW, Theriault A (2003) Role of tocotrienols in the prevention of cardiovascular disease and breast cancer. *Curr Topics Nutraceut Res*. 1, 121–136.

[16] McIntyre BS, Briski KP, Gapor A, Sylvester PW (2000) Antiproliferative and apoptotic effects of tocopherols and tocotrienols on preneoplastic and neoplastic mouse mammary epithelial cells. *Proc Soc Exp Biol Med*. 224, 292–301.

[17] McIntyre BS, Briski KP, Tirmenstein MA, Fariss MW, Gapor A, Sylvester PW (2000) Antiproliferative and apoptotic effects of tocopherols and tocotrienols on normal mouse mammary epithelial cells. *Lipids*. 35, 171–180.

[18] Tiwari RV, Parajuli P, Sylvester PW (2014) gamma-Tocotrienol-induced autophagy in malignant mammary cancer cells. *Exp Biol Med (Maywood)*. 239, 33–44.

[19] Akl MR, Ayoub NM, Abuasal BS, Kaddoumi A, Sylvester PW (2013) Sesamin synergistically potentiates the anticancer effects of gamma-tocotrienol in mammary cancer cell lines. *Fitoterapia*. 84, 347–359.

[20] Ayoub NM, Akl MR, Sylvester PW (2013) Combined gamma-tocotrienol and Met inhibitor treatment suppresses mammary cancer cell proliferation, epithelial-to-mesenchymal transition and migration. *Cell Prolif*. 46, 538–553.

[21] Malaviya A, Sylvester PW (2014) Synergistic antiproliferative effects of combined gamma -tocotrienol and PPAR gamma antagonist treatment are mediated through PPAR gamma-independent mechanisms in breast cancer cells. *PPAR Res*. 2014, 439146.

[22] Shirode AB, Sylvester PW (2010) Synergistic anticancer effects of combined gamma-tocotrienol and celecoxib treatment are associated with suppression in Akt and NFkappaB signaling. *Biomed Pharmacother*. 64, 327–332.

[23] Wali VB, Sylvester PW (2007) Synergistic antiproliferative effects of gamma-tocotrienol and statin treatment on mammary tumor cells. *Lipids*. 42, 1113–1123.

[24] Lee WK, Chakraborty PK, Roussa E, Wolff NA, Thevenod F (2012) ERK1/2-dependent bestrophin-3 expression prevents ER-stress-induced cell death in renal epithelial cells by reducing CHOP. *Biochim Biophys Acta*. 1823, 1864–1876.

[25] Lu Z, Xu S (2006) ERK1/2 MAP kinases in cell survival and apoptosis. *IUBMB Life*. 58, 621–631.

[26] Sun W, Wang Q, Chen B, Liu J, Liu H, Xu W (2008) Gamma-tocotrienol-induced apoptosis in human gastric cancer SGC-7901 cells is associated with a suppression in mitogen-activated protein kinase signalling. *Br J Nutr*. 99, 1247–1254.

[27] Shah S, Gapor A, Sylvester PW (2003) Role of caspase-8 activation in mediating vitamin E-induced apoptosis in murine mammary cancer cells. *Nutr Cancer*. 45, 236–246.

[28] Bachawal SV, Wali VB, Sylvester PW (2010) Enhanced antiproliferative and apoptotic response to combined treatment of gamma-tocotrienol with erlotinib or gefitinib in mammary tumor cells. *BMC Cancer*. 10, 84.

[29] Park SK, Sanders BG, Kline K (2010) Tocotrienols induce apoptosis in breast cancer cell lines via an endoplasmic reticulum stress-dependent increase in extrinsic death receptor signaling. *Breast Cancer Res Treat*. 124, 361–375.

[30] Wali VB, Bachawal SV, Sylvester PW (2009) Endoplasmic reticulum stress mediates gamma-tocotrienol-induced apoptosis in mammary tumor cells. *Apoptosis*. 14, 1366–1377.

[31] Jiang Q, Rao X, Kim CY, Freiser H, Zhang Q, Jiang Z, Li G (2012) Gamma-tocotrienol induces apoptosis and autophagy in prostate cancer cells by increasing intracellular dihydrosphingosine and dihydroceramide. *Int J Cancer*. 130, 685–693.

[32] Lekli I, Ray D, Mukherjee S, Gurusamy N, Ahsan MK, Juhasz B, Bak I, Tosaki A, Gherghiceanu M, Popescu LM, Das DK (2010) Co-ordinated autophagy with resveratrol and gamma-tocotrienol confers synergetic cardioprotection. *J Cell Mol Med*. 14, 2506–2518.

[33] Abelson PH (1990) Medicine from plants. *Science*. 247, 513.

[34] Fujita E, Nagao Y, Node M, Kaneko K, Nakazawa S, Kuroda H (1976) Antitumor activity of the Isodon diterpenoids: structural requirements for the activity. *Experientia*. 32, 203–206.

[35] Ren KK, Wang HZ, Xie LP, Chen DW, Liu X, Sun J, Nie YC, Zhang RQ (2006) The effects of oridonin on cell growth, cell cycle, cell migration and differentiation in melanoma cells. *J Ethnopharmacol*. 103, 176–180.

[36] Chen S, Gao J, Halicka HD, Huang X, Traganos F, Darzynkiewicz Z (2005) The cytostatic and cytotoxic effects of oridonin (Rubescenin), a diterpenoid from Rabdosia rubescens, on tumor cells of different lineage. *Int J Oncol*. 26, 579–588.

[37] Cheng Y, Qiu F, Ye YC, Tashiro S, Onodera S, Ikejima T (2009) Oridonin induces G2/M arrest and apoptosis via activating ERK-p53 apoptotic pathway and inhibiting PTK-Ras-Raf-JNK survival pathway in murine fibrosarcoma L929 cells. *Arch Biochem Biophys*. 490, 70–75.

[38] Hsieh TC, Wijeratne EK, Liang JY, Gunatilaka AL, Wu JM (2005) Differential control of growth, cell cycle progression, and expression of NF-kappaB in human breast cancer cells MCF-7, MCF-10A, and MDA-MB-231 by ponicidin and oridonin, diterpenoids from the Chinese herb Rabdosia rubescens. *Biochem Biophys Res Commun*. 337, 224–231.

[39] Cui Q, Tashiro S, Onodera S, Minami M, Ikejima T (2007) Autophagy preceded apoptosis in oridonin-treated human breast cancer MCF-7 cells. *Biol Pharm Bull*. 30, 859–864.

[40] Sylvester PW, Wali VB, Bachawal SV, Shirode AB, Ayoub NM, Akl MR (2011) Tocotrienol combination therapy results in synergistic anticancer response. *Front Biosci (Landmark Ed)*. 16, 3183–3195.

[41] Rickmann M, Vaquero EC, Malagelada JR, Molero X (2007) Tocotrienols induce apoptosis and autophagy in rat pancreatic stellate cells through the mitochondrial death pathway. *Gastroenterology*. 132, 2518–2532.

[42] Vaquero EC, Rickmann M, Molero X (2007) Tocotrienols: balancing the mitochondrial crosstalk between apoptosis and autophagy. *Autophagy*. 3, 652–654.

[43] Pattingre S, Tassa A, Qu X, Garuti R, Liang XH, Mizushima N, Packer M, Schneider MD, Levine B (2005) Bcl-2 antiapoptotic proteins inhibit Beclin 1-dependent autophagy. *Cell*. 122, 927–939.

[44] Han J, Back SH, Hur J, Lin YH, Gildersleeve R, Shan J, Yuan CL, Krokowski D, Wang S, Hatzoglou M, Kilberg MS, Sartor MA, Kaufman RJ (2013) ER-stress-induced transcriptional regulation increases protein synthesis leading to cell death. *Nat Cell Biol*. 15, 481–490.

[45] Schroder M, Kaufman RJ (2005) ER stress and the unfolded protein response. *Mutat Res*. 569, 29–63.

[46] Yoshida H (2007) ER stress and diseases. *FEBS J*. 274, 630–658.

[47] Lu PD, Harding HP, Ron D (2004) Translation reinitiation at alternative open reading frames regulates gene expression in an integrated stress response. *J Cell Biol*. 167, 27–33.

[48] Ohoka N, Yoshii S, Hattori T, Onozaki K, Hayashi H (2005) TRB3, a novel ER stress-inducible gene, is induced via ATF4-CHOP pathway and is involved in cell death. *EMBO J*. 24, 1243–1255.

[49] Siu F, Bain PJ, LeBlanc-Chaffin R, Chen H, Kilberg MS (2002) ATF4 is a mediator of the nutrient-sensing response pathway that activates the human asparagine synthetase gene. *J Biol Chem*. 277, 24120–24127.

[50] Su N, Kilberg MS (2008) C/EBP homology protein (CHOP) interacts with activating transcription factor 4 (ATF4) and negatively regulates the stress-dependent induction of the asparagine synthetase gene. *J Biol Chem*. 283, 35106–35117.

[51] Xu C, Bailly-Maitre B, Reed JC (2005) Endoplasmic reticulum stress: cell life and death decisions. *J Clin Invest*. 115, 2656–2664.

[52] Oh SH, Lim SC (2009) Endoplasmic reticulum stress-mediated autophagy/apoptosis induced by capsaicin (8-methyl-N-vanillyl-6-nonenamide) and dihydrocapsaicin is regulated by the extent of c-Jun NH_2-terminal kinase/extracellular signal-regulated kinase activation in WI38 lung epithelial fibroblast cells. *J Pharmacol Exp Ther*. 329, 112–122.

[53] Singh BN, Kumar D, Shankar S, Srivastava RK (2012) Rottlerin induces autophagy which leads to apoptotic cell death through inhibition of PI3K/Akt/mTOR pathway in human pancreatic cancer stem cells. *Biochem Pharmacol*. 84, 1154–1163.

[54] Tiwari RV, Parajuli P, Sylvester PW (2015) gamma-Tocotrienol-induced endoplasmic reticulum stress and autophagy act concurrently to promote breast cancer cell death. *Biochem Cell Biol*. 12, 1–15.

[55] Tanida I, Ueno T, Kominami E (2004) LC3 conjugation system in mammalian autophagy. *Int J Biochem Cell Biol*. 36, 2503–2518.

[56] Klionsky DJ (2005) The molecular machinery of autophagy: unanswered questions. *J Cell Sci*. 118, 7–18.

[57] Nesaretnam K, Stephen R, Dils R, Darbre P (1998) Tocotrienols inhibit the growth of human breast cancer cells irrespective of estrogen receptor status. *Lipids*. 33, 461–469.

[58] Yu W, Simmons-Menchaca M, Gapor A, Sanders BG, Kline K (1999) Induction of apoptosis in human breast cancer cells by tocopherols and tocotrienols. *Nutr Cancer*. 33, 26–32.

[59] Akl MR, Ayoub NM, Sylvester PW (2012) Mechanisms mediating the synergistic anticancer effects of combined gamma-tocotrienol and sesamin treatment. *Planta Med.* 78, 1731–1739.

[60] Amin A, Gali-Muhtasib H, Ocker M, Schneider-Stock R (2009) Overview of major classes of plant-derived anticancer drugs. *Int J Biomed Sci.* 5, 1–11.

[61] Kang N, Zhang JH, Qiu F, Chen S, Tashiro S, Onodera S, Ikejima T (2010) Induction of G(2)/M phase arrest and apoptosis by oridonin in human laryngeal carcinoma cells. *J Nat Prod.* 73, 1058–1063.

[62] Li CY, Wang EQ, Cheng Y, Bao JK (2011) Oridonin: an active diterpenoid targeting cell cycle arrest, apoptotic and autophagic pathways for cancer therapeutics. *Int J Biochem Cell Biol.* 43, 701–704.

[63] Zhang W, Huang Q, Hua Z-C (2010) Oridonin: a promising anticancer drug from China. *Front Biol.* 5, 540–545.

[64] Tiwari RV, Parajuli P, Sylvester PW (2015) Synergistic anticancer effects of combined gamma-tocotrienol and oridonin treatment is associated with the induction of autophagy. *Mol Cell Biochem.* 408, 123–137.

Chapter 6

TRAIL Induces Apoptosis and Autophagy

Zhenyu Yao and Dexian Zheng

Additional information is available at the end of the chapter

http://dx.doi.org/10.5772/64246

Abstract

It is known that tumor necrosis factor-related apoptosis-inducing ligand (TRAIL) could induce both apoptosis and autophagy. Here, we summarized the recent findings of the key regulators and the crosstalk pathway that highlights the intricate interplay between TRAIL-induced apoptosis and autophagy.

Keywords: apoptosis, autophagy, caspase-8, RIP1, TRAIL

1. Introduction

Tumor necrosis factor (TNF)-related apoptosis-inducing ligand (TRAIL), also known as Apo-2 ligand (Apo2L), is a multifunctional cytokine of the TNF superfamily (TNFSF) [1, 2]. TRAIL gained much attention due to its specific antitumor potential without toxic side effects [3], making TRAIL itself as well as agonists of its two receptors, which can submit an apoptotic signal, TRAIL-R1 (DR4) [4] and TRAIL-R2 (DR5) [5–8], promising novel biotherapeutics for cancer therapy [9–11]. Importantly, TRAIL can also induce autophagy, which has been linked to apoptosis, serving either a prosurvival or prodeath function [12, 13]. Recent findings reveal that the cellular contexts require a balanced interplay between apoptosis and autophagy. Here, we summarized the recent findings of the key regulator and the crosstalk pathway that highlights the intricate interplay between TRAIL-induced apoptosis and autophagy.

2. TRAIL-induced apoptosis and autophagy

2.1. TRAIL signaling

There are four TRAIL transmembrane receptors: TRAIL-R1 (DR4), TRAIL-R2 (DR5), TRAIL-R3, also known as decoy receptor 1 (DcR1), and TRAIL-R4 (DcR2), and a soluble receptor osteoprotegerin (OPG) [4, 7, 14]. Only TRAIL-R1 and TRAIL-R2 are able to induce apoptosis, whereas TRAIL-R3, TRAIL-R4, and OPG lack the intracellular functional domain, which is required for apoptosis induction [15, 16]. This domain is characteristic for all apoptosis-inducing members of the TNFR superfamily (SF) and is called the death domain (DD). TRAIL-R3 and TRAIL-R4 have been suggested to act as decoy receptors that inhibit apoptosis induction [17]. It has been delineated that TRAIL triggers two major apoptosis signaling pathways, the death receptor (extrinsic) and the mitochondrial (intrinsic) pathways. TRAIL triggers the extrinsic apoptosis pathway upon binding of the TRAIL trimer to TRAIL-R1 and/or TRAIL-R2, resulting in receptor trimerization, which in turn leads to recruitment of the adaptor protein Fas-associated DD (FADD). FADD in turn recruits procaspase-8 and procaspase-10 through homotypic interactions of death-effector domains (DED) presenting in FADD and caspase-8 and caspase-10, respectively. This multiprotein complex is called death-inducing signaling complex (DISC) [18–21]. The DISC is an aggregation of the intracellular death domain of the death receptor. In "type I" cells, the procaspase-8 and procaspase-10 form homodimers. This induces a conformational change that exposes their proteolytical active sites, resulting in autoactivation and subsequent cleavage of additional procaspase-8 and procaspase-10 molecules leading to activation of sufficient caspase-8 to stimulate effector caspase-3 to induce apoptosis [22–24]. However, "type II" cells generate less-active caspase-8 at the DISC. These cells induce apoptosis requiring further signal amplification by the intrinsic/mitochondrial pathway. In this situation, an intracellular complex is activated [25–27]. The next is triggered by caspase-8-mediated cleavage of Bid to truncated Bid (tBid) as the active fragment of this protein [28–31]. Subsequently, tBid activates the mitochondrial pathway eventually leading to mitochondrial outer membrane permeabilization (MOMP) and releasing of cytochrome C and Smac/DIABLO [30, 32]. In the cytosol, cytochrome c Apaf-1 and caspase-9v forms a multimeric complex called apoptosome. Activated caspase-9 as the initiator caspase cleaves and activates the effector caspases. Release of Smac augments apoptosis by antagonizing the inhibitor of apoptosis (IAP) proteins, a family of antiapoptotic proteins that block apoptosis by binding to and inhibiting effector caspases such as caspase-3 and caspase-7 [33, 34].

In the DISC, the main regulator protein is cellular FLICE-like inhibitory protein (cFLIP) and caspase-8, cFLIP contains a death domain, which allows them to interact with proteins of the TRAIL DISC, thereby blocking the transmission of the proapoptotic signal and preventing caspase-8 activation [35–37]. cFLIP closely resembles caspase-8 but lacks the protease activity required for apoptosis induction [38, 39]. Two main variants of cFLIP are expressed on the protein level: a short isoform (cFLIP-S) and a long isoform (cFLIP-L) [40]. The cFLIP-S isoform can inhibit caspase-8 activation in a dominant-negative manner by competing with it for binding to FADD. cFLIP-L can also completely prevent DR-induced apoptosis when it is expressed at high levels. Several studies have demonstrated that cancer cells exploit overex-

http://dx.doi.org/10.5772/64246

pression of cFLIP to evade TRAIL-induced apoptosis [41–43]. Overexpression of cFLIP is a frequent event in human cancers and has been correlated with resistance to the induction of apoptosis, including TRAIL-mediated cell death [36, 37]. Consequently, downregulation of cFLIP may sensitize certain cancers to TRAIL-induced apoptosis [44–46]. Another key regulator in the DISC is caspase-8 that, besides caspase-10, represents the initiator caspase that is engaged during TRAIL-induced apoptosis [47]. Hypermethylation of a regulatory motif that controls caspase-8 expression has been shown to be responsible for low or even absent caspase-8 expression in several cancer entities, resulting in resistance or decreased sensitivity to TRAIL-induced apoptosis [48–51]. Caspase-8 function can be suppressed in a dominant-negative manner by aberrant expression of a splice variant of caspase-8, that is, caspase-8 long (caspase-8L) [52, 53]. This variant of caspase-8 was detected in cancer cells. Caspase-8L interferes with caspase-8 activation by competing with wild-type caspase-8 for the recruitment into the TRAIL DISC. Additional regulatory mechanisms that control caspase-8 activity include post-translational alterations of caspase-8 such as phosphorylation. The tyrosine kinase Src has been reported to phosphorylate caspase-8 on one specific residue (tyrosine-308), which impairs the enzymatic function of caspase-8 [54]. These regulation factors can influence the activity of caspase-8 that causes the change of TRAIL-induced apoptosis.

Except from inducing apoptosis, TRAIL can also induce cell survival signaling such as proinflammatory pathways (through NF-kB, Akt, MAPK, and JNK activation). TRAIL can promote a variety of cell survival cascades leading, for example, to proliferation, migration, invasion, and even metastasis, especially in cancers in which the cell death signaling part of the signaling network is impaired [55–57]. The induction of pathways has been suggested to be mediated by the formation of a secondary complex containing FADD, caspase-8, cFLIP, RIP1, TRAF2, and NEMO [25, 58]. RIP1 is an important regulatory protein in the DISC that can activate NF-κB and caspase-8 and generate reactive oxygen species (ROS) [59–61]. RIP1 function is modulated by ubiquitination and phosphorylation [62, 63]; a previous report showed that in TNF-α-induced DISC, RIP1, and NEMO form a stable chain of linear ubiquitin. This complex is involved in determining cell survival, necrosis, and apoptosis [64].

2.2. The regulators and pathways in TRAIL-induced apoptosis and autophagy

Apoptosis and autophagy are evolutionarily conserved processes that regulate cell fate together. Although apoptosis and autophagy has obvious difference, but their regulation is closely related; they share the same regulator molecules and same pathway; however, these same regulators may determine a different cell fate.

Nowadays, most studies focused on the relationship between TRAIL sensitivity and autophagy [12, 65–68], TRAIL has been shown to induce apoptosis and autophagy in a number of cancer cell lines, including colon, glioma, bladder and prostate, and breast carcinoma. Han et al. first explained TRAIL-mediated cytoprotective autophagy in apoptosis-deficient tumor cells. They found that TRAIL can induce autophagic response in apoptosis-defective tumor cells (Hct116-FLIP or $Bax^{-/-}$ Hct116). Engineered apoptotic deficiencies included stable FLIP transfection, which is expected to block the TRAIL-apoptotic cascade at the DISC level, and Bax knockout demonstrated to block the TRAIL apoptotic response of colon carcinoma Hct116 cells despite the processing of caspase-8 upstream of the mitochondria. Inhibition of autophagy by the knockdown of Beclin 1, UVRAG, Vps34, or Atg7 allows for the induction of significant apoptosis in response to TRAIL [69]. The following work from this laboratory demonstrates that TRAIL-mediated autophagic response counterbalances the TRAIL-mediated apoptotic response by the continuous sequestration of the large caspase-8 subunit in autophagosomes and its subsequent elimination in lysosome [66]. Inhibition of autophagy induces caspase-8 activity; these findings provide evidence for regulation of caspase activity by autophagy. These results suggest that the regulators, such as Beclin 1 and caspase-8, play an important role in the regulation of TRAIL-induced apoptosis and autophagy.

He et al. demonstrate that TRAIL induced cytoprotective autophagy in different cancer cell lines. MAPK8/JNK activation mediated by TRAF2 and RIP1 is required for TRAIL-induced autophagy. Blocking MAPK8 but not NF-κB effectively blocked autophagy, suggesting that MAPK8 is the main pathway for TRAIL-induced autophagy. TRAF2 and RIP1 modulated TRAIL-induced and MAPK8-mediated autophagy. These results reveal that inhibiting MAPK8 pathway-mediated autophagy will increase TRAIL's anticancer activity in cancer cells [65]. Inhibition of antiapoptosis factors in the DISC (cIAP1, cIAP2, XIAP, and c-FLIP, and so on) increases TRAIL-induced apoptosis. Also, some autophagy-related pathways, such as AMPK and MAPK/JNK pathway, are involved in TRAIL-induced apoptosis [65, 70, 71]. These results suggest that there are some regulators and pathways that are necessary for autophagy involved in the regulation of TRAIL-induced apoptosis and autophagy.

Following these researches, some new regulators were found. Caspase-9 is a novel coregulator of apoptosis and autophagy. Han et al. demonstrate that caspase-9 facilitates the early events leading to autophagosome formation; that it forms a complex with Atg7, and Atg7 represses the apoptotic capability of caspase-9, whereas the latter enhances the Atg7-mediated formation of light chain 3-II. The repression of caspase-9 apoptotic activity is mediated by its direct interaction with Atg7, and it is not related to the autophagic function of Atg7. The Atg7 caspase-9 complex performs a dual function of linking caspase-9 to the autophagic process while keeping in check its apoptotic activity [72]. So far it has been found that many regulators such as Beclin 1 and caspase 8 IAPs XIAP in TRAIL induced apoptosis and autophagy in cancer cells. Caspase-8L, cFLIP-L, and cFLIP-S act not only as antiapoptotic factors but also as suppressors of autophagy. Inhibition of autophagy by gene silencing of these regulators or small compounds targets these regulators sensitizing TRAIL-resistant tumor cells to TRAIL-induced apoptosis. Taken together, these researches suggest some potential targets in the prediction of tumor resistance to DR-targeted therapies. Interestingly, a basal level of autophagy is needed for TRAIL-induced apoptosis [73].

In addition to cancer cells, TRAIL has been shown to induce apoptosis and autophagy in other cell lines such as U937 cell, Jurkat T cell, breast epithelial cells, and so on. We found that TRAIL induces both apoptosis and autophagy in human U937 cells [74]. Inhibition of autophagy facilitates TRAIL-induced apoptosis, suggesting that autophagy of macrophages protects against TRAIL-induced apoptosis. RIP1 ubiquitination rapidly increased in U937 cells treated with TRAIL, and RIP1 ubiquitination was significantly reduced in the presence of 3-MA in the cells treated with TRAIL. RIP1 expression was also distinctly decreased in the presence of 3-

MA in the cells treated with TRAIL. Furthermore, c-FLIP-L cleaved into the p43 variant caspase-8 was degraded into p43/41 while autophagy was suppressed by 3-MA in the cells treated with TRAIL. Knockdown of RIP1 suppresses autophagy in macrophage. These data demonstrate that RIP1 is essential for the regulation of death receptor-mediated apoptosis and autophagy in macrophage and suggest that the expression and ubiquitination of RIP1 regulate TRAIL-induced apoptosis and autophagy. The results in this study contribute to understanding the regulation of apoptosis and autophagy in macrophages, and sheds light on inflammation and autoimmune diseases [74].

Wang et al. in our group demonstrate that HTLV-1 (human T cell leukemia virus type 1) Tax protein increases autophagosome accumulation in human U251 astroglioma cells. In addition, HTLV-1 Tax deregulated the autophagy pathway, which plays a protective role during the death receptor-mediated apoptosis. Tax-induced c-FLIP expression also contributes to the resistance against death receptor-mediated apoptosis. Tax-induced c-FLIP expression correlated with the phosphorylation of IKK and the transcriptional activation of NF-κB. But Tax-triggered autophagy only depends on the activation of IKK but not on the activation of NF-kB. TRAIL-induced apoptosis is correlated with the degradation of Tax, which can be facilitated by the inhibitors of autophagy [75]. These results outline a complex regulatory network between apoptosis and autophagy, and Tax-induced autophagy represents a new potential target for therapeutic intervention for the HTVL-1-related diseases.

Herrero-Martin et al. demonstrate that TRAIL triggers cytoprotective autophagy in untransformed human epithelial cells by the AMP-activated protein kinase pathway. Transforming growth factor-b-activating kinase 1 (TAK1) and TAK1-binding subunit 2 mediate TRAIL-induced activation of AMPK and autophagy. These data have broad implications for understanding the cellular control of energy homoeostasis as well as the resistance of untransformed cells against TRAIL-induced apoptosis [71]. These studies of macrophage, Jurkat T cell, and breast epithelial cells have shown that some new regulators are involved in TRAIL-induced apoptosis and autophagy, and the expression and ubiquitination of RIP1, HTLV-1 Tax protein, and TAK1-AMPK pathway regulate the balance of TRAIL-induced apoptosis and autophagy in different extent.

3. Conclusion

Taken together, both the regulators in apoptosis pathway such as caspase-8 and caspase-9 and the key factors in autophagy such as Beclin 1 and ATG7 can regulate the TRAIL-induced apoptosis and autophagy [66, 72]. Moreover, some molecular switchers, like RIP1, regulate the balance between TRAIL-induced apoptosis and autophagy by dynamic expression and modification [65, 74]. They share the same regulators even pathways to control the complicated process (**Table 1**).

Both apoptosis and autophagy are important biological processes that play essential roles in the development of tissue homeostasis and disease. Interactions among components of the two pathways indicate a complex crosstalk. Insight into the complex network of TRAIL-induced apoptosis and autophagy contributes to the development of novel therapeutic strategies for the treatment of TRAIL-related diseases and deeply understand the molecular mechanism of apoptosis and autophagy.

Cell lines	Key regulators pathway	References
Hct116-FLIP or ($Bax^{-/-}$)Hct116	Beclin 1 and caspase-8	[66, 69]
UM-UC-3, PC-3, and A549	TRAF2 (RIP1)-MAPK8/JNK pathway	[65]
Hct116, HeLa, MB-MDA-231, and RKO	Atg7·caspase-9 complex	[72]
U937	RIP1 caspase-8 and cFLIP	[74]
U251	HTLV-1 Tax and cFLIP	[75]
MCF10A–eGFP–LC3	TAK1-AMPK pathway	[71]

Table 1. The regulators and pathways in TRAIL-induced apoptosis and autophagy in differences cell lines.

Author details

Zhenyu Yao[1,2*] and Dexian Zheng[1]

*Address all correspondence to: zhengdx@pumc.edu.cn; zy.yao@siat.ac.cn

1 Institute of Basic Medical Sciences, Chinese Academy of Medical Sciences and Peking Union Medical College, Beijing, China

2 Shenzhen Institutes of Advanced Technology, Chinese Academy of Sciences, Shenzhen, China

References

[1] Pitti RM, Marsters SA, Ruppert S, Donahue CJ, Moore A, Ashkenazi A. Induction of apoptosis by Apo-2 ligand, a new member of the tumor necrosis factor cytokine family. J Biol Chem 1996;271:12687–12690. DOI: 10.1074/jbc.271.22.12687

[2] Wiley SR, Schooley K, Smolak PJ, et al. Identification and characterization of a new member of the TNF family that induces apoptosis. Immunity 1995;3:673–682. DOI: 10.1016/1074-7613(95)90057-8

[3] Walczak H, Miller RE, Ariail K, et al. Tumoricidal activity of tumor necrosis factor-related apoptosis-inducing ligand in vivo. Nat Med 1999; 5:157–163. DOI:10.1038/5517

[4] Pan G, O'Rourke K, Chinnaiyan AM, et al. The receptor for the cytotoxic ligand TRAIL. Science 1997; 276:111–113. DOI: 10.1126/science.276.5309.111

[5] Schneider P, Bodmer JL, Thome M, Hofmann K, Holler N, Tschopp J. Characterization of two receptors for TRAIL. FEBS Lett 1997; 416:329–334. DOI: 10.1016/S0014-5793(97)01231-3

[6] MacFarlane M, Ahmad M, Srinivasula SM, Fernandes-Alnemri T, Cohen GM, Alnemri ES. Identification and molecular cloning of two novel receptors for the cytotoxic ligand TRAIL. J Biol Chem 1997;272:25417–25420. DOI: 10.1074/jbc.272.41.25417

[7] Chaudhary PM, Eby M, Jasmin A, Bookwalter A, Murray J, Hood L. Death receptor 5, a new member of the TNFR family, and DR4 induce FADD-dependent apoptosis and activate the NF-kappaB pathway. Immunity 1997; 7:821–830. DOI:10.1016/S1074-7613(00)80400-8

[8] Walczak H, Degli-Esposti MA, Johnson RS, et al. TRAIL-R2: a novel apoptosis-mediating receptor for TRAIL. EMBO J 1997; 16:5386–5397. DOI:10.1093/emboj/16.17.5386

[9] Micheau O, Shirley S, Dufour F. Death receptors as targets in cancer. Br J Pharmacol 2013; 169:1723–1744. DOI:10.1111/bph.12238

[10] Lim B, Allen JE, Prabhu VV, Talekar MK, Finnberg NK, El-Deiry WS. Targeting TRAIL in the treatment of cancer: new developments. Expert Opin Ther Targets 2015; 19:1171–1185. DOI:10.1517/14728222.2015.1049838

[11] Refaat A, Abd-Rabou A, Reda A. TRAIL combinations: the new 'trail' for cancer therapy (Review). Oncol Lett 2014; 7:1327–1332. DOI:10.3892/ol.2014.1922

[12] Mills KR, Reginato M, Debnath J, Queenan B, Brugge JS. Tumor necrosis factor-related apoptosis-inducing ligand (TRAIL) is required for induction of autophagy during lumen formation in vitro. Proc Natl Acad Sci U S A 2004; 101:3438–3443. DOI:10.1073/pnas.0400443101

[13] Park KJ, Lee SH, Kim TI, et al. A human scFv antibody against TRAIL receptor 2 induces autophagic cell death in both TRAIL-sensitive and TRAIL-resistant cancer cells. Cancer Res 2007; 67:7327–7334. DOI:10.1158/0008-5472.CAN-06-4766

[14] Emery JG, McDonnell P, Burke MB, et al. Osteoprotegerin is a receptor for the cytotoxic ligand TRAIL. J Biol Chem 1998; 273:14363–14367. DOI: 10.1074/jbc.273.23.14363

[15] Degli-Esposti MA, Smolak PJ, Walczak H, et al. Cloning and characterization of TRAIL-R3, a novel member of the emerging TRAIL receptor family. J Exp Med 1997; 186:1165–1170. DOI: 10.1084/jem.186.7.1165

[16] Degli-Esposti MA, Dougall WC, Smolak PJ, Waugh JY, Smith CA, Goodwin RG. The novel receptor TRAIL-R4 induces NF-kappaB and protects against TRAIL-mediated apoptosis, yet retains an incomplete death domain. Immunity 1997; 7:813–820. DOI: 10.1016/S1074-7613(00)80399-4

[17] Merino D, Lalaoui N, Morizot A, Schneider P, Solary E, Micheau O. Differential inhibition of TRAIL-mediated DR5-DISC formation by decoy receptors 1 and 2. Mol Cell Biol 2006;26:7046–7055. DOI:10.1128/MCB.00520-06

[18] Bodmer JL, Holler N, Reynard S, et al. TRAIL receptor-2 signals apoptosis through FADD and caspase-8. Nat Cell Biol 2000; 2:241–243. DOI:10.1038/35008667

[19] Kuang AA, Diehl GE, Zhang J, Winoto A. FADD is required for DR4- and DR5-mediated apoptosis: lack of trail-induced apoptosis in FADD-deficient mouse embryonic fibroblasts. J Biol Chem 2000; 275:25065–25068. DOI:10.1074/jbc.C000284200

[20] Peter ME. The TRAIL DISCussion: it is FADD and caspase-8! Cell Death Differ 2000; 7:759–760. DOI:10.1038/sj.cdd.4400735

[21] Kischkel FC, Lawrence DA, Chuntharapai A, Schow P, Kim KJ, Ashkenazi A. Apo2L/TRAIL-dependent recruitment of endogenous FADD and caspase-8 to death receptors 4 and 5. Immunity 2000; 12:611–620. DOI:10.1016/S1074-7613(00)80212-5

[22] Chen M, Wang J. Initiator caspases in apoptosis signaling pathways. Apoptosis 2002; 7:313–319. DOI:10.1023/A:1016167228059

[23] Muhlethaler-Mottet A, Flahaut M, Bourloud KB, et al. Individual caspase-10 isoforms play distinct and opposing roles in the initiation of death receptor-mediated tumour cell apoptosis. Cell Death Dis 2011; 2:e125. DOI:10.1038/cddis.2011.8

[24] Salvesen GS, Dixit VM. Caspase activation: the induced-proximity model. Proc Natl Acad Sci U S A 1999; 96:10964–10967. DOI:10.1016/j.cell.2004.06.007

[25] Varfolomeev E, Maecker H, Sharp D, et al. Molecular determinants of kinase pathway activation by Apo2 ligand/tumor necrosis factor-related apoptosis-inducing ligand. J Biol Chem 2005; 280:40599–40608. DOI:10.1074/jbc.M509560200

[26] Ozoren N, El-Deiry WS. Defining characteristics of Types I and II apoptotic cells in response to TRAIL. Neoplasia 2002; 4:551–557. DOI:10.1038/sj.neo.7900270

[27] Rudner J, Jendrossek V, Lauber K, Daniel PT, Wesselborg S, Belka C. Type I and type II reactions in TRAIL-induced apoptosis -- results from dose-response studies. Oncogene 2005; 24:130–140. DOI:10.1038/sj.onc.1208191

[28] Grinberg M, Sarig R, Zaltsman Y, et al. tBID Homooligomerizes in the mitochondrial membrane to induce apoptosis. J Biol Chem 2002; 277:12237–12245. DOI:10.1074/jbc.M104893200

[29] Li H, Zhu H, Xu CJ, Yuan J. Cleavage of BID by caspase 8 mediates the mitochondrial damage in the Fas pathway of apoptosis. Cell 1998; 94:491–501. DOI:10.1016/S0092-8674(00)81590-1

[30] Luo X, Budihardjo I, Zou H, Slaughter C, Wang X. Bid, a Bcl2 interacting protein, mediates cytochrome c release from mitochondria in response to activation of cell surface death receptors. Cell 1998; 94:481–490. DOI:10.1016/S0092-8674(00)81589-5

[31] Wei MC, Lindsten T, Mootha VK, et al. tBID, a membrane-targeted death ligand, oligomerizes BAK to release cytochrome c. Genes Dev 2000; 14:2060–2071. DOI:10.1101/gad.14.16.2060

[32] Baliga B, Kumar S. Apaf-1/cytochrome c apoptosome: an essential initiator of caspase activation or just a sideshow? Cell Death Differ 2003; 10:16–18. DOI:10.1038/sj.cdd.4401166

[33] Roy N, Deveraux QL, Takahashi R, Salvesen GS, Reed JC. The c-IAP-1 and c-IAP-2 proteins are direct inhibitors of specific caspases. EMBO J 1997; 16:6914–6925. DOI:10.1093/emboj/16.23.6914

[34] Deveraux QL, Takahashi R, Salvesen GS, Reed JC. X-linked IAP is a direct inhibitor of cell-death proteases. Nature 1997; 388:300–304. DOI:10.1038/40901

[35] Fulda S. Targeting c-FLICE-like inhibitory protein (CFLAR) in cancer. Expert Opin Ther Targets 2013; 17:195–201. DOI:10.1517/14728222.2013.736499

[36] Haag C, Stadel D, Zhou S, et al. Identification of c-FLIP(L) and c-FLIP(S) as critical regulators of death receptor-induced apoptosis in pancreatic cancer cells. Gut 2011; 60:225–237. DOI:10.1136/gut.2009.202325

[37] Longley DB, Wilson TR, McEwan M, et al. c-FLIP inhibits chemotherapy-induced colorectal cancer cell death. Oncogene 2006; 25:838–848. DOI:10.1038/sj.onc.1209122

[38] Irmler M, Thome M, Hahne M, et al. Inhibition of death receptor signals by cellular FLIP. Nature 1997; 388:190–195. DOI:10.1038/40657

[39] Tschopp J, Irmler M, Thome M. Inhibition of fas death signals by FLIPs. Curr Opin Immunol 1998; 10:552–558. DOI:10.1016/S0952-7915(98)80223-9

[40] Krueger A, Schmitz I, Baumann S, Krammer PH, Kirchhoff S. Cellular FLICE-inhibitory protein splice variants inhibit different steps of caspase-8 activation at the CD95 death-inducing signaling complex. J Biol Chem 2001; 276:20633–20640. DOI:10.1074/jbc.M101780200

[41] Burns TF, El-Deiry WS. Identification of inhibitors of TRAIL-induced death (ITIDs) in the TRAIL-sensitive colon carcinoma cell line SW480 using a genetic approach. J Biol Chem 2001; 276:37879–37886. DOI:10.1074/jbc.M103516200

[42] Riley JS, Hutchinson R, McArt DG, et al. Prognostic and therapeutic relevance of FLIP and procaspase-8 overexpression in non-small cell lung cancer. Cell Death Dis 2013; 4:e951. DOI:10.1038/cddis.2013.481

[43] Guseva NV, Rokhlin OW, Taghiyev AF, Cohen MB. Unique resistance of breast carcinoma cell line T47D to TRAIL but not anti-Fas is linked to p43cFLIP(L). Breast Cancer Res Treat 2008; 107:349–357. DOI:10.1007/s10549-007-9563-2

[44] Geserick P, Drewniok C, Hupe M, et al. Suppression of cFLIP is sufficient to sensitize human melanoma cells to TRAIL- and CD95L-mediated apoptosis. Oncogene 2008; 27:3211–3220. DOI:10.1038/sj.onc.1210985

[45] Balsas P, Lopez-Royuela N, Galan-Malo P, Anel A, Marzo I, Naval J. Cooperation between Apo2L/TRAIL and bortezomib in multiple myeloma apoptosis. Biochem Pharmacol 2009; 77:804–812. DOI:10.1016/j.bcp.2008.11.024

[46] Palacios C, Yerbes R, Lopez-Rivas A. Flavopiridol induces cellular FLICE-inhibitory protein degradation by the proteasome and promotes TRAIL-induced early signaling and apoptosis in breast tumor cells. Cancer Res 2006; 66:8858–8869. DOI:10.1158/0008-5472.CAN-06-0808

[47] Fulda S. Caspase-8 in cancer biology and therapy. Cancer Lett 2009; 281:128–133. DOI: 10.1016/j.canlet.2008.11.023

[48] Eggert A, Grotzer MA, Zuzak TJ, et al. Resistance to tumor necrosis factor-related apoptosis-inducing ligand (TRAIL)-induced apoptosis in neuroblastoma cells correlates with a loss of caspase-8 expression. Cancer Res 2001; 61:1314–1319. http://cancerres.aacrjournals.org/content/61/4/1314.full

[49] Fulda S, Debatin KM. 5-Aza-2'-deoxycytidine and IFN-gamma cooperate to sensitize for TRAIL-induced apoptosis by upregulating caspase-8. Oncogene 2006; 25:5125–5133. DOI:10.1038/sj.onc.1209518

[50] Fulda S, Poremba C, Berwanger B, et al. Loss of caspase-8 expression does not correlate with MYCN amplification, aggressive disease, or prognosis in neuroblastoma. Cancer Res 2006; 66:10016–10023. DOI:10.1158/0008-5472.CAN-05-4079

[51] Teitz T, Wei T, Valentine MB, et al. Caspase 8 is deleted or silenced preferentially in childhood neuroblastomas with amplification of MYCN. Nat Med 2000; 6:529–535. DOI:10.1038/75007

[52] Himeji D, Horiuchi T, Tsukamoto H, Hayashi K, Watanabe T, Harada M. Characterization of caspase-8L: a novel isoform of caspase-8 that behaves as an inhibitor of the caspase cascade. Blood 2002; 99:4070–4078. DOI: http://dx.doi.org/10.1182/blood.V99.11.4070

[53] Miller MA, Karacay B, Zhu X, O'Dorisio MS, Sandler AD. Caspase 8L, a novel inhibitory isoform of caspase 8, is associated with undifferentiated neuroblastoma. Apoptosis 2006; 11:15–24. DOI:10.1007/s10495-005-3258-0

[54] Cursi S, Rufini A, Stagni V, et al. Src kinase phosphorylates Caspase-8 on Tyr380: a novel mechanism of apoptosis suppression. EMBO J 2006; 25:1895–1905. DOI:10.1038/sj.emboj.7601085

[55] Lin Y, Devin A, Cook A, et al. The death domain kinase RIP is essential for TRAIL (Apo2L)-induced activation of IkappaB kinase and c-Jun N-terminal kinase. Mol Cell Biol 2000; 20:6638–6645. DOI: 10.1128/MCB.20.18.6638-6645.2000

[56] Morel J, Audo R, Hahne M, Combe B. Tumor necrosis factor-related apoptosis-inducing ligand (TRAIL) induces rheumatoid arthritis synovial fibroblast proliferation through mitogen-activated protein kinases and phosphatidylinositol 3-kinase/Akt. J Biol Chem 2005; 280:15709–15718. DOI:10.1074/jbc.M414469200

[57] Tran SE, Holmstrom TH, Ahonen M, Kahari VM, Eriksson JE. MAPK/ERK overrides the apoptotic signaling from Fas, TNF, and TRAIL receptors. J Biol Chem 2001; 276:16484–16490. DOI:10.1074/jbc.M010384200

[58] Grunert M, Gottschalk K, Kapahnke J, Gundisch S, Kieser A, Jeremias I. The adaptor protein FADD and the initiator caspase-8 mediate activation of NF-kappaB by TRAIL. Cell Death Dis 2012; 3:e414. DOI:10.1038/cddis.2012.154

[59] Galluzzi L, Kepp O, Kroemer G. RIP kinases initiate programmed necrosis. J Mol Cell Biol 2009; 1:8–10. DOI:10.1093/jmcb/mjp007

[60] Kelliher MA, Grimm S, Ishida Y, Kuo F, Stanger BZ, Leder P. The death domain kinase RIP mediates the TNF-induced NF-kappaB signal. Immunity 1998; 8:297–303. DOI: 10.1016/S1074-7613(00)80535-X

[61] Lin Y, Devin A, Rodriguez Y, Liu ZG. Cleavage of the death domain kinase RIP by caspase-8 prompts TNF-induced apoptosis. Genes Dev 1999; 13:2514–2526. DOI: 10.1101/gad.13.19.2514

[62] Cho YS, Challa S, Moquin D, et al. Phosphorylation-driven assembly of the RIP1-RIP3 complex regulates programmed necrosis and virus-induced inflammation. Cell 2009; 137:1112–1123. DOI:10.1016/j.cell.2009.05.037

[63] Ea CK, Deng L, Xia ZP, Pineda G, Chen ZJ. Activation of IKK by TNFalpha requires site-specific ubiquitination of RIP1 and polyubiquitin binding by NEMO. Mol Cell 2006; 22:245–257. DOI:10.1016/j.molcel.2006.03.026

[64] Gerlach B, Cordier SM, Schmukle AC, et al. Linear ubiquitination prevents inflammation and regulates immune signalling. Nature 2011; 471:591–596. DOI:10.1038/nature09816

[65] He W, Wang Q, Xu J, et al. Attenuation of TNFSF10/TRAIL-induced apoptosis by an autophagic survival pathway involving TRAF2- and RIPK1/RIP1-mediated MAPK8/JNK activation. Autophagy 2012; 8:1811–1821. DOI: 10.4161/auto.22145

[66] Hou W, Han J, Lu C, Goldstein LA, Rabinowich H. Autophagic degradation of active caspase-8: a crosstalk mechanism between autophagy and apoptosis. Autophagy 2010; 6:891–900. DOI:10.4161/auto.6.7.13038

[67] Niu TK, Cheng Y, Ren X, Yang JM. Interaction of Beclin 1 with survivin regulates sensitivity of human glioma cells to TRAIL-induced apoptosis. FEBS Lett 2010; 584:3519–3524. DOI:10.1016/j.febslet.2010.07.018

[68] Zhang Y, Zhang B. TRAIL resistance of breast cancer cells is associated with constitutive endocytosis of death receptors 4 and 5. Mol Cancer Res 2008; 6:1861–1871. DOI: 10.1158/1541-7786.MCR-08-0313

[69] Han J, Hou W, Goldstein LA, et al. Involvement of protective autophagy in TRAIL resistance of apoptosis-defective tumor cells. J Biol Chem 2008; 283:19665–19677. DOI: 10.1074/jbc.M710169200

[70] Chiacchiera F, Grossi V, Cappellari M, et al. Blocking p38/ERK crosstalk affects colorectal cancer growth by inducing apoptosis in vitro and in preclinical mouse models. Cancer Lett 2012; 324:98–108. DOI:10.1016/j.canlet.2012.05.006

[71] Herrero-Martin G, Hoyer-Hansen M, Garcia-Garcia C, et al. TAK1 activates AMPK-dependent cytoprotective autophagy in TRAIL-treated epithelial cells. EMBO J 2009; 28:677–685. DOI:10.1038/emboj.2009.8

[72] Han J, Hou W, Goldstein LA, Stolz DB, Watkins SC, Rabinowich H. A Complex between Atg7 and Caspase-9: a novel mechanism of cross-regulation between autophagy and apoptosis. J Biol Chem 2014; 289:6485–6497. DOI:10.1074/jbc.M113.536854

[73] Di X, Zhang G, Zhang Y, Takeda K, Rivera Rosado LA, Zhang B. Accumulation of autophagosomes in breast cancer cells induces TRAIL resistance through downregulation of surface expression of death receptors 4 and 5. Oncotarget 2013; 4:1349–1364. DOI:10.18632/oncotarget.1174

[74] Yao Z, Zhang P, Guo H, et al. RIP1 modulates death receptor mediated apoptosis and autophagy in macrophages. Mol Oncol 2015; 9:806–817. DOI:10.1016/j.molonc.2014.12.004

[75] Wang W, Zhou J, Shi J, et al. Human T-cell leukemia virus type 1 Tax-deregulated autophagy pathway and c-FLIP expression contribute to resistance against death receptor-mediated apoptosis. J Virol 2014; 88:2786–2798. DOI:10.1128/JVI.03025-13

Chapter 7

The Role of Ubiquitin System in Autophagy

Yi-Ting Wang and Guang-Chao Chen

Additional information is available at the end of the chapter

http://dx.doi.org/10.5772/64728

Abstract

Autophagy is a highly conserved lysosomal degradation pathway, which has been shown to play a pivotal role during normal physiological and pathological conditions. Many proteins and signaling pathways have been shown to regulate autophagy during different stages of the process. Modifying autophagy-related proteins (Atg) by posttranslational modification (PTM) is an important way to control proper autophagic activity. Ubiquitination is one of the PTM that has a crucial role in controlling protein stability and functions. Proteins can be conjugated with ubiquitin chains with different topologies that are associated with different outcomes. Many autophagy regulators are found to be substrates for ubiquitin E3 ligases or deubiquitinating enzymes (DUBs). Ubiquitination modifications of these autophagy regulators result in autophagy induction or termination. Moreover, ubiquitin is also involved in selective autophagy by acting as a degradation signal. Here, we are going to review how E3 ligases and DUBs function in autophagy regulation and discuss the recent findings about ubiquitination regulation in autophagy-related processes and diseases.

Keywords: autophagy, ubiquitin, E3 ligase, deubiquitinating enzyme

1. Introduction

Proteome dynamics and complexity are tightly regulated to maintain normal cellular function and homeostasis. The ubiquitin-proteasome system (UPS) and autophagy are the two main intracellular degradative machineries in eukaryotes [1, 2]. UPS mainly degrades specific short-lived proteins, whereas autophagy is responsible for the bulk degradation of long-lived proteins and damaged organelles. Autophagy is a lysosome-mediated catabolic process by which cytoplasmic components are degraded and recycled for cellular homeostasis [3, 4]. It is tightly controlled by complex signaling pathways and serves as a cytoprotective mechanism in response to environmental stresses such as nutrient deprivation, reactive oxygen species (ROS), and pathogen invasion [5]. Dysregulation of autophagy pathway has been implicated in various human diseases [6, 7], including myopathies, aging, neurodegeneration, and cancer, as well as in heart, liver, and kidney diseases. To date, more than 35 autophagy-related (Atg) genes have been implicated in regulating the autophagic process [8]. The core Atg proteins are highly conserved and can be assembled into different complexes such as the autophagy-initiating Atg1/Ulk protein kinase complex, Beclin1-class III PI3K complex, the Atg5-Atg12 and Atg8/LC3 ubiquitin-like conjugation systems, and the Atg9 recycling system [8, 9].

Protein posttranslational modification (PTM) plays a pivotal role in increasing proteome complexity and determining the fates of proteins [10–13]. It is a widespread mechanism that involves the addition of a functional group covalently to a protein. The major types of PTMs include phosphorylation, glycosylation, acetylation, ubiquitination, methylation, and lipidation. The diversity of PTM provides enormous flexibility for control of protein structure, localization, activity, and function; this modification can be reversible or irreversible. Recent studies have identified various forms of PTMs in the regulation of autophagy [14–16]. Some PTMs regulate autophagy by affecting the enzymatic activity of Atg proteins. For example, the Ulk1 kinase (the key initiator of autophagy) can be phosphorylated by upstream regulators such as AMP-activated protein kinase (AMPK) and mammalian target of rapamycin (mTOR), which result in the activation or inactivation of Ulk1 kinase activity, respectively [17–19]. Additionally, Ulk1 can also be acetylated at K162 and K606 by acetyl transferase KAT5/TIP60, and the PTM acetylation is vital to the activation of Ulk1 [20]. PTMs can also regulate autophagy by changing the interacting partners of Atg proteins [21, 22]. It has been shown that the posttranslational modification of Beclin1 affects its interaction with Vps34 complex. Zalckvar et al. showed that death-associated protein kinase (DAPk)-mediated phosphorylation of Beclin1 promotes the association between Beclin1 and Vps34 complex [23]. On the contrary, Beclin1 phosphorylation by CDK1 leads to the dissociation of Beclin1 from Vps34 [24].

Accumulating evidence indicates that protein ubiquitination and deubiquitination play multiple roles in the regulation of protein stability and signaling during autophagy [15, 16]. Several ubiquitin E3 ligases and deubiquitinating enzymes (DUBs) have been shown to regulate autophagy at different stages. However, the detailed mechanisms of the E3 ligases and the DUBs in controlling both "on" and "off" signals of autophagy remain unclear. The ubiquitin system is also essential for the recognition and removal of damaged organelles and invading pathogens during selective autophagy processes. Moreover, two ubiquitin-like conjugation systems are found to be crucial for the expansion of the elongation and expansion of autophagosomal membrane. Here, we will discuss recent advances on the role of ubiquitin systems in autophagy.

http://dx.doi.org/10.5772/64728

2. Ubiquitin modification and protein fate determination

2.1. Ubiquitin modification

Ubiquitination is an ATP-dependent enzymatic process that involves the covalent conjugate a highly conserved 8-kDa ubiquitin (Ub) peptide to lysine residues of target proteins [25]. The ubiquitination reaction requires three classes of enzyme: ubiquitin-activating enzymes (E1), ubiquitin-conjugating enzymes (E2), and substrate-specific ubiquitin ligases (E3) [25, 26]. In mammals, there are more than 500 E3 ligases, 30 E2-conjugating enzymes, and two E1-activating enzymes [26–28]. The E1 enzyme activates free ubiquitin by forming a thioester linkage between the C-terminal carboxyl group of ubiquitin and the E1 cysteine sulfhydryl group in an ATP-dependent manner. The activated ubiquitin is then transferred from E1 to the catalytic cysteine of E2, which can determine the type of ubiquitin chain formed. Finally, the E3 binds both the Ub-charged E2 and substrates to catalyze and transfers the C-terminus of Ub to the lysine residue of substrates. The E3 ligase transfers ubiquitin to specific protein targets and is critical for conferring the substrate specificity. Because of the large number of E2 and E3 enzymes, a broad range of substrates can be modified by distinct ubiquitin chain configurations.

Ubiquitination of a substrate can be classified according to the number of ubiquitins, linkage of ubiquitin chains, and chain length [27]. It has been shown that the attachment of an ubiquitin molecule at one site in the substrate causes monoubiquitination [28, 29]. Whereas monoubiquitin is conjugated on several lysine residues of the substrate results in multiubiquitination. Mono- and multiubiquitination regulate processes that range from histone modification to membrane-receptor endocytic trafficking [30]. In addition, the E2/E3 complexes can catalyze further cycles of ubiquitination on the substrate-conjugated ubiquitin, resulting in substrate polyubiquitination [28, 29]. Ubiquitin links to another ubiquitin molecule to form polyubiquitin chain via one of its seven lysines (K6, K11, K27, K29, K33, K48, and K63) or the N-terminal methionine residue (M1) [27, 31]. The polyubiquitin chain can elongate using the same lysine residue on each ubiquitin (homogeneous ubiquitin chain) or the polyubiquitin chain can form through conjugation with mixed topology (heterogeneous ubiquitin chain). The different ubiquitination modification patterns provide a way to increase the diversity of protein regulation and functions. For example, the polyubiquitin chain linked through Lys48 (K48-linked polyubiquitin) provides a signal for protein degradation by the 26S proteasome, thereby regulating the stability of proteins [32]. The K11-linked polyubiquitin can trigger the degradation of cell cycle regulators via the ubiquitin-proteasome pathway during mitosis [33, 34]. Moreover, ubiquitination modification also regulates protein-protein interaction, enzyme activity, and the cellular localization of proteins. The K63-linked polyubiquitin was reported to be involved in signal transduction [35], kinase activation [36, 37], and protein-protein interaction [21]. Together, these findings indicate that ubiquitin modification affects diverse cellular processes by regulating the stability and the function of proteins. The temporal and spatial control of ubiquitin signaling plays a pivotal role to maintain normal cellular functions.

2.2. Deubiquitinating enzymes (DUBs)

Protein ubiquitination is a reversible posttranslational modification process. The process to cleave ubiquitin from proteins and other molecules is called deubiquitination, and the process is catalyzed by a large group of ubiquitin-cleaving proteases, the deubiquitinating enzymes (DUBs) [38–41]. DUBs can be classified into five families: ubiquitin-specific proteases (USPs), ubiquitin C-terminal hydrolase (UCHs), ovarian tumor proteases (OTUs), Machado-Joseph disease domain proteases (MJDs), and Jab1/Mpn/Mov34 metalloenzymes (JAMMs) and play several critical roles in the ubiquitin pathway [38, 40]. First, they are responsible for processing inactive ubiquitin precursors. The ubiquitin is translated in the form of linear polyubiquitin chain or fusion with ribosomal proteins. DUBs are needed to generate free ubiquitin monomer. Second, DUBs can antagonize E3 ligases by removing the ubiquitin molecule or trimming the ubiquitin chain form substrates, thereby changing the ubiquitin signaling or stability of targeted proteins. Finally, DUBs are required for the recycling of ubiquitin molecule. The ubiquitin molecules cleaved from substrates or ubiquitin chains can re-enter to the free ubiquitin pool. The coordination of E3 ligases and DUBs leads to conjugating, trimming, and removing ubiquitin modification of target proteins for various biological processes.

3. The role of ubiquitin modifications in autophagy regulation

Recent investigations have implicated the involvement of complex signaling pathways during different stages of autophagic process. Modifying autophagy-related proteins (Atg) by posttranslational modification is one of the important mechanisms to control proper autopha-

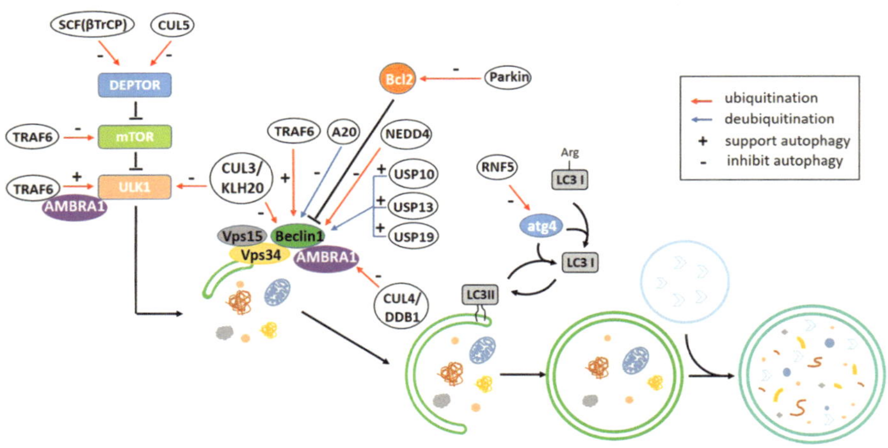

Figure 1. Functional role of ubiquitin-related enzymes (E3 ligases and DUBs) in autophagy regulation. An overview shows the involvement of the ubiquitination events during different steps of autophagy. Ubiquitination modifications by E3 ligases are shown in red arrows, and deubiquitination by DUBs are shown in blue arrows. "+" indicates that enzymes play positive regulatory roles (USP10, USP13, and USP19), and "–" indicates those thought to play negative roles (SCF$^{\beta TrCP}$, CUL5, CUL3/KLH20, A20, NEDD4, CUL4/DDB1, Parkin, and RNF5) in autophagy. TRAF6 can function as positive or negative regulator in autophagy via ubiquitination modification of different substrates.

gic activity [15, 16]. Many autophagy regulators are found to be substrates for ubiquitin E3 ligases or deubiquitinating enzymes (DUBs) [42, 43]. Ubiquitination modification of these autophagy regulators controls autophagy induction, nucleation, maturation, or termination. Here, we are going to review recent findings on the role of E3 ligases and DUBs in the regulation of autophagy (**Figure 1**).

3.1. Ubiquitin systems and autophagy initiation

Many signaling pathways participate in the induction of autophagy. The inhibition of mTOR function and the activation of Ulk1 complex are two major mechanisms in the initiation of autophagy [8]. Several ubiquitination enzymes have been shown to participate in the induction of autophagy by modifying the initiators of autophagy [44–47]. It has been reported that mTOR inhibitor, DEPTOR, can be ubiquitinated by $SCF^{\beta TrCP}$ E3 ubiquitin ligase and CUL5, and led to the degradation of DEPTOR. In response to growth signals, DEPTOR is phosphorylated by the downstream components of mTOR pathway such as RSK1 and S6K1, and the phosphorylated DEPTOR is then targeted for ubiquitination by $SCF^{\beta TrCP}$ E3 ligase for degradation [44, 45]. Interestingly, Antonioli et al. recently showed that CUL5 can also catalyze the DEPTOR ubiquitination and promote its degradation under normal conditions. The degradation of DEPTOR leads to the activation of mTOR, which acts as an inhibitor of autophagy. Upon autophagy stimulation, the CUL5-mediated degradation of DEPTOR is inhibited by Ambra1 in an Ulk1-dependent manner and promotes the onset of autophagy [46, 47].

The mTOR activity can also be regulated by ubiquitination through the TRAF6 E3 ligase [48]. Upon amino acids stimulation, TRAF6 is recruited to mTOR complex 1 (mTORC1) through p62 and catalyzes K63-linked polyubiquitination of mTOR which is required for mTORC1 translocation to the lysosome and its subsequent activation. Moreover, TRAF6 has also been shown to promote K63-linked polyubiquitination of Ulk1, which results in Ulk1 stabilization, self-association, and autophagy induction [49]. The TRAF6-mediated Ulk1 ubiquitination depends on Ambra1 which is also a substrate target of Ulk1 during autophagy induction. These findings together indicate that ubiquitin system and ubiquitin-related enzymes play a critical role in autophagy initiation.

3.2. Ubiquitin systems and autophagy nucleation

The Beclin1 and Vps34, a class III phosphoinositide 3-kinase (PI 3-kinase), are the key regulators in the nucleation step of autophagy [8]. Beclin1 acts as an adaptor, which recruits cellular components such as Ambra1 and UVRAG to form different Beclin1-Vps34 complexes that are responsible for modulating the activity of Vps34. Recent studies have shown that the interaction between Beclin1 and its binding partners can be regulated by PTMs including ubiquitination [21].

It was reported that, upon lipopolysaccharide (LPS) stimulation, TRAF6 catalyzes K63-linked polyubiquitination of Beclin1 at K117 [50]. The TRAF6-mediated Beclin1 ubiquitination leads to the disassociation of Beclin1 with Bcl2 and promotes autophagy. On the contrary, the deubiquitinating enzyme A20 antagonized the TRAF6-mediated Beclin1 ubiquitination and abrogated autophagy induction. Recently, Chen et al. showed that the E3 ubiquitin ligase Parkin, which is found to be involved in the neurodegenerative Parkinson's disease (PD), can also catalyze the monoubiquitination of Bcl2 [51]. Parkin-mediated Bcl2 ubiquitination increases the steady-state levels of Bcl2 and enhances the interactions between Bcl2 and Beclin1, leading to the inhibition of autophagy. Moreover, E3 ubiquitin ligase Nedd4 (neural precursor cell expressed developmentally down-regulated protein 4) promotes Beclin1 degradation through proteasomal system in the absence of Vps34 interaction [52]. Nedd4 controls the stability of Beclin1 via K11-linked polyubiquitination. Through the degradation of Beclin1, Nedd4 acts as a negative regulator of autophagy. Besides E3 ligases, Liu et al. showed that USP10 and USP13 DUBs also participate in the autophagy nucleation by regulating the stability of Beclin1-Vps34 complex components including Vps34, Beclein1, Vps15, and Atg14L [53]. Unexpectedly, Beclin1-Vps34 complex also promotes the stability and activity of USP10 and USP13 [53, 54]. Recently, the deubiquitinating enzyme USP19 was found to stabilize Beclin1 by removing the K11-linked ubiquitin chains of Beclin-1 at lysine 437 and act as a positive regulator of autophagy [55]. Moreover, USP19 inhibits RIG-I-mediated type I interferon (IFN) signaling and antiviral immune responses by blocking RIG-I-MAVS interaction in a Beclin-1-dependent manner. In addition, the deubiquitinating enzyme USP33 is also involved in autophagy induction by deubiquitinating the RAS-like GTPase RALB under starvation conditions [56]. RALB interacts with the exocyst components EXO84. Upon nutrient deprivation, the USP33-mediated deubiquitylation of RALB induces the assembly of RALB-EXO84-Beclin1 complex and the initiation of autophagy.

By interfering the interaction or by controlling the stability of the Vps34-Beclin1 complex components, the ubiquitin system provides a flexible and diverse way to regulate autophagy nucleation.

3.3. Ubiquitin systems and autophagosome elongation/expansion

The Atg8 (LC3 and GABARAP in mammals) and Atg12 ubiquitin-like conjugation systems are two major pathways involved in the regulation of autophagosomal elongation and expansion [57]. Similar to ubiquitination, Atg12 is conjugated to the lysine residue in Atg5 by the E1 enzyme Atg7 and the E2 enzyme Atg10. The Atg12-Atg5 conjugate subsequently forms a complex with Atg16 for phagophore membrane elongation. On the other hand, Atg8 is first processed at the C-terminus by the cysteine protease Atg4 and then activated by Atg7 (E1) and Atg3 (E2) for the conjugation of the lipid phosphatidylethanolamine (PE). Atg4 is also required for the cleavage of Atg8 from PE on the autophagic membrane after the completion of autophagosome formation. As Atg4 plays a critical role in the phagophore expansion and autophagosome completion, depletion of Atg4 inhibits the processing of Atg8 paralogues and autophagy. It has been shown that the membrane-associated E3 ligase RNF5 regulates autophagy by ubiquitinating Atg4b and promoting the proteolytic degradation of Atg4b [58].

3.4. Ubiquitin systems and autophagy termination

Like the initiation of autophagy, the termination of autophagy is also tightly regulated after completion of each run. The failure of autophagy termination under prolonged starvation will lead to unrestrained cellular degradation and cell death [59]. In addition to mTOR activation induced by the regeneration of intracellular nutrients, recent studies revealed that the proteasomal degradation of autophagy components also plays a critical role in controlling autophagy termination [46, 59]. It has been shown that CUL4 controls autophagy termination by promoting Ambra1 ubiquitination and regulating Ambra1 protein levels [47]. Under high nutrient conditions, DDB1/CUL4 mediates Ambra1 ubiquitination and maintains Ambra1 at low level. Upon starvation, CUL4 dissociates with Ambra1, and Ambra1 is stabilized by Ulk1 phosphorylation. The phosphorylated Ambra1 inhibits CUL5-mediated degradation of DEPTOR and further downregulating mTOR activity. Under prolonged stress conditions, DDB1/CUL4 re-establishes its interaction with Ambra1 and promotes the ubiquitination and degradation of Amba1, which in turn leads to autophagy termination [46].

Recently, Liu et al. reported that CUL3 also participates in the termination of autophagy [59]. KLHL20/CUL3 recruited autophosphorylated Ulk1 for ubiquitination and degradation under stress conditions. Moreover, KLHL20/CUL3 also promotes ubiquitination of phagophore-residing VPS34 and Beclin1, and the ubiquitination leads to their degradation. KLHL20/CUL3 plays a crucial role in autophagy termination by regulating the turnover of Ulk1 and VPS34 complex to restrain the amplitude and duration of autophagy.

To date, many ubiquitination events have been shown to participate in autophagy regulation. The ubiquitin-mediated modification functions at different steps of autophagy and targets at different substrates in response to distinct stress stimulation. Moreover, some of the ubiquitination modifications are antagonized by the deubiquitinating enzymes. Therefore, the ubiquitin system provides flexible, diverse, and effective ways to control the onset and the termination of autophagy.

4. Ubiquitination modification and selective autophagy

Although autophagy was originally thought to be a non-selective pathway which appears to randomly sequester cytosolic components for lysosomal degradation, it is now recognized that autophagy also acts in selective processes that involves specific receptors to target certain cargos [60, 61]. Accumulating evidence indicates that many intracellular degradation events are processed through selective autophagy, including the turnover of damaged organelles such as mitochondria (mitophagy) [62, 63] and peroxisomes (pexophagy) [64, 65], removal of protein aggregates (aggrephagy) [66], and elimination of intracellular pathogens (xenophagy) [67, 68].

Upon the induction of selective autophagy, phagophore is enriched with specific cargos in a process dependent on cargo receptors [61]. These cargo receptors can interact with both target proteins and the autophagic vesicle components such as LC3/Atg8 family proteins, which result in the enclosure of selective cargos to the autophagosome and promote the autophagic degradation of cargos. Like nonselective autophagy, selective autophagy also plays an important role in cellular homeostasis and has been associated with a variety of human diseases [63, 69].

4.1. The role of ubiquitin in selective autophagy

Ubiquitination has long been recognized as a key regulator to determine protein fate by tagging proteins for proteasomal degradation [60]. Ubiquitination of cargo proteins plays a crucial role in selective autophagy process. In selective autophagy, cargos are ubiquitinated and recognized by ubiquitin-binding receptors to transport cargos for lysosomal degradation [70]. Therefore, ubiquitin acts as a degradation signal for selective autophagy. Protein aggregates, damaged organelles, or pathogens can be tagged and targeted for degradation through the lysosome machinery to maintain cellular homeostasis. In this section, we will illustrate the mechanism and importance of ubiquitination in selective autophagy (**Figure 2**).

Recent studies have shown that selective autophagy is responsible for delivering a wide range of cargos to the lysosome for degradation [70–72]; however, the detailed mechanisms of selective degradation by lysosome remain largely unknown. Several types of adaptor proteins such as p62, NDP52, optineurin (OPTN), NBR1, and HDAC6, which contain the ubiquitin-binding motif, have been reported to target ubiquitinated cargos for lysosomal degradation under stress conditions [70, 71]. Besides the ubiquitin-binding motif, these cargo receptors often also contain a LC3-interacting region (LIR) or Atg8 interaction motif to interact with the LC3/Atg8 family members [60, 70]. Therefore, through binding to ubiquitinated cargos and LC3 simultaneously, these receptors can deliver selective cargos to the autophagosome and promotes the autophagic degradation.

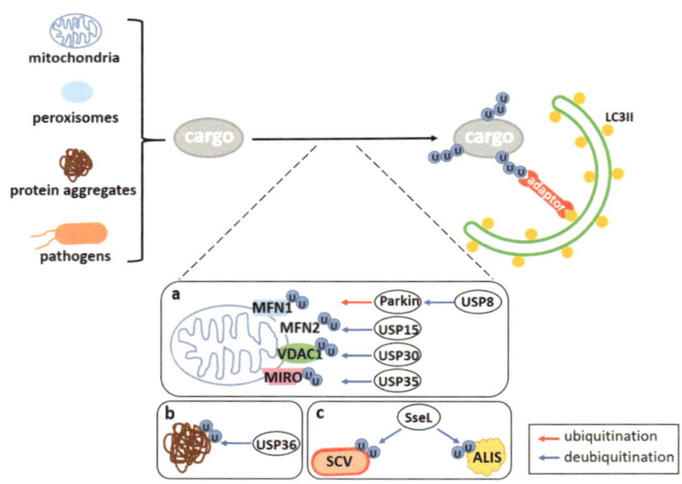

Figure 2. The involvement of ubiquitin-related enzymes (E3 ligases and DUBs) in selective autophagy. Selective autophagy is a process that depends on the ubiquitin signals and the ubiquitin recognition adaptor proteins. (a) The E3 ligase (Parkin) promotes mitophagy by catalyzing the ubiquitination of mitochondrial proteins, and DUBs (USP15, USP30, and USP35) inhibit mitophagy by removing the ubiquitin signals of mitochondrial proteins. USP8 participates in mitophagy by removing non-canonical K6-linked ubiquitin chains from Parkin, a process required for the efficient recruitment of Parkin to depolarized mitochondria. (b) USP36 removes the ubiquitin markers from protein aggregates, which inhibits aggrephagy. (c) The DUB SseL which is secreted by Salmonella can remove the ubiquitin tags on Salmonella-containing vacuole (SCV) and aggresome-like induced structures (ALIS) in order to escape xenophagy. "U" means ubiquitin (this involves different types of ubiquitination); the red arrows and the blue arrows indicate ubiquitination events and deubiquitination events, respectively.

Ubiquitinated cytosolic proteins can undergo degradation via proteasome or the lysosome. Proteins conjugated with K48-linked polyubiquitin chains often are recognized by UBD (ubiquitin-binding domain) containing proteasomal receptors and degraded by proteasome [73]. On the other hand, the ubiquitin-binding autophagy adaptors have been shown to interact with cargos containing K63-linked polyubiquitin chains [74–77]. Cargos modified with K63 polyubiquitination are preferentially targeted via the autophagy/lysosomal degradation pathway.

4.2. E3 ligases and DUBs in selective autophagy

Given that ubiquitin plays a critical role by acting as a tag for substrates recognition in selective autophagy, it is important to understand the regulatory mechanism of ubiquitin system during this process. E3 ubiquitin ligases and deubiquitiylating enzymes (DUBs) involved in the cargos ubiquitination are crucial in the selective autophagy regulation.

Autophagy of the mitochondria, also known as mitopahagy, depends on a set of ubiquitination modification on mitochondrial outer membrane proteins [63]. Upon the induction of mitophagy, Parkin, an E3 ubiquitin ligase that is also involved in the pathogenesis of Parkinson's disease, is recruited to the depolarized mitochondria [77, 78] to ubiquitinate several mitochondrial proteins, including MFN1, MFN2, VDAC1, and MIRO [79–81]. How is Parkin recruited to the damaged mitochondria? Recently, three studies showed that Pink1-mediated phosphorylation of ubiquitin at Ser65 activates Parkin [82–84]. The accumulation of ubiquitinated mitochondrial proteins then recruits the autophagy adaptors NDP52 and optineurin, which then promote the formation of mitophagy [85, 86]. Mitophagy is also modulated by a number of DUBs. It has been shown that USP15 [87], USP30 [88, 89], and USP35 [89] reduce the ubiquitin levels from the ubiquitinated mitochondrial proteins, thereby preventing the recognition by autophagy adaptors and blocking mitophagy. Moreover, TRAF6 [90] and USP8 [91] can also participate in mitophagy by regulating the ubiquitination of Parkin.

The selective degradation of protein aggregates requires aggregative proteins to be labeled with K63-linked ubiquitin chains which then are recognized by autophagy adaptors including p62 and NBR1 and HDAC6 [92]. Taillebourg et al. recently showed that DUB USP36 can act as a negative regulator to inhibit the selective autophagy of protein aggregates by removing the ubiquitin signals [93]. However, the specific ubiquitin ligases involved in aggrephagy remain to be identified.

The process of selective autophagy also plays a crucial role in host defense. It has been shown that the intracellular pathogen Salmonella Typhimurium can be eliminated by selective autophagy [94]. After infection, Salmonella Typhimurium grows in a membranous compartment, the Salmonella-containing vacuole (SCV). The bacterial infection often induces immune and nonimmune cells forming aggresome-like induced structures (ALIS). The host cell can eliminate SCV and ALIS by ubiquitination and xenophagy. However, S. Typhimurium can remove the ubiquitin signals by secreting the deubiquitinating enzyme SseL, which leads to lower autophagy flux due to the failure of autophagy-receptor recognition [94].

The ubiquitin-mediated selective autophagy plays an important role in maintaining cellular homeostasis and in the elimination of invading pathogens. Therefore, it is critical to further identify the E3 ubiquitin ligases and DUBs involved in selective autophagy under physiological and pathological conditions.

5. Ubiquitin system and autophagy in human health and disease

Like the ubiquitin-proteasome system, autophagy is a tightly regulated lysosomal degradation pathway which has been implicated in various human pathological and physiological processes. Basal autophagy is essential for removing misfolded proteins and damaged organelles; therefore, autophagy is also important for maintaining normal cellular processes in all tissues [95, 96]. Since the ubiquitin system serves as a central regulator to modify the autophagic activity and functions, it is no doubt that the protein modification by ubiquitination and deubiquitination also play crucial roles in autophagy-related diseases [97–100]. However, the relationship between the ubiquitin system and the autophagy-related pathological processes remain unclear. In this section, we will discuss the recent findings and progresses in the field.

5.1. Ubiquitin system and autophagy in cancer

Cancer is one of the first human diseases identified to be associated with autophagy malfunction [101]. Some autophagy genes mutation or deletion can lead to cancer. For instance, monoallelic deletion of Beclin1 gene has been detected in 40–75% of human breast, ovarian, and prostate cancer [102]. Besides Beclin1, many autophagy genes are found to be involved in human cancer, including UVRAG [103], Atg5, and Atg7 [104]. Accumulating evidence has indicated that autophagy also plays a crucial role in cancer cell progression. Autophagy likely plays distinct roles during different stages of cancer development [105]. It has been shown that autophagy has a preventive effect against tumorigenesis and the cancer occurrence during early cancer formation. However, autophagy provides a protective mechanism and supports the tumor growth once cancer progresses [105–107].

Although the role of autophagy in cancer progression remains elusive, several recent studies have shown that the inhibition of autophagic pathway can enhance the efficacy of anticancer drugs [108, 109]. Shao et al. showed that autophagy inhibitor-1 (spautin-1), an inhibitor of USP10 and USP13, can enhance Imatinib mesylate (IM)-induced cell death in chronic myeloid leukemia (CML) in a Beclin1-dependent manner [102]. Since autophagy plays a role in IM resistance and spautin-1 inhibits IM-induced autophagy in CML cells, inhibition of autophagy with the DUBs inhibitor spautin-1 may provide a promising approach to increase the efficacy of IM for patients with CML [110]. In another study, Yang et al. showed that knockdown of the regulator of CUL1 (ROC1) suppresses the growth of liver cancer cells through the induction of autophagy and senescence [103]. The triggering of autophagic response in ROC1 silencing cells is through the accumulation of the mTOR inhibitory protein DEPTOR [111]. Another link of ubiquitination regulation in autophagy and cancers is that the ubiquitin modification of

Beclin1 and p53 by E3 ligases or DUBs can balance the interaction between Beclin1 and p53, and their interaction is thought to regulate the cellular decision between apoptosis and autophagy in embryonal carcinoma cells [43].

As ubiquitination modification in autophagy regulation plays a notable role in cancer cells, it is crucial to further investigate the detailed mechanisms of how ubiquitin system regulates autophagic function during cancer progression. Findings from these studies will provide new insights into cancer biology as well as novel approaches in cancer prevention and treatment.

5.2. Ubiquitin system and autophagy in neurodegenerative diseases

In recent years, there is increased attention on the role of autophagy in neurodegenerative diseases, including Alzheimer's disease (AD), Parkinson's disease (PD), and Huntington's disease (HD) [112, 113]. These neurodegenerative disorders are usually characterized by the presence of protein inclusions and aggregates in neurons, which result from the failure of protein degradation, and these protein aggregates may be one cause of the progressive degeneration and/or death of neuronal cells [113, 114]. The ubiquitin-proteasome and autophagy-lysosome pathways are the two major pathways to degrade misfolded proteins and damaged organelles [60]. Accumulating evidence indicates that the dysfunction of autophagy may result in the accumulation of abnormally folded protein aggregates, which may contribute to neurodegenerative disorders [112, 114]. Although there are growing studies indicating the importance of autophagy in the neurodegenerative diseases, the molecular mechanisms of how autophagy or selective autophagy functions in these disorders are still not completely understood.

Ubiquitination of cellular proteins and organelles has been shown to promote the autophagic clearance of cargos associated with neurodegenerative diseases [76, 112, 115]. Parkin is a multifunctional ubiquitin ligase that has been found to be mutated in sporadic and familial early onset Parkinson's disease [116]. The involvement of Parkin and DUBs such as USP15, USP30, and USP35 in mitophagy has also been demonstrated to be critical in neurodegeneration related to PD [43, 116]. Besides its role in regulating mitochondrial homeostasis, it has been shown that Parkin catalyzes the ubiquitination modification of misfolded proteins, which then promotes the degradation of these substrates via proteasome or autophagy pathway [116]. The dysfunction of Parkin leads to the accumulation of protein aggregates and causes some neurodegenerative diseases. It was also reported that the C-terminus of Hsc70 interacting protein (CHIP) E3 ubiquitin ligase can promote the ubiquitination of denatured proteins and play an important role in neurodegeneration. By binding to different E2 enzymes, CHIP can catalyze K48- and K63-linked polyubiquitin chains which promote proteins degradation via chaperone-dependent ubiquitin-proteasome system and autophagy-lysosome pathway, respectively [117].

5.3. Ubiquitin system and autophagy in infectious diseases

Activation of autophagy provides a promising approach in the treatment of infectious diseases. Recent studies have shown that this cellular process can either selectively target microorganisms for lysosomal degradation (referred to as xenophagy) or promote the delivery of microbial nucleic acids and antigens to endo/lysosomal compartments for innate and adaptive immunity activation [118–120]. Accumulating evidence indicates that autophagy activity is higher upon the pathogen infection [119], and it is also known that autophagy can facilitate the intracellular antigen-processing events [118]. Moreover, the autophagy pathway can cross talk with immunity pathways [50, 120]. Interestingly, several reports also indicate that autophagy may provide a pathway for pathogens to escape from host defense and help them to invade host tissues [97].

The role of autophagy in immunity has been further confirmed by the finding of the connections between autophagy and several immune diseases. For example, Atg16 mutations are associated with increased risk of an inflammatory bowel disease, Crohn disease, which affects the gastrointestinal tract from the mouth to the anus [98]. Moreover, several studies have linked single-nucleotide polymorphisms (SNPs) in ATG5 to systemic lupus erythematosus (SLE) susceptibility [119]. Ubiquitination regulation of autophagy regulators was also found to participate in the infectious events. The ubiquitin E3 ligase TRAF6 has been shown to catalyze the K63-linked polyubiquitination of Beclin1 upon LPS stimulation and is critical for TLR4-triggered autophagy in macrophages [50]. And the deubiquitinating enzyme A20 antagonizes TRAF6-mediated ubiquitination of Beclin1 and limits the induction of autophagy in response to TLR signaling. The balanced activity of TRAF6 and A20 is required for the inflammatory response [50]. Another indication that ubiquitination modification regulates inflammation through autophagy involves the E3 ligase RNF216 (ring finger protein 216). RNF216 was reported to inhibit autophagy in macrophages by catalyzing K48-linked polyubiquitination of Beclin1, which induces the degradation of Beclin1 [121]. Manipulating RNF216 expression may provide a therapeutic approach for treatment of inflammatory diseases. In addition, the ubiquitination modification processes participated in xenophagy also plays important roles for the bacterial infection. The DUB SseL regulates the ubiquitin modification of SCVs and ALIS and is important for the removal of pathogens [115]. Kuang et al. showed that RNF5 promotes the ubiquitination and the degradation of Atg4b limits the basal levels of autophagy and influences susceptibility to bacterial infection [58].

In addition to the diseases discussed earlier, there are also disorders related to autophagy dysfunction, including developmental defect, muscle atrophy, heart diseases, liver disease, and aging [95]. However, it remains unclear whether the ubiquitin system also plays a role in these diseases.

6. Conclusion

Protein ubiquitination is considered as one of the most important reversible posttranslational modifications and has been implicates in various cellular signaling processes. Increasing evidence indicates that the ubiquitin system plays a pivotal role in the regulation of autophagy pathway. Recent studies have explored and highlighted the important functions of ubiquitin system in the pathogenesis of autophagy-related diseases such as tumorigenesis, neurodegeneration, and pathogen infection. Further investigations to identify novel E3 ligases and DUBs involved in autophagy and to determine their underlying mechanisms will not only contribute to our understanding on how autophagy is controlled by the ubiquitin system but also provide a rationale for novel therapeutic interventions in autophagy-related diseases.

Acknowledgements

This work was supported by grants from the Ministry of Science and Technology of Taiwan and the Academia Sinica Career Development Award (101CDA-L04).

Author details

Yi-Ting Wang[2] and Guang-Chao Chen[1,2*]

*Address all correspondence to: gcchen@gate.sinica.edu.tw

1 Institute of Biological Chemistry, Academia Sinica, Taipei, Taiwan

2 Institute of Biochemical Sciences, National Taiwan University, Taipei, Taiwan

References

[1] Ding, W.X. and X.M. Yin, Sorting, recognition and activation of the misfolded protein degradation pathways through macroautophagy and the proteasome. Autophagy, 2008. 4(2): p. 141–50.

[2] Wang, C. and X. Wang, The interplay between autophagy and the ubiquitin-proteasome system in cardiac proteotoxicity. Biochim Biophys Acta, 2015. 1852(2): p. 188–94.

[3] Klionsky, D.J., Autophagy: from phenomenology to molecular understanding in less than a decade. Nat Rev Mol Cell Biol, 2007. 8(11): p. 931–37.

[4] Klionsky, D.J., et al., Guidelines for the use and interpretation of assays for monitoring autophagy (3rd edition). Autophagy, 2016. 12(1): p. 1–222.

[5] Kroemer, G., G. Marino, and B. Levine, Autophagy and the integrated stress response. Mol Cell, 2010. 40(2): p. 280–93.

[6] Choi, A.M., S.W. Ryter, and B. Levine, Autophagy in human health and disease. N Engl J Med, 2013. 368(7): p. 651–62.

[7] Murrow, L. and J. Debnath, Autophagy as a stress-response and quality-control mechanism: implications for cell injury and human disease. Annu Rev Pathol, 2013. 8: p. 105–37.

[8] Feng, Y., et al., The machinery of macroautophagy. Cell Res, 2014. 24(1): p. 24–41.

[9] Xie, Z. and D.J. Klionsky, Autophagosome formation: core machinery and adaptations. Nat Cell Biol, 2007. 9(10): p. 1102–09.

[10] Jensen, O.N., Modification-specific proteomics: characterization of post-translational modifications by mass spectrometry. Curr Opin Chem Biol, 2004. 8(1): p. 33–41.

[11] Mann, M. and O.N. Jensen, Proteomic analysis of post-translational modifications. Nat Biotechnol, 2003. 21(3): p. 255–61.

[12] Deribe, Y.L., T. Pawson, and I. Dikic, Post-translational modifications in signal integration. Nat Struct Mol Biol, 2010. 17(6): p. 666–72.

[13] Khoury, G.A., R.C. Baliban, and C.A. Floudas, Proteome-wide post-translational modification statistics: frequency analysis and curation of the swiss-prot database. Sci Rep, 2011. 1, 90: p.1–5.

[14] McEwan, D.G. and I. Dikic, The three musketeers of autophagy: phosphorylation, ubiquitylation and acetylation. Trends Cell Biol, 2011. 21(4): p. 195–201.

[15] Xie, Y., et al., Posttranslational modification of autophagy-related proteins in macroautophagy. Autophagy, 2015. 11(1): p. 28–45.

[16] Wani, W.Y., et al., Regulation of autophagy by protein post-translational modification. Lab Invest, 2015. 95(1): p. 14–25.

[17] Kijanska, M., et al., Activation of Atg1 kinase in autophagy by regulated phosphorylation. Autophagy, 2010. 6(8): p. 1168–78.

[18] Yeh, Y.Y., K. Wrasman, and P.K. Herman, Autophosphorylation within the Atg1 activation loop is required for both kinase activity and the induction of autophagy in *Saccharomyces cerevisiae*. Genetics, 2010. 185(3): p. 871–82.

[19] Kim, J., et al., AMPK and mTOR regulate autophagy through direct phosphorylation of Ulk1. Nat Cell Biol, 2011. 13(2): p. 132–41.

[20] Lin, S.Y., et al., GSK3-TIP60-ULK1 signaling pathway links growth factor deprivation to autophagy. Science, 2012. 336(6080): p. 477–81.

[21] Abrahamsen, H., H. Stenmark, and H.W. Platta, Ubiquitination and phosphorylation of Beclin 1 and its binding partners: Tuning class III phosphatidylinositol 3-kinase activity and tumor suppression. FEBS Lett, 2012. 586(11): p. 1584–91.

[22] McKnight, N.C. and Y. Zhenyu, Beclin 1, an essential component and master regulator of PI3K-III in health and disease. Curr Pathobiol Rep, 2013. 1(4): p. 231–38.

[23] Zalckvar, E., et al., DAP-kinase-mediated phosphorylation on the BH3 domain of beclin 1 promotes dissociation of beclin 1 from Bcl-XL and induction of autophagy. EMBO Rep, 2009. 10(3): p. 285–92.

[24] Furuya, T., et al., Negative regulation of Vps34 by Cdk mediated phosphorylation. Mol Cell, 2010. 38(4): p. 500–11.

[25] Pickart, C.M. and M.J. Eddins, Ubiquitin: structures, functions, mechanisms. Biochim Biophys Acta, 2004. 1695(1–3): p. 55–72.

[26] Pickart, C.M., Mechanisms underlying ubiquitination. Annu Rev Biochem, 2001. 70: p. 503–33.

[27] Komander, D. and M. Rape, The ubiquitin code. Annu Rev Biochem, 2012. 81: p. 203–29.

[28] Xu, G. and S.R. Jaffrey, The new landscape of protein ubiquitination. Nat Biotechnol, 2011. 29(12): p. 1098–100.

[29] Sadowski, M., et al., Protein monoubiquitination and polyubiquitination generate structural diversity to control distinct biological processes. IUBMB Life, 2012. 64(2): p. 136–42.

[30] Schnell, J.D. and L. Hicke, Non-traditional functions of ubiquitin and ubiquitin-binding proteins. J Biol Chem, 2003. 278(38): p. 35857–60.

[31] Sadowski, M. and B. Sarcevic, Mechanisms of mono- and poly-ubiquitination: ubiquitination specificity depends on compatibility between the E2 catalytic core and amino acid residues proximal to the lysine. Cell Div, 2010. 5: p. 19.

[32] Ashizawa, A., et al., The ubiquitin system and Kaposi's sarcoma-associated herpesvirus. Front Microbiol, 2012. 3: p. 66.

[33] Jin, L., et al., Mechanism of ubiquitin-chain formation by the human anaphase-promoting complex. Cell, 2008. 133(4): p. 653–65.

[34] Matsumoto, M.L., et al., K11-linked polyubiquitination in cell cycle control revealed by a K11 linkage-specific antibody. Mol Cell, 2010. 39(3): p. 477–84.

[35] Passmore, L.A. and D. Barford, Getting into position: the catalytic mechanisms of protein ubiquitylation. Biochem J, 2004. 379(Pt 3): p. 513–25.

[36] Xia, Z.P., et al., Direct activation of protein kinases by unanchored polyubiquitin chains. Nature, 2009. 461(7260): p. 114–19.

[37] Hou, F., et al., MAVS forms functional prion-like aggregates to activate and propagate antiviral innate immune response. Cell, 2011. 146(3): p. 448–61.

[38] Reyes-Turcu, F.E., K.H. Ventii, and K.D. Wilkinson, Regulation and cellular roles of ubiquitin-specific deubiquitinating enzymes. Annu Rev Biochem, 2009. 78: p. 363–97.

[39] Amerik, A.Y. and M. Hochstrasser, Mechanism and function of deubiquitinating enzymes. Biochim Biophys Acta, 2004. 1695(1–3): p. 189–207.

[40] Komander, D., M.J. Clague, and S. Urbe, Breaking the chains: structure and function of the deubiquitinases. Nat Rev Mol Cell Biol, 2009. 10(8): p. 550–63.

[41] Nijman, S.M., et al., A genomic and functional inventory of deubiquitinating enzymes. Cell, 2005. 123(5): p. 773–86.

[42] Kuang, E., J. Qi, and Z. Ronai, Emerging roles of E3 ubiquitin ligases in autophagy. Trends Biochem Sci, 2013. 38(9): p. 453–60.

[43] Magraoui, F.E., et al., Autophagy-related deubiquitinating enzymes involved in health and disease. Cells, 2015. 4(4): p. 596–621.

[44] Zhao, Y., X. Xiong, and Y. Sun, DEPTOR, an mTOR inhibitor, is a physiological substrate of SCF(betaTrCP) E3 ubiquitin ligase and regulates survival and autophagy. Mol Cell, 2011. 44(2): p. 304–16.

[45] Gao, D., et al., mTOR drives its own activation via SCF(betaTrCP)-dependent degradation of the mTOR inhibitor DEPTOR. Mol Cell, 2011. 44(2): p. 290–303.

[46] Antonioli, M., et al., AMBRA1 interplay with cullin E3 ubiquitin ligases regulates autophagy dynamics. Dev Cell, 2014. 31(6): p. 734–46.

[47] McEwan, D.G. and I. Dikic, Cullins keep autophagy under control. Dev Cell, 2014. 31(6): p. 675–76.

[48] Linares, J.F., et al., K63 polyubiquitination and activation of mTOR by the p62-TRAF6 complex in nutrient-activated cells. Mol Cell, 2013. 51(3): p. 283–96.

[49] Nazio, F., et al., mTOR inhibits autophagy by controlling ULK1 ubiquitylation, self-association and function through AMBRA1 and TRAF6. Nat Cell Biol, 2013. 15(4): p. 406–16.

[50] Shi, C.S. and J.H. Kehrl, TRAF6 and A20 regulate lysine 63-linked ubiquitination of Beclin-1 to control TLR4-induced autophagy. Sci Signal, 2010. 3(123): p. ra42.

[51] Chen, D., et al., Parkin mono-ubiquitinates Bcl-2 and regulates autophagy. J Biol Chem, 2010. 285(49): p. 38214–23.

[52] Platta, H.W., et al., Nedd4-dependent lysine-11-linked polyubiquitination of the tumour suppressor Beclin 1. Biochem J, 2012. 441(1): p. 399–406.

[53] Liu, J., et al., Beclin1 controls the levels of p53 by regulating the deubiquitination activity of USP10 and USP13. Cell, 2011. 147(1): p. 223–34.

[54] Yuan, J., et al., USP10 regulates p53 localization and stability by deubiquitinating p53. Cell, 2010. 140(3): p. 384–96.

[55] Jin, S., et al., USP19 modulates autophagy and antiviral immune responses by deubiquitinating Beclin-1. EMBO J, 2016.

[56] Simicek, M., et al., The deubiquitylase USP33 discriminates between RALB functions in autophagy and innate immune response. Nat Cell Biol, 2013. 15(10): p. 1220–30.

[57] Chen, Y. and D.J. Klionsky, The regulation of autophagy-unanswered questions. J Cell Sci, 2011. 124(Pt 2): p. 161–70.

[58] Kuang, E., et al., Regulation of ATG4B stability by RNF5 limits basal levels of autophagy and influences susceptibility to bacterial infection. PLoS Genet, 2012. 8(10): p. e1003007.

[59] Liu, C.C., et al., Cul3-KLHL20 ubiquitin ligase governs the turnover of ULK1 and VPS34 complexes to control autophagy termination. Mol Cell, 2016. 61(1): p. 84–97.

[60] Schreiber, A. and M. Peter, Substrate recognition in selective autophagy and the ubiquitin-proteasome system. Biochim Biophys Acta, 2014. 1843(1): p. 163–81.

[61] Svenning, S. and T. Johansen, Selective autophagy. Essays Biochem, 2013. 55: p. 79–92.

[62] Novak, I., et al., Nix is a selective autophagy receptor for mitochondrial clearance. EMBO Rep, 2010. 11(1): p. 45–51.

[63] Kubli, D.A. and A.B. Gustafsson, Mitochondria and mitophagy: the yin and yang of cell death control. Circ Res, 2012. 111(9): p. 1208–21.

[64] Farre, J.C., et al., PpAtg30 tags peroxisomes for turnover by selective autophagy. Dev Cell, 2008. 14(3): p. 365–76.

[65] Till, A., et al., Pexophagy: the selective degradation of peroxisomes. Int J Cell Biol, 2012. 2012: p. 512721.

[66] Pankiv, S., et al., p62/SQSTM1 binds directly to Atg8/LC3 to facilitate degradation of ubiquitinated protein aggregates by autophagy. J Biol Chem, 2007. 282(33): p. 24131–45.

[67] Thurston, T.L., et al., Galectin 8 targets damaged vesicles for autophagy to defend cells against bacterial invasion. Nature, 2012. 482(7385): p. 414–18.

[68] Watson, R.O., P.S. Manzanillo, and J.S. Cox, Extracellular M. tuberculosis DNA targets bacteria for autophagy by activating the host DNA-sensing pathway. Cell, 2012. 150(4): p. 803–15.

[69] Mizumura, K., A.M. Choi, and S.W. Ryter, Emerging role of selective autophagy in human diseases. Front Pharmacol, 2014. 5: p. 244.

[70] Shaid, S., et al., Ubiquitination and selective autophagy. Cell Death Differ, 2013. 20(1): p. 21–30.

[71] Kirkin, V., et al., A role for ubiquitin in selective autophagy. Mol Cell, 2009. 34(3): p. 259–69.

[72] Kim, P.K., et al., Ubiquitin signals autophagic degradation of cytosolic proteins and peroxisomes. Proc Natl Acad Sci U S A, 2008. 105(52): p. 20567–74.

[73] Welchman, R.L., C. Gordon, and R.J. Mayer, Ubiquitin and ubiquitin-like proteins as multifunctional signals. Nat Rev Mol Cell Biol, 2005. 6(8): p. 599–609.

[74] van Wijk, S.J., et al., Fluorescence-based sensors to monitor localization and functions of linear and K63-linked ubiquitin chains in cells. Mol Cell, 2012. 47(5): p. 797–809.

[75] Lamark, T. and T. Johansen, Aggrephagy: selective disposal of protein aggregates by macroautophagy. Int J Cell Biol, 2012. 2012: p. 736905.

[76] Tan, J.M., et al., Lysine 63-linked ubiquitination promotes the formation and autophagic clearance of protein inclusions associated with neurodegenerative diseases. Hum Mol Genet, 2008. 17(3): p. 431–39.

[77] Jin, S.M. and R.J. Youle, PINK1- and Parkin-mediated mitophagy at a glance. J Cell Sci, 2012. 125(Pt 4): p. 795–99.

[78] Vives-Bauza, C., et al., PINK1-dependent recruitment of Parkin to mitochondria in mitophagy. Proc Natl Acad Sci U S A, 2010. 107(1): p. 378–83.

[79] Gegg, M.E., et al., Mitofusin 1 and mitofusin 2 are ubiquitinated in a PINK1/parkin-dependent manner upon induction of mitophagy. Hum Mol Genet, 2010. 19(24): p. 4861–70.

[80] Geisler, S., et al., PINK1/Parkin-mediated mitophagy is dependent on VDAC1 and p62/SQSTM1. Nat Cell Biol, 2010. 12(2): p. 119–31.

[81] Chan, N.C., et al., Broad activation of the ubiquitin-proteasome system by Parkin is critical for mitophagy. Hum Mol Genet, 2011. 20(9): p. 1726–37.

[82] Koyano, F., et al., Ubiquitin is phosphorylated by PINK1 to activate parkin. Nature, 2014. 510(7503): p. 162–66.

[83] Kane, L.A., et al., PINK1 phosphorylates ubiquitin to activate Parkin E3 ubiquitin ligase activity. J Cell Biol, 2014. 205(2): p. 143–53.

[84] Kazlauskaite, A., et al., Parkin is activated by PINK1-dependent phosphorylation of ubiquitin at Ser65. Biochem J, 2014. 460(1): p. 127–39.

[85] Lazarou, M., et al., The ubiquitin kinase PINK1 recruits autophagy receptors to induce mitophagy. Nature, 2015. 524(7565): p. 309–14.

[86] Heo, J.M., et al., The PINK1-PARKIN mitochondrial ubiquitylation pathway drives a program of OPTN/NDP52 recruitment and TBK1 activation to promote mitophagy. Mol Cell, 2015. 60(1): p. 7–20.

[87] Cornelissen, T., et al., The deubiquitinase USP15 antagonizes Parkin-mediated mitochondrial ubiquitination and mitophagy. Hum Mol Genet, 2014. 23(19): p. 5227–42.

[88] Bingol, B., et al., The mitochondrial deubiquitinase USP30 opposes parkin-mediated mitophagy. Nature, 2014. 510(7505): p. 370–75.

[89] Wang, Y., et al., Deubiquitinating enzymes regulate PARK2-mediated mitophagy. Autophagy, 2015. 11(4): p. 595–606.

[90] Murata, H., et al., SARM1 and TRAF6 bind to and stabilize PINK1 on depolarized mitochondria. Mol Biol Cell, 2013. 24(18): p. 2772–84.

[91] Durcan, T.M., et al., USP8 regulates mitophagy by removing K6-linked ubiquitin conjugates from parkin. EMBO J, 2014. 33(21): p. 2473–91.

[92] Knaevelsrud, H. and A. Simonsen, Fighting disease by selective autophagy of aggregate-prone proteins. FEBS Lett, 2010. 584(12): p. 2635–45.

[93] Taillebourg, E., et al., The deubiquitinating enzyme USP36 controls selective autophagy activation by ubiquitinated proteins. Autophagy, 2012. 8(5): p. 767–79.

[94] Mesquita, F.S., et al., The Salmonella deubiquitinase SseL inhibits selective autophagy of cytosolic aggregates. PLoS Pathog, 2012. 8(6): p. e1002743.

[95] Klionsky, D.J., The autophagy connection. Dev Cell, 2010. 19(1): p. 11–2.

[96] Badadani, M., Autophagy mechanism, regulation, functions, and disorders. ISRN Cell Biol, 2012. 2012: p. 11.

[97] Cheng, Y., et al., Therapeutic targeting of autophagy in disease: biology and pharmacology. Pharmacol Rev, 2013. 65(4): p. 1162–97.

[98] Huang, J. and D.J. Klionsky, Autophagy and human disease. Cell Cycle, 2007. 6(15): p. 1837–49.

[99] Klionsky, D.J. and S.D. Emr, Autophagy as a regulated pathway of cellular degradation. Science, 2000. 290(5497): p. 1717–21.

[100] Mizushima, N., et al., Autophagy fights disease through cellular self-digestion. Nature, 2008. 451(7182): p. 1069–75.

[101] Kenific, C.M. and J. Debnath, Cellular and metabolic functions for autophagy in cancer cells. Trends Cell Biol, 2015. 25(1): p. 37–45.

[102] Qu, X., et al., Promotion of tumorigenesis by heterozygous disruption of the beclin 1 autophagy gene. J Clin Invest, 2003. 112(12): p. 1809–20.

[103] Liang, C., et al., Autophagic and tumour suppressor activity of a novel Beclin1-binding protein UVRAG. Nat Cell Biol, 2006. 8(7): p. 688–99.

[104] Takamura, A., et al., Autophagy-deficient mice develop multiple liver tumors. Genes Dev, 2011. 25(8): p. 795–800.

[105] White, E. and R.S. DiPaola, The double-edged sword of autophagy modulation in cancer. Clin Cancer Res, 2009. 15(17): p. 5308–16.

[106] Guo, J.Y., B. Xia, and E. White, Autophagy-mediated tumor promotion. Cell, 2013. 155(6): p. 1216–19.

[107] White, E., Deconvoluting the context-dependent role for autophagy in cancer. Nat Rev Cancer, 2012. 12(6): p. 401–10.

[108] Abedin, M.J., et al., Autophagy delays apoptotic death in breast cancer cells following DNA damage. Cell Death Differ, 2007. 14(3): p. 500–10.

[109] Janku, F., et al., Autophagy as a target for anticancer therapy. Nat Rev Clin Oncol, 2011. 8(9): p. 528–39.

[110] Shao, S., et al., Spautin-1, a novel autophagy inhibitor, enhances imatinib-induced apoptosis in chronic myeloid leukemia. Int J Oncol, 2014. 44(5): p. 1661–68.

[111] Yang, D., et al., Induction of autophagy and senescence by knockdown of ROC1 E3 ubiquitin ligase to suppress the growth of liver cancer cells. Cell Death Differ, 2013. 20(2): p. 235–47.

[112] Nixon, R.A., The role of autophagy in neurodegenerative disease. Nat Med, 2013. 19(8): p. 983–97.

[113] Pan, T., et al., The role of autophagy-lysosome pathway in neurodegeneration associated with Parkinson's disease. Brain, 2008. 131(Pt 8): p. 1969–78.

[114] Rubinsztein, D.C., The roles of intracellular protein-degradation pathways in neurodegeneration. Nature, 2006. 443(7113): p. 780–86.

[115] Yamano, K., N. Matsuda, and K. Tanaka, The ubiquitin signal and autophagy: an orchestrated dance leading to mitochondrial degradation. EMBO Rep, 2016. 17(3): p. 300–16.

[116] Zhang, C.W., et al., Parkin regulation and neurodegenerative disorders. Front Aging Neurosci, 2015. 7: p. 248.

[117] Ciechanover, A. and Y.T. Kwon, Degradation of misfolded proteins in neurodegenerative diseases: therapeutic targets and strategies. Exp Mol Med, 2015. 47: p. e147.

[118] Rubinsztein, D.C., P. Codogno, and B. Levine, Autophagy modulation as a potential therapeutic target for diverse diseases. Nat Rev Drug Discov, 2012. 11(9): p. 709–30.

[119] Levine, B., N. Mizushima, and H.W. Virgin, Autophagy in immunity and inflammation. Nature, 2011. 469(7330): p. 323–35.

[120] Shibutani, S.T., et al., Autophagy and autophagy-related proteins in the immune system. Nat Immunol, 2015. 16(10): p. 1014–24.

[121] Xu, C., et al., Regulation of autophagy by E3 ubiquitin ligase RNF216 through BECN1 ubiquitination. Autophagy, 2014. 10(12): p. 2239–50.

Section 3

Autophagy and biological barriers: defense, offense, survival or death

Chapter 8

HIV-1, Drug Addiction, and Autophagy

Ming-Lei Guo, Palsamy Periyasamy and Shilpa Buch

Additional information is available at the end of the chapter

http://dx.doi.org/10.5772/63980

Abstract

Despite the dramatic success of combined antiretroviral therapies (cART) in controlling peripheral virus replication, the prevalence of HIV-1-associated neurocognitive disorders (HAND) is on a rise as infected individuals continue to live longer. Almost half of the infected individuals on ART develop HAND, out of which at least 30% suffer from a comorbid condition of substance abuse. Involvement of autophagy has been implicated not only in HIV-1 infection of the CNS but also in CNS cells exposed to drugs such as amphetamine, opiates, and cocaine, contributing in turn, to cellular dysfunction. HIV-1 is known to interfere with the autophagy pathway, resulting in turn to upregulation of HIV-1 replication. Specifically, different HIV-1 proteins such as TAT, gp120, and Nef have been shown to act on various stages of autophagy such as initiation and maturation and to affect overall autophagy levels. Whether or not abused drugs and HIV-1 can cooperate to dysregulate autophagy, however, remains unclear. This chapter is focused on identifying the molecular mechanism(s) underlying HIV-1 (proteins) and cocaine, opiate, methamphetamine-mediated impairment of autophagy. Such effects could underlie the synergistic effects of HIV-1 and abused drugs in exacerbating symptoms of HAND.

Keywords: HIV-1, TAT, gp120, drug addiction, cART, autophagy

1. Introduction

Since the advent of combined antiretroviral therapy (cART), human immunodeficiency virus (HIV-1) infection has transformed into more manageable and controllable chronic disease analogues to diabetes [1]. HIV-1-infected individuals on cART are living longer compared to HIV-1-seronegative controls. Despite the dramatic success of cART in controlling viremia and increased longevity, various comorbidities involving multiple organs including the brain are on a rise in the infected individuals [2, 3]. In the context of CNS, there is an increased prevalence of HIV-1-associated neurocognitive disorders (HAND) due to the extended life expectancy. HAND is composed of various entities of neurocognitive impairments ranging from asymptomatic to milder forms of motor disorders. Increased activation of microglia and astrocytes (neuroinflammation) and synaptodendritic injury are the emerging hallmark features of HAND. Although multiple factors and pathways have been proposed in the development of HAND, the exact molecular mechanisms underlying the pathogenesis of HAND in the era of cART remain elusive.

Autophagy is a highly conserved cellular process that is present in all eukaryotic cells. Autophagy allows the orderly degradation and recycling of cellular components [4]. During this process, targeted cytoplasmic constituents are isolated from the rest of the cell within a double-membraned vesicle known as the autophagosome. The autophagosome fuses with a lysosome and the contents are degraded and recycled. Autophagy can be broadly divided into several stages: initiation, elongation, autophagosome formation, and maturation. Each stage is tightly and sequentially regulated by multiple autophagy-related proteins, to ensure the completion of whole process (autophagy flux).

It is well recognized that drug abuse serves as a significant risk factor for acquiring HIV-1 infection and has been implicated in worsening the symptoms of HAND [5, 6]. HIV-1-positive individuals with a history of abused drug exposure exhibit severe cognitive and behavioral dysfunction compared to those without exposure to abused drugs. Clinical evidence points to increased neuroinflammation, severe neuronal injury, and increased viral loads in the brains of HIV-1-positive patients with a history of drug addiction compared to the brains of infected individuals that were drug naive [5, 6]. Abused drugs including cocaine, methamphetamine, and opiates have been shown to interact with the autophagy pathway in various kinds of cells including microglia, astrocytes, and neurons to alter their homeostasis. Dysfunction of these three cell types plays critical roles in the pathogenesis of HAND.

In this chapter, we focus on the intermingled relationship between HIV-1, abused drugs, and autophagy. HIV-1/HIV-1 proteins and drugs of abuse, that have been shown to interact with various stages of autophagy pathway, can synergistically lead to autophagy dysregulation, which, in turn, contributes to accelerated pathogenesis of HAND. Understanding and clarification of the complex interplay of HIV-1, abused drugs and HAND could set a stage for future development of novel target(s) as alternative therapeutic approaches to ameliorate HAND in HIV-1-infected individuals on cART.

2. HIV-1, HAND, and cART

In 1983, HIV-1 was first isolated and identified as a lentivirus (a subgroup of retrovirus) that was the causative agent for acquired immune deficiency syndrome (AIDS). In the pre-antiretroviral therapy era, the average survival time following infection with HIV-1 was estimated to be 9–11 years, depending on the HIV-1 subtype [7]. In 1996, the introduction of combined antiretroviral therapy (cART) changed the course of the HIV-1 epidemic. In the current era of cART, HIV-1 infection has become a more controllable and manageable disease, with increased patient longevity [1, 8]. Clinical studies have shown that in patients with well-controlled viremia, the CD4 numbers are relatively maintained in a normal range (400–1600/mm^3) and also that the function of immune system is well preserved with extended periods of undetectable viral loads.

HAND is composed of neurological complications that range from asymptomatic neurocognitive impairment (ANI) to milder forms of cognitive impairment (MNI). While severe complications of CNS such as HIV-associated dementia (HAD) have declined in the post-cART era, emergence of milder cognitive motor disorders is actually on rise. In fact, as HIV-1-infected individuals continue to live longer lives, it is estimated that up to 50% of those individuals still continue to display varying degrees of neurocognitive impairment, which could impact their quality of life, leading to increased healthcare burden.

Various theories are extant about the mediators that regulate pathogenesis of HAND. For example, it has been implicated that direct injury caused by HIV-1 and its associated proteins such as the envelope gp120, transactivator of transcription (TAT), and Vpr can mediate damage in the CNS. Furthermore, indirect damage caused by proinflammatory cytokines/chemokines and chronic, sustained immune activation in the CNS can all play key roles in the pathogenesis and progression of HAND [9, 10]. It is now well accepted that events secondary to HIV-1 infection (inflammation) are critical for CNS damage associated with HAND [10]. Other factors have also been shown to contribute to HAND development. For example, factors such as inefficient CNS penetration of cART and/or emergence of latent virus reservoirs can also be attributable factors to the pathogenesis of HAND [11, 12]. Additionally, it is now becoming well recognized that HAND is prevalent in infected individuals abusing recreational drugs. In fact, substance abuse and co-infection with hepatitis C are recognized comorbidities of HIV infection. In summary, while treatment with cART has reduced the incidence of the severe form of CNS impairment such as HAD, low-level ongoing inflammation in the periphery and in the CNS and HAND is emerging as newer comorbidity of HIV infection.

3. HIV-1 and autophagy

Autophagy and the innate immune responses are highly conserved evolutionarily processes in virtually all the eukaryotic cells [4]. Autophagy is an important player in various diseases such as cancer and neurodegenerative disorders [4]. At the cellular level, HIV-1 can infect dendritic cells (DC), monocytes, and CD4$^+$ T cells in the periphery and microglia and astrocytes (limited infection) in the CNS. Depending on the cell type, HIV-1 can either usurp certain autophagy-related proteins or block autophagy flux to invade the host immune response, thereby exacerbating HIV-1 replication and transmission [13].

In DCs, HIV-1 impairs autophagy to subvert host immunity and enhance trans-infection. The efficiency of DCs being infected by HIV-1 heavily depends on their origin and activation state [14]. Previous reports have demonstrated that immature, resting DCs are more readily infected with HIV-1 than mature, activated DCs that are induced either by TNF-α or by poly I:C [14, 15]. Of note, while immature DCs seem to be more easily infected with HIV-1, mature DCs are more efficient at mediating trans-infection to CD4⁺ T cells without themselves getting infected [14]. The idea that autophagy is a mechanism that protects against HIV-1 infection originated from the fact that treatment of DCs with either TNF-α or poly (I:C) led to the activation and upregulation of autophagy suggests thereby that elevated levels of autophagy degrade the incoming virions in mature DCs [15]. Another study indicated that viral Env downregulated autophagy through mTOR activation, thereby protecting the virus from autophagy-mediated degradation, ultimately leading to increased trans-infection to $CD4^+$ T cells [16]. These results reveal a mechanism by which HIV-1 limits autophagy in DCs to target immune function leading, in turn, to increased transmission.

HIV-1 is known to infect macrophages, without a negligible change in cell numbers [17]. The role of autophagy in viral replication in the macrophages, however, remains poorly explored. In one study, it was shown that autophagic vacuoles were increased in infected macrophages. The mechanism underlying this phenomenon involved inhibition of the maturation stage of autophagy by HIV-1 factor Nef, via its association with the autophagy regulator Beclin-1 [18]. Thus, HIV-1 can exploit early stages of autophagic signaling in the macrophages for biogenesis and egress, while concomitantly also inhibiting the maturation stages of autophagy to prevent its own degradation. One of the key hallmark features of HIV-1 infection is a significant depletion of $CD4^+$ T cell in patients that are treatment naive [19, 20]. Previous investigations have addressed the effects of HIV-1 protein Env on bystander T-cell death [20]. Although autophagic vacuoles (AV) were increased in these bystander T cells, the exact role of autophagy contributing to cell death still remains elusive. Since the autophagic process involves the capture and degradation of activated inflammasome complexes, increased numbers of AV in the bystander $CD4^+$ T cells could thus imply an attempt by the cell to upregulate autophagy as part of an anti-inflammatory response. The failure of autophagy to protect these bystander cells from inflammasome activation could eventually lead to cell death. Thus, while autophagy has no direct role in $CD4^+$ T-cell death, it can be envisioned as a secondary, compensatory mechanism that is an attempt at salvaging the cells from apoptosis [21].

Microglia is the primary immune-competent cells of the brain. Similar to peripheral immune cells, HIV-1 and viral proteins also modulate autophagy levels in microglia. A recent study has demonstrated that expression levels of autophagy markers such as Beclin-1, ATG5, LC3-II, and p62 were significantly altered in microglia from HIV-1-infected individuals with NCI ± HIV-1 encephalitis (HIVE) [22]. Autophagy dysregulation could thus be associated with the microglial activation status. Another investigation provided a more direct evidence focusing on the role of HIV-1 infection on microglial autophagy [23]. In this study, it was shown that in primary human microglia infected with macrophage-tropic HIV-1SF162 strain, there was increased protein expression of Beclin-1 and LC3-II. In these infected cells, accumulation of LC3 reporter RFP^+ GFP^+ (yellow) puncta suggested that HIV-1 infection triggered autophagosome formation without promoting protein degradation by the lysosomes. These findings imply that HIV-1 can usurp autophagy pathway to promote its own replication. Increasing autophagy flux could thus serve as a potential therapeutic approach against HIV-1 infection in microglia. Similar to microglia, dysregulation of autophagy in HIV-1-infected astrocytes has also been reported [24].

http://dx.doi.org/10.5772/63980

Unlike the microglia and astrocytes, HIV-1 does not infect the neurons, primarily due to lack of CD4⁺ receptor in these latter cells. Neurons, however, are susceptible to the toxic effects of HIV-1 proteins TAT and gp120, which have been shown to dysregulate autophagy, leading, in turn, to neurotoxicity and/or neuronal injury including synaptodendritic injury. A seminal report [25] investigated the role of autophagy in microglia-induced neurotoxicity in primary rodent neurons. These authors demonstrated that conditioned media from simian immunodeficiency virus (SIV)-infected microglia inhibited autophagy in rodent neurons, leading to decreased neuronal survival. It is likely that a combination of HIV-1 proteins (TAT/gp120) and/or multiple inflammatory mediators released from the infected cells could be contributing to the dysregulation of autophagy. Another study has provided evidence that HIV-1 protein TAT can impair the endolysosome structure and function and ensuing autophagy in the neurons [26]. In this study, it was shown that following the treatment of primary cultured rat hippocampal neurons with HIV-1 TAT, neuronal viability was significantly decreased. The authors reported significant changes in the structure and membrane integrity of endolysosomes, endolysosome pH, and autophagy flux, which were responsible for neuronal death. In agreement with these findings, report by Fields et al. also demonstrated that TAT altered neuronal autophagy by modulating autophagosome fusion to the lysosome [27]. In this study, TAT exposure resulted in increased numbers of LC3-II puncta and autophagosomes in a neuronal-derived cell line *in vitro*. Similarly, *in vivo* studies in GFAP-TAT transgenic mice showed increased autophagosome accumulation in the neurons that was accompanied with altered LC3-II levels and neuron dysfunction. The findings from this study implicate that therapies targeting TAT-mediated autophagy alterations could mitigate neurodegeneration in HIV-1-infected individuals with HAND.

4. Cocaine, HIV-1, and autophagy

Cocaine, one of most abused drugs, is known to lead to drug addiction. At the molecular level, cocaine binds to dopamine transporter to block dopamine uptake by the neurons, leading subsequently to elevated synaptic dopamine concentrations. Increased synaptic dopamine levels can, in turn, overstimulate the dopamine receptors located in the post-synaptic membrane resulting in enhanced neuronal excitability in the brain striatum [28]. Emerging evidence also demonstrates that cocaine-mediated microglial activation can contribute to the development of drug addiction [29]. In this study, it was shown that cocaine directly interacts with the Toll-like receptor 4-mediated pathway to activate microglia, resulting in enhanced microglial-neuronal cross talk and enhanced synaptic dopamine concentrations. Inhibition of microglial activation through pharmacological or genetic approaches was shown to block cocaine-mediated reward-related behavioral changes.

Compared to other abused drugs such as methamphetamine and morphine that have been implicated to dysregulate autophagy, reports on cocaine-mediated induction of autophagy in cells of the CNS are relatively scant. In a recent study, it was shown that [30] cocaine exposure resulted in induction of autophagy both in microglia and in astrocytes and also *in vivo* in mice administered cocaine. Cocaine exposure can lead to increased expression of various autophagy markers such as Beclin-1, ATG5, and LC3-II in both a dose- and time-dependent manner. In this study, upstream activation of ER stress was shown to mediate induction of autophagy. It was shown that increased autophagy contributed to cocaine-mediated activation of microglia since pre-treatment of cells with wortmannin resulted in decreased expression and release of inflammatory factors (TNF-α, IL-1β, IL-6, and CCL2). In another report, it was demonstrated that cocaine induced autophagic death in astrocytes, a process involving activation of sigma 1 receptor, PI3K, and mTOR pathway [31]. In another recent report, cocaine exposure of neurons was also shown to elicit autophagic cytotoxicity via a nitric oxide-GAPDH signaling cascade [32]. In this study, cocaine exposure markedly increased levels of LC3-II with a concomitant depletion of p62. Pharmacologic inhibition of autophagy protected neurons against cocaine-induced cell death.

Cocaine abuse is one of the comorbidities of HIV-1 infection. Cocaine abuse has been demonstrated to increase HIV-1 infection rate and to accelerate HAND pathogenesis [33, 34]. HIV-1-infected individuals abusing cocaine exhibited increased virus loads in the brain with associated neuroinflammation and neuronal injury compared to HIV-1-infected individuals that were drug naïve. Clinical evidence showed that cocaine use resulted in lack of virologic suppression and accelerated decline of $CD4^+$ T cells even among ART-adherent patients [35] which was consistent with findings that cocaine enhanced HIV-1 replication, as demonstrated in multiple *in vitro* and *in vivo* studies [36, 37]. In addition to its effects on virus replication, cocaine has also been shown to disrupt the integrity of brain-blood barrier, resulting in enhanced macrophage/monocyte influx into the brain, leading to increased neuroinflammation [38]. Additionally, cocaine has also been shown to potentiate toxicity of HIV-1 proteins such as gp120 in the brain [39]. Similar to gp120, cocaine and TAT exerted a synergistic neurotoxic effect in rat primary hippocampal neurons [40]. In these cells, cocaine exposure exacerbated TAT-induced mitochondrial depolarization and generation of intracellular ROS [40]. Cocaine can thus modulate the activation status of microglia and astrocytes through multiple mechanisms including ROS, ER stress, and autophagic pathways [30, 41].

5. Methamphetamine (METH), HIV-1, and autophagy

METH is a psychostimulant drug which is extensively abused for its stimulant, euphoric, empathogenic, and hallucinogenic properties [42]. METH predominantly disrupts the monoamine neurotransmitter system of the brain leading to persistent damage in the serotonin and dopamine neurotransmission, a mechanism also accountable for the motor deficiencies observed in Parkinson's disease [43]. While METH-mediated impairment in neurotransmission results in nigrostriatal denervation, accumulation of ubiquitin-positive neuronal inclusions, and striatal dopamine loss resulting in long-term neurotoxicity [42], METH does not affect dopaminergic neurons in the substantia nigra pars compacta. Although nigral cell bodies are largely preserved following exposure to METH, cytoplasmic features do reveal the presence of autophagic-like vacuolization and accumulation of α-synuclein, ubiquitin, and parkin-positive inclusion-like bodies [44], as is the feature of several neurodegenerative diseases [45]. Chronic use of METH is closely coupled with cognitive deficits ranging from impaired impulse control, attentional problems, working memory, and decision making to motor coordination, including inhibitory control [46]. Apart from this, chronic METH abuse also leads to a limited, but persistent, loss of dopamine and serotonin transporters in the striatum, cortex, and hippocampal areas. METH abuse also induces basal ganglia-mediated behavioral deficits, which is secondary to compromised dopamine neurotransmission in the moderately denervated striatum [47].

Emerging studies also correlate the use of METH with the activation of autophagy in both the CNS and periphery [48]. The effects of METH on the autophagic pathway are likely to depend on the superfluous free radicals and oxidative species generated within the cells. These effects become predominant within the dopaminergic neurons, which is likely due to the specificity of METH for dopamine targets and the ability of dopamine to self-oxidize and produce free radicals [49]. Indeed, the mechanisms of action of METH involve release of cytosolic dopamine and are based on various molecular targets such as the dopamine transporter, the vesicular monoamine transporter-2, and monoamine oxidase, which are involved in the uptake, storage, and release of dopamine [50]. By acting on the dopamine vesicular storage, METH interrupts the physiological gradient, with the diffusion of large amount of dopamine into the cytosol, where the blockage of monoamine oxidase type A abrogates its physiological metabolism, leaving the cytosolic dopamine vulnerable to self-oxidize to form dopamine-quinones and cysteinyl aggregates, thereby promoting neurotoxicity [49].

In addition to its direct neurotoxic effects, the dysfunction of blood-brain barrier (BBB) is also a feature of METH-induced neurotoxicity [51]. METH has been shown to induce the impairment of GLUT1 at the brain endothelium thereby contributing the energy-associated disruption of tight junction assembly and loss of BBB integrity [52]. Moreover, METH acts directly on cultured rat brain microvascular endothelial cells to compromise the BBB via the involvement of eNOS/NO-mediated transcytosis [51]. Additionally, METH mediates a transient increase in the permeability of the BBB in the hippocampus compared with the frontal cortex and striatum, via alterations in the expression levels of tight junction proteins and matrix metalloproteinase-9 [53]. Low doses of METH (1 μM) have been shown to induce endothelial cell barrier dysfunction, thereby underscoring the role of METH in BBB compromise [51]. Interestingly, exposure of low doses of METH also induces induction of autophagy in two dopaminergic neuronal-derived cell lines such as the rat pheochromocytoma PC12 and the human neuroblastoma SH-SY5Y through the phosphatidylinositide 3-kinase III signaling. In contrast, caspase-dependent neuronal cell death involves inhibition of the autophagy maturation process despite the aggregation of α-synuclein and damaged mitochondria in the cytoplasm of METH-exposed cells [44]. Similarly, METH exposure also leads to the induction of the autophagic sequestering process (by inducing LC3-II expression) in human dopaminergic SK-N-SH cell line by ameliorating the mTOR activity on its downstream phosphorylation of the target eukaryotic translation initiation factor 4E-binding protein 1, eventually resulting in decreased cell viability [54]. METH exposure of SK-N-SH cells also induces autophagy by blocking both the dissociation of the Bcl-2/Beclin-1 complex and upstream activation of c-Jun N-terminal kinase 1 signaling [55]. Acute METH exposure also has been shown to induce autophagy as an early protective response. Chronic METH exposure on the other hand exacerbates the progression of autophagic flux leading, in turn, to apoptotic cell death in primary human brain microvascular endothelial cells and human umbilical vein endothelial cells, through the Kappa opioid receptor [56]. Recent investigations have also identified the potentiation of METH-mediated neurotoxicity by caffeine through inhibition of autophagy via the protein kinase A activation pathway and via enhancement of LC3-II phosphorylation, thereby abrogating incorporation of LC3-II into the autophagosomes [57]. In cardiomyocytes and dopaminergic neurons, METH exposure also stimulates the damage-inducible transcript 4 expression leading to induction of autophagy and apoptosis [58, 59].

METH use has been associated with higher risk-taking behaviors that set drug abusers at a higher risk of exposure to infections such as HIV-1 and hepatitis C, each, in turn, contributing eventually to CNS dysfunction. METH abuse has also been shown to accelerate the onset and severity of HAND. Neurotoxic consequences of METH abuse and HIV-1 infection include brain hyperthermia, release of inflammatory cytokines and reactive oxygen species, excitotoxicity, and astrogliosis [60]. METH abuse in combination with HIV-1 infection leads to notable variations in the functioning of dopaminergic neurons. Role of HIV-1 TAT protein that is actively released by infected glial and lymphoid cells has been well documented in the pathogenesis of HAND [61–63]. Intrastriatal administrations of HIV-1 TAT combined with gp120 protein have been shown to damage both efferent and afferent neurons in the striatum [64], including the nigrostriatal DA neurons [65], likely underlie the motor abnormalities observed in HIV-1-infected individuals. Co-exposure of HIV-1 TAT and METH cross-amplifies their deleterious cellular effects through oxidative stress-mediated inflammatory mediators such as TNF-α, IL-β, and ICAM-1 in distinct regions of the mice brain, with implications for CNS complications in HIV-1-infected individuals abusing drugs. Additionally, mice administered HIV-1 TAT and METH were shown to exhibit enhanced DNA-binding activities of transcription factors such as NF-kB, AP-1, and CREB in the frontal cortex and hippocampal regions compared with mice administered either HIV-1 TAT or METH alone [60]. These findings likely suggest that HIV-1 TAT and METH synergistically decrease striatal dopamine release and content, which likely leads to increased risk for basal ganglia dysfunction and cognitive impairment in METH abusers that are infected with HIV-1 [66].

Interestingly, METH abuse is closely associated with higher viral loads in ART-receiving HIV-1-positive individuals [67]. The combination of METH and HIV-1 leads to increased neurocognitive deficits and neuropathology compared with the either agent alone [68]. The potential mechanistic interactions of virus, antiviral treatment, and the psychostimulant drug, however, remain largely unknown. Furthermore, METH-mediated dopamine accumulation in the synapse also increases the levels of free radicals in the neurons, thereby promoting protein damage as well as protein dysfunction, leading to upregulation of autophagy [48]. It has also been shown that exposure of neurons to varying combinations of cART and METH, along with HIV-1-envelope gp120, compromised cellular ATP homeostasis in association with activation of both AMP-activated protein kinase (AMPK) and autophagy [68]. Overall, METH-mediated neurotoxicity is mediated by induction of a specific cellular pathway that is activated when dopamine is not effectively sequestered in the synaptic vesicles, thereby producing oxidative stress, autophagy, and eventually neurite degeneration.

6. Morphine, HIV-1, and autophagy

Opiates are the most potent and popular compounds known to control pain and are also among the most common drugs of abuse. Heroin is one such highly addictive, illegal opioid drug. Since heroin is converted to morphine in the brain, morphine has been the preferred opiate of choice in most studies. Morphine is a natural opiate that has a clinically widespread use for pain management in cancer patients. Despite its beneficial effects, chronic use of opiates elicits adverse side effects, such as memory impairment, tolerance, dependence, drug addiction, and neural injury [69]. Morphine exposure also significantly alters the immune system by modifying the functions of a range of immune cells such as phagocytes, T cells, and dendritic cells [70, 71]. Morphine induces cellular impairment via inhibition of the central cholinergic system, altered expression of μ-opioid receptor, attenuation of long-term potentiation in the hippocampus caused by accumulation of extracellular adenosine [72], increased glucocorticoid concentration in plasma [73], and inhibition of nitric oxide synthesis in the brain [74].

Morphine induces autophagy through sustained activation of μ-opioid receptor in rat hippocampal neurons and in neuroblastoma SH-SY5Y cells resulting in neuronal injury [75]. Morphine-mediated induction of autophagy in these cells is operated primarily by PTX-sensitive G protein-coupled receptor signaling. Following its binding to μ-opioid receptor, there is an increase in Beclin-1 protein levels and a significant decrease in the association of Beclin-1 with Bcl-2 leading, in turn, to dissociation of Beclin-1 from its pro-autophagic events. It has been reported that Bcl-2 overexpression remarkably impedes morphine-mediated autophagy induction and that genetic silencing of Beclin-1 or another autophagy marker ATG5 inhibits morphine-mediated autophagy. Long-term exposure of morphine has been shown to stimulate neuronal cell death, an effect that is exacerbated by genetic silencing of Beclin-1. Taken together, this study for the first time identified key roles of Beclin-1 and ATG5 in morphine-mediated induction of neuronal injury [75].

In another study, long-term morphine exposure was shown to induce autophagy in the superficial layer of the spinal cord through upregulation of Beclin-1, LC3-II, and cathepsin B in the GABAergic interneurons, resulting in the development of antinociceptive tolerance. Blockade of either cathepsin B or autophagy notably suppressed morphine-mediated antinociceptive tolerance [76]. Morphine abuse was also closely associated with decreased mitochondrial DNA copy number both in the hippocampus and in peripheral blood and was linked to drug addiction. These findings were also corroborated in heroin addicts [77]. Morphine exposure also has been shown to potentiate LPS-induced autophagy initiation and block the fusion of autophagosomes in macrophages, leading to defective bacterial clearance and increased bacterial load via TLR4-dependent and TLR4-independent pathways. Morphine exposure thus increases the susceptibility to infection as well as prevalence of persistent infection in drug addicts [78]. Morphine also increases autophagic flux both in the hippocampal CA1 neurons and in microglia, leading to increased neuronal cell death in the CA1 and CA3 regions and escalated inflammation in the hippocampal microglia, ultimately resulting in morphine-mediated spatial memory impairment [79].

Opiates abuse and HIV-1 are interlinked epidemics, and opiates such as heroin exacerbate the neuropathogenesis of HIV-1 with rapid disease progression [80]. Long-term opiate abuse in the pre-cART era was demonstrated to be directly associated with increased progression to HIVE and exacerbated neuropathology in patients on cART [81]. Morphine exacerbates HIV-1 toxicity through distinct pathways in neurons and in glia, primarily through the μ-opioid receptors. Acute exposure of morphine leads to increased HIV-1 replication in the infected microglia [82]. Morphine potentiates the deleterious effects of HIV-1 TAT via dysregulation of intracellular calcium homeostasis, leading to decreased buffering of extracellular glutamate in the astrocytes. This dysfunction in turn leads to decreased excitotoxicity threshold of neurons, resulting in increased production of reactive species and proinflammatory mediators, which ultimately damage the neurons [83]. Morphine exposure has also been shown to enhance both HIV-1 replication and inflammatory response through the Beclin-1-independent mechanism in microglial cells, implicating the connections between autophagy and HIV-1 pathogenesis [23]. In another study, chronic morphine exposure of HIV-1-infected human monocyte-derived macrophages led to significant alterations in the secretion of IL-6 and monocyte chemoattractant protein 2, thereby suggesting enhanced CNS inflammation in HIV-1-infected opiate abusers [84].

7. Conclusion

Overall, this chapter sheds light on HIV-1, drugs of abuse such as cocaine, methamphetamine, and morphine-mediated autophagy primarily in the CNS and periphery. Such effects could underlie the synergistic effects of HIV-1 and abused drugs in exacerbating symptoms of HAND. Further studies, however, are warranted to unravel the mechanistic roles of autophagy in CNS cells and periphery in HIV-1-infected individuals with a history of drug addiction. Primarily, how autophagy might be involved in these cells and its relative contribution to immunopathogenesis, particularly HAND, has yet to be determined. Investigating autophagy in various cell types involved in the pathogenesis of HIV-1 infection in the context of drug abuse could provide a basis for future development of novel therapeutic strategies aimed at treating HIV-1-infected individuals that abuse drugs.

Conflict of interests

There are no conflicts of interest for any of the authors.

Acknowledgements

This work was supported by NIH grants: DA033614, DA035203, DA036157, DA033150, DA040397, and MH106425.

Author details

Ming-Lei Guo, Palsamy Periyasamy and Shilpa Buch*

*Address all correspondence to: sbuch@unmc.edu

Department of Pharmacology and Experimental Neuroscience, University of Nebraska Medical Center, Omaha, Nebraska, USA

References

[1] Winston A, Underwood J. Emerging concepts on the use of antiretroviral therapy in older adults living with HIV infection. Curr Opin Infect Dis. 2015;28(1):17–22. doi:10.1097/QCO.0000000000000117

[2] Cai Y, Yang L, Callen S, Buch S. Multiple faceted roles of cocaine in potentiation of HAND. Curr HIV Res. DOI: 10.2174/1570162X14666160324125158.

[3] Kurapati KR, Atluri VS, Samikkannu T, Garcia G, Nair MP. Natural products as anti-HIV agents and role in HIV-associated neurocognitive disorders (HAND): a brief overview. Front Microbiol. 2015;6:1444. doi:10.3389/fmicb.2015.01444

[4] Mizushima N, Komatsu M. Autophagy: renovation of cells and tissues. Cell. 2011;147(4):728–741. doi:10.1016/j.cell.2011.10.026

[5] Tyagi M, Bukrinsky M, Simon GL. Mechanisms of HIV transcriptional regulation by drugs of abuse. Curr HIV Res. DOI: 10.2174/1570162X14666160324124736.

[6] Maubert ME, Pirrone V, Rivera NT, Wigdahl B, Nonnemacher MR. Interaction between tat and drugs of abuse during HIV-1 infection and central nervous system disease. Front Microbiol. 2015;6:1512. doi:10.3389/fmicb.2015.01512

[7] Brown T, Salomon JA, Alkema L, Raftery AE, Gouws E. Progress and challenges in modelling country-level HIV/AIDS epidemics: the UNAIDS Estimation and Projection Package 2007. Sex Transm Infect. 2008;84 Suppl 1:i5–i10. doi:10.1136/sti.2008.030437

[8] Brew BJ, Chan P. Update on HIV dementia and HIV-associated neurocognitive disorders. Curr Neurol Neurosci Rep. 2014;14(8):468. doi:10.1007/s11910–014-0468-2

[9] Elbirt D, Mahlab-Guri K, Bezalel-Rosenberg S, Gill H, Attali M, Asher I. HIV-associated neurocognitive disorders (HAND). Isr Med Assoc J. 2015;17(1):54–59.

[10] Rumbaugh JA, Tyor W. HIV-associated neurocognitive disorders: five new things. Neurol Clin Pract. 2015;5(3):224–231. doi:10.1212/CPJ.0000000000000117

[11] Simoes E, Justino JD. HIV-associated neurocognitive disorders: a review for NPs. Nurse Pract. 2015;40(7):1–7. doi:10.1097/01.NPR.0000466501.42049.99

[12] Watkins CC, Treisman GJ. Cognitive impairment in patients with AIDS—prevalence and severity. HIV AIDS (Auckl). 2015;7:35–47. doi:10.2147/HIV.S39665

[13] Dinkins C, Pilli M, Kehrl JH. Roles of autophagy in HIV infection. Immunol Cell Biol. 2015;93(1):11–17. doi:10.1038/icb.2014.88

[14] Alexaki A, Liu Y, Wigdahl B. Cellular reservoirs of HIV-1 and their role in viral persistence. Curr HIV Res. 2008;6(5):388–400.

[15] Delgado M, Singh S, De Haro S, Master S, Ponpuak M, Dinkins C, et al. Autophagy and pattern recognition receptors in innate immunity. Immunol Rev. 2009;227(1):189–202. doi:10.1111/j.1600-065X.2008.00725.x

[16] Blanchet FP, Moris A, Nikolic DS, Lehmann M, Cardinaud S, Stalder R, et al. Human immunodeficiency virus-1 inhibition of immunoamphisomes in dendritic cells impairs early innate and adaptive immune responses. Immunity. 2010;32(5):654–669. doi: 10.1016/j.immuni.2010.04.011

[17] Manches O, Frleta D, Bhardwaj N. Dendritic cells in progression and pathology of HIV infection. Trends Immunol. 2014;35(3):114–122. doi:10.1016/j.it.2013.10.003

[18] Kyei GB, Dinkins C, Davis AS, Roberts E, Singh SB, Dong C, et al. Autophagy pathway intersects with HIV-1 biosynthesis and regulates viral yields in macrophages. J Cell Biol. 2009;186(2):255–268. doi:10.1083/jcb.200903070

[19] Lackner AA, Lederman MM, Rodriguez B. HIV pathogenesis: the host. Cold Spring Harb Perspect Med. 2012;2(9):a007005. doi:10.1101/cshperspect.a007005

[20] Espert L, Varbanov M, Robert-Hebmann V, Sagnier S, Robbins I, Sanchez F, et al. Differential role of autophagy in CD4 T cells and macrophages during X4 and R5 HIV-1 infection. PLoS One. 2009;4(6):e5787. doi:10.1371/journal.pone.0005787

[21] Doitsh G, Galloway NL, Geng X, Yang Z, Monroe KM, Zepeda O, et al. Cell death by pyroptosis drives CD4 T-cell depletion in HIV-1 infection. Nature. 2014;505(7484):509–514. doi:10.1038/nature12940

[22] Dever SM, Rodriguez M, Lapierre J, Costin BN, El-Hage N. Differing roles of autophagy in HIV-associated neurocognitive impairment and encephalitis with implications for morphine co-exposure. Front Microbiol. 2015;6:653. doi:10.3389/fmicb.2015.00653

[23] El-Hage N, Rodriguez M, Dever SM, Masvekar RR, Gewirtz DA, Shacka JJ. HIV-1 and morphine regulation of autophagy in microglia: limited interactions in the context of HIV-1 infection and opioid abuse. J Virol. 2015;89(2):1024–1035. doi:10 1128/JVI. 02022-14

[24] Mehla R, Chauhan A. HIV-1 differentially modulates autophagy in neurons and astrocytes. J Neuroimmunol. 2015;285:106–118. doi:10.1016/j.jneuroim.2015.06.001

[25] Alirezaei M, Kiosses WB, Flynn CT, Brady NR, Fox HS. Disruption of neuronal autophagy by infected microglia results in neurodegeneration. PLoS One. 2008;3(8):e2906. doi:10.1371/journal.pone.0002906

[26] Hui L, Chen X, Haughey NJ, Geiger JD. Role of endolysosomes in HIV-1 Tat-induced neurotoxicity. ASN Neuro. 2012;4(4):243–252. doi:10.1042/AN20120017

[27] Fields J, Dumaop W, Eleuteri S, Campos S, Serger E, Trejo M, et al. HIV-1 Tat alters neuronal autophagy by modulating autophagosome fusion to the lysosome: implications for HIV-associated neurocognitive disorders. J Neurosci. 2015;35(5):1921–1938. doi:10.1523/JNEUROSCI.3207-14.2015

[28] Broseus J, Gentile N, Esseiva P. The cutting of cocaine and heroin: a critical review. Forensic Sci Int. 2016;262:73–83. doi:10.1016/j.forsciint.2016.02.033

[29] Northcutt AL, Hutchinson MR, Wang X, Baratta MV, Hiranita T, Cochran TA, et al. DAT isn't all that: cocaine reward and reinforcement require Toll-like receptor 4 signaling. Mol Psychiatry. 2015;20(12):1525–1537. doi:10.1038/mp.2014.177

[30] Guo ML, Liao K, Periyasamy P, Yang L, Cai Y, Callen SE, et al. Cocaine-mediated microglial activation involves the ER stress-autophagy axis. Autophagy. 2015;11(7): 995–1009. doi:10.1080/15548627.2015.1052205

[31] Cao L, Walker MP, Vaidya NK, Fu M, Kumar S, Kumar A. Cocaine-mediated autophagy in astrocytes involves sigma 1 receptor, PI3K, mTOR, Atg5/7, beclin-1 and induces type II programed cell death. Mol Neurobiol. 2015. doi:10.1007/s12035–015-9377-x

[32] Guha P, Harraz MM, Snyder SH. Cocaine elicits autophagic cytotoxicity via a nitric oxide-GAPDH signaling cascade. Proc Natl Acad Sci USA. 2016;113(5):1417–1422. doi: 10.1073/pnas.1524860113

[33] Dash S, Balasubramaniam M, Villalta F, Dash C, Pandhare J. Impact of cocaine abuse on HIV pathogenesis. Front Microbiol. 2015;6:1111. doi:10.3389/fmicb.2015.01111

[34] Dahal S, Chitti SV, Nair MP, Saxena SK. Interactive effects of cocaine on HIV infection: implication in HIV-associated neurocognitive disorder and neuroAIDS. Front Microbiol. 2015;6:931. doi:10.3389/fmicb.2015.00931

[35] Carrico AW, Johnson MO, Morin SF, Remien RH, Riley ED, Hecht FM, et al. Stimulant use is associated with immune activation and depleted tryptophan among HIV-positive persons on anti-retroviral therapy. Brain Behav Immun. 2008;22(8):1257–1262. doi: 10.1016/j.bbi.2008.07.010

[36] Dhillon NK, Williams R, Peng F, Tsai YJ, Dhillon S, Nicolay B, et al. Cocaine-mediated enhancement of virus replication in macrophages: implications for human immunodeficiency virus-associated dementia. J Neurovirol. 2007;13(6):483–495. doi: 10.1080/13550280701528684

[37] Napuri J, Pilakka-Kanthikeel S, Raymond A, Agudelo M, Yndart-Arias A, Saxena SK, et al. Cocaine enhances HIV-1 infectivity in monocyte derived dendritic cells by

suppressing microRNA-155. PLoS One. 2013;8(12):e83682. doi:10.1371/journal.pone.0083682

[38] Sharma HS, Muresanu D, Sharma A, Patnaik R. Cocaine-induced breakdown of the blood-brain barrier and neurotoxicity. Int Rev Neurobiol. 2009;88:297–334. doi:10.1016/S0074-7742(09)88011-2

[39] Yang Y, Yao H, Lu Y, Wang C, Buch S. Cocaine potentiates astrocyte toxicity mediated by human immunodeficiency virus (HIV-1) protein gp120. PLoS One. 2010;5(10):e13427. doi:10.1371/journal.pone.0013427

[40] Aksenov MY, Aksenova MV, Nath A, Ray PD, Mactutus CF, Booze RM. Cocaine-mediated enhancement of Tat toxicity in rat hippocampal cell cultures: the role of oxidative stress and D1 dopamine receptor. Neurotoxicology. 2006;27(2):217–228. doi:10.1016/j.neuro.2005.10.003

[41] Liao K, Guo M, Niu F, Yang L, Callen SE, Buch S. Cocaine-mediated induction of microglial activation involves the ER stress-TLR2 axis. J Neuroinflammation. 2016;13(1):33. doi:10.1186/s12974-016-0501-2

[42] Yu S, Zhu L, Shen Q, Bai X, Di X. Recent advances in methamphetamine neurotoxicity mechanisms and its molecular pathophysiology. Behav Neurol. 2015;2015:103969. doi:10.1155/2015/103969

[43] Riddle EL, Fleckenstein AE, Hanson GR. Mechanisms of methamphetamine-induced dopaminergic neurotoxicity. AAPS J. 2006;8(2):E413–418.

[44] Castino R, Lazzeri G, Lenzi P, Bellio N, Follo C, Ferrucci M, et al. Suppression of autophagy precipitates neuronal cell death following low doses of methamphetamine. J Neurochem. 2008;106(3):1426–1439. doi:10.1111/j.1471-4159.2008.05488.x

[45] Fornai F, Lenzi P, Gesi M, Soldani P, Ferrucci M, Lazzeri G, et al. Methamphetamine produces neuronal inclusions in the nigrostriatal system and in PC12 cells. J Neurochem. 2004;88(1):114–123. doi:10.1046/j.1471-4159.2003.02137.x

[46] Simon SL, Domier CP, Sim T, Richardson K, Rawson RA, Ling W. Cognitive performance of current methamphetamine and cocaine abusers. J Addict Dis. 2002;21(1):61–74. doi:10.1300/J069v21n01_06

[47] Howard CD, Pastuzyn ED, Barker-Haliski ML, Garris PA, Keefe KA. Phasic-like stimulation of the medial forebrain bundle augments striatal gene expression despite methamphetamine-induced partial dopamine denervation. J Neurochem. 2013;125(4):555–565. doi:10.1111/jnc.12234

[48] Pasquali L, Lazzeri G, Isidoro C, Ruggieri S, Paparelli A, Fornai F. Role of autophagy during methamphetamine neurotoxicity. Ann N Y Acad Sci. 2008;1139:191–196. doi:10.1196/annals.1432.016

[49] Gibb JW, Johnson M, Hanson GR. Neurochemical basis of neurotoxicity. Neurotoxicology. 1990;11(2):317–321.

[50] Lohr KM, Stout KA, Dunn AR, Wang M, Salahpour A, Guillot TS, et al. Increased vesicular monoamine transporter 2 (VMAT2; Slc18a2) protects against methamphetamine toxicity. ACS Chem Neurosci. 2015;6(5):790–799. doi:10.1021/acschemneuro.5b00010

[51] Martins T, Burgoyne T, Kenny BA, Hudson N, Futter CE, Ambrosio AF, et al. Methamphetamine-induced nitric oxide promotes vesicular transport in blood-brain barrier endothelial cells. Neuropharmacology. 2013;65:74–82. doi:10.1016/j.neuropharm.2012.08.021

[52] Abdul Muneer PM, Alikunju S, Szlachetka AM, Murrin LC, Haorah J. Impairment of brain endothelial glucose transporter by methamphetamine causes blood-brain barrier dysfunction. Mol Neurodegener. 2011;6:23. doi:10.1186/1750-1326-6-23

[53] Martins T, Baptista S, Goncalves J, Leal E, Milhazes N, Borges F, et al. Methamphetamine transiently increases the blood-brain barrier permeability in the hippocampus: role of tight junction proteins and matrix metalloproteinase-9. Brain Res. 2011;1411:28–40. doi:10.1016/j.brainres.2011.07.013

[54] Kongsuphol P, Mukda S, Nopparat C, Villarroel A, Govitrapong P. Melatonin attenuates methamphetamine-induced deactivation of the mammalian target of rapamycin signaling to induce autophagy in SK-N-SH cells. J Pineal Res. 2009;46(2):199–206. doi:10.1111/j.1600-079X.2008.00648.x

[55] Nopparat C, Porter JE, Ebadi M, Govitrapong P. The mechanism for the neuroprotective effect of melatonin against methamphetamine-induced autophagy. J Pineal Res. 2010;49(4):382–389. doi:10.1111/j.1600-079X.2010.00805.x

[56] Ma J, Wan J, Meng J, Banerjee S, Ramakrishnan S, Roy S. Methamphetamine induces autophagy as a pro-survival response against apoptotic endothelial cell death through the Kappa opioid receptor. Cell Death Dis. 2014;5:e1099. doi:10.1038/cddis.2014.64

[57] Pitaksalee R, Sanvarinda Y, Sinchai T, Sanvarinda P, Thampithak A, Jantaratnotai N, et al. Autophagy inhibition by caffeine increases toxicity of methamphetamine in SH-SY5Y neuroblastoma cell line. Neurotox Res. 2015;27(4):421–429. doi:10.1007/s12640-014-9513-9

[58] Li B, Chen R, Chen L, Qiu P, Ai X, Huang E, et al. Effects of DDIT4 in methamphetamine-induced autophagy and apoptosis in dopaminergic neurons. Mol Neurobiol. doi:10.1007/s12035-015-9637-9

[59] Chen R, Wang B, Chen L, Cai D, Li B, Chen C, et al. DNA damage-inducible transcript 4 (DDIT4) mediates methamphetamine-induced autophagy and apoptosis through mTOR signaling pathway in cardiomyocytes. Toxicol Appl Pharmacol. 2016;295:1–11. doi:10.1016/j.taap.2016.01.017

[60] Flora G, Lee YW, Nath A, Hennig B, Maragos W, Toborek M. Methamphetamine potentiates HIV-1 Tat protein-mediated activation of redox-sensitive pathways in discrete regions of the brain. Exp Neurol. 2003;179(1):60–70.

[61] Chang HC, Samaniego F, Nair BC, Buonaguro L, Ensoli B. HIV-1 Tat protein exits from cells via a leaderless secretory pathway and binds to extracellular matrix-associated heparan sulfate proteoglycans through its basic region. AIDS. 1997;11(12):1421–1431.

[62] Tardieu M, Hery C, Peudenier S, Boespflug O, Montagnier L. Human immunodeficiency virus type 1-infected monocytic cells can destroy human neural cells after cell-to-cell adhesion. Ann Neurol. 1992;32(1):11–17. doi:10.1002/ana.410320104

[63] Ensoli B, Buonaguro L, Barillari G, Fiorelli V, Gendelman R, Morgan RA, et al. Release, uptake, and effects of extracellular human immunodeficiency virus type 1 Tat protein on cell growth and viral transactivation. J Virol. 1993;67(1):277–287.

[64] Bansal AK, Mactutus CF, Nath A, Maragos W, Hauser KF, Booze RM. Neurotoxicity of HIV-1 proteins gp120 and Tat in the rat striatum. Brain Res. 2000;879(1–2):42–49.

[65] Zauli G, Secchiero P, Rodella L, Gibellini D, Mirandola P, Mazzoni M, et al. HIV-1 Tat-mediated inhibition of the tyrosine hydroxylase gene expression in dopaminergic neuronal cells. J Biol Chem. 2000;275(6):4159–4165.

[66] Langford D, Adame A, Grigorian A, Grant I, McCutchan JA, Ellis RJ, et al. Patterns of selective neuronal damage in methamphetamine-user AIDS patients. J Acquir Immune Defic Syndr. 2003;34(5):467–474.

[67] Hinkin CH, Barclay TR, Castellon SA, Levine AJ, Durvasula RS, Marion SD, et al. Drug use and medication adherence among HIV-1 infected individuals. AIDS Behav. 2007;11(2):185–194. doi:10.1007/s10461-006-9152-0

[68] Sanchez AB, Varano GP, de Rozieres CM, Maung R, Catalan IC, Dowling CC, et al. Antiretrovirals, methamphetamine, and HIV-1 envelope protein gp120 compromise neuronal energy homeostasis in association with various degrees of synaptic and neuritic damage. Antimicrob Agents Chemother. 2015;60(1):168–179. doi:10.1128/AAC.01632-15

[69] Nestler EJ, Hope BT, Widnell KL. Drug addiction: a model for the molecular basis of neural plasticity. Neuron. 1993;11(6):995–1006.

[70] Reddy PV, Pilakka-Kanthikeel S, Saxena SK, Saiyed Z, Nair MP. Interactive effects of morphine on HIV infection: role in HIV-associated neurocognitive disorder. AIDS Res Treat. 2012;2012:953678. doi:10.1155/2012/953678

[71] Messmer D, Hatsukari I, Hitosugi N, Schmidt-Wolf IG, Singhal PC. Morphine reciprocally regulates IL-10 and IL-12 production by monocyte-derived human dendritic cells and enhances T cell activation. Mol Med. 2006;12(11–12):284–290. doi:10.2119/2006-00043.Messmer

[72] Lu G, Zhou QX, Kang S, Li QL, Zhao LC, Chen JD, et al. Chronic morphine treatment impaired hippocampal long-term potentiation and spatial memory via accumulation of extracellular adenosine acting on adenosine A1 receptors. J Neurosci. 2010;30(14): 5058–5070. doi:10.1523/JNEUROSCI.0148-10.2010

[73] Mesripour A, Hajhashemi V, Rabbani M. Metyrapone and mifepristone reverse recognition memory loss induced by spontaneous morphine withdrawal in mice. Basic Clin Pharmacol Toxicol. 2008;102(4):377–381. doi:10.1111/j.1742-7843.2007.00183.x

[74] Babaei R, Javadi-Paydar M, Sharifian M, Mahdavian S, Almasi-Nasrabadi M, Norouzi A, et al. Involvement of nitric oxide in pioglitazone memory improvement in morphine-induced memory impaired mice. Pharmacol Biochem Behav. 2012;103(2):313–321. doi: 10.1016/j.pbb.2012.08.018

[75] Zhao L, Zhu Y, Wang D, Chen M, Gao P, Xiao W, et al. Morphine induces Beclin 1- and ATG5-dependent autophagy in human neuroblastoma SH-SY5Y cells and in the rat hippocampus. Autophagy. 2010;6(3):386–394.

[76] Hayashi Y, Koga Y, Zhang X, Peters C, Yanagawa Y, Wu Z, et al. Autophagy in superficial spinal dorsal horn accelerates the cathepsin B-dependent morphine antinociceptive tolerance. Neuroscience. 2014;275:384–394. doi:10.1016/j.neuroscience.2014.06.037

[77] Feng YM, Jia YF, Su LY, Wang D, Lv L, Xu L, et al. Decreased mitochondrial DNA copy number in the hippocampus and peripheral blood during opiate addiction is mediated by autophagy and can be salvaged by melatonin. Autophagy. 2013;9(9):1395–1406. doi: 10.4161/auto.25468

[78] Wan J, Ma J, Anand V, Ramakrishnan S, Roy S. Morphine potentiates LPS-induced autophagy initiation but inhibits autophagosomal maturation through distinct TLR4-dependent and independent pathways. Acta Physiol (Oxf). 2015;214(2):189–199. doi: 10.1111/apha.12506

[79] Pan J, He L, Li X, Li M, Zhang X, Venesky J, et al. Activating autophagy in hippocampal cells alleviates the morphine-induced memory impairment. Mol Neurobiol. doi: 10.1007/s12035-016-9735-3

[80] Hauser KF, Fitting S, Dever SM, Podhaizer EM, Knapp PE. Opiate drug use and the pathophysiology of neuroAIDS. Curr HIV Res. 2012;10(5):435–452.

[81] Smith DB, Simmonds P, Bell JE. Brain viral burden, neuroinflammation and neurodegeneration in HAART-treated HIV positive injecting drug users. J Neurovirol. 2014;20(1):28–38. doi:10.1007/s13365-013-0225-3

[82] El-Hage N, Dever SM, Podhaizer EM, Arnatt CK, Zhang Y, Hauser KF. A novel bivalent HIV-1 entry inhibitor reveals fundamental differences in CCR5-mu-opioid receptor interactions between human astroglia and microglia. AIDS. 2013;27(14):2181–2190. doi: 10.1097/QAD.0b013e3283639804

[83] Podhaizer EM, Zou S, Fitting S, Samano KL, El-Hage N, Knapp PE, et al. Morphine and gp120 toxic interactions in striatal neurons are dependent on HIV-1 strain. J Neuroimmune Pharmacol. 2012;7(4):877–891. doi:10.1007/s11481-011-9326-z

[84] Dave RS. Morphine affects HIV-induced inflammatory response without influencing viral replication in human monocyte-derived macrophages. FEMS Immunol Med Microbiol. 2012;64(2):228–236. doi:10.1111/j.1574-695X.2011.00894.x

Chapter 9

Autophagic Flux Failure in Neurodegeneration: Identifying the Defect and Compensating Flux Offset

Claudia Ntsapi, Chrisna Swart,
Dumisile Lumkwana and Ben Loos

Additional information is available at the end of the chapter

http://dx.doi.org/10.5772/64197

Abstract

Protein degradation through autophagy is one of the key pathways that maintain proteostasis and neuronal viability. Dysregulation in autophagy has been associated with a number of major protein aggregation storage disorders that are characterized by increased cellular vulnerability and susceptibility to undergo cell death. Although the molecular machinery, the proteome, and the regulation of the autophagy system are becoming increasingly clear, the specific nature of its dysfunction in the context of neuronal disease pathogenesis remains largely unclear. Moreover, although the intricate network of autophagy regulatory proteins with key metabolic checkpoints is increasingly being revealed, the relationship between autophagy dysfunction, the changing rate of protein degradation in the specific pathology, and the aggregate prone behavior of specific candidate proteins remains less understood. Many questions remain and deserve urgent attention. When does a neuron respond with heightened autophagic activity and When does the system fail to degrade autophagy cargo? This book chapter will focus on some of the main challenges in the field of autophagy research, the identity, and nature of autophagic flux failure in neurodegeneration, current means to discern and measure autophagic flux dysfunction in neuronal tissue, and recent advances in compensating the flux offset. Specifically, the role of both macroautophagy and chaperone-mediated autophagy in neuronal function and dysfunction and the spatiotemporal changes in their rates of protein degradation will be discussed and their molecular interplay highlighted. Finally, current advances in the use of autophagy modulators to better control autophagy activity will be stressed and contextualized within the framework of re-establishing neuronal proteostasis to favorably control cellular fate.

Keywords: autophagic flux, proteotoxicity, Alzheimer's disease, cell death onset, neurodegeneration, autophagosome, lysosome

1. Introduction

Maintaining the delicate balance between protein synthesis and the degradation of cytotoxic aggregate-prone proteins is crucial for sustained cell growth and development [1]. In neurons, the continuous removal of deleterious intracellular components, including aberrant proteinaceous species and irreversibly damaged organelles, is governed by the machinery of two proteolytic systems: the ubiquitin proteasome system (UPS) and the autophagy-lysosome pathways [2]. While the activity of the UPS is limited to the degradation of short-lived cytosolic and nuclear proteins, the autophagy pathways are responsible for the bulk sequestration, degradation, and recycling of long-lived or misfolded cytosolic proteins and damaged organelles [3]. Defects in the autophagic pathways are particularly detrimental to neuronal cells, with heightened vulnerability to the accumulation of toxic cytoplasmic components [4, 5]. Autophagy is a highly conserved and tightly regulated pathway that is constitutively active in all cell types and is markedly induced under stress conditions [6]. Depending on the cargo targeted, and the mode of cargo delivery to the lysosome, autophagy is generally classified into macroautophagy, chaperone-mediated autophagy (CMA), and microautophagy [7]. All three autophagic pathways usually coexist in the same cell, but only macroautophagy (henceforth referred to as autophagy) and CMA have been implicated in the central nervous system and associated with specific neurodegenerative diseases [8]. In contrast to the distinctive vesicular formation, and the indiscriminate bulk degradation of cytoplasmic materials by autophagy, CMA's characteristic feature is selectivity, whereby cytosolic proteins containing a pentapeptide motif (KFERQ) are targeted and bound by the cytosolic chaperone heat-shock cognate protein of 70 kDa (hsc70) and its cochaperones [9]. An estimated 30% of all cytosolic proteins are thought to contain the KFERQ-like targeting motif, but this number is likely underrated given that post-translational modifications can also render proteins amiable to CMA-mediated degradation [9]. Similar to autophagy, CMA is constitutively active in all cell types studied thus far and upregulated in response to various stressors [10]. Importantly, the inability to upregulate CMA has been shown to render cells more vulnerable to cell death onset [11]. Neurons are highly efficient in autophagic cargo degradation [12], which contributes to their heightened vulnerability when autophagic flux, that is the rate of protein degradation though autophagy, is impaired. Changes in autophagic flux alter the cell's susceptibility to undergo cell death and it is becoming increasingly clear that the autophagic machinery is anchored within an energetic feedback loop that includes metabolic checkpoints that govern cell survival [13] (**Figure 1**). It is therefore critical in our understanding of autophagic flux deviation or dysfunction to reliably and robustly quantify this process, in vitro as well as in vivo. Autophagic flux is defined as the rate of cargo degradation within autophagosomes through autophagy [4]. Transmission electron microscopy (TEM), Western blotting, and fluorescence microscopy, all of which have been extensively described elsewhere [14], are widely used in this context. In brief, TEM remains a most powerful technique for assessing autophagy, as it allows the identification of the autophagic machinery structures at nm range [15]. Western blotting monitors endogenous microtubule-associated protein 1 light chain 3 (MAP1-LC3/Atg8/LC3) [16] as well as p62 (sequestosome/SQSTM1) degradation [17] as an indicator of autophagic flux. Importantly, the

http://dx.doi.org/10.5772/64197

amount of LC3 II correlates with the number of autophagosomes. p62/SQSTM1 is responsible for selecting cargo and to deliver proteins for degradation. It binds directly to LC3 and is codegraded by autophagy [18]. Therefore, the total amount of p62 expressed in a given cell inversely correlates with autophagic flux and provides an indication for the autophagic flux status. Fluorescence microscopy-based analysis techniques enable the counting of LC3 and p62 punctate as well as the quantification of the fluorescence signal at a single-cell level [19]. However, although above techniques are valuable in assessing whether autophagic flux has changed, they are most powerful when complemented with single-cell measurements that allow the assessment of the organelle pool size of autophagosomes (nA), autophagolysosomes (nAL), and lysosomes (nL), thus enabling to report on autophagic flux (J) and transition time (τ). Currently, there has been a major progression using such single cell-based assays to quantify flux using combinations of live cell imaging and photoswitchable fluorochromes [19–21]. These techniques are highly aligned with measuring the rate of cargo degradation, that is degradation per hour, and hence autophagic flux.

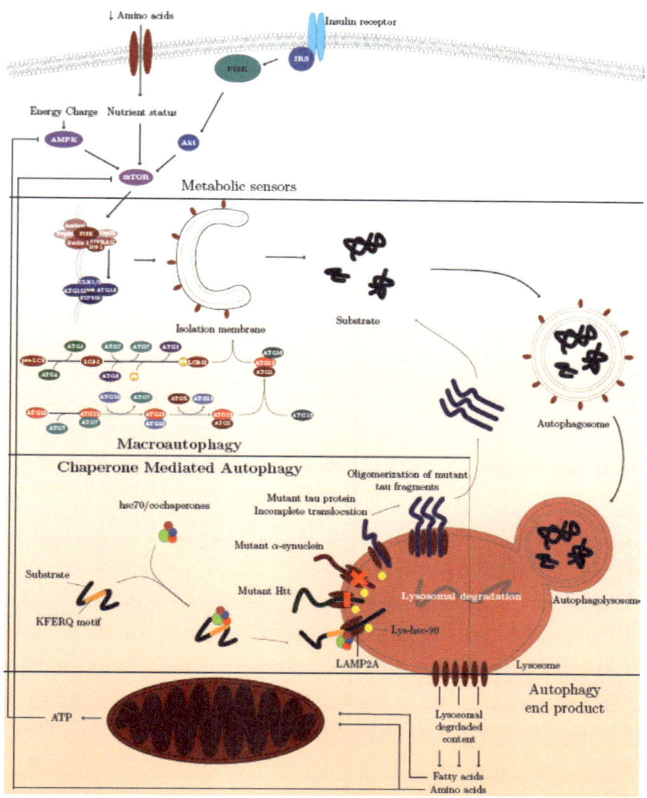

Figure 1. Changes in macroautophagy (MA) and chaperone-mediated autophagy (CMA) impact on cell vulnerability. The autophagic machinery is anchored within an energetic feedback loop that includes key metabolic checkpoints governing proteostasis and cell survival.

2. Macroautophagy and chaperone-mediated autophagy in neurodegeneration: from spatiotemporal changes to complete pathway failure

A unifying theme in neurodegenerative diseases is the failure of the proteolytic systems to adequately dispose unwanted, deleterious proteins [22]. The first pathological evidence of dysfunctional autophagy related to neurodegeneration came from electron microscopy studies of the Alzheimer's disease (AD) brain showing amyloid plaque-associated dystrophic neurites displaying massive autophagic vacuole (AV) accumulation [23]. Similar observations have been made in multiple animal models of AD [24] as well as in brains of patients with Parkinson's disease (PD) [25] and Huntington's disease (HD) [26]. Although the exact mechanism underlying autophagic dysfunction in neurodegeneration remains unclear, AV buildup may result from increased autophagic induction, impairment of downstream degradative processes in the autophagic pathway, or a decreased rate of autophagosome formation combined with insufficient lysosomal fusion [27]. The role of chaperone-mediated autophagy (CMA) in neurodegeneration is twofold: On the one hand, CMA contributes to the removal of pathogenic proteins, but, on the other hand, CMA itself becomes functionally affected by the toxicity of abnormal proteins [28] (**Figure 1**). In the following section, we will focus specifically on the three candidate pathologies with emphasis on the variability of autophagy and CMA dysfunction.

2.1. Alzheimer's disease

Evidence indicates that abnormal autophagy at the level of induction or autophagosome formation may contribute to AD pathogenesis as the expression of Beclin 1, an essential initiator of autophagy, was found to be decreased in AD patients [29, 30], possibly due to an increase in caspase 3-mediated cleavage of Beclin 1 [31]. However, a genome-wide study reported an upregulation of autophagy in AD due to transcriptional upregulation of positive regulators of autophagy as well as reactive oxygen species (ROS)-dependent activation of type III PI3 kinase, a critical kinase for the initiation of autophagy [32]. Furthermore, accumulation of electron-dense autolysosomes in the AD brain indicates lysosomal proteolytic failure [23, 33]. The morphology of such accumulated AVs resembles those resulting from selectively blocking lysosomal proteolysis through deletion of specific cathepsins or addition of lysosomal inhibitors [34, 35]. The most common cause of early-onset, familial AD is autosomal-dominant mutation in presenilin 1 (PS1) and PS2 [36], which enhance the disproportionate release of aggregation prone Aβ. However, not all AD-linked PS mutations manifest with this effect. Apart from its role in the cleavage of γ-secretase, PS1 was suggested to function in calcium homeostasis [37]. Calcium flux regulates both autophagic induction and lysosomal fusion, and PS mutations appear to aggravate this dysfunction [24] and may represent a mechanistic link unifying these pathologies [38]. Therefore, presenilins may affect autophagic flux by facilitating two crucial aspects, firstly, vesicle fusion and secondly, lysosomal function [39]. In fact, PS1 is involved in lysosomal acidification and autophagosome-lysosome fusion, and recent findings demonstrated its association with defective proteolysis of autophagic substrates in

AD patients [39, 40]. It was suggested that familial AD-linked PS1 mutations may have a loss-of-function effect on lysosomal proteolysis that leads to AV accumulation and impaired autophagic substrate turnover in AD [41]. Moreover, defective axonal transport of AVs is being implicated in AD pathogenesis. Under normal conditions, immature AVs are transported retrogradely toward the soma for degradation, but in the AD brain, a significant buildup of AVs is found within dystrophic neurites, an event that could be mimicked by inhibiting autophagosome delivery to lysosomes in healthy cells [34]. The exact molecular defects underlying axonal transport failure remain, however, largely unclear. Neuronal damage may be further inferred via inflammatory reactions generated by brain amyloid deposits. Such reactions may affect both neuronal and glial functions [42], with glial autophagy specifically affecting amyloid processing during the advanced stages of the disease [43].

Moreover, pathogenic variants of proteins, such as mutant tau associated with AD and other proteinopathies, block CMA leading to increased levels of neurofibrillary tangles [44]. When mutant tau binds to the lysosomal surface protein LAMP-2A, it is only partially internalized and the portion that gains entry is trimmed resulting in smaller amyloidogenic tau fragments at the lysosomal membrane [44]. Tau fragment oligomerization disrupts lysosomal membrane integrity and blocks CMA function. In addition, tau oligomers released from the lysosomes upon membrane rupture may act as a nucleating agent to further seed tau aggregation. It was suggested that alterations in mTOR signaling and autophagy occur at early stages of the disease [45]. A significant increase in Aβ (1–42) levels associated with a reduction in autophagy (Beclin 1 and LC3) was observed in postmortem tissue from the inferior parietal lobule of AD, amnestic mild cognitive impairment (MCI), and preclinical AD (PCAD) subjects. Hyperactivation of the PI3K/Akt/mTOR was evident in MCI and AD subjects, but not in PCAD subjects, indicating that autophagy is dynamically altered early on in the disease pathogenesis of AD.

2.2. Parkinson's disease

Faulty CMA has been widely reported in both familial [28] and sporadic PD [46]. An important role for CMA in familial PD was indicated by sequence analysis showing the presence of CMA-targeting motifs in the majority of PD-related proteins. The two most predominantly mutated proteins affected in PD, α-synuclein, and leucine-rich repeat kinase 2 (LRRK2) have been shown to undergo lysosomal degradation through CMA [28]. Mutant variants of these proteins fail to reach the lysosomal lumen despite recognition by cytosolic hsc70 and successful delivery to the lysosomal membrane [28, 47]. Aberrant interactions of these toxic proteins with lysosomal surface protein LAMP-2A obstruct internalization [28]. Importantly, such toxic interactions not only impede the degradation of these proteins but also obstruct the degradation of other CMA substrates [28, 47]. In sporadic PD, post-translational modifications caused by environmental or cellular stressors may reduce dopamine-modified α-synuclein susceptibility to CMA degradation in a manner similar to mutant α-synuclein [46]. Moreover, the persistent binding of modified forms of α-synuclein to the lysosomal membrane promotes the formation of highly toxic α-synuclein oligomers or protofibrils. Studies show that an increase in the cellular levels of either α-synuclein [28] or LRRK2 [47] beyond a tolerable threshold has similar inhibitory effects on CMA activity even in the absence of modifications. Aberrant α-

synuclein not only inhibits CMA but also inhibits autophagy [48], while the overexpression of α-synuclein blocks autophagosome formation [49]. The block in autophagy through α-synuclein overexpression presents early, prior to autophagosome formation, suggesting an effect on Atg9, the only transmembrane autophagic protein.

Several genes related to PD participate in the removal of damaged mitochondria via the specialized form of autophagy, termed mitophagy [49]. In nearly 50% of autosomal recessive PD, and about 15% of sporadic early-onset PD cases, the PARK2 gene is mutated. The gene product of PARK2, PARKIN, is a ubiquitin E3 ligase containing a ubiquitin-like domain, two RING finger domains, and a conserved region between the RING domains [50]. PARKIN, a cytosolic protein, plays an important role in eliminating dysfunctional mitochondria [51]. It is recruited to the membrane of damaged mitochondria and promotes their autophagic degradation [52]. Degradation of mitochondria is both dependent on the expression of PARKIN and the presence of Atgs. Another PD-related protein, PTEN-induced kinase 1 (PINK1), interacts with PARKIN. p62 connects ubiquitinated proteins to LC3 for degradation via the autophagic pathway [18] and the loss of mitochondrial membrane potential promotes p62 accumulation on clustered mitochondria in a PARKIN-dependent fashion. It remains, however, less clear whether p62 is required for mitophagy [53, 54].

2.3. Huntington's disease

Wild-type Huntingtin protein (Htt) is a short-lived, regulatory protein usually degraded through the ubiquitin proteasome system (UPS) [55]. In HD, the long polyQ may affect the UPS by obstructing the system with mutant Htt (mHtt) [55, 56]; however, the exact affliction remains less clear. In HD, a unique situation arises compared to other neurodegenerative proteinopathies: Apart from autophagy being dysfunctional, wild-type Huntingtin protein (Htt) plays multiple roles in regulating the dynamics of the autophagic process [57]. mHtt contributes toward the induction of autophagy through mTOR sequestering and inactivation [58]. Importantly, the autophagosomes detected, while increased in abundance, appear devoid of contents indicating cargo recognition failure [59]. Hence, a situation arises where aggregated proteins and damaged organelles are not readily degraded despite the increase in autophagic induction. The presence of mHtt results in defective autophagy, leading to increased accumulation of protein aggregates, which in turn leads to compensatory upregulation of autophagy, resulting in accumulation of mHtt and subsequent toxicity [57]. mHtt affects autophagosome motility and prevents their fusion with lysosomes, further contributing to the heightened autophagosome pool size [60]. However, the exact point in disease pathogenesis during which the specific molecular defects manifests remains elusive. Fusion dynamics may be affected early on in the disease leading to compensation through alternative pathways followed by autophagic failure to recognize mHtt and subsequent toxicity, or *vice versa* [57]. In order to implement a successful autophagic therapeutic strategy in neurodegeneration, such defects need to be precisely mapped and quantified, in order to correct and offset a specific flux deviation.

3. Spatiotemporal changes of MA and CMA flux in the pathogenesis of neurodegeneration

Functional autophagic flux involves both the execution of autophagosome formation and lysosomal clearance, and dual evaluation is required when studying disease pathology [38]. The presence of autophagosomes alone is not a measure of functional autophagy and autophagic flux; the net rate of autophagosome content degradation [19, 61] reflects the efficiency of the process. Many neurodegenerative diseases have been characterized by a low autophagic flux leading to accumulation of diseased proteins and neurotoxicity [62]. Reports on autophagic flux are often contradictory as dysfunction in multiple steps of the pathway may be implicated. In the case of HD, for example, human and rodent samples have been reported to display increased numbers of autophagosomes while, at the same time, maintaining basal, or even increased, levels of autophagic flux compared to wild-type controls [59, 63]. In AD, decreased expression of autophagic induction proteins and increased activity of autophagy-suppressing molecules indicate impaired autophagic induction [30, 64]. However, accumulating intermediate AVs containing partially digested cargo indicates intact autophagic induction and failure instead of substrate clearance [23, 30]. Given the number of pathological events occurring in the lysosomal network of AD neurons, such changes in autophagic status are likely to reflect different stages of AD progression. During normal ageing, autophagy is downregulated; however, transcriptional regulation thereof seems to be upregulated in AD brains [32, 65]. This upregulation may represent a compensatory attempt to increase flux affected by the defective autophagosome maturation that occurs in AD neurons [38]. Impaired autophagy was suggested to occur early in the onset of AD and the dysregulated overcompensation in the advanced stages instead. It becomes clear that a fine dissection and quantification of autophagic flux [19] are required to better elucidate the extent of pathway failure and to better align autophagy modulating drugs to compensate for the existing offset.

4. Autophagy biomarkers?

There is currently an urgent need for validated biomarkers to guide clinical diagnosis in the early stages of neurodegenerative disease progression, to estimate disease risk, to evaluate disease stages, and to monitor progression and/or response to therapy before the brain is irreversibly damaged [66]. Some of the earliest pathogenic events in AD have also been linked to the Aβ clearance systems, which consists of an interconnected vesicular network of endosomes, lysosomes, and autophagosomes [67]. These alterations are followed by an increase in lysosomal biogenesis, autophagy impairment, and loss of function in genes and proteins related to the lysosomal system in AD [23]. A recent study investigated whether alterations in the lysosomal system are mirrored in the CSF of AD patients and found that the lysosomal proteins LAMP-1 and LAMP-2 were significantly upregulated in the CSF of AD patients [68]. Moreover, strongly reduced BECN1 levels have been observed in the affected brain regions of presymptomatic AD patients compared with controls [69]. APP-transgenic mice with a homozygous BECN1 deletion ($BECN1^{-/-}$) died during embryogenesis [70], whereas mice containing a heterozygous deletion, that is BECN1$^{+/-}$, revealed increased Aβ plaque deposition, neuronal loss, and prominent accumulation of dysfunctional lysosomes containing electron-dense material [30].

These data indicate that the autophagy profile changes substantially in the disease pathogenesis, increasing the complexity of treating neuronal autophagy dysfunction. Brain imaging studies of AD disease progression have previously been monitored by the presence of tangle-bearing neurons in selective brain areas classified into Braak stages 0–VI [71, 72]. In Braak stages V and VI, the clinical diagnosis of dementia is made as NFC-associated neuropathology is spread throughout most parts of the neocortex [71, 72]. However, few studies have investigated the alterations in gene expression patterns throughout the entire course of AD progression. These and above data strongly highlight that an assessment of the autophagy proteome, autophagic as well as CMA flux parameters, and a correlation with clinical data or Braak stages would be highly beneficial in advancing successful implementation of autophagy modulation in the clinical scenario.

5. MA and CMA in disease-specific target protein clearance

Although knowledge of how autophagy and CMA are linked is limited, these two pathways have been shown to provide an integrated cytoprotective response against various proteotoxic challenges [11]. Indeed, experimental inhibition of either pathway has been shown to result in compensatory upregulation of the other, revealing a close "cross talk" between these systems [11, 73] (**Figure 2**). For example, blockage of CMA through Lamp-2A silencing in cultured cells not only leads to the constitutive upregulation of autophagy [11, 73] but also sensitizes cells to various stressors, such as oxidative stress [11, 74]. The autophagy–CMA compensatory response appears to be sequential rather than simultaneous, further stressing the need for a time-dependent flux profile assessment in the disease pathogenesis. For example, autophagy is rapidly upregulated as a transitory response to starvation [75], while CMA is sequentially upregulated in response to long-term starvation following the downregulation of autophagy [76]. In some instances, autophagy and CMA have been shown to degrade the same substrate proteins, but to varying degree. For example, wild-type α-synuclein [28], mutant HTT [77], and mutant tau protein [44] are all degraded by autophagy and CMA (**Figure 2**). Therefore, it is possible that the compensatory upregulation of these pathways may attenuate a specific disease pathogenesis by preferentially targeting and eliminating a specific candidate mutant protein aggregate. Indeed, autophagy has been shown to serve as the primary route for mutant HTT degradation [58, 78] and to eliminate both soluble and mutant tau protein aggregates in vitro and in vivo models [79, 80]. It would therefore be expected that, in the presence of CMA dysfunction, autophagy would be upregulated, thereby enabling CMA-defective cells to maintain their normal protein degradative capacity to sustain cell viability. However, in cortical neurons and differentiated SHSY5Y cells, CMA blockage due to the overexpression of mutant α-synuclein was not found to result in the compensatory upregulation of autophagic activity [81]. Instead, it led to the accumulation of autophagosomes, cytoplasmic release of vacuolar hydrolases, and eventually induced autophagic cell death of primary cortical

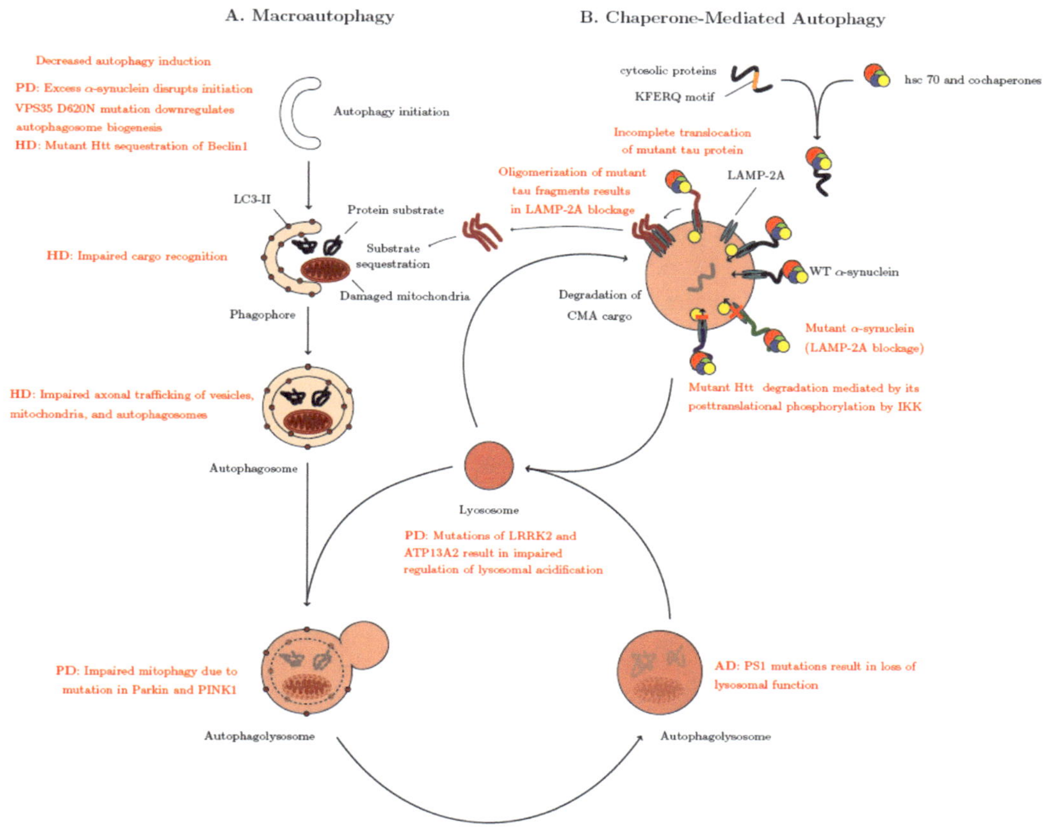

Figure 2. Substantial cross talk exists between macroautophagy (MA) and chaperone mediated autophagy (CMA), with precision defect localization. This calls for the need to precisely map and quantify both MA and CMA fluxes, in order to correct and offset the pathological flux deviation, re-establishing proteostasis.

neurons [82, 83]. In this regard, the interaction between autophagy and CMA may be a detrimental, calling for the need to accurately determine how these pathways are sequentially activated and why the molecular interplay does not always operate functionally.

6. Flux modulation and future outlook

Autophagy can be modulated through mammalian target of rapamycin (mTOR)-dependent and mammalian target of rapamycin (mTOR)-independent pathways using either pharmacological agents or lifestyle interventions. Autophagy upregulation has been shown to clear various types of aggregate prone proteins in vitro [84–86] as well as in vivo [58, 87, 88].Agents such as rapamycin, rilmenidine, lithium, and trehalose have been used in various disease models of AD, PD, and HD and have been shown to reduce the disease pathology (**Table 1**). However, the application of these drugs in different cell types, at different concentrations, or varying time durations makes conclusive flux modulation challenging. In addition, autophagic flux was not always assessed, as the techniques and the approach to accurately measure autophagy activity have often evolved in parallel. For example, rapamycin has been used to induce autophagy and has been shown to protect against AD and HD when administered in COS-7 cells for 15 and 48 h [84, 86] and in PD models when given at 2 μg/ml in PC12 cells for 48 h [85] (**Table 1**). The same can be reported in in vivo studies where rapamycin has been used at different concentration and durations, for example 2.24or 14 mg/kg for 13 weeks or 16 months, respectively [89, 90]. Rilmenidine has been shown to reduce aggregate prone proteins associated with HD when administered at 1 μM incortical neurons and PC12 for 8 and 24 h, respectively [87]. Lithium has been shown to increase the clearance of mutant Huntingtin and α-synuclein when given at 10 nM [91] and 15 mM [92] in Hela cells (**Table 1**).

Intervention	Pathology	Pathology specificity	Model system	Concentration applied	Duration applied	Mode of flux assessment	References
Rapamycin	AD	Autophagy induction and autophagosome clearance	COS7	0.2 μg/ml	48 h	–	Berger et al. [86]
	HD	Cargo recognition	CO7	0.2 μg/ml	15 h	–	Ravikumar et al. [84]
	PD	Cargo recognition and autophagy induction	PC12	0.2 μg/ml	48 h	–	Webb et al. [85]
	AD	Autophagy induction and autophagosome clearance	Mouse	2.24 mg/kg	1X per day	WB (LC3II, p62), FM (LC3 II)	Spilman et al. [89]
Rilmenidine	HD			1 μM	8 and 24 h	FM & WB (LC3 II)	Rose et al. [87]
Lithium	HD		Hela	15 mM	48 h/5 days	FM (LC3) WB (p62)0	Wu et al. [92]

The application of these drugs in different cell types, at different concentrations or varying durations applied, makes conclusive flux modulation challenging and calls for enhanced method standardization.

EM, electron microscopy; FM, fluorescence microscopy; and WB, Western blotting.

Table 1. Autophagic flux modulators in key model systems of neurodegenerative disease.

Although it becomes clear that major promise exists to achieve favorable therapeutic effects through autophagy upregulation, it remains largely unclear what the concentration or dose

http://dx.doi.org/10.5772/64197

and the duration of exposure should be. In addition, AD, PD, and HD affect the autophagic pathway in different compartments and subtypes of autophagy, changing autophagic flux distinctively. Increased autophagic induction prior to developing AD-like pathology in 3xTg-AD mice reduces levels of soluble Aβ and tau, but induction after formation of mature plaques and tangles has no effect on either pathology or cognition [90]. In a scenario where the lysosomal clearance of autophagosomes is halted, activation of autophagy will result in an increase in the harmful accumulation of intermediate AVs [93]. In the case of Aβ, it was found that autophagosomes in AD brains may be a major reservoir of Aβ [94]; therefore, enhancement of new autophagosome formation without the parallel increase in their degradation may lead to an increase in Aβ production and subsequent toxicity [95]. Ideally, modification of autophagic failure should improve autophagosome clearance via the lysosome. Thus, restoring normal lysosomal proteolysis may hold a key to optimal therapeutic interventions against AD [33]. Currently, such therapeutic compounds are not yet available. With regard to the role of Htt in regulating autophagy, it is necessary to identify therapeutic targets that are able to both restore Htt function and normalize defects associated with key autophagic processes [57]. CMA regulation also represents a potential therapeutic target given the cross talk that exists between autophagic pathways [96]. Currently, it is, however, unclear to what extent autophagic flux is being affected. This demands a better quantitative assessment of autophagic flux as well as subsequent improved alignment of autophagy modulators, to allow for precision in compensating flux offset. Taken together, upregulation of autophagy may be beneficial, especially in the early stages of disease pathogenesis; however, the precise molecular target within the autophagy machinery as well as the approach and timing of the intervention has to be strongly aligned with the particular disease specific autophagic flux deviation. Future studies will undoubtedly better address these challenges, thereby impacting on the therapeutic success brought about by autophagic flux control.

Acknowledgements

The authors acknowledge financial support from the South African National Research Foundation (NRF), the Medical Research Council as well as the Cancer Association of South Africa (CANSA). We also wish to thank Andre du Toit, Stellenbosch University, South Africa, for assistance in figure development.

Author details

Claudia Ntsapi, Chrisna Swart, Dumisile Lumkwana and Ben Loos*

*Address all correspondence to: bloos@sun.ac.za

Department of Physiological Sciences, Faculty of Natural Sciences, University of Stellenbosch, Stellenbosch, South Africa

References

[1] Ryazanov AG, Nefsky BS. Protein turnover plays a key role in aging. Mech. Ageing Dev. 2002;123:207–213. doi:10.1016/S0047-6374(01)00337-2

[2] Goldberg AL. Protein degradation and protection against misfolded or damaged proteins. Nature. 2003;426:895–899. doi:10.1038/nature02263

[3] Rubinsztein DC. The roles of intracellular protein-degradation pathways in neurodegeneration. Nature. 2006;443: 780–786. doi:10.1038/nature05291

[4] Nixon R A. Autophagy induction and autophagosome clearance in neurons: relationship to autophagic pathology in Alzheimer's disease. J. Neurosci. Off. J. Soc. Neurosci. 2008;28:6926–6937. doi:10.1523/JNEUROSCI.0800-08.2008

[5] Winslow AR, Rubinsztein DC. Autophagy in neurodegeneration and development. Biochim. Biophys. Acta. 2008; 1782:723–729. doi:10.1016/j.bbadis.2008.06.010

[6] Mizushima N, Levine B, Cuervo AM, Klionsky DJ. Autophagy fights disease through cellular self-digestion. Nature. 2008; 451:1069–1075. doi:10.1038/nature06639

[7] Mizushima N. Autophagy: process and function. Genes Dev. 2007; 21:2861–2873. doi: 10.1101/gad.1599207

[8] Cuervo AM. Chaperone-mediated autophagy: selectivity pays off. Trends Endocrinol. Metab. 2010; TEM 21: 142–150. doi:10.1016/j.tem.2009.10.003

[9] Dice JF. Peptide sequences that target cytosolic proteins for lysosomal proteolysis. Trends Biochem. Sci. 1990; 15: 305–309. doi:10.1016/0968-0004(90)90019-8

[10] Kiffin R, Christian C, Knecht E, Cuervo, AM. Activation of chaperone-mediated autophagy during oxidative stress. Mol. Biol. Cell. 2004; 15: 4829–4840. doi:10.1091/mbc.E04-06-0477

[11] Massey AC, Kaushik S, Sovak G, Kiffin R, Cuervo AM. Consequences of the selective blockage of chaperone-mediated autophagy. Proc. Natl. Acad. Sci. USA. 2006b; 103: 5805–5810. doi:10.1073/pnas.0507436103

[12] Mizushima N, Yamamoto A, Matsui M, Yoshimori T, and Ohsumi Y. In vivo analysis of autophagy in response to nutrient starvation using transgenic mice expressing a fluorescent autophagosome marker. Mol. Biol. Cell 2004;15: 1101–1111. doi:10.1091/mbc.E03-09-0704

[13] Loos B, Lockshin R, Klionsky DJ, Engelbrecht AM and Zakheri Z. On the variability of autophagy and cell death susceptibility. Autophagy. 2013; 9(9): 1270–1285. doi:10.4161/auto.25560

[14] Klionsky DJ, Cuervo AM, Seglen PO. Methods for monitoring autophagy from yeast to human. Autophagy. 2007; 3:181–206. doi:10.4161/auto.3678

[15] Klionsky DJ, Abdelmohsen K, Abe A, Abedin MJ, Abeliovich H, Acevedo-Arozena A, Adach H, Adams CM, Adam PD, Adeli K and Adhihetty PJ. Guidelines for the use and interpretation of assays for monitoring autophagy. Autophagy 2016; 12(1): 1–222. doi: 10.1080/15548627.2015.1100356

[16] Rubinsztein DC, Cuervo AM, Ravikumar B, Sarkar S, Korolchuk VI, Kaushik S, Klionsky DJ. In search of an "autophagomometer". Autophagy 2009; 5:585–589. doi: 10.4161/auto.5.5.8823

[17] Bjørkøy G, Lamark T, Brech A, Outzen H, Perander M, Øvervatn A, Stenmark H, Johansen T. p62/SQSTM1 forms protein aggregates degraded by autophagy and has a protective effect on huntingtin-induced cell death. J Cell Biol 2005; 171:603-14. doi: 10.1083/jcb.200507002

[18] Pankiv, S., Clausen, T.H., Lamark, T., Brech, A., Bruun, J.-A., Outzen, H., Øvervatn, A., Bjørkøy, G., and Johansen, T. p62/SQSTM1 binds directly to Atg8/LC3 to facilitate degradation of ubiquitinated protein aggregates by autophagy. J. Biol. Chem. 2007;282: 24131–24145. doi:10.1074/jbc.M702824200

[19] Loos B, du Toit A and Hofmeyr JHS. Defining and measuring autophagosome flux—concept and reality. Autophagy. 2014;10: 2087–2096. doi:10.4161/15548627.2014.973338

[20] Koga H, Martinez-Vicente M, Arias E, Kaushik S, Sulzer D and Cuervo AM. Constitutive upregulation of chaperone-mediated autophagy in Huntington's disease. J. Neurosci. Off. J. Soc. Neurosci. 2011;31: 18492–18505. doi:10.1523/JNEUROSCI.3219-11.2011

[21] Tsvetkov AS, Arrasate M, Barmada S, Ando DM, Sharma P, Shaby BA and Finkbeiner S. Proteostasis of polyglutamine varies among neurons and predicts neurodegeneration. Nat Chem Biol 2013; 9: 586–592. doi:10.1038/nchembio.1308

[22] Morimoto RI, Driessen AJM, Hegde RS and Langer T. The life of proteins: the good, the mostly good and the ugly. Nat. Struct. Mol. Biol. 2011; 18: 1–4. doi: 10.1038/nsmb0111-1

[23] Nixon RA, Wegiel J, Kumar A, Yu WH, Peterhoff C, Cataldo A and Cuervo AM. Extensive involvement of autophagy in Alzheimer disease: an immuno-electron microscopy study. J. Neuropathol. Exp. Neurol. 2005; 64: 113–122. doi:10.1093/jnen/64.2.113 113-122

[24] Cataldo AM, Peterhoff CM, Schmidt SD, Terio NB, Duff K, Beard M, Mathews PM and Nixon RA. Presenilin mutations in familial Alzheimer disease and transgenic mouse models accelerate neuronal lysosomal pathology. J. Neuropathol. Exp. Neurol. 2004; 63: 821–830. doi:10.1074/jbc.M801279200

[25] Anglade P, Vyas S, Javoy-Agid F, Herrero MT, Michel PP, Marquez J, Mouatt-Prigent A, Ruberg M, Hirsch EC and Agid Y. Apoptosis and autophagy in nigral neurons of

patients with Parkinson's disease. Histol. Histopathol. 1997; 12: 25–31. doi:10.1111/jnc.13266

[26] Petersén A, Larsen KE, Behr GG, Romero N, Przedborski S, Brundin P and Sulzer D. Expanded CAG repeats in exon 1 of the Huntington's disease gene stimulate dopamine-mediated striatal neuron autophagy and degeneration. Hum. Mol. Genet. 2001; 10: 1243–1254. doi:10.1093/hmg/10.12.1243

[27] Barnett A and Brewer GJ. Autophagy in aging and Alzheimer's disease: pathologic or protective? J. Alzheimers Dis. JAD 2011; 25: 385–394. doi:10.3233/JAD-2011-101989

[28] Cuervo AM, Stefanis L, Fredenburg R, Lansbury PT and Sulzer D. Impaired degradation of mutant alpha-synuclein by chaperone-mediated autophagy. Science. 2004; 305: 1292–1295. doi:10.1126/science.1101738

[29] Liang XH, Jackson S, Seaman M, Brown K, Kempkes B, Hibshoosh H and Levine B. Induction of autophagy and inhibition of tumorigenesis by beclin 1. Nature. 1999; 402: 672–676. doi:10.1038/45257

[30] Pickford F, Masliah E, Britschgi M, Lucin K, Narasimhan R, Jaeger PA, Small S, Spencer B, Rockenstein E, Levine B, et al. The autophagy-related protein beclin 1 shows reduced expression in early Alzheimer disease and regulates amyloid β accumulation in mice. J. Clin. Invest. 2008; 118: 2190–2199. doi:10.1172/JCI33585

[31] Rohn TT, Wirawan E, Brown RJ, Harris JR, Masliah E and Vandenabeele P. Depletion of Beclin-1 due to proteolytic cleavage by caspases in the Alzheimer's disease brain. Neurobiol. Dis. 2011; 43: 68–78. doi:10.1016/j.nbd.2010.11.003

[32] Lipinski MM, Zheng B, Lu T, Yan Z, Py BF, Ng A, Xavier RJ, Li C, Yankner BA, Scherzer CR., et al. Genome-wide analysis reveals mechanisms modulating autophagy in normal brain aging and in Alzheimer's disease. Proc. Natl. Acad. Sci. USA. 2010; 107: 14164–14169. doi:10.1073/pnas.1009485107

[33] Yang DS, Stavrides P, Mohan PS, Kaushik S, Kumar A, Ohno M, Schmidt SD, Wesson DW, Bandyopadhyay U, Jiang Y., et al. Therapeutic effects of remediating autophagy failure in a mouse model of Alzheimer disease by enhancing lysosomal proteolysis. Autophagy 2011b; 7: 788–789. doi:10.4161/auto.7.7.15596

[34] Boland B, Kumar A, Lee S, Platt FM, Wegiel J, Yu WH and Nixon RA. Autophagy induction and autophagosome clearance in neurons: relationship to autophagic pathology in Alzheimer's disease. J. Neurosci. Off. J. Soc. Neurosci. 2008; 28: 6926–6937. doi:10.1523/JNEUROSCI.0800-08.2008

[35] Koike M, Nakanishi H, Saftig P, Ezaki J, Isahara K, Ohsawa Y, Schulz-Schaeffer W, Watanabe T, Waguri S, Kametaka S., et al. Cathepsin D deficiency induces lysosomal storage with ceroid lipofuscin in mouse CNS neurons. J. Neurosci. Off. J. Soc. Neurosci. 2000;20: 6898–6906. doi:10.1016/S0168-0102(00)81020-2

[36] Querfurth HW and LaFerla FM. Alzheimer's disease. N. Engl. J. Med. 2010;362: 329–344. doi:10.1056/NEJMra0909142

[37] Supnet C and Bezprozvanny I. Presenilins function in ER calcium leak and Alzheimer's disease pathogenesis. Cell Calcium. 2011;50: 303–309. doi:10.1016/j.ceca.2011.05.013

[38] Orr ME and Oddo S. Autophagic/lysosomal dysfunction in Alzheimer's disease. Alzheimers Res. Ther. 2013;5: 53. doi:10.1186/alzrt217

[39] Neely KM, Green KN and LaFerla FM. Presenilin is necessary for efficient proteolysis through the autophagy-lysosome system in a γ-secretase-independent manner. J. Neurosci. Off. J. Soc. Neurosci. 2011;31: 2781–2791. doi:10.1523/JNEUROSCI.5156-10.2010

[40] Esselens C, Oorschot V, Baert V, Raemaekers T, Spittaels K, Serneels L, Zheng H, Saftig P, De Strooper B, Klumperman J., et al. Presenilin 1 mediates the turnover of telencephalin in hippocampal neurons via an autophagic degradative pathway. J. Cell Biol. 2004;166: 1041–1054. doi:10.1083/jcb.200406060

[41] Lee JH, Yu WH, Kumar A, Lee S, Mohan PS, Peterhoff CM, Wolfe DM, Martinez-Vicente M, Massey AC, Sovak G., et al. Lysosomal proteolysis and autophagy require presenilin 1 and are disrupted by Alzheimer-related PS1 mutations. Cell. 2010;141: 1146–1158. doi:10.1016/j.cell.2010.05.008

[42] Mosher KI and Wyss-Coray T. Microglial dysfunction in brain aging and Alzheimer's disease. Biochem. Pharmacol. 2014;88: 594–604. doi:10.1016/j.bcp.2014.01.008

[43] Pomilio C, Pavia P, Gorojod RM, Vinuesa A, Alaimo A, Galvan V, Kotler ML, Beauquis J and Saravia F. Glial alterations from early to late stages in a model of Alzheimer's disease: evidence of autophagy involvement in Aβ internalization. Hippocampus. 2016;26: 194–210. doi:10.1002/hipo.22503

[44] Wang Y, Martinez-Vicente M, Krüger U, Kaushik S, Wong E, Mandelkow EM, Cuervo AM and Mandelkow E. Tau fragmentation, aggregation and clearance: the dual role of lysosomal processing. Hum. Mol. Genet. 2009;18: 4153–4170. doi:10.1093/hmg/ddp367

[45] Tramutola A, Triplett JC, Di Domenico F, Niedowicz DM, Murphy MP, Coccia R, Perluigi M and Butterfield DA. Alteration of mTOR signaling occurs early in the progression of Alzheimer disease (AD): analysis of brain from subjects with pre-clinical AD, amnestic mild cognitive impairment and late-stage AD. J. Neurochem. 2015;133: 739–749. doi:10.1111/jnc.13037

[46] Martinez-Vicente M, Talloczy Z, Kaushik S, Massey AC, Mazzulli J, Mosharov EV, Hodara R, Fredenburg R, Wu DC, Follenzi A., et al. Dopamine-modified alpha-synuclein blocks chaperone-mediated autophagy. J. Clin. Invest. 2008;118: 777–788. doi:10.1172/JCI32806

[47] Orenstein SJ, Kuo SH, Tasset I, Arias E, Koga H, Fernandez-Carasa I, Cortes E, Honig LS, Dauer W, Consiglio A., et al. Interplay of LRRK2 with chaperone-mediated autophagy. Nat. Neurosci. 2013;16: 394–406. doi:10.1038/nn.3350

[48] Winslow AR, Chen CW, Corrochano S, Acevedo-Arozena A, Gordon DE, Peden AA, Lichtenberg M, Menzies FM, Ravikumar B, Imarisio S., et al. α-Synuclein impairs macroautophagy: implications for Parkinson's disease. J. Cell Biol. 2010;190: 1023–1037. doi:10.1083/jcb.201003122

[49] Lynch-Day MA, Mao K, Wang K, Zhao M and Klionsky DJ. The role of autophagy in Parkinson's disease. Cold Spring Harb. Perspect. Med. 2012;2: a009357. doi:10.1101/cshperspect.a009357

[50] Schapira AHV. Mitochondria in the aetiology and pathogenesis of Parkinson's disease. Lancet Neurol. 2008;7: 97–109. doi:10.1002/ana.410440714

[51] Narendra D, Tanaka A, Suen DF and Youle RJ. Parkin is recruited selectively to impaired mitochondria and promotes their autophagy. J. Cell Biol. 2008; 183: 795–803. doi:10.1083/jcb.200809125

[52] Gasser T. Mendelian forms of Parkinson's disease. Biochim. Biophys. Acta 2009; 1792: 587–596. doi:10.1016/j.bbadis.2008.12.007

[53] Geisler S, Holmström KM, Skujat D, Fiesel FC, Rothfuss OC, Kahle PJ and Springer W. PINK1/Parkin-mediated mitophagy is dependent on VDAC1 and p62/SQSTM1. Nat. Cell Biol. 2010; 12: 119–131. doi:10.1038/ncb2012

[54] Narendra D, Kane LA, Hauser DN, Fearnley IM and Youle RJ. p62/SQSTM1 is required for Parkin-induced mitochondrial clustering but not mitophagy; VDAC1 is dispensable for both. Autophagy 2010; 6: 1090–1106. doi:10.4161/auto.6.8.13426

[55] Schipper-Krom S, Juenemann K, Reits EAJ, Schipper-Krom S, Juenemann K and Reits EAJ. The ubiquitin-proteasome system in Huntington's disease: are proteasomes impaired, initiators of disease, or coming to the rescue? Biochem. Res. Int. Biochem. Res. Int. 2012; e837015. doi:10.1155/2012/837015

[56] Kisselev AF, Akopian TN, Woo KM and Goldberg AL. The sizes of peptides generated from protein by mammalian 26 and 20 S proteasomes. Implications for understanding the degradative mechanism and antigen presentation. J. Biol. Chem. 1999; 274 (6): 3363–3371. doi:10.1074/jbc.274.6.3363

[57] Martin DDO, Ladha S, Ehrnhoefer DE and Hayden MR. Autophagy in Huntington disease and Huntingtin in autophagy. Trends Neurosci. 2015; 38: 26–35. doi:10.1016/j.tins.2014.09.003

[58] Ravikumar B, Vacher C, Berger Z, Davies JE, Luo S, Oroz LG, Scaravilli F, Easton DF, Duden R, O'Kane CJ., et al. Inhibition of mTOR induces autophagy and reduces toxicity of polyglutamine expansions in fly and mouse models of Huntington disease. Nat. Genet. 2004; 36: 585–595. doi:10.1038/ng1362

[59] Martinez-Vicente M, Talloczy Z, Wong E, Tang G, Koga H, Kaushik S, de Vries R, Arias E, Harris S, Sulzer D., et al. Cargo recognition failure is responsible for inefficient autophagy in Huntington's disease. Nat. Neurosci. 2010;13: 567–576. doi:10.1038/nn.2528

[60] Wong YC and Holzbaur ELF. The regulation of autophagosome dynamics by huntingtin and HAP1 is disrupted by expression of mutant huntingtin, leading to defective cargo degradation. J. Neurosci. Off. J. Soc. Neurosci. 2014; 34: 1293–1305. doi:10.1523/JNEUROSCI.1870-13.2014

[61] Klionsky DJ, Abdalla FC, Abeliovich H, Abraham RT, Acevedo-Arozena A, Adeli K, Agholme L, Agnello M, Agostinis P, Aguirre-Ghiso JA, et al. Guidelines for the use and interpretation of assays for monitoring autophagy. Autophagy. 2012; 8: 445–544. doi:10.4161/auto.19496

[62] Menzies FM, Moreau K and Rubinsztein DC. Protein misfolding disorders and macroautophagy. Curr. Opin. Cell Biol. 2011; 23: 190–197. doi:10.1016/j.ceb.2010.10.010

[63] Kegel KB, Kim M, Sapp E, McIntyre C, Castaño JG, Aronin N and DiFiglia M. Huntingtin expression stimulates endosomal-lysosomal activity, endosome tubulation, and autophagy. J. Neurosci. Off. J. Soc. Neurosci. 2000; 20: 7268–7278. doi:10.1016/j.toxlet.2014.01.039

[64] Funderburk SF, Wang QJ and Yue Z. The Beclin 1–VPS34 complex—at the crossroads of autophagy and beyond. Trends Cell Biol. 2010; 20: 355–362. doi:10.1016/j.tcb.2010.03.002

[65] Ginsberg SD, Mufson EJ, Counts SE, Wuu J, Alldred MJ, Nixon RA and Che S. Regional selectivity of rab5 and rab7 protein upregulation in mild cognitive impairment and Alzheimer's disease. J. Alzheimers Dis. JAD 2010; 22: 631–639. doi:10.3233/JAD-2010-101080

[66] Blennow K, Hampel H, Weiner M, Zetterberg H. Cerebrospinal fluid and plasma biomarkers in Alzheimer disease. Nat. Rev. Neurol. 2010; 6: 131–144. doi:10.1038/nrneurol.2010.4

[67] Ihara Y, Morishima-Kawashima M, Nixon R. The ubiquitin-proteasome system and the autophagic-lysosomal system in Alzheimer disease. Cold Spring Harb. Perspect. Med. 2012; 2: 1–27. doi:10.1101/cshperspect.a006361

[68] Armstrong A, Mattsson N, Appelqvist H, Janefjord C, Sandin L, Agholme L, Olsson B, Svensson S, Blennow K, Zetterberg H, Kågedal K. Lysosomal network proteins as potential novel CSF biomarkers for Alzheimer's disease. Neuromolecular Med. 2014; 16: 150–160. doi:10.1007/s12017-013-8269-3

[69] Small SA, Kent K, Pierce A, Leung C, Kang MS, Okada H, Honig L, Vonsattel JP, Kim TW. Model-guided microarray implicates the retromer complex in Alzheimer's disease. Ann. Neurol. 2005; 58: 909–919. doi:10.1002/ana.20667

[70] Qu X, Yu J, Bhagat G, Furuya N, Hibshoosh H, Troxel A, Rosen J, Eskelinen EL, Mizushima N, Ohsumi Y, Cattoretti G, Levine B. Promotion of tumorigenesis by heterozygous disruption of the beclin 1 autophagy gene. J. Clin. Invest. 2003; 112: 1809–1820. doi:10.1172/JCI20039

[71] Braak H and Braak E. Neuropathological stageing of Alzheimer-related changes. Acta Neuropathol. (Berl.) 1991; 82: 239–259. doi:10.1007/BF00308809

[72] Braak H, Alafuzoff I, Arzberger T, Kretzschmar H, Del Tredici K.. Staging of Alzheimer disease-associated neurofibrillary pathology using paraffin sections and immunocytochemistry. Acta Neuropathol. (Berl.) 2006; 112: 389–404. doi:10.1007/s00401-006-0127-z

[73] Vogiatzi T, Xilouri M, Vekrellis K and Stefanis L. Wild type alpha-synuclein is degraded by chaperone-mediated autophagy and macroautophagy in neuronal cells. J. Biol. Chem. 2008;283: 23542–23556. doi:10.1074/jbc.M801992200

[74] Kaushik S, Massey AC, Mizushima N, Cuervo AM. Constitutive Activation of chaperone-mediated autophagy in cells with impaired macroautophagy. Mol. Biol. Cell. 2008; 19: 2179–2192. doi:10.1091/mbc.E07-11-1155

[75] Massey AC, Kaushik S, Cuervo AM. Lysosomal chat maintains the balance. Autophagy. 2006a; 2: 325–327. doi:10.4161/auto.3090

[76] Cuervo AM, Knecht E, Terlecky SR, Dice JF. Activation of a selective pathway of lysosomal proteolysis in rat liver by prolonged starvation. Am. J. Physiol. 1995; 269: C1200–1208.

[77] Bauer PO, Goswami A, Wong HK, Okuno M, Kurosawa M, Yamada M, Miyazaki H, Matsumoto G, Kino Y, Nagai Y, Nukina N. Harnessing chaperone-mediated autophagy for the selective degradation of mutant Huntingtin protein. Nat. Biotechnol. 2010;.28: 256–263. doi:10.1038/nbt.1608

[78] Sarkar S, Perlstein EO, Imarisio S, Pineau S, Cordenier A, Maglathlin RL, Webster JA, Lewis TA, O'Kane CJ, Schreiber SL, Rubinsztein DC. Small molecules enhance autophagy and reduce toxicity in Huntington's disease models. Nat. Chem. Biol. 2007;.3: 331–338. doi:10.1038/nchembio883

[79] Krüger U, Wang Y, Kumar S, Mandelkow EM. Autophagic degradation of tau in primary neurons and its enhancement by trehalose. Neurobiol. Aging. 2012; 33: 2291–2305. doi:10.1016/j.neurobiolaging.2011.11.009

[80] Rodríguez-Navarro JA, Rodríguez L, Casarejos MJ, Solano RM, Gómez A, Perucho J, Cuervo AM, García de Yébenes J, Mena MA. Trehalose ameliorates dopaminergic and tau pathology in parkin deleted/tau overexpressing mice through autophagy activation. Neurobiol. Dis. 2010; 39: 423–438. doi:10.1016/j.nbd.2010.05.014

[81] Xilouri M, Vogiatzi T, Vekrellis K, Park D and Stefanis L. Abberant α-synuclein confers toxicity to neurons in part through inhibition of chaperone-mediated autophagy. PLoS One. 2009;4: e5515. doi:10.1371/journal.pone.0005515

[82] Bains M, Florez-McClure ML, Heidenreich KA. Insulin-like growth factor-I prevents the accumulation of autophagic vesicles and cell death in Purkinje neurons by increasing the rate of autophagosome-to-lysosome fusion and degradation. J. Biol. Chem. 2009; 284: 20398–20407. doi:10.1074/jbc.M109.011791

[83] Kroemer G and Jäättelä M. Lysosomes and autophagy in cell death control. Nat. Rev. Cancer. 2005; 5: 886–897. doi:10.1038/nrc1738

[84] Ravikumar B, Duden R and Rubinsztein DC. Aggregate-prone proteins with polyglutamine and polyalanine expansions are degraded by autophagy. Hum. Mol. Genet. 2002;11: 1107–1117. doi:10.1093/hmg/11.9.1107

[85] Webb JL, Ravikumar B, Atkins J, Skepper JN, Rubinsztein DC. Alpha-synuclein is degraded by both autophagy and the proteasome. J. Biol. Chem. 2003; 278: 25009–25013. doi:10.1074/jbc.M300227200

[86] Berger Z. Ravikumar B, Menzies FM, Oroz LG, Underwood BR, Pangalos MN, Schmitt I, Wullner U, Evert BO, O'Kane CJ, Rubinsztein DC. Rapamycin alleviates toxicity of different aggregate-prone proteins. Hum Mol Genetics. 2006;15(3): 433–442. doi: 10.1093/hmg/ddi458

[87] Rose C, Menzies FM, Renna M, Acevedo-Arozena A, Corrochano S, Sadiq O, Brown SD and Rubinsztein DC. Rilmenidine attenuates toxicity of polyglutamine expansions in a mouse model of Huntington's disease. Hum Mol Genetics. 2010; 19(11): 2144–2153. doi:10.1093/hmg/ddq093

[88] Hebron ML, Lonskaya I and Moussa CEH. Nilotinib reverses loss of dopamine neurons and improves motor behavior via autophagic degradation of α-synuclein in Parkinson's disease models. Hum Mol Genetics. 2013. 22(16): 3315–28. doi: 10.1093/hmg/ddt192

[89] Spilman P, Podlutskaya N, Hart MJ, Debnath J, Gorostiza O, Bredesen D, Richardson A, Strong R and Galvan V. Inhibition of mTOR by rapamycin abolishes cognitive deficits and reduces amyloid-βlevels in a mouse model of Alzheimer's disease. PLoS One. 2010;5: e9979. doi:10.1371/journal.pone.0009979

[90] Majumder S, Richardson A, Strong R and Oddo S. Inducing Autophagy by rapamycin before, but not after, the formation of plaques and tangles ameliorates cognitive deficits. PLoS One 2011;6: e25416. doi:10.1371/journal.pone.0025416

[91] Sarkar S, Krishna G, Imarisio S, Saiki S, O'Kane CJ and Rubinsztein DC. A rational mechanism for combination treatment of Huntington's disease using lithium and rapamycin. Hum. Mol. Genet. 2008;17: 170–178. doi:10.1093/hmg/ddm294

[92] Wu S, Zheng SD, Huang HL, Yan LC, Yin XF, Xu HN, Zhang KJ, Gui JH, Chu L, Liu XY. Lithium down-regulates histone deacetylase 1 (HDAC1) and induces degradation of mutant huntingtin. J. Biol. Chem. 2013; 288 (49): 35500–35510. doi:10.1074/jbc.M113.479865

[93] Liang JH and Jia JP. Dysfunctional autophagy in Alzheimer's disease: pathogenic roles and therapeutic implications. Neurosci. Bull. 2014; 30: 308–316. doi:10.1007/s12264-013-1418-8

[94] Yu WH, Cuervo AM, Kumar A, Peterhoff CM, Schmidt SD, Lee JH, Moha PS, Mercken M, Farmery MR, Tjernberg LO, et al. Macroautophagy - a novel Beta-amyloid peptide-generating pathway activated in Alzheimer's disease. J. Cell Biol. 2005;171: 87–98. doi: 10.1083/jcb.200505082

[95] Nixon RA. Autophagy, amyloidogenesis and Alzheimer disease. J. Cell Sci. 2007;120: 4081–4091. doi:10.1242/jcs.019265

[96] Bejarano E and Cuervo AM. Chaperone-mediated autophagy. Proc. Am. Thorac. Soc. 2010;7: 29–39. doi:10.1513/pats.200909-102JS.

Chapter 10

Role of Autophagy in Burn Wound Progression and Wound Healing

Ligen Li and Mengjing Xiao

Additional information is available at the end of the chapter

http://dx.doi.org/10.5772/63711

Abstract

Background: Burn wound progression refers to the phenomenon of continued tissue loss following abatement of an initial thermal insult, which makes the treatment more difficult, prolongs hospital stay, increases medical costs, and raises the likelihood of scar formation. Autophagy is a highly conserved pathway that delivers intracellular macromolecule waste to lysosomes, where they are degraded into biologically active monomers, such as amino acids, that are subsequently reused to maintain cellular metabolic turnover and homeostasis. We aimed to observe the expression of autophagy in burn wounds in a deep second-degree rodent burn model and further investigate the role of autophagy on burn wound progression and wound healing.

Methods: A rat deep second-degree burn model was established by placing a 100°C brass rod to the shaved skin of rats for 6 s. Rats were randomly divided into sham burn, burn, burn treated with rapamycin, and burn treated with vehicle. Immediately after creating the thermal injury, rats in the treatment group were given rapamycin 1 mg/kg by intraperitoneal injection, and the rats in the vehicle group were injected with equivalent carrier solution. Skin tissue specimens from burn wounds were taken for assessment of autophagy, apoptosis, myeloperoxidase activity, methanedicarboxylicaldehyde contents, laser Doppler flow values, H&E, and Masson staining.

Results: Reduced levels of autophagy, impaired blood flow, together with increased levels of apoptosis, inflammation, and myeloperoxidase activity during the early course of burn wound progression were observed. However, enhanced autophagy in the deep dermal layers was observed, which may function as a prosurvival mechanism against inflammation and ischemia.

Rapamycin enhanced the level of autophagy in burn wounds, ameliorated the early progression of burn depth, and promoted burn wound healing, possibly by protecting the zone of stasis from further necrosis and by reducing the extent or level of apoptosis commonly seen in burn wounds.

Conclusion: Autophagy protects from burn wound progression and promotes burn wound healing through protecting the zone of stasis from further necrosis, which is probably mediated by phagocytizing damaged organelles to sustain homeostatic response and remodeling in the injured tissue.

Keywords: Autophagy, Burn, injury progression, wound healing, rapamycin

1. Introduction

After the initial thermal insult, tissue necrosis is continued in burn wound, which is referred to as the phenomenon of "burn wound progression" [1]. The progression of burn wound is an important problem in clinic, largely determining the morbidity and mortality of burn patients. Generally, the burn wound can be divided into three concentric zones [2]: the zone of coagulation, the zone of stasis, and the zone of hyperemia. While the central zone of coagulation is distinguished by irreversible tissue necrosis and the outer zone of hyperemia can always naturally recover, the intermediate zone of stasis may either progress into necrosis or can be salvaged if optimal treatments be available [3].

It has been demonstrated that burn wound progression is attributed to many mechanisms, including tissue hypoperfusion, prolonged inflammation, edema, hypercoagulability, and free radical damage [4]. Many studies have aimed to ameliorate burn wound progression by intervening these above mechanisms, and certain advance has already been achieved. However, further studies are still needed to help us better understand this complicated process.

Autophagy is a lysosomal-dependent degradation pathway, which is indispensable for survival, development, differentiation, and cellular homeostasis [5]. Autophagy has been reported to protect against ischemic injuries, inflammatory diseases, and other disease models [6]. To understand the role of autophagy in burn wound progression, firstly, we designed an animal study to examine the expressing pattern of autophagy in burn wounds. We found that autophagy level in the zone of stasis was increased when comparing to the normal unburned tissue. We speculated that the increase of autophagy level in the zone of stasis was a survival mechanism against tissue ischemia, excessive inflammatory response, and oxidative stress in burn wounds [7].

Rapamycin is a macrolide antibiotic agent, which was initially used as an antifungal drug, and **has potent** anti-proliferative and immunosuppressive properties. Rapamycin is now a commonly used autophagy-enhancing agent and it can induce autophagy by inhibiting mTOR kinase activity [8]. In the second part of the work, we used a rat model of second degree burns and selected rapamycin as an autophagic-enhancing agent to determine whether it could enhance autophagy in burn wounds and ameliorate burn wound progression while promoting wound healing. We found that rapamycin enhanced the level of autophagy in burn wounds, ameliorated the early progression of burn depth and promoted burn wound healing, possibly by protecting the zone of stasis from further necrosis and by reducing the extent or level of apoptosis commonly seen in burn wounds [9].

2. Materials and methods

See details at Refs. [7, 9].

3. Results

3.1. Expression rule of autophagy proteins in burn wounds

Levels of LC3 and Beclin-1 protein were maintained at a certain degree in normal skin tissue. After burn was inflicted, levels of these proteins declined continuously in wound tissue until 24 h after the burn. Levels then increased slightly, but remained far lower than in normal skin samples. Quantitative analysis show that autophagy level decreased about fourfold over 24 h, and then began to increase but still could not reach their normal level (**Figure 1**).

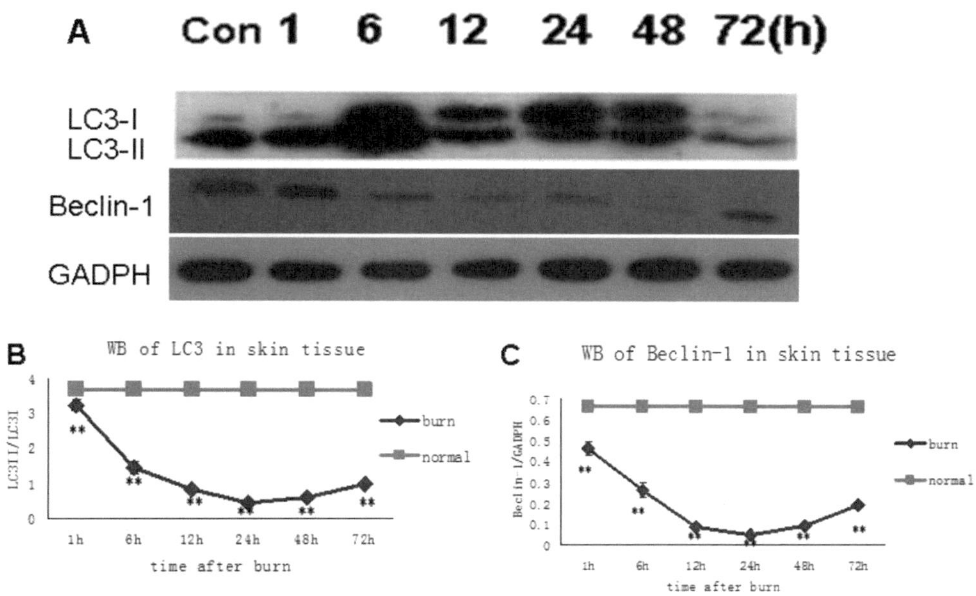

Figure 1. (A) Western blot analyses of LC3 and Beclin-1 in wound and normal unburned skin at different time points after burn. (B, C) Quantitative analysis of LC3II/I and Beclin-1 in burn wounds. B and C show that the levels of these proteins declined continuously in wound tissue until 24 h after the burn. Levels then increased slightly, but remained far lower than in skin samples. The times trends for LC3II/I and Beclin-1 change were almost identical. Data are presented as means ± SD (*$P < 0.05$, **$P < 0.01$, n = 8 per group).

3.2. Change of apoptosis level in burn wounds

To investigate the apoptosis level of burn wounds, we conducted TUNEL staining to see how the apoptosis level in burn wounds would change after burn was inflicted. We found more cells undergoing apoptosis in burn wounds than control. The apoptotic rates in burn wounds at 6, 24, 48, and 72 h are 4.56 ± 0.27, 11.76 ± 0.7, 13.16 ± 0.65, and 8.14 ± 0.62%, respectively, versus 2.23 ± 0.23% of controls (mean ± SD, $P < 0.05$ at 6 h, $P < 0.01$ at 24, 48, 72 h) (**Figure 2**).

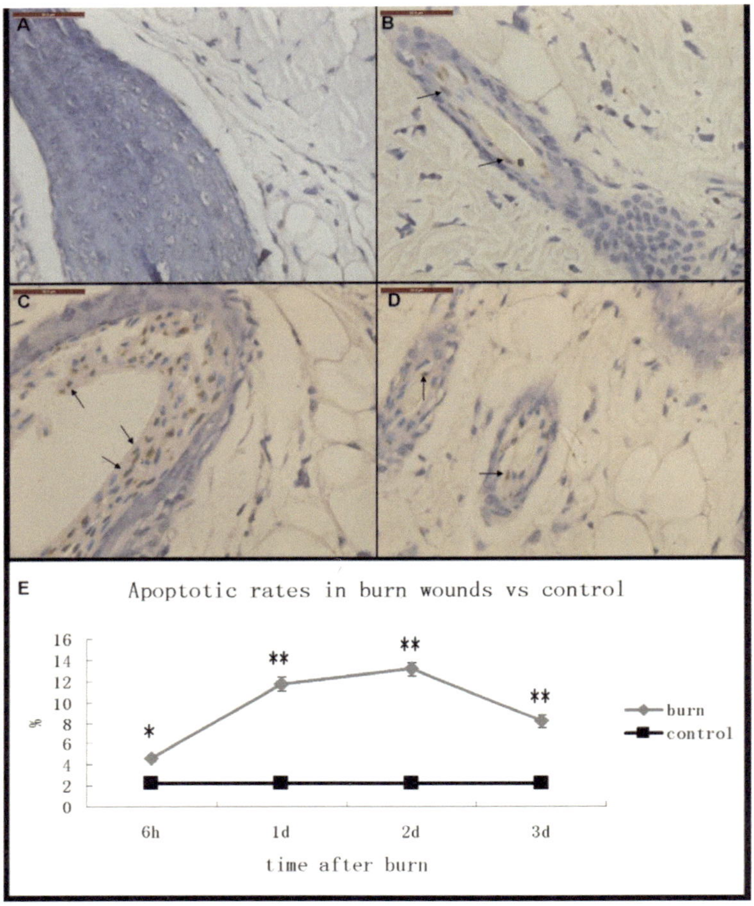

Figure 2. TUNEL staining of normal skin (A) and burn wound tissue 1 day (B), 2 days (C), and 3 days (D) post burn. Cells with shrunken brown stained nuclei were considered positive (black arrows). E: The p values for 6h, 1d, 2d and 3d are 0.028, 0.005, 0.003 and 0.007 respectively. Scale bars = 50 um, $^{*}P < 0.05$, $^{**}P < 0.01$, n = 8 per group.

3.3. Change of Laser Doppler Flowmetry (LDF) value and Myeloperoxidase (MPO) activity in burn wounds

LDF value in burn wounds is an indicator of burn wound blood flow, while MPO activity reflects the inflammatory response in the burn wound. We found that the LDF values decreased fourfold over 12 h after burn and then began to increase but still far lower normal level. The MPO activity increased about 5.3-fold over 48 h after burn and then decreased slightly, yet still significantly higher than control (**Figure 3**).

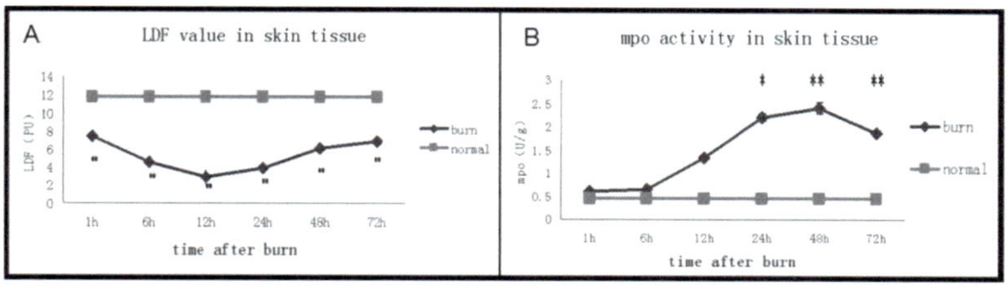

Figure 3. LDF value and MPO activity in burn wounds and normal skin. (A) LDF values in the burn wounds were significantly decreased as compared to the control group. LDF values in the burn wounds decreased fourfold over 12 h and then began to rise, yet still could not reach their normal level. (B) MPO activities in burn wounds were significantly increased than the control. The levels of MPO activity in burn wounds were increased 5.3-fold over 48 h, and then decreased but still far higher than normal. Alterations in the LDF value were significant at all the time points post burn. Changes in the levels of MPO activity between burn and control groups were significant at the time points of 1, 2, and 3 days post burn. The values described herein were mean ± SD (burn versus control group, *$P < 0.05$, **$P < 0.01$, n = 8 per group).

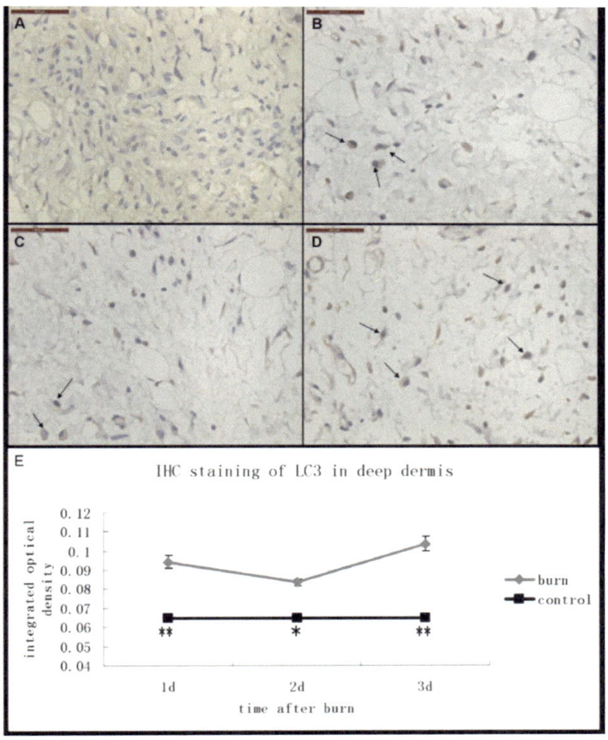

Figure 4. Immunohistochemical staining of LC3 (black arrows, cells with brown cytoplasm) in deep dermis of normal skin (A) and burn wounds (B–D). (B) 1 day post burn; (C) 2 days post burn; (D) 3 days post burn. At all time points after burn, the cells expressing LC3 were increased in the deep dermis of burn wound tissue as compared to that of control. Scale bars = 50 um. (E) Semi-quantitative analysis of the immunohistochemical staining results. The expression of LC3 in the deep dermis of burn wounds was significantly higher than that of the control at all time points. The values herein were mean ± SD (*$P < 0.05$, **$P < 0.01$, n = 8 per group).

3.4. LC3 staining in the zone of stasis

We found that LC3-positive cells in the deep dermis adjacent to the subcutaneous tissue, which in this deep second-degree burn model, may represent the zone of stasis, which were much more than that of control. We conducted a quantitative analysis and found that the increase of LC3-positive cells in the deep dermis was significant at 6, 24, and 48 h post burn (**Figure 4**).

3.5. Rapamycin enhanced the autophagic level in burn wounds

Western bolt analysis showed that the protein levels of two autophagy markers, LC3 and Beclin-1, were significantly augmented in the thermal burn wounds of the rapamycin-treated group as compared with the levels seen in the control group (**Figure 5**). The expression of both proteins in burn wounds were at their lowest 24 h post-thermal burn injury and then gradually increased with time. The patterns of LC3-II/LC3-I and Beclin-1 expression were almost identical (**Figure 5**). Moreover, the increases in expression of these proteins were statistically significant at 6 h ($p < 0.05$) and at 24 and 48 h post-thermal burn injury ($p < 0.01$).

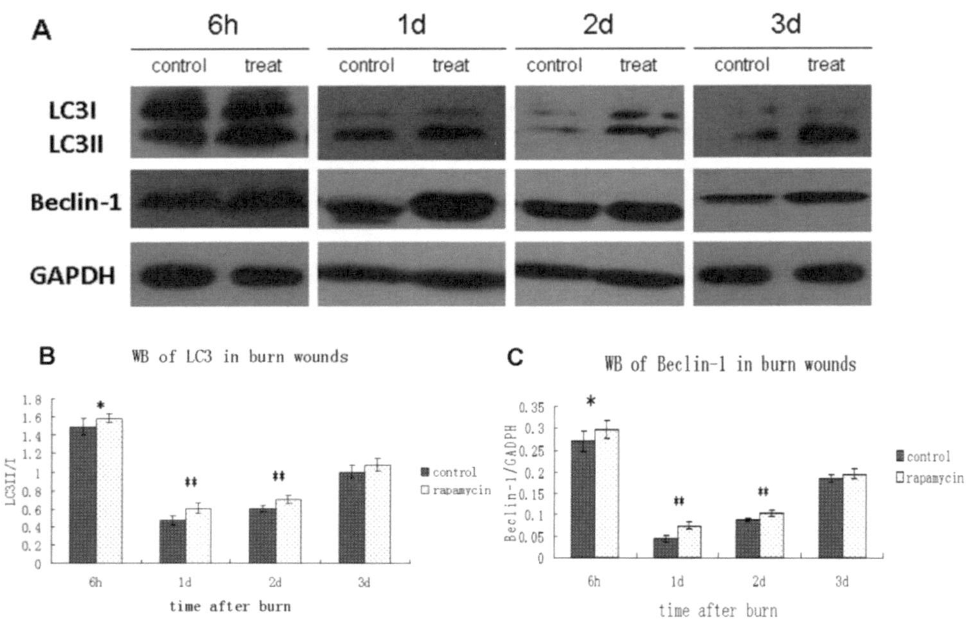

Figure 5. Western immunoblot assay of LC3 and Beclin-1 expression in burn wounds of control and treatment groups (A–C). The values herein were mean ± SD (*$p < 0.05$, **$p < 0.01$, n = 8 per group).

3.6. Rapamycin treatment inhibited apoptosis in burn wounds

To further investigate if the enhanced autophagy had any effects on apoptosis of burn wounds, we conducted TUNEL staining to see if the apoptosis level in burn wounds of treated rats were different from controls. Cells with shrunken brown stained nuclei were considered positive.

We found fewer cells undergoing apoptosis in rats treated with rapamycin. The apoptotic rates in burn wounds of treated rats at 6, 24, 48, and 72 h are 2.99 ± 0.33, 8.85 ± 0.6, 12.48 ± 0.67, and 6.7 ± 0.31%, respectively, versus 4.56 ± 0.27, 11.76 ± 0.7, 13.16 ± 0.65, and 8.14 ± 0.62% of controls (mean ± SD, $p < 0.05$ at 6, 24, 72 h). Besides, we found that the TUNEL-positive cells were mainly observed in the epithelium of hair follicles or blood vessels (**Figure 6**).

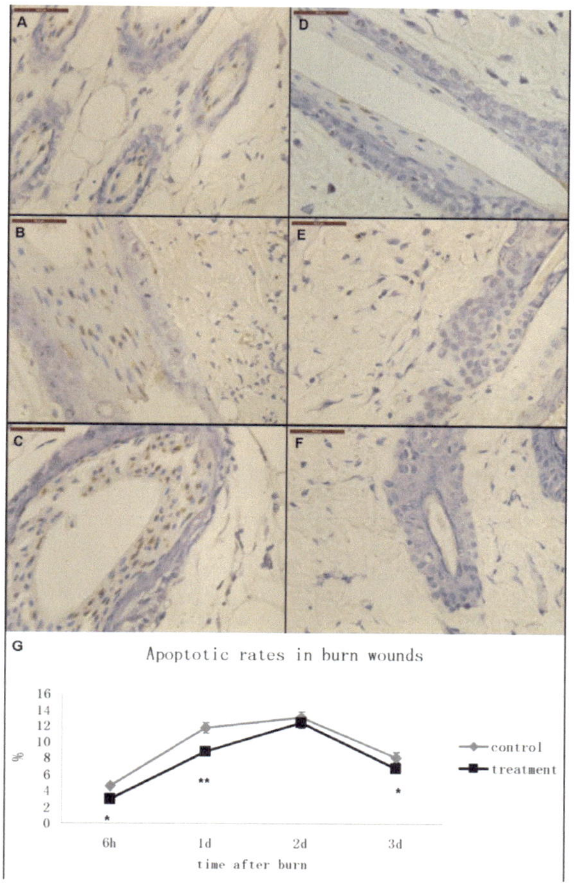

Figure 6. TUNEL staining of burn wounds in control (A–C) and treatment (D–F). (A, D) 6 h post burn; (B, E) 24 h post burn; (C, F) 72 h post burn. TUNEL positive cells were characterized by shrunken nucleus with brown staining. Scale bars = 50 um. (G) Quantitative analyses of TUNEL staining results. The values herein were mean ± SD ($*p < 0.05$, $**p < 0.01$, n = 6 per group).

3.7. Rapamycin treatment reduced inflammatory infiltration and restored blood perfusion in burn wounds

The LDF values and Na/K-ATPase activities in burn wounds of the treatment group were significantly increased when compared to those of the control group (**Figure 7A, B**). Furthermore, both LDF values and Na/K-ATPase activities in burn wounds were lower than those of normal unburned skin. The IL-8 level, MPO activity, and MDA content in burn wounds of the treatment group were significantly decreased when compared to those of the control group. In addition, these measurements were almost all higher than those of normal unburned skin (**Figure 7C–E**). Changes in the activities of Na/K-ATPase and MPO between the treatment and control groups were significantly different at the time points of 1 day and 2 days post-thermal burn injury. The changes in LDF, MDA, and IL-8 content between the treatment and control groups were also significant at the time points of 1, 2, and 3 days post-thermal burn injury.

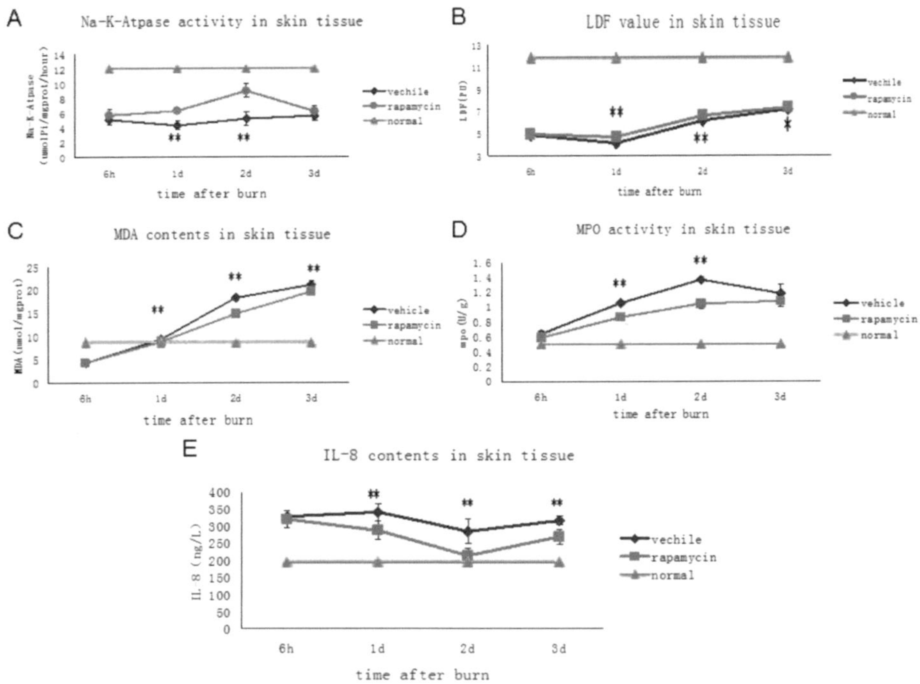

Figure 7. Na/K-ATPase activities, LDF values, MDA contents, MPO activities, and IL-8 contents in burn wounds of the treatment group, vehicle control, and normal unburned control. A: The p values for 1d and 2d are 0.008 and 0.006. B: The p values for 1d, 2d and 3d are 0.007, 0.007 and 0.043, respectively. C: The p values for 1d, 2d and 3d are 0.009, 0.006 and 0.008, respectively. D: The p values for 1d and 2d are 0.005 and 0.003. E: The p values for 1d, 2d and 3d are 0.006, 0.004 and 0.006, respectively. The values described herein were mean ± SD (treatment versus control group, $*p < 0.05$, $**p < 0.01$, n = 8 per group).

3.8. Rapamycin treatment reduced burn wound progression

Histopathological examination demonstrated that the burn wound progression was reduced in the treatment group when compared to the controls (**Figure 8**). Fewer residual hair follicles,

http://dx.doi.org/10.5772/63711

deeper tissue necrosis, more severe collagen denaturation, and more inflammatory infiltrations in burn wounds of the control group were seen when compared to those of the treatment group. Masson staining showed that less intense staining with Masson red were seen in the treated rats when compared to control, which may indicate less tissue necrosis in the treated rats (**Figure 8A–F**).

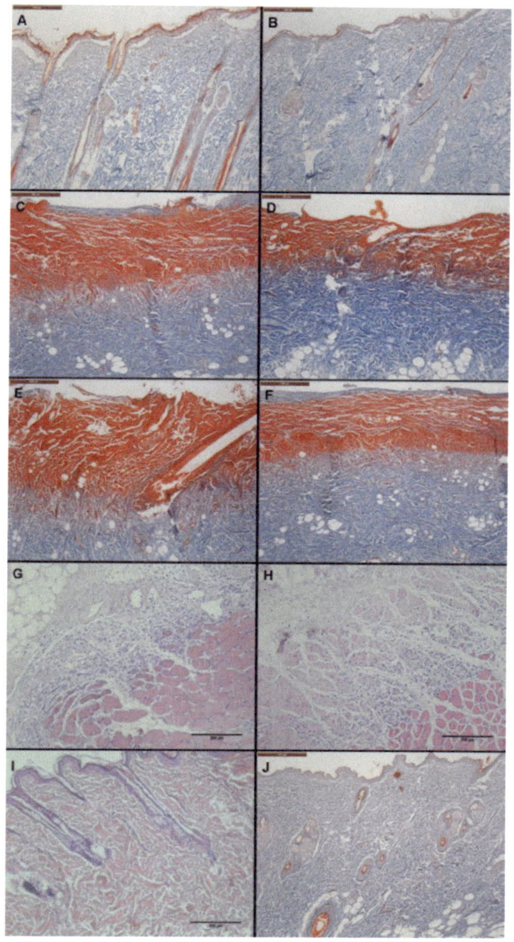

Figure 8. H&E and Masson staining of normal skin tissue and burn wounds. (A–F) Masson staining of burn wounds of 1-day control (A), 1 day following treatment (B), burn wounds of 2-day control (C), 2 days following treatment (D), 3-day control (E), and 3 days following treatment (F). (G, H) H&E staining of 3-day control (G), and 3 days following treatment (H). (I, J) H&E and Masson staining of normal skin tissue. Scale bars = 200 um.

3.9. Rapamycin treatment promoted burn wound healing

Finally, the time to wound re-epithelialization was shorter in treated rats as compared with controls (22.5 ± 1.4 vs. 24.8 ± 1.3 days, respectively). (mean ± SD, p<0.01, **Figure 9**).

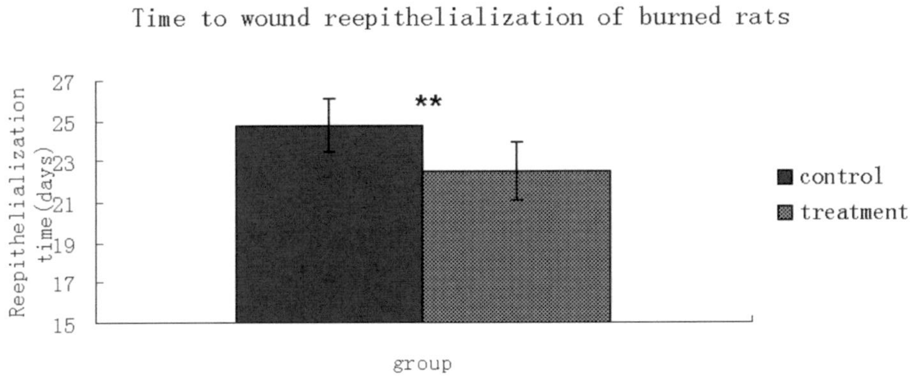

Figure 9. Time to wound re-epithelialization. The time to wound re-epithelialization was significantly shorter in treated rats compared with controls.

4. Discussion

The phenomenon of burn wound progression has long been considered a complex and challenging clinical and theoretical challenge. For over half a century, investigators in the field have been attempting to study the mechanisms of burn wound penetration of the epidermal and dermal layers of the skin with the objective of gaining greater insights into the pathology. Several studies [10] have demonstrated that a number of factors contribute to the invasive deepening of thermal burn wounds including a deficiency in local burn wound perfusion, edema, thrombosis, free radical damage, accumulation of factors that contribute to cell-mediated toxicity and inflammatory cell infiltration.

The first part of our study showed that after the occurrence of burn injury, laser Doppler flow values and Na/K-ATPase enzyme activities, which collectively represent the burn wound blood circulation and index of energy metabolism, decreased significantly. Moreover, both MPO and MDA activities, which respectively represent wound inflammation and free radical generation, increased significantly. Moreover, the autophagy level was at first reduced and then increased, yet remained far below those in normal skin. We speculate that at the early stage when the tissue necrosis due to thermal injury was inevitable, autophagy may not be a principal cellular activity as necrosis did in burn wounds and which is why autophagy level decreased at first. At the latter stage, when the inevitable tissue necrosis ceased, the remain viable tissue adjacent to coagulation was subject to ischemic and inflammatory damage among others, and then autophagy may protect against these stress stimuli as a survival mechanism which account for the latter rise of autophagy level. The prosurvival role of autophagy is supported by our another finding in IHC staining, which demonstrated significantly activated autophagy in the deep dermis which in this burn model, and represents the zone of stasis (or ischemia). In addition, at the latter stage (48–72 h) after burn, tissue perfusion as determined by LDF was restored and inflammatory infiltration in burn wounds as shown by MPO activity was reduced, in coincidence with the increase of autophagy level at the same time, which may add to evidence for the pro-survival roles of autophagy in this burn model.

To further dissect the role of autophagy in burn wound progression, we used rapamycin as treatment intervention in the second part of the study. The results demonstrated that the autophagic levels in burn wounds were enhanced following treatment with rapamycin. Furthermore, both LDF values and Na/K-ATPase activities (indicators of blood perfusion and energy metabolism) in the wounds of treatment group were significantly increased as compared to those of the control group. However, the IL-8 level, MPO activities, and MDA content (representative of inflammatory reaction and free radial damage) in the wounds of rapamycin-treated group were significantly decreased as compared to those of the control group [11]. Therefore, these results suggested that enhanced autophagy restored blood perfusion and energy metabolism, and inhibited inflammatory reaction and oxidative damage in burn wounds. Thus, it can be inferred that autophagy maintained cellular metabolic turnover and homeostasis in this experimental burn model [12].

Histopathological results showed that the burn wound depth of the rapamycin-treated group was less remarkable and "more superficial" than that seen in the control group, particularly at 2 and 3 days following thermal tissue injury. Additionally, as compared with the burn wounds of the treatment group, there were fewer residual hair follicles, more severe denatured collagen events recorded, evidence of deeper tissue necrosis, and more obvious inflammatory infiltrates in the burn wounds of the control group. These results suggested that burn wound progression in the treated rats was ameliorated. At last, burn wounds in treatment group re-epithelialized faster than those of control group, further indicating that prevention of burn wound progression were beneficial to burn wound healing.

5. Conclusion

Autophagy protects from burn wound progression and promotes burn wound healing through protecting the zone of stasis from further necrosis, which is probably mediated by phagocytizing damaged organelles to sustain homeostatic response and remodeling in the injured tissue.

Author details

Ligen Li* and Mengjing Xiao

*Address all correspondence to: liligen@126.com

The Department of Burn and Plastic Surgery, Burn Institute, First Affiliated Hospital of General Hospital of PLA, Beijing, China

References

[1] Shupp JW, Nasabzadeh TJ, Rosenthal DS, Jordan MH, Fidler P, Jeng JC. A review of the local pathophysiologic bases of burn wound progression. J Burn Care Res. 2010;31:849–73.

[2] Jackson DM. [The diagnosis of the depth of burning]. Br J Surg. 1953;40:588–96.

[3] Jackson DM. Second thoughts on the burn wound. J Trauma. 1969;9:839–62.

[4] Singh V, Devgan L, Bhat S, Milner SM. The pathogenesis of burn wound conversion. Ann Plast Surg. 2007;59:109–15.

[5] Levine B, Kroemer G. Autophagy in the pathogenesis of disease. Cell. 2008;132:27–42.

[6] Deretic V, Saitoh T, Akira S. Autophagy in infection, inflammation and immunity. Nat Rev Immunol. 2013;13:722–37.

[7] Xiao M, Li L, Li C, et al. Role of autophagy and apoptosis in wound tissue of deep second-degree burn in rats. Acad Emerg Med. 2014;21:383–91.

[8] Vignot S, Faivre S, Aguirre D, Raymond E. mTOR-targeted therapy of cancer with rapamycin derivatives. Ann Oncol. 2005;16:525–37.

[9] Xiao M, Li L, Hu Q, et al. Rapamycin reduces burn wound progression by enhancing autophagy in deep second-degree burn in rats. Wound Repair Regen. 2013;21:852–9.

[10] Tobalem M, Harder Y, Schuster T, Rezaeian F, Wettstein R. Erythropoietin in the prevention of experimental burn progression. Br J Surg. 2012;99:1295–303.

[11] Schmauss D, Rezaeian F, Finck T, Machens HG, Wettstein R, Harder Y. Treatment of secondary burn wound progression in contact burns—a systematic review of experimental approaches. J Burn Care Res. 2015;36:e176–89.

[12] Levine B, Mizushima N, Virgin HW. Autophagy in immunity and inflammation. Nature. 2011;469:323–35.

Chapter 11

Autophagy in Model Organisms: Insights into Cancer

Elite Possik and Arnim Pause

Additional information is available at the end of the chapter

http://dx.doi.org/10.5772/64541

Abstract

Autophagy is an evolutionarily conserved process utilized by most organisms to clear cellular damage and recycle building blocks for energy production. In this chapter, we emphasize the importance of genetic model organisms, including yeast, nematodes, flies, and mammals in the discovery and understanding of the autophagy process. We highlight the important roles of autophagy in aging, stress tolerance, neuronal health, organismal development, and pathogen resistance in invertebrate and vertebrate model organisms. We provide examples on how the same autophagy-related pathways that increase stress response and longevity in lower organisms could be utilized by cancer cells to survive harsh microenvironments, proliferate, and metastasize, with emphasis on the dual role of autophagy in cancer: an antitumorigenic or a protumorigenic process.

Keywords: autophagy, model organisms, stress tolerance, aging, organismal development, cancer

1. Introduction to autophagy

Autophagy is an evolutionarily conserved "self-degradation" process through which cytosolic compartments and organelles are delivered to the lysosome for degradation [1]. Autophagy exists in three forms: microautophagy where cytosolic components are directly engulfed in lysosomes, chaperone-mediated autophagy through which designated proteins are selectively targeted to the lysosomes, and macroautophagy (noted herein as autophagy) where cytosolic material is enclosed in a double-membrane autophagosomal structure that is delivered to lysosomes for degradation by acidic hydrolases [1]. Autophagy is selectively activated to remove cellular damage or is non-selectively activated under stress situations to supply energy and sustain cellular/organismal viability.

The autophagy machinery components and the physiology of this process are highly conserved across evolution from yeast to mammals. The autophagy-related genes (ATGs) have been initially identified in yeast *Saccharomyces cerevisiae* by pioneering genetic screens [2–7]. Later, their orthologues in other organisms have been determined, which led to the assessment of the functional roles of autophagy. ATG proteins form distinct autophagic complexes that function upon phagophore biogenesis, autophagosome formation, and maturation. The autophagy process comprises several steps. First, it starts with the nucleation and formation of the phagophore, which elongates and closes to form the double-membrane autophagosome, engulfing material to be recycled. Then, the autophagosome fuses with the lysosome to form the autolysosome where the material is digested by hydrolases [8–12]. The autophagy proteins are classified into six functional groups: the Atg1 autophagy initiation complex, the autophagy-specific phosphatidylinositol PI 3-kinase complex, the Atg12 the Atg2-Atg18 complex, the Atg9 transmembrane protein, the Atg12 autophagy conjugation system, and the Atg8 lipid conjugation system [8, 9]. The autophagic components of every group, their functions, and homologues in yeast, *Drosophila*, and the nematode *Caenorhabditis elegans* are described in **Table 1**.

	Yeasts	***Caenorhabditis elegans***	***Drosophila melanogaster***	**Mammals**
Regulation of induction	yTOR	*let-363*	dTOR	*mTOR*
	Snf1	*aak-1, aak-2*	AMPK	*AMPK*
Atg1/ULK autophagy initiation complex	Atg1	*unc-51*	Atg1	*ULK1, ULK2*
	Atg13	*atg-13*	Atg13	*ATG13*
	Fip200	*atg-11*	Fip200	*ATG17*
	Atg101	epg-9	Atg101	*ATG101*
Class III PI3K complex	Vps34	*vps-34*	Vps34	*VPS34*
	Vps15	*vps-15*	Vps15	*VPS15*
	Atg6	*Bec-1*	Atg6	*ATG6*
	Atg14	*epg-8*	Atg14	*ATG14L*
Atg2-Atg18 conjugation complex	Atg2	*atg-2*	Atg2	*ATG2*
	Atg18	*atg-18, epg-6*	Atg18a, Atg18b	*WIPI1, WIPI2, WIPI3, WIPI4*
Atg 9 transmembrane	Atg9	*atg-9*	Atg9	*ATG9A, ATG9B*
Atg12 conjugation system	Atg12	*lgg-3*	Atg12	*ATG12*
	Atg5	*atg-5*	Atg5	*ATG5*
	Atg10	*atg-10*	Atg10	*ATG10*
	Atg16	*atg-16.1, atg-16.2*	Atg16	*ATG16L1, ATG16L2*
	Atg7	*atg-7*	Atg7	*ATG7*
Atg8 conjugation system	Atg8	*lgg-1, lgg-2*	Atg8a, Atg8b	*GABARAP, LC3, GABARAPL1, GABARAPL2*
	Atg3	*atg-3*	Atg3	*ATG3*
	Atg4	*atg-4.1, atg-4.2*	Atg4a, Atg4b	*ATG4A, ATG4B, ATG4C, ATG4D*
	Atg7	*atg-7*	Atg7	*ATG7*

Table 1. Conserved autophagy genes in yeast, nematodes, flies, and mammals.

This review focuses on the multifaceted roles of autophagy in model organisms and how these conserved pathways could be adopted by cancer cells to suppress or promote tumorigenesis.

2. The importance of invertebrate model organisms

Although mammalian model organisms such as mice and rats are highly advantageous to study disease-related biological processes in humans due to the close anatomical and physiological similarities between systems, they have disadvantages including space, cost, and time-consuming transgenic technologies. Yeast models including budding yeast *Saccharomyces cerevisiae* (*S. cerevisiae*) and fission yeast *Schizosaccharomyces pombe* (*S. pombe*), the fruit fly *Drosophila melanogaster* (*D. melanogaster*), the nematode *Caenorhabditis elegans* (*C. elegans*), and other invertebrate models have emerged as excellent model organisms to study conserved signaling pathways. Many biological processes including autophagy are highly evolutionary conserved such that findings in these models are often applicable to humans.

First, yeasts, flies, and nematodes are characterized by their short lifespans and rapid reproductive lifecycles. Second, their genomes are fully sequenced [13–15] and well annotated, and a large number of tools and resources are available in accessible bioinformatics databases specific to every model (Yeast: www.yeastgenome.org; *Drosophila:* www.flybase.org; *C. elegans:* www.wormbase.org). Third, a high percentage of genes in invertebrate model organisms is homologous to disease-associated genes in humans. Fourth, several tools have been invented and developed in these systems including microscopy, transgenic techniques, biochemical methods, and others, rendering them attractive models to study genetically signaling pathways linked to diseases in humans including autophagy.

Although autophagy has been first observed by electron microscopy in mammalian cells in the 1950s [16], more than 30 autophagy genes have been discovered using genetic screens in yeast, and many of them have homologues in humans [2–7]. The rapid reproductive life cycles and short lifespans, the massive generation of tools to study autophagy, and the ease with which researchers pursue genetics work *in vivo* emphasize the importance of these models to study not only the molecular basics of the autophagic process but also the multifaceted roles of autophagy in organismal aging, stress tolerance, neuronal health, metabolism, pathogen infection, and others.

Despite the large advantages of invertebrate model organisms, they also have many limitations. The anatomy and physiology of the organismal systems, including immune, circulatory, respiratory, and nervous systems, largely differ from that of humans. Therefore, the importance of mammalian *in vitro* and *in vivo* models in studying autophagy is also unquestionable.

3. Methods to monitor autophagy in model organisms

Similar methods to study autophagy have been used in invertebrate model organisms and mammalian systems with the employment of the benefits of every system. These methods are recently reviewed in detail for yeast [17–20], *C. elegans* [21–25], flies [26–30], and mammalian systems [31–36].

Despite its complexity and difficulty to pursue, electron microscopy is one of the most reliable methods to visualize autophagic structures and has been used to monitor autophagy in many model organisms. However, since it requires a substantial specialized expertise, most researchers currently rely on light microscopic and biochemical methods, which are more accessible and easier to perform in most organisms. The fluorescent image analysis of autophagic components using reporters of tagged autophagic proteins has been widely used. LC3/ATG8 exists in two forms: LC3-I is cytosolic and soluble, and LC3-II is conjugated with phosphatidylethanolamine and is bound to the autophagosomal membranes. When autophagy is induced, the conjugation reaction can be monitored using the LC3:GFP reporter and the change between the diffuse localization of LC3 into autophagosomal puncta structures reflects the autophagic activity. This reporter is one of the most popular with its orthologues in *C. elegans* (LGG1:GFP) [23, 24] and in *Drosophila* and yeast (ATG8:GFP) [18, 29, 30]. The autophagic activity has been also assessed using Western blotting of the LC3:GFP protein extracts with or without inhibitors to determine the conversion of LC3-I to LC3-II. Moreover, previous studies in yeast, *C. elegans*, and mammalian cells have demonstrated that LC3-II is degraded inside the autolysosomes and that the GFP fragment is resistant to degradation and accumulates when autophagy is induced [37–40]. Therefore, researchers have used Western blot analysis on protein extracts to assess the levels of GFP and cleavage of GFP-LC3-I.

Since autophagic proteins also accumulate upon defective autophagy, researchers have monitored the degradation of cargo proteins such as p62 in most model organisms as well [24, 25, 28, 41, 42]. Furthermore, autophagy inhibitors have been used to determine whether the accumulation of autophagosomes is due to impaired autophagy or to a heightened autophagic flux. The most recent studies employ the mRFP-GFP-LC3, which enables the distinction between heightened autophagic flux and impaired autophagy. In this method, mCherry and GFP have been used as red and green fluorescent protein markers, respectively, to trace the autophagic protein LC3. Upon physiological pH in newly formed autophagosomes or when autophagy is impaired, both GFP and mCherry colocalize in puncta leading to yellow puncta structures, whereas upon lysosomal fusion and acidification, the GFP signal is lost and only mCherry is detected.

High-resolution live-cell imaging to visualize the dynamics of autophagy has been also employed and reviewed in detail [36].

4. Autophagy-related biological roles in model organisms

Despite the anatomical, morphological, and physiological differences between model organisms, autophagy appears to play similar important roles across evolution. In this section, we review the major autophagy-associated roles at the cellular and organismal levels in invertebrate and mammalian model systems.

4.1. Stress tolerance

In most organisms, autophagy is activated by different stresses including nutrient deprivation, oxidative stress, hypoxia, temperature shifts, and others, to eliminate damaged macromolecules and to produce energy

In yeast, mutation of *Atg1*, *Atg2*, *Atg4*, *Atg7*, or *Atg8* genes increases sensitivity to the oxidative stressor paraquat [43]. In *C. elegans*, starvation, oxidative stress, and hypoxia stresses induce autophagy in multiple tissues of the animal as monitored by the number of positive GFP:LGG-1 puncta [44–47]. The increased autophagy levels induced by stress are essential for organismal survival to stressful conditions. In addition, the inhibition of autophagy genes causes defects in the formation of the *C. elegans* dauer animals, a static larval stage adapted to survive prolonged starvation [45]. Furthermore, autophagy is required for the survival of *C. elegans* nematodes to starvation [47, 48], hypoxic environments [44], oxidative stress [46], and hyperosmotic stress [49].

In *Drosophila*, *Atg7* mutant flies are hypersensitive to complete starvation, sugar-only diets, and oxidative stress [50, 51]. Moreover, JNK signaling induces the transcription of autophagy genes to help protect flies from oxidative stress [52]. Specifically, mutation of *Atg1* and *Atg6* in young adult flies overexpressing JNK signaling suppressed their increased resistance to the oxidative stressor paraquat [52]. Consistently, the spermidine-induced autophagy is required for the resistance of *Drosophila* animals to paraquat [53].

The role of autophagy in stress resistance has been demonstrated not only in invertebrate models but also with mammalian cell culture and *in vivo* models. For example, in mice, ATG5 overexpression induces autophagy, increases oxidative stress resistance, and extends lifespan [54]. Additionally, autophagy is significantly induced following the early starvation-associated postnatal period in mouse neonates and is required for their survival until supply with milk nutrients [55]. Several studies also reported that following ischemic injuries, autophagy is activated and contributes to neuroprotection by delaying neuronal cell death in rats [56–58]. Collectively, these studies demonstrate an evolutionarily conserved role of autophagy in stress tolerance. However, how autophagy mediates stress tolerance is still unclear. While many studies highlight the important role of autophagy in the clearance of stress-induced damaged organelles, others claim that the stress resistance is due to the role of autophagy in sustaining energy levels and providing building blocks for mitochondrial energy production.

4.2. Extension of lifespan

Accumulating evidence demonstrates that longevity pathways converge on autophagic processes in many organisms to regulate diverse cellular functions including the clearance of damaged proteins and organelles and the remodeling of cellular metabolism. In *C. elegans*, multiple genetic or pharmacological manipulations extend lifespan [59]. For instance, mutations of genes in the insulin-signaling pathway, including *daf-2* and *age-1*, which are orthologues of the insulin signaling receptor and PI3K, respectively, deficiency in target of rapamycin (TOR) signaling, overexpression of activated protein kinase (AMPK) signaling, mutation of mitochondrial genes, dietary restriction through mutation of *eat-2*, mutation in sitruin-1, are all genetic alterations that extend lifespan in *C. elegans* [59]. Pharmacological alterations, such as spermidine, resveratrol, and w-6 polyunsaturated fatty acids treatment also prolong lifespan in *C. elegans* [60, 61]. Importantly, autophagy is induced in most of the above-mentioned longevity pathways and contributes to the lifespan extension phenotypes in *C. elegans*. For example, the inhibition of the autophagy gene *bec-1* suppresses the increased lifespan mediated by caloric restriction in *eat-2* mutant animals or by TOR inhibition [62]. Furthermore, the inhibition of *bec-1* in *daf-2* long-lived *C. elegans* mutants severely reduces their lifespan [45]. In addition, autophagy is highly induced in calcineurin *C. elegans* mutant animals and its inhibition by RNAi feeding against *bec-1* or *atg-7* abolishes the increased longevity phenotype [63]. Moreover, the mutation of *cep-1*, the worm orthologue of P53 promotes an autophagy-dependent lifespan extension [64]. Additionally, both spermidine and resveratrol extend *C. elegans* lifespan by inducing autophagy [60, 65]. Mitophagy also contributes to the extension of lifespan upon low insulin signaling and mitochondrial mutations [66].

HLH-30 is the worm homologue of transcription factor EB (TFEB), a master transcriptional regulator of lysosomal and autophagic pathways [67, 68]. The overexpression of HLH-30 increases lifespan in *C. elegans* [67]. Furthermore, the impairment of the production of the yolk lipoprotein vitellogenin extends lifespan in *C. elegans* [69]. Importantly, autophagy and HLH-30 are both induced by the reduction in vitellogenesis and contribute to the extension of lifespan in vitellogenesis-defective *vit* mutant animals [69].

In *Drosophila*, mutations in *Atg7* and *Atg8* genes shorten lifespan [50, 51]. In addition, mutation of the autophagic protein FIP200, a component of the Atg1 autophagy initiation complex, leads to neuronal degeneration and shortens lifespan [70]. The administration of phosphatidylethanolamine enhances autophagic flux and increases lifespan in yeast, *Drosophila*, and mammalian cells in culture [71].

In yeast, the role of autophagy in aging seems to be context-dependent. Autophagy has been shown to be required for the extension of chronological lifespan by low doses of the mammalian target of rapamycin (mTOR) inhibitor rapamycin [72, 73], methionine limitation [74], and calorie restriction [75]. In contrast, Tang et al., 2008 claim that autophagy genes may be required or not for the lifespan extension by calorie restriction depending on their role in the autophagy process. Specifically, they show that the deletion of genes involved in autophagosome formation including *Atg1*, *Atg6*, *Atg7*, and *Atg8* did not affect lifespan of budding yeast upon calorie restriction [76]. However, the deletion of *Atg15*, *Atg17*, or other genes involved in vacuole-vacuole fusion reduced the lifespan extension promoted by calorie restriction [76].

In mammals, the link between autophagy and the organismal extension of lifespan has not been clearly established. A few studies support the role of autophagy in promoting longevity in mammals. For instance, ATG5 overexpression has been shown to extend lifespan by 17.2% in mice [54]. Interestingly, rapamycin feeding of mice at their old age extends their lifespan, which could be due to autophagy activation [77]. While rapamycin is a strong mTOR inhibitor and autophagy inducer, the link between rapamycin feeding and increased autophagy has not been made, and therefore, the extension of lifespan by administration of rapamycin in mice may not be due to autophagy activation per se but to other mechanisms [77].

Although the role of autophagy in mammalian organismal lifespan is still not clearly elucidated, many studies demonstrate an important role for autophagy in delaying the acquisition of aging features of multiple cells and tissues. Numerous studies also claim a decline in the autophagic activity in many mammalian organs upon aging [78–83]. For example, autophagy genes *Atg5*, *Becn1*, and *Atg7* are significantly downregulated in the human aging brain [84]. Cardiac-specific *Atg5* deficiency in mice leads to cardiac abnormalities after 6 months of age and early death [85]. Consistently, cardiac-specific overexpression of *Atg7* increased autophagic flux and improved cardiac function in desmin-related cardiomyopathies in mice [86]. Furthermore, the hyperactivation of chaperone-mediated autophagy in aging livers maintains hepatic function in old mice to a level comparable to that reported in young mice [87]. Recently, autophagy inhibition has been shown to increase aging features in macrophages including the reduction in phagocytosis and nitrite burst and increased inflammatory response [78]. Numerous studies have also linked autophagy to improved neuronal health in mice and protection from age-associated neurological disorders [58, 81, 88–93]. This is further detailed in the neuronal health section of this chapter. Moreover, the role of autophagy in suppressing tumor initiation is well described at the end of this chapter. Therefore, although it is not clear whether autophagy extends organismal lifespan in mammals, collective evidence supports its implication in the extension of healthy living or health span and the delay of the appearance of age-associated diseases.

4.3. Resistance to pathogen infection

The induction of autophagy has been widely shown to contribute to the organismal survival to infection with pathogens. In *C. elegans*, autophagy genes are required for survival to infection with pathogens, including *Pseudomonas aeruginosa*, *Salmonella typhimurium*, and other pathogens [68, 94, 95]. Using the GFP:LGG1 reporter, autophagy has been shown to be induced in the hypodermal seam cells and intestinal cells of wild-type animals following infection with *Pseudomonas aeruginosa* [95] and *Staphylococcus aureus* [68]. Importantly, inhibition of autophagy genes suppresses not only the resistance of wild-type animals but also the resistance of highly stress-resistant strains including *daf-2* mutant animals and *daf-16* overexpressing animals to infection with *Salmonella typhimurium* [94].

In *Drosophila*, IRD1 is the fly homologue of mammalian VPS15, an important autophagic serine/threonine kinase implicated in phagosome maturation. IRD1 plays an important role in antibacterial immune responses in *Drosophila* [96]. *Ird1* mutant flies are incapable of expressing antimicrobial peptide genes upon infection [96]. In addition, the conditional inactivation of autophagy genes *Atg5*, *Atg7*, *Atg12* in *Drosophila* reduces survival of the animals upon infection with *Escherichia Coli* [97]. Moreover, autophagy genes *Atg5*, *Atg8a*, and *Atg18* are also required to limit the infection of *Drosophila* cells with the *Vesicular stomatisis* virus [98]. Furthermore, the inhibition of *Atg5* using RNAi in flies increased the susceptibility of the animals to infection with *Listeria monocytogenes* [99].

The transcriptional upregulation of autophagy genes by TFEB has been also associated with increased resistance to pathogens. Upon infection with *Staphylococcus aureus*, HLH-30 rapidly translocates to the nucleus and activates the expression of autophagy genes, lysosomal genes, and antimicrobial peptide genes in both *C. elegans* [68]. In murine cell lines, TFEB translocates to the nucleus following infection and induces the transcription of chemokines and cytokines [68]. Mitophagy is also another mechanism of defense against invasion with *P. aeruginosa* [100].

How autophagy mediates resistance to pathogens is still not clear. Xenophagy (eating the pathogen) is a cellular defense mechanism through which cells direct autophagy to degrade the invading pathogens. Autophagy genes restrict *Salmonella* bacterial replication in both hosts, the unicellular organism *Dictyostelium discoideum* and in *C. elegans* [94]. However, autophagy only increased resistance of *C. elegans* to *Pseudomonas aeruginosa* and to *Staphylococcus aureus* without decreasing bacterial load suggesting that xenophagy is not the only defense mechanism attributed to autophagy [68, 95].

In mammalian cells, autophagy also plays an essential role in the protection against invading pathogens, including *Streptococcus*, *Shigella flexneri*, *Mycobacterium tuberculosis*, and *Toxoplasma gondii* [12, 101–103]. Autophagy has also been shown to protect against toxins released by bacterial pathogens [37]. In mice, recent work demonstrates the involvement of autophagy in the clearance of pathogens, including *Listeria monocytogenes*, and moreover, IRF8 directs stress-induced autophagy in macrophages and promotes clearance of *L. monocytogenes* [104] and *Staphylococcus aureus* [105, 106] and *Mycobacterium tuberculosis* [107, 108]. However, recent work demonstrates a unique role of ATG5 in the resistance of mice to *Mycobacterium tuberculosis* infection distinct from autophagy in contrast to previous reports. ATG5 prevents polymorphonuclear cell-mediated immunopathology enhancing resistance to *Mycobacterium tuberculosis* infection [109].

4.4. Organismal development

Accumulating evidence highlights an important role for autophagy during organismal development. Deletion of autophagy genes leads to severe defects and causes early lethality in many organisms. For example, *bec-1* mutation leads to severe defects during embryogenesis in *C. elegans* and mutant animals display a highly penetrant lethal phenotype where only few animals are capable of reaching adulthood [110]. The *unc-51/atg-1 C. elegans* mutant animals exhibit an uncoordinated movement and paralysis. Moreover, autophagy is highly induced at several stages during *C. elegans* development and a genome-wide genetic screen has identified signaling pathways that regulate this process in *C. elegans* [111]. In *Drosophila*, mutations in *Atg1* are pupal lethal [112] and strong hypomorphic mutations in *Atg8* lead to a semi-lethality phenotype [50, 51]. Autophagy is also induced during the development of Zebrafish larvae and the knockdown of autophagy genes *Atg5*, *Beclin1*, and *Atg7* results in aberrant cardiac morphogenesis and reduced survival in Zebrafish [113]. ATG5 deficiency in Zebrafish impairs nervous system development, specifically brain morphogenesis [114]. Additionally, AMBRA1 (autophagy/Beclin 1 regulator 1) is an evolutionary conserved positive regulator of BECN1 and is essential for proper autophagic activity. The inhibition of AMBRA1 in Zebrafish leads to incomplete organogenesis and defects in skeletal muscle development [115, 116].

In mice, *Becn1* homozygous deletion leads to embryonic lethality [117], while *Atg7* and *Atg5* null mice are born alive but die soon after birth. Similarly to what has been reported in

Zebrafish, *Atg5* is required for the proper cardiac development [113] and cortical astrocyte differentiation [118] during embryogenesis in mice. Autophagy is also involved in chondrocyte differentiation and bone formation through fibroblast growth factor (FGF) signaling in mice [119]. FIP200 is an important autophagic protein that interacts with ULK1 in the autophagy initiation complex. Homozygous deletion of FIP200 in mice leads to embryonic lethality due to heart failure and severe hepatic defects [120]. Other than its important role in the heart and liver, FIP200 plays a central role in the differentiation of neural stem cells and is essential for maintenance and function of fetal hematopoietic stem cells [121]. Supporting the role of autophagy in stem cell differentiation during development, a recent study reports a retardation in stem cell differentiation during the embryonic development of mice hypomorphic for Atg16l1 [122].

The discovery that autophagy is involved in the degradation of the paternal mitochondria is another important aspect during development. In most eukaryotes, the maternal mitochondrial genome is believed to be the one inherited and thus the degradation of the sperm-inherited mitochondrial genome is essential. In *C. elegans*, autophagosomes engulf the paternal mitochondria and target them to the lysosomes for degradation during embryonic development [123]. Similarly, paternal mitochondria are also destroyed by endocytic and autophagic pathways in *Drosophila* [124]. However, in mammalian zygotes, the degradation of the paternally inherited mitochondria requires the ubiquitin proteasome system rather than autophagy [125]. Therefore, autophagy plays central role (s) in organismal development across evolution, which includes key checkpoints during embryogenesis, cellular differentiation, and tissue organization.

4.5. Neuronal health

The accumulation of autophagosomes has been observed in the neurons of individuals affected with neurodegenerative diseases including Alzheimer's disease, Parkinson's disease, Huntington's disease, and amyotrophic lateral sclerosis. Autophagy improves neuronal health by degrading damaged proteins, specifically mutant proteins associated with neurological disorders and toxic aggregation-prone proteins [88, 91, 126–128]. Non-mammalian model systems are excellent to study protein homeostasis in regard to fatal neurological disorders. In addition, C. elegans [129–133] and flies [134–139] researchers have generated transgenic animals that express polyglutamine repeats, beta-amyloid peptides, and the αsynuclein protein, to mimic the pathologies of Huntington's disease, Alzheimer's disease, and Parkinson's disease, respectively. Using electron microscopy and the LGG-1:GFP reporter, the expression of human beta-amyloid (1–42) in *C. elegans* muscles resulted in the accumulation of autophagic vacuoles. Autophagy contributes to the degradation of the Beta-amyloid peptide in *daf-2* mutant nematodes [129]. In *Drosophila*, inhibition of autophagy genes increases neuronal toxicity of amyloid beta 1–42 peptides [140]. In *C. elegans*, inactivation of autophagy genes *atg-18* and *atg-7* accelerates the accumulation of polyQ40:YFP protein aggregates in the body wall muscles of the animals over time [133]. In *C. elegans*, the *unc-51 (atg-1)* gene is essential for normal axonal elongation and structure [141].

In *Drosophila*, mutation of *Atg7* or *Atg8* genes enhanced the accumulation of insoluble poly-ubiquitinated proteins with age as determined by Western blot analysis using fly head extracts [50, 51]. Consistently, the overexpression of *Atg8* in the central nervous system of adult flies reduced the accumulation of insoluble ubiquitinated proteins [51]. The *Drosophila* homologue of P62, refractory to sigma P (Ref(2)P), a scaffold protein with diverse biological roles, marks ubiquitinated protein aggregates for degradation [142, 143]. Ref(2)P acts as a receptor for selective autophagic degradation. In flies, p62-tagged ubiquitinated protein aggregates accumulate in the brains of older animals as compared to young animals [143]. The accumulation of protein aggregates correlates not only with poor autophagic functions with age but also with a decline in fly behaviors and aging [144]. While the accumulation of Ref(2)P-linked protein aggregates is enhanced in *Drosophila Atg8* [143] and *Vps15* [145] mutant flies, Ref(2)P is also required to form these aggregates [143]. In *Drosophila, Atg17/FIP200* localizes to Ref(2)P protein aggregates proximate to the lysosomes and interacts with the autophagy-activating protein ATG1 to promote autophagy [146]. By sustaining autophagy, the N-ethyl-maleimide-sensitive fusion protein (NSF1) protects dopaminergic neurons from degeneration and promotes longevity in *Drosophila* [147]. Also, the inhibition of the ectopic P-granules autophagy protein 5 (*Epg5*) in the retina of adult *Drosophila* animals leads to the degeneration of photoreceptor neurons and loss of the retina [148] mirroring the genetic neurological disorders of EPG5-related Vici Syndrome in humans. Mutation of the autophagy gene-related proteinase ATG4D in dogs has been recently associated with a novel neurodegenerative disorder in the Lagotto Romagnolo dog breed [149]. Accordingly, knockdown of Atg4D in Zebrafish also leads to neurodegeneration of the central nervous system [149].

Consistently with what has been observed in *C. elegans* and *Drosophila*, the induction of autophagy by starvation [150] or by rapamycin [151] reduced the amount of poly-ubiquitinated proteins [150] or α-synuclein [151] protein aggregates in yeast. However, yeast *Atg8* mutants displayed an accumulation of ubiquitinated aggregate-prone proteins upon starvation and high temperature stresses [150]. Moreover, the mutation of *Atg1* or *Atg7* delayed the clearance of α-synuclein aggregates in yeast [152, 153].

Numerous studies highlight an important role for autophagy in mammalian neurogenesis and neuronal "maintenance." Several neurological disorders in humans are associated with impaired autophagy and defects in the clearance of damaged organelles and proteins [154, 155]. Among several examples, mutations in *WDR45*, one of the mammalian homologues of yeast Atg18, cause encephalopathy in children and neurodegeneration in adults [156]. Importantly, V471A polymorphism in the *Atg7* gene in human patients, mostly of Italian origin, has been strongly correlated with an earlier onset of Huntington's disease [157, 158]. In mice, lack of autophagy genes *Atg7* and *Atg5* in the neurons promotes the accumulation of poly-ubiquitinated aggregation-prone proteins leading to neuronal degeneration [92, 93]. The specific knockout of *Atg7* in the Purkinje cells of mice leads to neurodegeneration and destabilization of axonal homeostasis [159]. Moreover, the induction of autophagy in neuronal of amyotrophic lateral sclerosis models decreases proteotoxicity by enhancing TDP43 turnover and neuronal survival [89]. An increasing number of studies support the correlation between autophagy deficiency and neurodegeneration in mammals. Here, we only listed few examples to support this idea. For detailed reviews, please see [154, 155].

4.6. Autophagic cell death and clearance of cellular corpses

Apoptosis or programmed cell death is a fundamental component in the development of *C. elegans* nematodes [160]. Pioneering studies in *C. elegans* led to the discovery of evolutionarily conserved key players implicated in this important biological process. There are two types of programmed cell death in *C. elegans*: "developmental cell death," which occurs in the somatic tissues throughout worm development, and "germ cell death," which takes place in the gonads of adult hermaphrodites [160–165]. During the embryonic and postembryonic stages of *C. elegans* development, only 131 cells of 1090 cells undergo apoptosis to form the adult hermaphrodite [160–165]. The morphological changes in apoptotic *C. elegans* cells are similar to those of mammalian cells and include DNA fragmentation, chromatin condensation, and changes in mitochondrial and plasma membrane potentials [160]. Autophagy plays a major role in the clearance of apoptotic corpses generated during both the developmental cell death and germ cell death [166–169]. Specifically, the number of embryonic apoptotic corpses is significantly increased in nine *C. elegans* strains harboring mutations in essential genes of the autophagic pathway [167]. Autophagy proteins LGG-1, ATG-18, and EPG-5 are recruited to engulfed apoptotic corpses and are essential for the degradation inside the phagocyte [169].

In *Drosophila*, several studies have reported the requirement of autophagy in the death and clearance of specific cells throughout the fly development. In contrast to the role of autophagy in mediating cellular survival, autophagy contributes to fly development by killing particular cells in specific tissues. For instance, autophagy genes are required for the killing and clearance of cells in the salivary glands, ovary, intestine, and embryonic serosa membranes [170–175]. Autophagy also occurs in dying midgut cells and is essential for the clearance of this tissue.

In mice, autophagy contributes to the programmed cell death-mediated clearance of apoptotic cell corpses. Lack of *Atg5* leads to defective apoptotic corpses engulfment in the developing mice embryos [176]. Autophagy is also required for the clearance of cell corpses in the retinal neuroepithelium of developing chick embryos [177]. Therefore, the role of autophagy in the clearance of corpses is evolutionarily conserved and essential for the proper organogenesis and development in most animals.

4.7. Metabolism

In invertebrates, the storage and biosynthesis of energy reserves, including yolk particles, lipids, and glycogen, play a crucial role in development during early embryogenesis and later during adulthood [178]. In *C. elegans*, the yolk particles accumulate with age and are synthesized in the intestine and transported later to the pseudoceolom (body cavity) of *C. elegans* animals. These granules are essential to survival upon starvation during L1 diapause [178, 179]. In *Drosophila*, the yolk particles are also important for embryonic development. Importantly, ATG1 is required for the catabolism of yolk particles in *Drosophila* [180].

The role of autophagy in lipid metabolism has been reported in many organisms. In *C. elegans,* inhibition of autophagy genes leads to a decline in organismal lipid content supporting an essential role for autophagy in lipid metabolism [181]. Moreover, autophagy and lipolysis work inter-dependently to promote longevity in germline-less *C. elegans* strains [181, 182]. The role of autophagy in the degradation of lipid droplets has not been clearly elucidated in *C. elegans*. The fact that autophagy mutants display reduced lipid contents in *C. elegans* could be due to the role of autophagy in the restoration of energy levels and storage in the form of yolk, glycogen, and fat. To determine whether autophagy plays a role in lipid degradation in nematodes, both wild-type and autophagy mutant *C. elegans* strains should be subjected to an energy depletion stress that induces lipid degradation and the difference in the efficiency of degradation should be investigated. A similar experiment has been conducted upon loss of HLH-30, the TFEB homologue in *C. elegans*. In this case, *hlh-30* mutant animals displayed a less efficient degradation of lipid content upon starvation in comparison with the wild-type animals supporting a potential role of autophagy in the mobilization of lipids upon stress in *C. elegans* [183]. This role of HLH-30 is evolutionarily conserved. In fact, TFEB has been also shown to prevent diet-induced obesity in mice [183].

Following stress and energy depletion, the mobilization of "energy-rich" intracellular contents is essential. The autophagic degradation of lipids has been reported throughout evolution. In contrast to what has been observed in *C. elegans,* where the inhibition of autophagy leads to a decrease in lipid content, autophagic pathways are important for targeting lipid droplets for lysosomal degradation in yeast [184, 185]. In mammalian systems, autophagy has been linked to lipid metabolism but with opposite effects depending on the context. In hepatocytes, the pharmacological or genetic inhibition of autophagy increases triglyceride content supporting an important role of autophagy in lipid breakdown. Consistently, lipid content is significantly increased in *Atg7* liver-specific knockout mice as compared to the controls [186]. However, knockdown of *Atg7*, *Atg5*, or the pharmacological inhibition of autophagy in 3T3-L1 pre-adipocytes reduced lipid accumulation [187]. This is in accordance with the observation that the mass of white adipose tissue decreased significantly in *Atg7* adipocyte-specific knockout mice in comparison with the control [187]. The connection between autophagy and lipid metabolism is reviewed in detail in Ref. [188].

In accordance with the role of autophagy in lipid metabolism, autophagy also plays an important role in glycogen metabolism. In *Drosophila,* the inhibition of autophagy in the fly skeletal muscles using chloroquine reduced the efficiency of glycogen degradation [189]. Using electron microscopy, the same group has revealed glycogen as electron dense material inside the double membrane structures of the autophagosomes [189]. Importantly, *Vps15* deficiency led to the accumulation of glycogen in murine skeletal muscles, whereas the overexpression of *Vps34/Vps15* in myoblasts from Danon autophagic vacuolar myopathy patients decreased glycogen storage [190]. In humans, the impairment of lysosomal and autophagic functions is associated with glycogen storage diseases and is linked to muscle atrophy and neurodegeneration [191–194]. Altogether, accumulating evidence supports the role of autophagy in the degradation of lipids and glycogen across evolution.

http://dx.doi.org/10.5772/64541

5. From model organisms to cancer in humans

Genetic pathways that alter autophagy in model organisms are often linked to cancer in humans. For instance, AMPK, TOR, Insulin, SKN-1/NRF2, CEP-1/p53, FLCN-1, and other signaling pathways modulate autophagy in model organisms and are associated with cancer initiation and progression in humans. Two major kinases are important in stress sensing and autophagy regulation: the mammalian target of rapamycin (mTOR) and the 5′ AMP-activated protein kinase (AMPK). TOR is a serine/threonine kinase that is activated during nutrient-rich conditions and is inhibited by starvation. In *S. cerevisiae*, *D. melanogaster*, and mammalian systems, TOR has been linked to autophagy through the regulation of the autophagy initiation complex ULK1/ATG1 [112, 195–199]. AMPK is activated upon starvation and drives autophagy in mammalian cells and in invertebrate model organisms. In yeast, ATG1 and ATG13 have been found as potential genetic interactors and downstream effectors of *SNF1*, the yeast AMPK homologue [200]. In mammals, two groups reported the ability of AMPK to induce autophagy through ULK1/ATG1 activation [195, 201]. In this section of this chapter, we will emphasize the dual role of autophagy in cancer.

Autophagy deregulation has been widely reported in human cancers. This is reviewed in detail in Refs. [202, 203]. Whether autophagy plays a tumor-suppressing role or a tumor-promoting role is still controversial since both cases have been reported. Although autophagy protects against tumorigenesis since it plays a central role in the clearance of damaged cellular macromolecules and organelles, increasing evidence suggests that autophagy could also acquire tumor-promoting functions. By supplying cancer cells with energy, autophagy may promote their survival because they are often exposed to nutrient deprivation and hypoxia due to lack of blood vessels.

5.1. Autophagy as a tumor-suppressing mechanism

The observation that autophagy gene ATG6/BECN1 is monoallelically lost in a large number of prostate, breast, and ovarian cancers supported the tumor suppression role of autophagy at first [117, 204–206]. Consistently, autophagy genes are frequently downregulated in tumors. In mice, homozygous deletion of *Becn1* leads to embryonic lethality. However, *Becn1* heterozygous mice exhibit a high frequency of spontaneous tumors that still express the wild-type *Becn1* mRNA and protein supporting a role of *Becn1* as a haploinsufficient tumor suppressor gene [117, 206]. Moreover, BIF-1 and UVRAC, which are essential components of the Beclin1/class III PI3K complex, also contribute to the control of proliferation and suppression of tumor growth [207]. Furthermore, the deficiency in autophagy genes *Atg5*, *Atg7*, and *Becn1* in mice leads to benign hepatic tumors [208].

How autophagy acts as a tumor suppressor is not clear yet. A plausible explanation could be that loss of autophagy increases oxidative stress, which leads to the accumulation of damaged macromolecular cellular components [209, 210]. This is supported by the fact that impaired autophagy increases genomic instability presumably through lack of degradation of damaged mitochondria and an intracellular increase in the levels of reactive oxygen species (ROS) [211, 212]. The selective degradation of damaged mitochondria by autophagy has been shown to protect against oxidative stress and mitochondrial dysfunction [213]. Autophagy deficiency has been shown to contribute to the tumorigenesis induced by oncogene activation or by chemical carcinogens. Deletion of *Atg7* in mice drives early tumorigenesis induced by BRAF ^{V600E} activation [214], supporting the tumor suppression function of autophagy in the initiation of tumorigenesis. However, *Atg7* deletion also abrogated the ability of the BRAF ^{V600E}-driven tumors to progress into a more malignant phenotype [214]. Also, Atg4C/autophagin3 knock-out mice exhibited an increased susceptibility to develop fibrosarcomas induced by chemical carcinogens [215].

Autophagy has been recently shown to mediate cellular senescence through the degradation of nuclear lamina upon oncogenic events, suggesting that this guardian role of autophagy might prevent tumorigenesis [216].

5.2. Autophagy as a tumor-promoting mechanism

The balance between autophagy and apoptosis is a key factor in the cellular decision between life and death. These two pathways are connected, and deregulation in this balance is a main factor in carcinogenesis. Upon cellular exposure to stress, when the damage cannot be repaired, cells normally undergo programmed cell death to eliminate them. When cells escape these control mechanisms and are unable to die, resistant clones emerge which could lead to cancer. Therefore, mechanisms of resistance to stress are often utilized by cancer cells to survive and proliferate. Autophagy is induced in hypoxic and highly nutrient-stressed tumor microenvironments [211, 212]. Autophagy is also required to promote tumorigenesis by activating mutations of multiple oncogenes, including *Kras*^{G12D} [217–219] and Braf^{V600E} [214]. In fact, *Atg7* deletion in mice extends the lifespan of mice carrying an activating mutation in *Braf*^{V600E} that drives lung tumor growth and impairs mitochondrial metabolism and survival to starvation [214]. Similarly, the inhibition of autophagy using the autophagy inhibitor chloroquine abrogates the growth of lymphoma tumors induced by Myc activation. Additionally, deletion of the autophagic component FIP200 in mammary epithelial cells in mice suppressed mammary tumor growth in the MMTV-PyMT mouse model of human breast cancer [220].

The role of P62/SQSTM1 in tumorigenesis is controversial and context-dependent. While autophagy suppresses tumorigenesis by eliminating P62, recent findings demonstrate that P62 synergizes with autophagy to promote tumor growth *in vivo* [221].

Several tumor suppressor genes are associated with aberrant autophagic flux. Mutation in the tumor suppressor gene *Flcn* in humans, responsible for the Birt-Hogg-Dubé neoplastic syndrome, increases the predisposition to renal cysts and tumors [222, 223]. Importantly, autophagy is required for survival to oxidative and nutrient deprivation stresses of FLCN-deficient cells and for the FLCN-driven tumorigenesis [46, 224]. A similar role for VHL, another renal tumor suppressor, in the regulation of autophagic events in renal cell carcinomas has also been described [225]. Autophagy inhibition by MiR-204 suppressed the tumor growth in VHL-deficient cells and the inhibition of LC3B/ATG5 suppressed the development of VHL-deficient renal cell carcinomas in nude mice [225]. Autophagy also contributes to the tumorigenesis induced by loss of the tumor suppressor tuberous sclerosis complex TSC2 [226].

Recently, ATG7 has been shown to cooperate with loss of PTEN to drive tumorigenesis in prostate cancer [227].

Autophagy also plays a critical role in sustaining cancer cell viability and promoting tumor growth in pancreatic ductal adenocarcinoma [228]. MiT/TFE-dependent transcriptional activation of the lysosomal-autophagic pathway is essential for metabolic reprogramming in pancreatic ductal adenocarcinomas and drives aggressive malignancies [229].

6. Conclusion and perspectives

The autophagy-associated pathways that alter lifespan, stress tolerance, neuronal health, resistance to pathogens, and metabolism in lower organisms are highly evolutionarily conserved and are associated with tumorigenesis in mammals. Although the autophagic process does not change between cells/tissues/organisms, its roles are diverse and depend on the context. The important role of autophagy as a "guardian" of cellular integrity by clearing damaged components helps protect organisms against many diseases, including neurological disorders and cancer. Moreover, the important role of autophagy in energy supply and survival to harsh environmental conditions could be employed by cancer cells to survive hypoxic tumor microenvironments. Due to the fact that the molecular and functional basis of autophagic processes are highly conserved between organisms, it is of great interest to use these organisms to link autophagy to important disease-associated signaling pathways. Finding pathways that alter autophagic activities is essential and could help the development of cures for multiple diseases with the common denominator: autophagy. Performing such assays in invertebrate models is an advantageous fast, inexpensive, and a reliable method that has great potential and value for the understanding and treatment of human diseases linked to autophagy including cancer.

Author details

Elite Possik and Arnim Pause*

*Address all correspondence to: arnim.pause@mcgill.ca

Goodman Cancer Research Centre and Biochemistry Department, McGill University, Montreal, QC, Canada

References

[1] Kaur J, Debnath J. Autophagy at the crossroads of catabolism and anabolism. Nat Rev Mol Cell Biol. 2015;16(8):461–72.

[2] Tsukada M, Ohsumi Y. Isolation and characterization of autophagy-defective mutants of *Saccharomyces cerevisiae*. FEBS Lett. 1993;333(1–2):169–74.

[3] Thumm M, Egner R, Koch B, Schlumpberger M, Straub M, Veenhuis M, et al. Isolation of autophagocytosis mutants of *Saccharomyces cerevisiae*. FEBS Lett. 1994;349(2):275–80.

[4] Harding TM, Morano KA, Scott SV, Klionsky DJ. Isolation and characterization of yeast mutants in the cytoplasm to vacuole protein targeting pathway. J Cell Biol. 1995;131(3): 591–602.

[5] Hutchins MU, Klionsky DJ. Vacuolar localization of oligomeric alpha-mannosidase requires the cytoplasm to vacuole targeting and autophagy pathway components in *Saccharomyces cerevisiae*. J Biol Chem. 2001;276(23):20491–8.

[6] Klionsky DJ, Cregg JM, Dunn WA, Jr., Emr SD, Sakai Y, Sandoval IV, et al. A unified nomenclature for yeast autophagy-related genes. Dev Cell. 2003;5(4):539–45.

[7] Wang CW, Klionsky DJ. The molecular mechanism of autophagy. Mol Med. 2003;9(3–4):65–76.

[8] Bento CF, Renna M, Ghislat G, Puri C, Ashkenazi A, Vicinanza M, et al. Mammalian autophagy: how does it work? Annu Rev Biochem. 2016.

[9] Noda NN, Inagaki F. Mechanisms of autophagy. Annu Rev Biophys. 2015;44:101–22.

[10] Mizushima N. Autophagy: process and function. Genes Dev. 2007;21(22):2861–73.

[11] Xie Z, Klionsky DJ. Autophagosome formation: core machinery and adaptations. Nat Cell Biol. 2007;9(10):1102–9.

[12] Nakatogawa H, Suzuki K, Kamada Y, Ohsumi Y. Dynamics and diversity in autophagy mechanisms: lessons from yeast. Nat Rev Mol Cell Biol. 2009;10(7):458–67.

[13] Goffeau A, Barrell BG, Bussey H, Davis RW, Dujon B, Feldmann H, et al. Life with 6000 genes. Science. 1996;274(5287):546, 63–7.

[14] Wood V, Gwilliam R, Rajandream MA, Lyne M, Lyne R, Stewart A, et al. The genome sequence of *Schizosaccharomyces pombe*. Nature. 2002;415(6874):871–80.

[15] Adams MD, Celniker SE, Holt RA, Evans CA, Gocayne JD, Amanatides PG, et al. The genome sequence of *Drosophila melanogaster*. Science. 2000;287(5461): 2185–95.

[16] Klionsky DJ. Autophagy: from phenomenology to molecular understanding in less than a decade. Nat Rev Mol Cell Biol. 2007;8(11):931–7.

[17] Guimaraes RS, Delorme-Axford E, Klionsky DJ, Reggiori F. Assays for the biochemical and ultrastructural measurement of selective and nonselective types of autophagy in the yeast *Saccharomyces cerevisiae*. Methods. 2015;75:141–50.

[18] Delorme-Axford E, Guimaraes RS, Reggiori F, Klionsky DJ. The yeast *Saccharomyces cerevisiae*: an overview of methods to study autophagy progression. Methods. 2015;75:3–12.

[19] Klionsky DJ, Abdalla FC, Abeliovich H, Abraham RT, Acevedo-Arozena A, Adeli K, et al. Guidelines for the use and interpretation of assays for monitoring autophagy. Autophagy. 2012;8(4):445–544.

[20] Nakatogawa H, Ohsumi Y. SDS-PAGE techniques to study ubiquitin-like conjugation systems in yeast autophagy. Methods Mol Biol. 2012;832:519–29.

[21] Palmisano NJ, Melendez A. Detection of autophagy in *Caenorhabditis elegans*. Cold Spring Harb Protoc. 2016;2016(2):pdb top070466.

[22] Palmisano NJ, Melendez A. Detection of autophagy in *Caenorhabditis elegans* by western blotting analysis of LGG-1. Cold Spring Harb Protoc. 2016;2016(2):pdb prot086512.

[23] Palmisano NJ, Melendez A. Detection of autophagy in *Caenorhabditis elegans* using GFP::LGG-1 as an autophagy marker. Cold Spring Harb Protoc. 2016;2016(1):pdb prot086496.

[24] Zhang H, Chang JT, Guo B, Hansen M, Jia K, Kovacs AL, et al. Guidelines for monitoring autophagy in *Caenorhabditis elegans*. Autophagy. 2015;11(1):9–27.

[25] Jenzer C, Simionato E, Legouis R. Tools and methods to analyze autophagy in *C. elegans*. Methods. 2015;75:162–71.

[26] DeVorkin L, Gorski SM. Genetic manipulation of autophagy in the *Drosophila ovary*. Cold Spring Harb Protoc. 2014;2014(9):973–9.

[27] DeVorkin L, Gorski SM. LysoTracker staining to aid in monitoring autophagy in *Drosophila*. Cold Spring Harb Protoc. 2014;2014(9):951–8.

[28] DeVorkin L, Gorski SM. Monitoring autophagic flux using Ref(2)P, the *Drosophila* p62 ortholog. Cold Spring Harb Protoc. 2014;2014(9):959–66.

[29] DeVorkin L, Gorski SM. Monitoring autophagy in *Drosophila* using fluorescent reporters in the UAS-GAL4 system. Cold Spring Harb Protoc. 2014;2014(9):967–72.

[30] Mauvezin C, Ayala C, Braden CR, Kim J, Neufeld TP. Assays to monitor autophagy in *Drosophila*. Methods. 2014;68(1):134–9.

[31] Ligeon LA, Barois N, Werkmeister E, Bongiovanni A, Lafont F. Structured illumination microscopy and correlative microscopy to study autophagy. Methods. 2015;75:61–8.

[32] Tooze SA, Dooley HC, Jefferies HB, Joachim J, Judith D, Lamb CA, et al. Assessing mammalian autophagy. Methods Mol Biol. 2015;1270:155–65.

[33] Joachim J, Jiang M, McKnight NC, Howell M, Tooze SA. High-throughput screening approaches to identify regulators of mammalian autophagy. Methods. 2015;75:96–104.

[34] Esteban-Martinez L, Boya P. Autophagic flux determination *in vivo* and *ex vivo*. Methods. 2015;75:79–86.

[35] Seglen PO, Luhr M, Mills IG, Saetre F, Szalai P, Engedal N. Macroautophagic cargo sequestration assays. Methods. 2015;75:25–36.

[36] Karanasios E, Ktistakis NT. Live-cell imaging for the assessment of the dynamics of autophagosome formation: focus on early steps. Methods. 2015;75:54–60.

[37] Gutierrez MG, Saka HA, Chinen I, Zoppino FC, Yoshimori T, Bocco JL, et al. Protective role of autophagy against *Vibrio cholerae* cytolysin, a pore-forming toxin from *V. cholerae*. Proc Natl Acad Sci USA. 2007;104(6):1829–34.

[38] Hosokawa N, Hara Y, Mizushima N. Generation of cell lines with tetracycline-regulated autophagy and a role for autophagy in controlling cell size. FEBS Lett. 2006;580(11): 2623–9.

[39] Shintani T, Klionsky DJ. Cargo proteins facilitate the formation of transport vesicles in the cytoplasm to vacuole targeting pathway. J Biol Chem. 2004;279(29):29889–94.

[40] Djeddi A, Michelet X, Culetto E, Alberti A, Barois N, Legouis R. Induction of autophagy in ESCRT mutants is an adaptive response for cell survival in *C. elegans*. J Cell Sci. 2012;125(Pt 3):685–94.

[41] Jiang P, Mizushima N. LC3- and p62-based biochemical methods for the analysis of autophagy progression in mammalian cells. Methods. 2015;75:13–8.

[42] Bjorkoy G, Lamark T, Pankiv S, Overvatn A, Brech A, Johansen T. Monitoring autophagic degradation of p62/SQSTM1. Methods Enzymol. 2009;452:181–97.

[43] Mikawa T, Kanoh J, Ishikawa F. Fission yeast Vps1 and Atg8 contribute to oxidative stress resistance. Genes Cells. 2010;15(3):229–42.

[44] Samokhvalov V, Scott BA, Crowder CM. Autophagy protects against hypoxic injury in *C. elegans*. Autophagy. 2008;4(8):1034–41.

[45] Melendez A, Talloczy Z, Seaman M, Eskelinen EL, Hall DH, Levine B. Autophagy genes are essential for dauer development and life-span extension in *C. elegans*. Science. 2003;301(5638):1387–91.

[46] Possik E, Jalali Z, Nouet Y, Yan M, Gingras MC, Schmeisser K, et al. Folliculin regulates ampk-dependent autophagy and metabolic stress survival. PLoS Genet. 2014;10(4):e1004273.

[47] Kang C, You YJ, Avery L. Dual roles of autophagy in the survival of *Caenorhabditis elegans* during starvation. Genes Dev. 2007;21(17):2161–71.

[48] Gomez TA, Banfield KL, Trogler DM, Clarke SG. The L-isoaspartyl-O-methyltransferase in *Caenorhabditis elegans* larval longevity and autophagy. Dev Biol. 2007;303(2): 493–500.

[49] Possik E, Ajisebutu A, Manteghi S, Gingras MC, Vijayaraghavan T, Flamand M, et al. FLCN and AMPK confer resistance to hyperosmotic stress via remodeling of glycogen stores. PLoS Genet. 2015;11(10):e1005520.

[50] Juhasz G, Erdi B, Sass M, Neufeld TP. Atg7-dependent autophagy promotes neuronal health, stress tolerance, and longevity but is dispensable for metamorphosis in *Drosophila*. Genes Dev. 2007;21(23):3061–6.

[51] Simonsen A, Cumming RC, Brech A, Isakson P, Schubert DR, Finley KD. Promoting basal levels of autophagy in the nervous system enhances longevity and oxidant resistance in adult *Drosophila*. Autophagy. 2008;4(2):176–84.

[52] Wu H, Wang MC, Bohmann D. JNK protects *Drosophila* from oxidative stress by transcriptionally activating autophagy. Mech Dev. 2009;126(8–9):624–37.

[53] Minois N, Carmona-Gutierrez D, Bauer MA, Rockenfeller P, Eisenberg T, Brandhorst S, et al. Spermidine promotes stress resistance in *Drosophila melanogaster* through autophagy-dependent and -independent pathways. Cell Death Dis. 2012;3:e401.

[54] Pyo JO, Yoo SM, Ahn HH, Nah J, Hong SH, Kam TI, et al. Overexpression of Atg5 in mice activates autophagy and extends lifespan. Nat Commun. 2013;4:2300.

[55] Kuma A, Hatano M, Matsui M, Yamamoto A, Nakaya H, Yoshimori T, et al. The role of autophagy during the early neonatal starvation period. Nature. 2004;432(7020):1032–6.

[56] Ginet V, Puyal J, Clarke PG, Truttmann AC. Enhancement of autophagic flux after neonatal cerebral hypoxia-ischemia and its region-specific relationship to apoptotic mechanisms. Am J Pathol. 2009;175(5):1962–74.

[57] Yan W, Zhang H, Bai X, Lu Y, Dong H, Xiong L. Autophagy activation is involved in neuroprotection induced by hyperbaric oxygen preconditioning against focal cerebral ischemia in rats. Brain Res. 2011;1402:109–21.

[58] Sheng R, Zhang LS, Han R, Liu XQ, Gao B, Qin ZH. Autophagy activation is associated with neuroprotection in a rat model of focal cerebral ischemic preconditioning. Autophagy. 2010;6(4):482–94.

[59] Lapierre LR, Hansen M. Lessons from *C. elegans*: signaling pathways for longevity. Trends Endocrinol Metab. 2012.Dec; 23(12): 637–644.

[60] Morselli E, Galluzzi L, Kepp O, Criollo A, Maiuri MC, Tavernarakis N, et al. Autophagy mediates pharmacological lifespan extension by spermidine and resveratrol. Aging (Albany NY). 2009;1(12):961–70.

[61] O'Rourke EJ, Kuballa P, Xavier R, Ruvkun G. omega-6 Polyunsaturated fatty acids extend life span through the activation of autophagy. Genes Dev. 2013;27(4):429–40.

[62] Hansen M, Chandra A, Mitic LL, Onken B, Driscoll M, Kenyon C. A role for autophagy in the extension of lifespan by dietary restriction in *C. elegans*. PLoS Genet. 2008;4(2):e24.

[63] Dwivedi M, Song HO, Ahnn J. Autophagy genes mediate the effect of calcineurin on life span in *C. elegans*. Autophagy. 2009;5(5):604–7.

[64] Tavernarakis N, Pasparaki A, Tasdemir E, Maiuri MC, Kroemer G. The effects of p53 on whole organism longevity are mediated by autophagy. Autophagy. 2008;4(7):870–3.

[65] Morselli E, Maiuri MC, Markaki M, Megalou E, Pasparaki A, Palikaras K, et al. Caloric restriction and resveratrol promote longevity through the Sirtuin-1-dependent induction of autophagy. Cell Death Dis. 2010;1:e10.

[66] Palikaras K, Lionaki E, Tavernarakis N. Coordination of mitophagy and mitochondrial biogenesis during ageing in *C. elegans*. Nature. 2015;521(7553):525–8.

[67] Lapierre LR, De Magalhaes Filho CD, McQuary PR, Chu CC, Visvikis O, Chang JT, et al. The TFEB orthologue HLH-30 regulates autophagy and modulates longevity in *Caenorhabditis elegans*. Nat Commun. 2013;4:2267.

[68] Visvikis O, Ihuegbu N, Labed SA, Luhachack LG, Alves AM, Wollenberg AC, et al. Innate host defense requires TFEB-mediated transcription of cytoprotective and antimicrobial genes. Immunity. 2014;40(6):896–909.

[69] Seah NE, de Magalhaes Filho CD, Petrashen AP, Henderson HR, Laguer J, Gonzalez J, et al. Autophagy-mediated longevity is modulated by lipoprotein biogenesis. Autophagy. 2016;12(2):261–72.

[70] Kim M, Park HL, Park HW, Ro SH, Nam SG, Reed JM, et al. *Drosophila* Fip200 is an essential regulator of autophagy that attenuates both growth and aging. Autophagy. 2013;9(8):1201–13.

[71] Rockenfeller P, Koska M, Pietrocola F, Minois N, Knittelfelder O, Sica V, et al. Phosphatidylethanolamine positively regulates autophagy and longevity. Cell Death Differ. 2015;22(3):499–508.

[72] Alvers AL, Fishwick LK, Wood MS, Hu D, Chung HS, Dunn WA, Jr., et al. Autophagy and amino acid homeostasis are required for chronological longevity in *Saccharomyces cerevisiae*. Aging Cell. 2009;8(4):353–69.

[73] Alvers AL, Wood MS, Hu D, Kaywell AC, Dunn WA, Jr., Aris JP. Autophagy is required for extension of yeast chronological life span by rapamycin. Autophagy. 2009;5(6):847–9.

[74] Ruckenstuhl C, Netzberger C, Entfellner I, Carmona-Gutierrez D, Kickenweiz T, Stekovic S, et al. Lifespan extension by methionine restriction requires autophagy-dependent vacuolar acidification. PLoS Genet. 2014;10(5):e1004347.

[75] Aris JP, Alvers AL, Ferraiuolo RA, Fishwick LK, Hanvivatpong A, Hu D, et al. Autophagy and leucine promote chronological longevity and respiration proficiency during calorie restriction in yeast. Exp Gerontol. 2013;48(10):1107–19.

[76] Tang F, Watkins JW, Bermudez M, Gray R, Gaban A, Portie K, et al. A life-span extending form of autophagy employs the vacuole-vacuole fusion machinery. Autophagy. 2008;4(7):874–86.

[77] Harrison DE, Strong R, Sharp ZD, Nelson JF, Astle CM, Flurkey K, et al. Rapamycin fed late in life extends lifespan in genetically heterogeneous mice. Nature. 2009;460(7253):392–5.

[78] Stranks AJ, Hansen AL, Panse I, Mortensen M, Ferguson DJ, Puleston DJ, et al. Autophagy controls acquisition of aging features in macrophages. J Innate Immun. 2015;7(4):375–91.

[79] Yang F, Chu X, Yin M, Liu X, Yuan H, Niu Y, et al. mTOR and autophagy in normal brain aging and caloric restriction ameliorating age-related cognition deficits. Behav Brain Res. 2014;264:82–90.

[80] Zhao S, Lin L, Kan G, Xu C, Tang Q, Yu C, et al. High autophagy in the naked mole rat may play a significant role in maintaining good health. Cell Physiol Biochem. 2014;33(2):321–32.

[81] Kaushik S, Arias E, Kwon H, Lopez NM, Athonvarangkul D, Sahu S, et al. Loss of autophagy in hypothalamic POMC neurons impairs lipolysis. EMBO Rep. 2012;13(3): 258–65.

[82] Vittorini S, Paradiso C, Donati A, Cavallini G, Masini M, Gori Z, et al. The age-related accumulation of protein carbonyl in rat liver correlates with the age-related decline in liver proteolytic activities. J Gerontol A Biol Sci Med Sci. 1999;54(8):B318–23.

[83] Cuervo AM, Dice JF. Age-related decline in chaperone-mediated autophagy. J Biol Chem. 2000;275(40):31505–13.

[84] Lipinski MM, Zheng B, Lu T, Yan Z, Py BF, Ng A, et al. Genome-wide analysis reveals mechanisms modulating autophagy in normal brain aging and in Alzheimer's disease. Proc Natl Acad Sci USA. 2010;107(32):14164–9.

[85] Taneike M, Yamaguchi O, Nakai A, Hikoso S, Takeda T, Mizote I, et al. Inhibition of autophagy in the heart induces age-related cardiomyopathy. Autophagy. 2010;6(5):600–6.

[86] Bhuiyan MS, Pattison JS, Osinska H, James J, Gulick J, McLendon PM, et al. Enhanced autophagy ameliorates cardiac proteinopathy. J Clin Invest. 2013;123(12): 5284–97.

[87] Zhang C, Cuervo AM. Restoration of chaperone-mediated autophagy in aging liver improves cellular maintenance and hepatic function. Nat Med. 2008;14(9): 959–65.

[88] Webb JL, Ravikumar B, Atkins J, Skepper JN, Rubinsztein DC. Alpha-Synuclein is degraded by both autophagy and the proteasome. J Biol Chem. 2003;278(27):25009–13.

[89] Barmada SJ, Serio A, Arjun A, Bilican B, Daub A, Ando DM, et al. Autophagy induction enhances TDP43 turnover and survival in neuronal ALS models. Nat Chem Biol. 2014;10(8):677–85.

[90] Park HK, Chu K, Jung KH, Lee ST, Bahn JJ, Kim M, et al. Autophagy is involved in the ischemic preconditioning. Neurosci Lett. 2009;451(1):16–9.

[91] Cuervo AM, Stefanis L, Fredenburg R, Lansbury PT, Sulzer D. Impaired degradation of mutant alpha-synuclein by chaperone-mediated autophagy. Science. 2004;305(5688): 1292–5.

[92] Komatsu M, Waguri S, Chiba T, Murata S, Iwata J, Tanida I, et al. Loss of autophagy in the central nervous system causes neurodegeneration in mice. Nature. 2006;441(7095): 880–4.

[93] Hara T, Nakamura K, Matsui M, Yamamoto A, Nakahara Y, Suzuki-Migishima R, et al. Suppression of basal autophagy in neural cells causes neurodegenerative disease in mice. Nature. 2006;441(7095):885–9.

[94] Jia K, Thomas C, Akbar M, Sun Q, Adams-Huet B, Gilpin C, et al. Autophagy genes protect against *Salmonella typhimurium* infection and mediate insulin signaling-regulated pathogen resistance. Proc Natl Acad Sci USA. 2009;106(34): 14564–9.

[95] Zou CG, Ma YC, Dai LL, Zhang KQ. Autophagy protects *C. elegans* against necrosis during *Pseudomonas aeruginosa* infection. Proc Natl Acad Sci USA. 2014;111(34): 12480–5.

[96] Wu J, Randle KE, Wu LP. ird1 is a Vps15 homologue important for antibacterial immune responses in *Drosophila*. Cell Microbiol. 2007;9(4):1073–85.

[97] Ren C, Finkel SE, Tower J. Conditional inhibition of autophagy genes in adult *Drosophila* impairs immunity without compromising longevity. Exp Gerontol. 2009;44(3):228–35.

[98] Shelly S, Lukinova N, Bambina S, Berman A, Cherry S. Autophagy is an essential component of *Drosophila* immunity against vesicular stomatitis virus. Immunity. 2009;30(4):588–98.

[99] Yano T, Mita S, Ohmori H, Oshima Y, Fujimoto Y, Ueda R, et al. Autophagic control of listeria through intracellular innate immune recognition in *Drosophila*. Nat Immunol. 2008;9(8):908–16.

[100] Kirienko NV, Ausubel FM, Ruvkun G. Mitophagy confers resistance to siderophore-mediated killing by *Pseudomonas aeruginosa*. Proc Natl Acad Sci USA. 2015;112(6): 1821–6.

[101] Ogawa M, Yoshimori T, Suzuki T, Sagara H, Mizushima N, Sasakawa C. Escape of intracellular Shigella from autophagy. Science. 2005;307(5710):727–31.

[102] Gutierrez MG, Master SS, Singh SB, Taylor GA, Colombo MI, Deretic V. Autophagy is a defense mechanism inhibiting BCG and *Mycobacterium tuberculosis* survival in infected macrophages. Cell. 2004;119(6):753–66.

[103] Ling YM, Shaw MH, Ayala C, Coppens I, Taylor GA, Ferguson DJ, et al. Vacuolar and plasma membrane stripping and autophagic elimination of *Toxoplasma gondii* in primed effector macrophages. J Exp Med. 2006;203(9):2063–71.

[104] Gupta M, Shin DM, Ramakrishna L, Goussetis DJ, Platanias LC, Xiong H, et al. IRF8 directs stress-induced autophagy in macrophages and promotes clearance of *Listeria monocytogenes*. Nat Commun. 2015;6:6379.

[105] Maurer K, Torres VJ, Cadwell K. Autophagy is a key tolerance mechanism during *Staphylococcus aureus* infection. Autophagy. 2015;11(7):1184–6.

[106] Maurer K, Reyes-Robles T, Alonzo F, 3rd, Durbin J, Torres VJ, Cadwell K. Autophagy mediates tolerance to *Staphylococcus aureus* alpha-toxin. Cell Host Microbe. 2015;17(4): 429–40.

[107] Singh SB, Davis AS, Taylor GA, Deretic V. Human IRGM induces autophagy to eliminate intracellular mycobacteria. Science. 2006;313(5792):1438–41.

[108] Vergne I, Singh S, Roberts E, Kyei G, Master S, Harris J, et al. Autophagy in immune defense against *Mycobacterium tuberculosis*. Autophagy. 2006;2(3):175–8.

[109] Kimmey JM, Huynh JP, Weiss LA, Park S, Kambal A, Debnath J, et al. Unique role for ATG5 in neutrophil-mediated immunopathology during *M. tuberculosis* infection. Nature. 2015;528(7583):565–9.

[110] Takacs-Vellai K, Vellai T, Puoti A, Passannante M, Wicky C, Streit A, et al. Inactivation of the autophagy gene bec-1 triggers apoptotic cell death in *C. elegans*. Curr Biol. 2005;15(16):1513–7.

[111] Guo B, Huang X, Zhang P, Qi L, Liang Q, Zhang X, et al. Genome-wide screen identifies signaling pathways that regulate autophagy during *Caenorhabditis elegans* development. EMBO Rep. 2014;15(6):705–13.

[112] Scott RC, Schuldiner O, Neufeld TP. Role and regulation of starvation-induced autophagy in the *Drosophila* fat body. Dev Cell. 2004;7(2):167–78.

[113] Lee E, Koo Y, Ng A, Wei Y, Luby-Phelps K, Juraszek A, et al. Autophagy is essential for cardiac morphogenesis during vertebrate development. Autophagy. 2014;10(4):572–87.

[114] Hu Z, Zhang J, Zhang Q. Expression pattern and functions of autophagy-related gene atg5 in zebrafish organogenesis. Autophagy. 2011;7(12):1514–27.

[115] Benato F, Skobo T, Gioacchini G, Moro I, Ciccosanti F, Piacentini M, et al. Ambra1 knockdown in zebrafish leads to incomplete development due to severe defects in organogenesis. Autophagy. 2013;9(4):476–95.

[116] Skobo T, Benato F, Grumati P, Meneghetti G, Cianfanelli V, Castagnaro S, et al. Zebrafish ambra1a and ambra1b knockdown impairs skeletal muscle development. PLoS One. 2014;9(6):e99210.

[117] Yue Z, Jin S, Yang C, Levine AJ, Heintz N. Beclin 1, an autophagy gene essential for early embryonic development, is a haploinsufficient tumor suppressor. Proc Natl Acad Sci USA. 2003;100(25):15077–82.

[118] Wang S, Li B, Qiao H, Lv X, Liang Q, Shi Z, et al. Autophagy-related gene Atg5 is essential for astrocyte differentiation in the developing mouse cortex. EMBO Rep. 2014;15(10):1053–61.

[119] Cinque L, Forrester A, Bartolomeo R, Svelto M, Venditti R, Montefusco S, et al. FGF signalling regulates bone growth through autophagy. Nature. 2015;528(7581):272–5.

[120] Gan B, Peng X, Nagy T, Alcaraz A, Gu H, Guan JL. Role of FIP200 in cardiac and liver development and its regulation of TNFalpha and TSC-mTOR signaling pathways. J Cell Biol. 2006;175(1):121–33.

[121] Liu F, Guan JL. FIP200, an essential component of mammalian autophagy is indispensible for fetal hematopoiesis. Autophagy. 2011;7(2):229–30.

[122] Wu X, Fleming A, Ricketts T, Pavel M, Virgin H, Menzies FM, et al. Autophagy regulates Notch degradation and modulates stem cell development and neurogenesis. Nat Commun. 2016;7:10533.

[123] Sato M, Sato K. Degradation of paternal mitochondria by fertilization-triggered autophagy in *C. elegans* embryos. Science. 2011;334(6059):1141–4.

[124] Politi Y, Gal L, Kalifa Y, Ravid L, Elazar Z, Arama E. Paternal mitochondrial destruction after fertilization is mediated by a common endocytic and autophagic pathway in *Drosophila*. Dev Cell. 2014;29(3):305–20.

[125] Song WH, Ballard JW, Yi YJ, Sutovsky P. Regulation of mitochondrial genome inheritance by autophagy and ubiquitin-proteasome system: implications for health, fitness, and fertility. Biomed Res Int. 2014;2014:981867.

[126] Liberski PP, Sikorska B, Bratosiewicz-Wasik J, Gajdusek DC, Brown P. Neuronal cell death in transmissible spongiform encephalopathies (prion diseases) revisited: from apoptosis to autophagy. Int J Biochem Cell Biol. 2004;36(12):2473–90.

[127] Ravikumar B, Duden R, Rubinsztein DC. Aggregate-prone proteins with polyglutamine and polyalanine expansions are degraded by autophagy. Hum Mol Genet. 2002;11(9):1107–17.

[128] Shibata M, Lu T, Furuya T, Degterev A, Mizushima N, Yoshimori T, et al. Regulation of intracellular accumulation of mutant Huntingtin by Beclin 1. J Biol Chem. 2006;281(20):14474–85.

[129] Florez-McClure ML, Hohsfield LA, Fonte G, Bealor MT, Link CD. Decreased insulin-receptor signaling promotes the autophagic degradation of beta-amyloid peptide in *C. elegans*. Autophagy. 2007;3(6):569–80.

[130] Lakso M, Vartiainen S, Moilanen AM, Sirvio J, Thomas JH, Nass R, et al. Dopaminergic neuronal loss and motor deficits in *Caenorhabditis elegans* overexpressing human alpha-synuclein. J Neurochem. 2003;86(1):165–72.

[131] Cao S, Gelwix CC, Caldwell KA, Caldwell GA. Torsin-mediated protection from cellular stress in the dopaminergic neurons of *Caenorhabditis elegans*. J Neurosci. 2005;25(15):3801–12.

[132] Karpinar DP, Balija MB, Kugler S, Opazo F, Rezaei-Ghaleh N, Wender N, et al. Pre-fibrillar alpha-synuclein variants with impaired beta-structure increase neurotoxicity in Parkinson's disease models. EMBO J. 2009;28(20):3256–68.

[133] Jia K, Hart AC, Levine B. Autophagy genes protect against disease caused by poly-glutamine expansion proteins in *Caenorhabditis elegans*. Autophagy. 2007;3(1):21–5.

[134] Ye Y, Fortini ME. Apoptotic activities of wild-type and Alzheimer's disease-related mutant presenilins in *Drosophila melanogaster*. J Cell Biol. 1999;146(6):1351–64.

[135] Wittmann CW, Wszolek MF, Shulman JM, Salvaterra PM, Lewis J, Hutton M, et al. Tauopathy in *Drosophila*: neurodegeneration without neurofibrillary tangles. Science. 2001;293(5530):711–4.

[136] Feany MB, Bender WW. A *Drosophila* model of Parkinson's disease. Nature. 2000;404(6776):394–8.

[137] Auluck PK, Chan HY, Trojanowski JQ, Lee VM, Bonini NM. Chaperone suppression of alpha-synuclein toxicity in a *Drosophila* model for Parkinson's disease. Science. 2002;295(5556):865–8.

[138] Jackson GR, Salecker I, Dong X, Yao X, Arnheim N, Faber PW, et al. Polyglutamine-expanded human huntingtin transgenes induce degeneration of *Drosophila* photoreceptor neurons. Neuron. 1998;21(3):633–42.

[139] Warrick JM, Paulson HL, Gray-Board GL, Bui QT, Fischbeck KH, Pittman RN, et al. Expanded polyglutamine protein forms nuclear inclusions and causes neural degeneration in *Drosophila*. Cell. 1998;93(6):939–49.

[140] Omata Y, Lim YM, Akao Y, Tsuda L. Age-induced reduction of autophagy-related gene expression is associated with onset of Alzheimer's disease. Am J Neurodegener Dis. 2014;3(3):134–42.

[141] Ogura K, Wicky C, Magnenat L, Tobler H, Mori I, Muller F, et al. *Caenorhabditis elegans* unc-51 gene required for axonal elongation encodes a novel serine/threonine kinase. Genes Dev. 1994;8(20):2389–400.

[142] Saitoh Y, Fujikake N, Okamoto Y, Popiel HA, Hatanaka Y, Ueyama M, et al. p62 plays a protective role in the autophagic degradation of polyglutamine protein oligomers in polyglutamine disease model flies. J Biol Chem. 2015;290(3):1442–53.

[143] Nezis IP, Simonsen A, Sagona AP, Finley K, Gaumer S, Contamine D, et al. Ref(2)P, the *Drosophila melanogaster* homologue of mammalian p62, is required for the formation of protein aggregates in adult brain. J Cell Biol. 2008;180(6):1065–71.

[144] Ratliff EP, Mauntz RE, Kotzebue RW, Gonzalez A, Achal M, Barekat A, et al. Aging and autophagic function influences the progressive decline of adult *Drosophila* behaviors. PLoS One. 2015;10(7):e0132768.

[145] Lindmo K, Brech A, Finley KD, Gaumer S, Contamine D, Rusten TE, et al. The PI 3-kinase regulator Vps15 is required for autophagic clearance of protein aggregates. Autophagy. 2008;4(4):500–6.

[146] Nagy P, Karpati M, Varga A, Pircs K, Venkei Z, Takats S, et al. Atg17/FIP200 localizes to perilysosomal Ref(2)P aggregates and promotes autophagy by activation of Atg1 in *Drosophila*. Autophagy. 2014;10(3):453–67.

[147] Babcock DT, Shen W, Ganetzky B. A neuroprotective function of NSF1 sustains autophagy and lysosomal trafficking in *Drosophila*. Genetics. 2015;199(2):511–22.

[148] Byrne S, Jansen L, JM UK-I, Siddiqui A, Lidov HG, Bodi I, et al. EPG5-related Vici syndrome: a paradigm of neurodevelopmental disorders with defective autophagy. Brain. 2016;139(Pt 3):765–81.

[149] Kyostila K, Syrja P, Jagannathan V, Chandrasekar G, Jokinen TS, Seppala EH, et al. A missense change in the ATG4D gene links aberrant autophagy to a neurodegenerative vacuolar storage disease. PLoS Genet. 2015;11(4):e1005169.

[150] Lu K, Psakhye I, Jentsch S. Autophagic clearance of polyQ proteins mediated by ubiquitin-Atg8 adaptors of the conserved CUET protein family. Cell. 2014;158(3):549–63.

[151] Zabrocki P, Pellens K, Vanhelmont T, Vandebroek T, Griffioen G, Wera S, et al. Characterization of alpha-synuclein aggregation and synergistic toxicity with protein tau in yeast. FEBS J. 2005;272(6):1386–400.

[152] Petroi D, Popova B, Taheri-Talesh N, Irniger S, Shahpasandzadeh H, Zweckstetter M, et al. Aggregate clearance of alpha-synuclein in *Saccharomyces cerevisiae* depends more on autophagosome and vacuole function than on the proteasome. J Biol Chem. 2012;287(33):27567–79.

[153] Tenreiro S, Reimao-Pinto MM, Antas P, Rino J, Wawrzycka D, Macedo D, et al. Phosphorylation modulates clearance of alpha-synuclein inclusions in a yeast model of Parkinson's disease. PLoS Genet. 2014;10(5):e1004302.

[154] Nixon RA. The role of autophagy in neurodegenerative disease. Nat Med. 2013;19(8): 983–97.

[155] Menzies FM, Fleming A, Rubinsztein DC. Compromised autophagy and neurodegenerative diseases. Nat Rev Neurosci. 2015;16(6):345–57.

[156] Saitsu H, Nishimura T, Muramatsu K, Kodera H, Kumada S, Sugai K, et al. De novo mutations in the autophagy gene WDR45 cause static encephalopathy of childhood with neurodegeneration in adulthood. Nat Genet. 2013;45(4):445–9, 9e1.

[157] Metzger S, Saukko M, Van Che H, Tong L, Puder Y, Riess O, et al. Age at onset in Huntington's disease is modified by the autophagy pathway: implication of the V471A polymorphism in Atg7. Hum Genet. 2010;128(4):453–9.

[158] Metzger S, Walter C, Riess O, Roos RA, Nielsen JE, Craufurd D, et al. The V471A polymorphism in autophagy-related gene ATG7 modifies age at onset specifically in Italian Huntington disease patients. PLoS One. 2013;8(7):e68951.

[159] Komatsu M, Wang QJ, Holstein GR, Friedrich VL, Jr., Iwata J, Kominami E, et al. Essential role for autophagy protein Atg7 in the maintenance of axonal homeostasis and the prevention of axonal degeneration. Proc Natl Acad Sci USA. 2007;104(36): 14489–94.

[160] Lettre G, Hengartner MO. Developmental apoptosis in *C. elegans*: a complex CEDnario. Nat Rev Mol Cell Biol. 2006;7(2):97–108.

[161] Sulston JE, Brenner S. The DNA of *Caenorhabditis elegans*. Genetics. 1974;77(1):95–104.

[162] Sulston J, Dew M, Brenner S. Dopaminergic neurons in the nematode *Caenorhabditis elegans*. J Comp Neurol. 1975;163(2):215–26.

[163] Sulston JE, Schierenberg E, White JG, Thomson JN. The embryonic cell lineage of the nematode *Caenorhabditis elegans*. Dev Biol. 1983;100(1):64–119.

[164] Sulston JE, Horvitz HR. Post-embryonic cell lineages of the nematode, *Caenorhabditis elegans*. Dev Biol. 1977;56(1):110–56.

[165] Kimble J, Hirsh D. The postembryonic cell lineages of the hermaphrodite and male gonads in *Caenorhabditis elegans*. Dev Biol. 1979;70(2):396–417.

[166] Huang S, Jia K, Wang Y, Zhou Z, Levine B. Autophagy genes function in apoptotic cell corpse clearance during *C. elegans* embryonic development. Autophagy. 2013;9(2):138–49.

[167] Cheng S, Wu Y, Lu Q, Yan J, Zhang H, Wang X. Autophagy genes coordinate with the class II PI/PtdIns 3-kinase PIKI-1 to regulate apoptotic cell clearance in *C. elegans*. Autophagy. 2013;9(12):2022–32.

[168] Zou W, Wang X, Vale RD, Ou G. Autophagy genes promote apoptotic cell corpse clearance. Autophagy. 2012;8(8):1267–68.

[169] Li W, Zou W, Yang Y, Chai Y, Chen B, Cheng S, et al. Autophagy genes function sequentially to promote apoptotic cell corpse degradation in the engulfing cell. J Cell Biol. 2012;197(1):27–35.

[170] Berry DL, Baehrecke EH. Growth arrest and autophagy are required for salivary gland cell degradation in *Drosophila*. Cell. 2007;131(6):1137–48.

[171] Nezis IP, Shravage BV, Sagona AP, Lamark T, Bjorkoy G, Johansen T, et al. Autophagic degradation of dBruce controls DNA fragmentation in nurse cells during late *Drosophila melanogaster* oogenesis. J Cell Biol. 2010;190(4):523–31.

[172] Nezis IP, Lamark T, Velentzas AD, Rusten TE, Bjorkoy G, Johansen T, et al. Cell death during *Drosophila melanogaster* early oogenesis is mediated through autophagy. Autophagy. 2009;5(3):298–302.

[173] Mohseni N, McMillan SC, Chaudhary R, Mok J, Reed BH. Autophagy promotes caspase-dependent cell death during *Drosophila* development. Autophagy. 2009;5(3): 329–38.

[174] Lee CY, Cooksey BA, Baehrecke EH. Steroid regulation of midgut cell death during *Drosophila* development. Dev Biol. 2002;250(1):101–11.

[175] Denton D, Shravage B, Simin R, Mills K, Berry DL, Baehrecke EH, et al. Autophagy, not apoptosis, is essential for midgut cell death in *Drosophila*. Curr Biol. 2009;19(20):1741–46.

[176] Qu X, Zou Z, Sun Q, Luby-Phelps K, Cheng P, Hogan RN, et al. Autophagy gene-dependent clearance of apoptotic cells during embryonic development. Cell. 2007;128(5):931–46.

[177] Mellen MA, de la Rosa EJ, Boya P. The autophagic machinery is necessary for removal of cell corpses from the developing retinal neuroepithelium. Cell Death Differ. 2008;15(8):1279–90.

[178] Braeckman BP. Intermediary metabolism. In: Koen Houthoofd JRV, editor. 16 February 2009 ed. wormbook2009. http://www.wormbook.org/chapters/www_intermetabolism/intermetabolism.html

[179] Chotard L, Skorobogata O, Sylvain MA, Shrivastava S, Rocheleau CE. TBC-2 is required for embryonic yolk protein storage and larval survival during L1 diapause in *Caenorhabditis elegans*. PLoS One. 2010;5(12):e15662.

[180] Kuhn H, Sopko R, Coughlin M, Perrimon N, Mitchison T. The Atg1-Tor pathway regulates yolk catabolism in *Drosophila* embryos. Development. 2015;142(22):3869–78.

[181] Lapierre LR, Silvestrini MJ, Nunez L, Ames K, Wong S, Le TT, et al. Autophagy genes are required for normal lipid levels in *C. elegans*. Autophagy. 2013;9(3):278–86.

[182] Lapierre LR, Melendez A, Hansen M. Autophagy links lipid metabolism to longevity in *C. elegans*. Autophagy. 2012;8(1):144–6.

[183] Settembre C, De Cegli R, Mansueto G, Saha PK, Vetrini F, Visvikis O, et al. TFEB controls cellular lipid metabolism through a starvation-induced autoregulatory loop. Nat Cell Biol. 2013;15(8):1016.

[184] Toulmay A, Prinz WA. Direct imaging reveals stable, micrometer-scale lipid domains that segregate proteins in live cells. J Cell Biol. 2013;202(1):35–44.

[185] van Zutphen T, Todde V, de Boer R, Kreim M, Hofbauer HF, Wolinski H, et al. Lipid droplet autophagy in the yeast *Saccharomyces cerevisiae*. Mol Biol Cell. 2014;25(2):290–301.

[186] Singh R, Kaushik S, Wang Y, Xiang Y, Novak I, Komatsu M, et al. Autophagy regulates lipid metabolism. Nature. 2009;458(7242):1131–5.

[187] Singh R, Xiang Y, Wang Y, Baikati K, Cuervo AM, Luu YK, et al. Autophagy regulates adipose mass and differentiation in mice. J Clin Invest. 2009;119(11):3329–39.

[188] Singh R, Cuervo AM. Lipophagy: connecting autophagy and lipid metabolism. Int J Cell Biol. 2012;2012:282041.

[189] Zirin J, Nieuwenhuis J, Perrimon N. Role of autophagy in glycogen breakdown and its relevance to chloroquine myopathy. PLoS Biol. 2013;11(11):e1001708.

[190] Nemazanyy I, Blaauw B, Paolini C, Caillaud C, Protasi F, Mueller A, et al. Defects of Vps15 in skeletal muscles lead to autophagic vacuolar myopathy and lysosomal disease. EMBO Mol Med. 2013;5(6):870–90.

[191] Duran J, Gruart A, Garcia-Rocha M, Delgado-Garcia JM, Guinovart JJ. Glycogen accumulation underlies neurodegeneration and autophagy impairment in Lafora disease. Hum Mol Genet. 2014;23(12):3147–56.

[192] Criado O, Aguado C, Gayarre J, Duran-Trio L, Garcia-Cabrero AM, Vernia S, et al. Lafora bodies and neurological defects in malin-deficient mice correlate with impaired autophagy. Hum Mol Genet. 2012;21(7):1521–33.

[193] Nascimbeni AC, Fanin M, Masiero E, Angelini C, Sandri M. Impaired autophagy contributes to muscle atrophy in glycogen storage disease type II patients. Autophagy. 2012;8(11):1697–700.

[194] Nascimbeni AC, Fanin M, Masiero E, Angelini C, Sandri M. The role of autophagy in the pathogenesis of glycogen storage disease type II (GSDII). Cell Death Differ. 2012;19(10):1698–708.

[195] Kim J, Kundu M, Viollet B, Guan KL. AMPK and mTOR regulate autophagy through direct phosphorylation of Ulk1. Nat Cell Biol. 2011;13(2):132–41.

[196] Kamada Y, Yoshino K, Kondo C, Kawamata T, Oshiro N, Yonezawa K, et al. Tor directly controls the Atg1 kinase complex to regulate autophagy. Mol Cell Biol. 2010;30(4):1049–58.

[197] Nazio F, Strappazzon F, Antonioli M, Bielli P, Cianfanelli V, Bordi M, et al. mTOR inhibits autophagy by controlling ULK1 ubiquitylation, self-association and function through AMBRA1 and TRAF6. Nat Cell Biol. 2013;15(4):406–16.

[198] Alers S, Loffler AS, Wesselborg S, Stork B. Role of AMPK-mTOR-Ulk1/2 in the regulation of autophagy: cross talk, shortcuts, and feedbacks. Mol Cell Biol. 2012;32(1):2–11.

[199] Scott RC, Juhasz G, Neufeld TP. Direct induction of autophagy by Atg1 inhibits cell growth and induces apoptotic cell death. Curr Biol. 2007;17(1):1–11.

[200] Wang Z, Wilson WA, Fujino MA, Roach PJ. Antagonistic controls of autophagy and glycogen accumulation by Snf1p, the yeast homolog of AMP-activated protein kinase, and the cyclin-dependent kinase Pho85p. Mol Cell Biol. 2001;21(17):5742–52.

[201] Egan D, Kim J, Shaw RJ, Guan KL. The autophagy initiating kinase ULK1 is regulated via opposing phosphorylation by AMPK and mTOR. Autophagy. 2011;7(6):643–4.

[202] Wu WK, Coffelt SB, Cho CH, Wang XJ, Lee CW, Chan FK, et al. The autophagic paradox in cancer therapy. Oncogene. 2012;31(8):939–53.

[203] White E. The role for autophagy in cancer. J Clin Invest. 2015;125(1):42–6.

[204] Liang XH, Jackson S, Seaman M, Brown K, Kempkes B, Hibshoosh H, et al. Induction of autophagy and inhibition of tumorigenesis by beclin 1. Nature. 1999;402(6762):672–6.

[205] Aita VM, Liang XH, Murty VV, Pincus DL, Yu W, Cayanis E, et al. Cloning and genomic organization of beclin 1, a candidate tumor suppressor gene on chromosome 17q21. Genomics. 1999;59(1):59–65.

[206] Qu X, Yu J, Bhagat G, Furuya N, Hibshoosh H, Troxel A, et al. Promotion of tumorigenesis by heterozygous disruption of the beclin 1 autophagy gene. J Clin Invest. 2003;112(12):1809–20.

[207] Takahashi Y, Coppola D, Matsushita N, Cualing HD, Sun M, Sato Y, et al. Bif-1 interacts with Beclin 1 through UVRAG and regulates autophagy and tumorigenesis. Nat Cell Biol. 2007;9(10):1142–51.

[208] Takamura A, Komatsu M, Hara T, Sakamoto A, Kishi C, Waguri S, et al. Autophagy-deficient mice develop multiple liver tumors. Genes Dev. 2011;25(8):795–800.

[209] Abedin MJ, Wang D, McDonnell MA, Lehmann U, Kelekar A. Autophagy delays apoptotic death in breast cancer cells following DNA damage. Cell Death Differ. 2007;14(3):500–10.

[210] Katayama M, Kawaguchi T, Berger MS, Pieper RO. DNA damaging agent-induced autophagy produces a cytoprotective adenosine triphosphate surge in malignant glioma cells. Cell Death Differ. 2007;14(3):548–58.

[211] Degenhardt K, Mathew R, Beaudoin B, Bray K, Anderson D, Chen G, et al. Autophagy promotes tumor cell survival and restricts necrosis, inflammation, and tumorigenesis. Cancer Cell. 2006;10(1):51–64.

[212] Karantza-Wadsworth V, Patel S, Kravchuk O, Chen G, Mathew R, Jin S, et al. Autophagy mitigates metabolic stress and genome damage in mammary tumorigenesis. Genes Dev. 2007;21(13):1621–35.

[213] Lemasters JJ. Selective mitochondrial autophagy, or mitophagy, as a targeted defense against oxidative stress, mitochondrial dysfunction, and aging. Rejuvenation Res. 2005;8(1):3–5.

[214] Strohecker AM, Guo JY, Karsli-Uzunbas G, Price SM, Chen GJ, Mathew R, et al. Autophagy sustains mitochondrial glutamine metabolism and growth of BrafV600E-driven lung tumors. Cancer Discov. 2013;3(11):1272–85.

[215] Marino G, Salvador-Montoliu N, Fueyo A, Knecht E, Mizushima N, Lopez-Otin C. Tissue-specific autophagy alterations and increased tumorigenesis in mice deficient in Atg4C/autophagin-3. J Biol Chem. 2007;282(25):18573–83.

[216] Dou Z, Xu C, Donahue G, Shimi T, Pan JA, Zhu J, et al. Autophagy mediates degradation of nuclear lamina. Nature. 2015;527(7576):105–9.

[217] Guo JY, Chen HY, Mathew R, Fan J, Strohecker AM, Karsli-Uzunbas G, et al. Activated Ras requires autophagy to maintain oxidative metabolism and tumorigenesis. Genes Dev. 2011;25(5):460–70.

[218] Lock R, Roy S, Kenific CM, Su JS, Salas E, Ronen SM, et al. Autophagy facilitates glycolysis during Ras-mediated oncogenic transformation. Mol Biol Cell. 2011;22(2): 165–78.

[219] Guo JY, Karsli-Uzunbas G, Mathew R, Aisner SC, Kamphorst JJ, Strohecker AM, et al. Autophagy suppresses progression of K-ras-induced lung tumors to oncocytomas and maintains lipid homeostasis. Genes Dev. 2013;27(13):1447–61.

[220] Wei H, Wei S, Gan B, Peng X, Zou W, Guan JL. Suppression of autophagy by FIP200 deletion inhibits mammary tumorigenesis. Genes Dev. 2011;25(14):1510–27.

[221] Wei H, Wang C, Croce CM, Guan JL. p62/SQSTM1 synergizes with autophagy for tumor growth *in vivo*. Genes Dev. 2014;28(11):1204–16.

[222] Linehan WM, Srinivasan R, Schmidt LS. The genetic basis of kidney cancer: a metabolic disease. Nature Reviews Urology. 2010;7(5):277–85.

[223] Tee AR, Pause A. Birt-Hogg-Dube: tumour suppressor function and signalling dynamics central to folliculin. Fam Cancer. 2013;12(3):367–72.

[224] Yan M, Gingras MC, Dunlop EA, Nouet Y, Dupuy F, Jalali Z, et al. The tumor suppressor folliculin regulates AMPK-dependent metabolic transformation. J Clin Invest. 2014; 126(6):2640–2650.

[225] Mikhaylova O, Stratton Y, Hall D, Kellner E, Ehmer B, Drew AF, et al. VHL-regulated MiR-204 suppresses tumor growth through inhibition of LC3B-mediated autophagy in renal clear cell carcinoma. Cancer Cell. 2012;21(4):532–46.

[226] Parkhitko A, Myachina F, Morrison TA, Hindi KM, Auricchio N, Karbowniczek M, et al. Tumorigenesis in tuberous sclerosis complex is autophagy and p62/sequestosome 1 (SQSTM1)-dependent. Proc Natl Acad Sci USA. 2011;108(30):12455–60.

[227] Santanam U, Banach-Petrosky W, Abate-Shen C, Shen MM, White E, DiPaola RS. Atg7 cooperates with Pten loss to drive prostate cancer tumor growth. Genes Dev. 2016;30(4): 399–407.

[228] Yang S, Wang X, Contino G, Liesa M, Sahin E, Ying H, et al. Pancreatic cancers require autophagy for tumor growth. Genes Dev. 2011;25(7):717–29.

[229] Perera RM, Stoykova S, Nicolay BN, Ross KN, Fitamant J, Boukhali M, et al. Transcriptional control of autophagy-lysosome function drives pancreatic cancer metabolism. Nature. 2015;524(7565):361–5.

Chapter 12

Autophagy in Plant Pathogenic Fungi

Huan-Bin Shi, Shuang Liang, Lu-Yao Huang,
Xiao-Hong Liu, Xue-Ming Zhu, Ya-Hui Zhao and
Fu-Cheng Lin

Additional information is available at the end of the chapter

http://dx.doi.org/10.5772/64634

Abstract

Autophagy is a ubiquitous and conserved process in eukaryotic cells from yeasts to mammals. It also appears to play vital roles in plant pathogenic fungi, impacting growth, morphology, development, and pathogenicity. In this chapter, we have introduced a new concept to delineate the role of autophagy in homeostasis of plant pathogenic fungi and in their interaction with host cells, in breach of host barrier, and in the mechanisms of plant fungal infection.

Keywords: plant pathogenic fungi, autophagy, selective autophagy, process and function, molecular mechanism

1. Introduction of plant pathogenic fungi

Filamentous fungi play important roles in health care, agriculture, and bioprocessing. There are thousands of species of plant pathogenic fungi that collectively are responsible for 70% of all known plant diseases. An improved understanding of the molecular mechanisms of plant fungal pathogenesis will ultimately lead to better control of plant fungal diseases. However, until recently, the complexity of fungal pathogen-plant host interactions made for slow progress in understanding both the mechanisms of plant host resistance and fungal pathogenesis. Over the last 40 years, the development and application of productive approaches based on genetics and molecular biology has led to the cloning and analysis of many plant resistance genes and fungal pathogenesis determinants to reveal a plethora of fungus-plant molecular interactions [1–3]. Diverse plant organs present different obstacles to infection by potential fungal pathogens, and therefore successful pathogens have evolved specific strategies, especially infection structures, that are able to break through host plant roots, stems, leaves, flowers, or other special tissues. These infection structures are usually modified hyphae specialized for the invasion of the target host plant tissue(s).

The typical life cycle of plant pathogenic fungi includes an asexual and a sexual stage. In the asexual stage, pathogens often repeatedly produce many asexual spores during the crop growing season; these play important roles in the spread of plant disease. The asexual spores germinate into filamentous hyphae, which can differentiate into a series of complex infection structures such as infection cushions, haustoria, appressoria, penetration pegs, and others [4]. This stage plays an important role in the spread of plant disease. When the vegetative growth stage reaches a certain period of time, fungi begin to enter the sexual stage to form various sexual spores, which are the initial sources of infection during the next disease cycle, in addition to giving rise to offspring and aiding pathogenic fungi in adverse environments. In favor of pathogens undergoing successful sporulation and infection, autophagy plays important roles in nutrient homeostasis. However, studies examining the role of autophagy are still not advanced in plant pathogenic fungi. With the publication of the genome sequence of *Saccharomyces cerevisiae* in 1996, the use of yeast as a model system to study the molecular mechanisms of autophagy came to the fore. The noteworthy discovery of the role of autophagy, following its induction by nutrient starvation, in the development of turgor in the appressorium represents a milestone emphasizing the importance of recognizing the role of autophagy in the formation of infection structures in plant pathogenic fungi. This achievement adds to our fundamental knowledge of both plant fungal pathogenesis and the biological roles of autophagy [5]. In the last two decades, our knowledge has advanced remarkably and autophagy has been examined in organisms from yeast to plant pathogenic fungi, such as *Magnaporthe oryzae*, *Colletortrichum* spp., *Fusarium* spp., and *Ustilago maydis*.

The rice blast fungus, *M. oryzae*, a filamentous ascomycete fungus, is the causal agent of rice blast disease, the most destructive disease of rice worldwide [6]. *M. oryzae* differentiates a special infection structure, the appressorium, to rupture the strong cuticular layer, which is the first defense barrier of the plant host, and gain entry to colonize into plant tissue [4, 7]. The appressorium is a flattened, hyphal structure that is used to enter host cells during infection. It generates colossal intracellular turgor pressure (as much as 8.0 MPa), allowing it to penetrate the leaf cuticle. The mechanical forces that generate a mature appressorium and deliver its penetration peg have been confirmed by researchers worldwide. This enormous turgor in the appressorium is a consequence of the accumulation of very large quantities of glycerol in the cell, and potential sources for glycerol biosynthesis are lipids and glycogen, as well as sugars, trehalose, and mannitol, in the conidium [8]. The sequenced genomes of *M. oryzae* have been provided online [9].

Colletotrichum is the causal agent of anthracnose and other diseases on leaves, stems, and fruits of numerous plant species, including several important crops. The damage caused by *Colletotrichum* spp. extends to important staple food crops, including bananas, cassava, and sorghum. In addition, it is particularly successful as a postharvest pathogen because latent infections, which are initiated before harvest, do not become active until the fruit has been stored or appears on the market shelf. *Colletotrichum* is highly significant as an experimental model in studies of fungal development, infection processes, host resistance, signal transduction, and the molecular biology of plant-pathogen interactions. *Colletotrichum* can differentiate the specialized infection structure, the appressorium, as well. Dozens of laboratories are studying the biology and pathology of various species of *Colletotrichum* all over the world. The sequenced genomes of *C. graminicola* and *C. higginsianum* have been provided online [10].

The ascomycete *Fusarium graminearum* (teleomorph *Gibberella zeae*) is a highly destructive pathogen of all cereal species. It causes *Fusarium* head blight, a devastating disease on wheat and barley. Infection causes shifts in the amino acid composition of wheat, resulting in shriveled kernels and contamination of the remaining grain with mycotoxins, primarily deoxynivalenol. Mycotoxins in grain can affect human and animal health when they enter the food chain. The other important *Fusarium* species, *F. oxysporum*, can induce susceptibility in more than 120 types of plants such as cotton, tomato, and banana. The infection process begins when the pathogen reaches the surface of plant roots, subsequently resulting in the penetration and colonization of this fungus. Ultimately, the pathogens will diffuse in the xylem vessels, which can result in both local and systemic defense responses in the host plant. The typical disease symptoms in the infected plant cells are slow growth, browning, wilting, and finally the death of the host. In addition to the economic importance of *Fusarium* spp., *Fusarium* species also serve as key model organisms for biological and evolutionary researches [11].

U. maydis, a member of the smut fungi, infects certain important crops such as wheat, maize, and barley. Smut fungi are obligate parasite pathogens, thus they live in their host plants to obtain necessary nutrition for their sexual life cycle. The haploid yeast-like form of *U. maydis* can be propagated on artificial media. However, this form is unable to cause disease. Two compatible haploid strains fuse and generate a dikaryotic filament to cause infection. The dikaryotic filament penetrates the host cell via invagination of the plasma membrane. Unlike the systemic infections caused by other smut fungi, *U. maydis* is unable to cause systemic disease and can only infect the above-ground portions of the host plant maize [12].

In addition, the genetic models *Aspergillus* spp., *Podospora anserina*, and *Sordaria macrospore* are discussed in this chapter. We enumerate progress in studying autophagy and describe the many differences between single cell yeasts and multiple cell fungi in the context of this process.

2. Autophagy: functional roles in eukaryotes

Autophagy is an intracellular degradation system that is highly conserved in eukaryotic cells, which degrade proteins and organelles in the vacuole/lysosome. Studies examining the functions of autophagy have increased significantly in the last decade. Autophagy has a wide variety of functions in eukaryotic cells and intensive studies have shown that autophagy is not only involved in nutrient recycling but also in other cellular processes such as cellular differentiation, growth, and pathogenicity [13–18].

- Autophagy is a general term for the degradation of cytoplasmic components within lysosomes. This process is quite distinct from general endocytosis-mediated lysosomal degradation of extracellular and plasma membrane proteins [19]. There are three types of autophagy that are classified based on the different ways in which a substrate in the cytoplasm can be transferred into the vacuole/lysosome: macroautophagy, microautophagy, and chaperone-mediated autophagy (CMA). In general, we often consider the term "autophagy" to refer to macroautophagy. In recent years, given the selectivity of substrates for degradation, autophagy has been divided into two types: selective autophagy and nonselective autophagy [16–18, 20]. The unique organelle, called the autophagosome, mediates autophagy but where and how autophagosomes emerge has been a major question. In yeast, more than 38 Atg (autophagy-related) proteins have been identified, and most of them gather at a site that can be identified by fluorescence microscopy as a punctate spot very close to the vacuolar membrane. As autophagosomes are generated from this site, it is called the "preautophagosomal structure" (PAS) [21, 22]. However, the structure of the PAS has not been characterized until now.

2.1. Autophagy/autolysosomal events

In mammals, the regulation of autophagy is highly complicated. Limited numbers of studies examining autophagy in filamentous fungi have extended the knowledge gleaned from *S. cerevisiae,* the model organism for the study of autophagy [23]. Generally, autophagy consists of four sequential steps: (1) induction of autophagy; (2) recruitment of ATG proteins by phagophore assembly site proteins, the rapid formation of two-layer autophagosomal membrane structures, and isolation of the cytoplasm and organelles; (3) fusion of autophagosomes with lysosomes/vacuoles, in which the inner membrane of the autophagosome and the cytoplasm-derived materials are contained in the autophagosome; and (4) degradation of autophagic bodies in the vacuole into macromolecules that will be recycled. Autophagy facilitates the recycling of cytoplasmic components as nutrients to support cell survival. This feature is remarkably different from the ubiquitin-proteasome system, which only can specifically recognize ubiquitinated proteins for proteasomal degradation. Autophagy can selectively or nonselectively degrade proteins and fragments of organisms to maintain essential activity and viability in response to nutrient limitation.

2.2. The induction and inhibition of autophagy

Like yeast or mammals, fungal autophagy is typically induced by nutrition (e.g., carbon and nitrogen) starvation [24]. Conversely, autophagy is inhibited by the mammalian target of rapamycin (mTOR), a central cell growth regulator that is connected with growth factor and nutrient sensing. It has been demonstrated that the molecular mechanism is regulated by the mammalian autophagy-initiating kinase Ulk1, a homolog of yeast Atg1 [25]. Under nutrient-rich conditions, TOR kinase is activated and phosphorylates Atg13. Phosphorylated Atg13 does not possess sufficient affinity for the Atg1 kinase and cannot form a complex with Atg1. Thus, the process of autophagy is inhibited [26, 27].

Rapamycin is an inducer of autophagy as inhibition of mTOR mimics cellular starvation by blocking signals required for cell growth and proliferation. Fungi can also be induced by treatment with rapamycin, often with concentrations between 200 and 500 ng/ml in fungal culture [23, 24, 28, 29].

2.3. Autophagy in nutrient recycling

Under nutrient-limited conditions, autophagy can serve as a nutrient recycling pathway to support cell survival [30]. This role was demonstrated in the autophagy-deficient *Aspergillus fumigatus* Δ*Afatg1* mutant strain that experienced limited growth on a nutrient starvation medium (i.e., water-agarose), while the wild-type strain could maintain growth on the same medium. After reconstituting *AfATG1* back into the mutant, the complemented strain exhibited the same phenotype as the wild-type strain, suggesting that autophagy can recycle nutrients for growth [31]. For fungi, the colony margin mycelium is more active than the central mycelium and the inner hyphal network recycles nutrients to complement the growing tips [32]. When autophagy was blocked in Δ*Moatg1* mutants of *M. oryzae*, growth of the Δ*Moatg1* mutants was assessed on minimal media lacking either of the two crucial nutrients, nitrogen and carbon, and Δ*Moatg1* mutant colonies spread more slowly than those of the wild-type strain in both the cases [5]. These studies provide evidence that autophagy is required to support growth.

2.4. Autophagy in cellular degradation

Filamentous fungi have no lysosomes, like mammals, but the vacuoles play a similar role as degradative organelles. Hyphal vacuolation has been shown to increase rapidly in nutrient-starved *Aspergillus oryzae* mycelia [33]. It has been demonstrated that autophagic bodies can be visualized in vacuoles under starvation conditions; however, autophagic bodies cannot be detected in autophagy-blocked mutants, such as in the *A. fumigatus* Δ*Afatg1* mutant or in the *M. oryzae* Δ*Moatg1*, *4*, *5*, and *9* mutants [5, 31, 34–36]. Mon1 or Ypt7 are essential for vesicle fusion and vacuole morphology in yeast, and deleting the homologous genes in *M. oryzae* results in autophagy blockage in the mutants [37, 38].

2.5. Autophagy in cellular differentiation

In eukaryotic cells, autophagy is involved in cellular differentiation and development [14]. In plant pathogenic fungi, blocked autophagy impacts the phenotypes and morphologies of fungi. In the *M. oryzae* autophagy-deficient mutant Δ*Moatg8*, sucrose or glucose supplementation suppresses conidiation defects but the appressorium loses the ability to penetrate [39]. Deletion of *MoATG4* or *MoATG8* results in nuclear degeneration during appressorium formation and a drastic reduction in asexual conidiospore formation [34, 39]. Deletion of the lipase gene *ATG15* in *F. graminearum* results in reduced degradation of storage lipids [40]. Deletion of autophagy genes also inhibits the formation of sexual reproductive organs such as protoperithecia in *P. anserina* and perithecia in *M. oryzae* [5, 41]. These findings strongly suggest that autophagy is needed for proper cellular differentiation in several species of filamentous fungi. However, autophagy does not always appear to be involved in cellular differentiation.

For example, hyphal differentiation in Δ*atg9* mutant of the filamentous yeast *Candida albicans* does not affect the formation of chlamydospores [42]. Thus, there may be diverse roles for autophagy in the fungal differentiation of different species.

3. Macroautophagy-related genes studied in plant pathogenic fungi

Macroautophagy is the most widely studied type of autophagy. Many *ATG* genes involved in this process have been identified and characterized in yeast, plant, and mammals [43]. Additionally, it has been demonstrated that autophagy is a conserved catabolic pathway in plant pathogenic fungi and plays vital roles in development and differentiation among plant pathogens [44]. In contrast to yeast, many plant pathogenic fungi form special infection structures that can rupture the plant cuticle to gain entry to internal tissue, ultimately causing plant disease. The appressorium is known as a typical feature of some of the most important cereal pathogens such as the devastating rice blast disease-causing fungus, *M. oryzae*. Macro-autophagy-related genes have been studied extensively in the plant pathogens *M. oryzae*, *Aspergillus* spp., *Colletotrichum* spp., *Fusarium* spp., and *U. maydis*.

3.1. The Atg1 kinase complex

In *S. cerevisiae*, Atg1, which encodes a serine/threonine kinase, forms a complex with Atg13, Atg17, Atg29, and Atg31 that are required for PAS formation. The Atg1 complex initiates the formation of the autophagosome by forming a scaffold to recruit other complexes [45]. Homologs of the yeast *ATG1* gene have been identified and characterized in many filamentous fungi including *Colletotrichum lindemuthianum* [46], *M. oryzae* [5], and *P. anserina* [41]. Random insertional mutagenesis in *C. lindemuthianum* clones produced mutations in a putative serine/threonine protein kinase named Clk1 that is associated with pathogenicity in the common bean. Clk1 is homologous to Atg1 in *S. cerevisiae*, although its role in autophagy has not been reported [46]. Disruption of the *ATG1* gene results in blockage of the autophagy process as confirmed by a defect in autophagosome formation and the absence of autophagic bodies in the vacuole in *P. anserina* [47]. In *M. oryzae*, deletion of *MgATG1*(*MoATG1*) causes reduced conidiation, delayed spore germination, and loss of pathogenicity on barley and rice due to interrupted autophagy [5]. Similarly, deletion of *BcATG1* inhibits autophagosome accumulation in the vacuoles of nitrogen-starved cells. A null mutant of *BcATG1* is defective in vegetative growth, conidiation, sclerotial formation, and appressorium formation [48]. Functional analysis of *AoATG1* in *A. oryzae* shows that conidiation and aerial hyphae are reduced significantly. Overexpression of AoAtg1 leads to decreased conidiation and excessive development of aerial hyphae [49]. These data are consistent with the role of *S. cerevisiae ATG1* in the induction of autophagy.

Upon starvation induction, Atg13 is rapidly dephosphorylated and subsequently activates Atg1 kinase activity. Atg13 binds to the C-terminus of Atg1 and this interaction occurs in a constitutive manner [50]. Data from *M. oryzae* show that *MoATG13* and *MoATG29* are not essential for turgor generation and rice blast formation [51]. Although the *AoATG13* deletion mutant exhibits a reduction in conidiation and aerial hyphae, the mutant still exhibits autophagic activity in *A. oryzae* [52]. Thus, unlike *S. cerevisiae* Atg13, there may be another Atg13 in these plant pathogens.

3.2. Atg9-mediated vesicle transport

In yeast, Atg9 is a transmembrane protein consisting of six transmembrane domains with the amino and carboxyl termini exposed in the cytosol. Atg9 is positioned on the PAS and peripheral sites, shuttling and delivering membrane to form the phagophore. Atg1-Atg13 and Atg2-Atg18 facilitate Atg9-mediated vesicle transport from the PAS to peripheral sites [53], and the return of Atg9-mediated vesicle transport depends on Atg11, Atg23, and Atg27 [54, 55]. A new report indicates that Atg9 is a direct target of Atg1 kinase. Phosphorylated Atg9 is essential for the recruitment of Atg8 and Atg18 to the PAS and extension of the isolation membrane [56].

The functional characterization of Atg9 homologs in filamentous fungi has been documented. In *M. oryzae*, knockout of 23 autophagy-related genes confirmed that *MoATG9* is essential for rice blast formation [51]. Dong et al. also characterized the MoAtg9-mediated trafficking process in detail by observing the fluorescent localization of MoAtg9 in the wild-type strain and null mutants of *MoATG1*, 2, 13, and 18. MoAtg9 cycling depends on MoAtg1, 2, and 18 but not on MoAtg13. The null mutant of *MoATG9* exhibits similar phenotypes to the *MoATG1* deletion mutant, such as poor sporulation and appressorium formation, blockage of autophagy, and lack of pathogenicity on susceptible rice [35].

3.3. The two ubiquitin-like systems in autophagy

The process of autophagy involves two ubiquitin-like systems, the Atg12 and Atg8 conjugation systems. Both play key roles in bending and extension of the autophagosome membrane [45]. In the Atg12-Atg5-Atg16 ubiquitin-like system, Atg5 can directly bind the membrane, a process that is negatively regulated by Atg12 and positively regulated by Atg16 [57]. Identification of the proteins associated with the autophagosome has been performed in filamentous fungi. Disruption of the gene *BbATG5* resulted in abnormal conidia and reduction in growth, germination, blastospore formation, conidiation, and virulence in *Beauveria bassiana* [58]. Disruption of *MoATG5* in *M. oryzae* caused loss of pathogenicity, reduced conidiation, and perithecia formation [34]. In *Trichoderma reesei*, *TrATG5* gene knockout resulted in reduced conidiation and abnormal conidiophores [59]. Null mutants *Moatg12* and *Moatg16* also result in loss of pathogenicity in *M. oryzae* [51]. We can conclude that the Atg12 ubiquitin-like system is associated with conidiation and pathogenicity in plant pathogens.

Atg8 is another ubiquitin-like protein associated with autophagosome formation throughout the autophagy process and is well conserved in most model organisms and higher eukaryotes. It is associated with the autophagosome membrane and has been used as a marker for autophagy [60]. The amount of Atg8 can regulate the volume of autophagosomes by controlling phagophore expansion [61, 62]. The Atg12-Atg5-Atg16 complex acts like an E3 ligase, catalyzing Atg8 to phosphatidylethanolamine (PE) on the membrane [57].

Deletion of the *ATG8* gene in fungi detrimentally inhibits autophagy and affects cellular growth and differentiation [63]. Deletion of *FgATG8* in the necrotrophic plant pathogen *F. graminearum* results in loss of perithecia, reduced conidiation, and collapse of the aerial hyphae [64]. Lipid utilization is dependent on autophagy in *F. graminearum* for providing nutrients for the nonassimilating portion of the fungi. Although the ability to infect plants is not affected in the null mutant *Fgatg8*, plant colonization cannot be seen from spikelet to spikelet as with the wild-type fungus [64]. In *F. oxysporum*, the autophagy-related gene *FoATG8* is involved in nuclear degradation after hyphal fusion and the control of nuclear distribution [65]. The null mutant *Smatg8* in *S. macrospore* results in fruiting body loss, impaired ascospore germination, and the ability to undergo hyphal fusion [66]. In *M. oryzae*, loss of *MoATG8* blocks autophagic conidial cell death leading to impaired appressorium formation, loss of pathogenicity, and reduced conidiation [67]. Consistent with its function in *M. oryzae*, *CoATG8* is also involved in normal conidiation, appressorium formation, and pathogenicity in *C. orbiculare* [68]. Disruption of the *AoATG8* gene causes severe defects in the formation of aerial hyphae and conidia by affecting the autophagy process in *A. oryzae* [69].

The Atg12 and Atg8, two ubiquitin-like conjugating systems, share the same E1-like activating enzyme, Atg7, but have different E2-like conjugating enzymes: Atg10 and Atg3, respectively [45]. In *M. oryzae*, *MoATG7*, *MoATG10*, and *MoATG3* are all associated with pathogenicity. Deletion of any of these causes a loss of pathogenicity in *M. oryzae* [51]. Atg4, a cysteine protease, is responsible for the first cleavage of Atg8 to expose its C-terminal glycine and the second cleavage from PE to recycle Atg8 [47]. In *M. oryzae*, the cleavage event mediated by MoAtg4 is also conserved as observed in *in vitro* assays. Disruption of MoAtg4 blocks autophagy and causes defects like other null mutants of autophagy-related genes in *M. oryzae* [36]. In *Sordaria macrospora*, SmAtg4 is also capable of cleaving the SmAtg8 precursor [66]. Deletion of *AoATG4* results in the loss of aerial hyphae and reduced conidiation, resulting from the destruction of autophagy in *A. oryzae* [52].

3.4. The PI3K complex

Another protein complex required for autophagy in yeast is the phosphatidylinositol 3-kinase (PI3K) complex that involves Vps15, Vps34, Atg6, and Atg14 [45]. All of these proteins, with the exception of Atg14, are conserved in filamentous fungi. In *S. cerevisiae*, this complex is not specific to autophagy and is also involved in the cytoplasm-to-vacuole targeting pathway (CVT) pathway and a fairly diverse array of signaling and membrane transport events including Golgi to vacuole transport [61]. Thus, it is likely that autophagy may be impacted by other cellular signaling events. There have been few studies examining this complex in plant pathogens except in *M. oryzae*. A null mutant of *MoATG6* resulted in the loss of pathogenicity [50].

3.5. Membrane fusion

The membrane fusion of the autophagosome and vacuole requires many proteins. The autophagosome first docks with the surface of a vacuole, and then the outer membrane fuses with the vacuolar membrane. The proteins involved in this step are mostly those proteins which are implicated in membrane fusion, such as the SNARE family proteins and the homotypic fusion and vacuolar protein sorting (HOPS) tethering complex [70, 71]. Once inside the vacuolar lumen, the single-membrane vesicle turns into the autophagic body.

Many SNARE proteins have been characterized in plant pathogens especially in *M. oryzae* and *F. graminearum*. In *M. oryzae*, MoVam7 is essential for vacuolar membrane fusion and vacuolar maturation [72]. MoSec22, an R-SNARE protein, is involved in endocytosis. Both MoVam7 and MoSec22 are required for growth, conidiation, and pathogenicity in *M. oryzae* [73]. Recently, another syntaxin protein, MoSyn8, has been identified for its role in regulating intracellular trafficking in *M. oryzae* [74]. In *F. graminearum*, FgVam7, a homolog of Vam7, plays an important role in regulating cellular differentiation and virulence [75]. Roles for these SNARE proteins in fungal development have been explored in depth but their functions in the autophagy process must be further evaluated. It has been reported that the HOPS subunit MoVps39 is crucial for pathogenicity in *M. oryzae* due to its role in anchoring G protein signaling. Rab GTPases also function in vesicle-vacuolar fusion [76]. In *M. oryzae*, MoYpt7, a homolog of Ypt7 in yeast, has been confirmed to be required for autophagy by affecting membrane fusion and assembly of the mature vacuole [37]. MoMon1, an ortholog of *S. cerevisiae* Mon1, is also required for membrane fusion in mature vacuolar formation [38]. Consistent with the results in *M. oryzae*, it has also been demonstrated that the *F. graminearum* homologs of Ypt7 and Mon1 are conserved in vacuole fusion and autophagy [77, 78].

3.6. Degradation of the autophagic body

The last step during autophagy is the degradation of the autophagic body in the vacuoles and recycling of cellular macromolecules to supply nutrient. Two autophagy-related proteins, Atg15 and Atg22, have been identified in yeast and are involved in this process [45]. Atg15 is a putative lipase involved in the lysis of CVT bodies, autophagic bodies, and incorporated peroxisomes in the vacuole [79]. Atg15 is well conserved in yeast and filamentous fungi, but it contains a distinct repeat motif at its C-terminus in front of the Ser/Thr-rich region [80]. In *A. oryzae*, AoAtg15 is required for autophagosome formation and the lysis of autophagic bodies. A null mutant of *Aoatg15* exhibits differentiation defects for the aerial hyphae and conidia [52]. Disruption of *FgATG15* causes a delay in lipid body degradation and utilization and abnormal development of the conidia and aerial hyphae formation [81]. In *M. oryzae*, *MoATG15* is also involved in pathogenicity [53]. On the other hand, Atg22 is an integral vacuolar transmembrane protein with structural similarity to permeates. It may function as a transporter in the export of recycled amino acids from the vacuole to the cytosol [82]. In *A. nidulans* and many other filamentous fungi, multiple paralogs of Atg22-like proteins have been identified [80]. In *P. anserina*, the *pspA/idi-6* gene encoding serine protease A is the functional ortholog of the *S. cerevisiae* vacuolar protease B (Prb1). Both of these proteases are involved in autophagy [47].

4. Selective autophagy in plant pathogenic fungi

With the exception of bulk autophagy, which is nonselective, eukaryotic cells possess several types of selective autophagy to maintain homeostasis during stress. These processes include the CVT pathway (which is only found in *S. cerevisiae*), pexophagy (which targets peroxisomes), mitophagy (the specific elimination of mitochondria), nucleophagy (which targets the nucleus), reticulophagy (which mediates the turnover of the endoplasmic reticulum), and ribophagy (the specific elimination of ribosomes) [83]. Here, we provide a brief introduction to the mechanisms that have been identified and the development of selective autophagy in plant pathogenic fungi.

4.1. Pexophagy

Peroxisomes share enzymes involved in the β-oxidation of fatty acids and the production and degradation of H_2O_2 or other reactive oxygen species (ROS); thus, homeostasis in peroxisomes plays an important role in survival and development. Researchers have found that peroxisomes exhibit high variability under changing circumstances because their numbers can be rapidly increased when their functions are required but then can be quickly recycled when they are not essential to avoid wasting energy [84]. There are two biological processes that explain this phenomenon: the process of peroxisome growth and division and the specific degradation of peroxisomes by autophagy, known as pexophagy. Studies of methylotrophic yeasts have shown that two distinct, selective modes are exploited for pexophagy: macropexophagy and micropexophagy. It appears that high levels of ATP activate micropexophagy, while lower levels induce macropexophagy. Macropexophagy is initiated at a specific PAS (different from the CVT pathway); newly synthesized membrane wraps around and sequesters the peroxisomes one by one and then forms a double membrane pexophagosome, which is ultimately delivered to the vacuole. In contrast to macropexophagy, a cluster of peroxisomes is swallowed by vacuolar sequestering membranes (VSMs) in micropexophagy. Meanwhile, the micropexophagy-specific membrane apparatus (MIPA), which mediates fusion between the tips of the invagination vacuoles, extends from the PAS. Finally, membrane scission occurs on the inner side of the vacuolar membrane and the peroxisomes are cracked [85].

Among 38 *ATG* genes that have been identified in yeasts, 16 Atg proteins constitute the core machinery mediating the formation of the autophagosome for almost all types of autophagy. While the specialization of any selective autophagy pathway is determined by the selective cargo recognized and engulfed by the autophagosomes, this requires the help of other selective autophagy-specific *ATG* genes. In *Pichia pastoris*, phosphorylated Atg30 physically interacts directly with Atg11 and Atg17. These two proteins act as scaffolds at the PAS to recruit other proteins. In *Hansenula polymorpha*, Atg25 is essential to connect Atg11 and Atg17. In contrast, Atg30 is required for macropexophagy in *S. cerevisiae*. During the final stages of pexophagy in *P. pastoris* and *S. cerevisiae*, Atg24 is involved in the regulation of membrane fusion at the vacuolar surface. Atg26 is a sterol glucosyltransferase that synthesizes sterol glucoside and is necessary for both modes of pexophagy but not bulk autophagy. In *P. pastoris*, the protein is associated with the MIPA during micropexophagy [86].

In the cucumber anthracnose fungus *C. orbiculare*, pexophagy is required for phytopathogenicity. To penetrate host epidermal cells and cause infection, *C. orbiculare* forms a specific structure, the appressorium. Researchers showed that while appressoria could still form in the *Coatg26* deletion mutant, the infection process was severely delayed. Furthermore, deletion of *Coatg8* impaired formation of the appressoria, suggesting the important role of autophagy during infection. A GFP fusion protein was peroxisomally expressed in *C. orbiculare* to monitor peroxisome conditions. Massive fluorescent dots could be observed in the wild-type strain and *Coatg26* mutant showing that the Atg26 null mutant still possessed a normal ability to perform peroxisome biosynthesis. The authors also observed that CoAtg8-tagged phagophores could swallow the peroxisomes through diffusion of GFP inside the vacuoles in the wild-type strain. However, bright peroxisomal dots were present in the appressoria of the *Coatg26* mutant. This indicated that appressoria pexophagy of the *Coatg26* mutant was significantly impaired compared with the wild-type strain. Taken together, nonselective general autophagy is essential for early stage pathogen development and Atg26-dependent selective pexophagy is essential for later stages of infection [68].

4.2. Mitophagy

Mitochondria are the sites of oxidative metabolism in eukaryotic organisms and are the places where the energy is released by the final oxidation of carbohydrates, fats, and amino acids. Reactive oxygen species are a side-product of the mitochondria. ROS release damages mitochondrial DNA and proteins and other cellular compartments, e.g., nuclear DNA. Thus, mitochondrial homeostasis is critical for organisms. Mitochondria degradation is mediated through a selective type of autophagy, called mitophagy. In *S. cerevisiae*, Atg32 (a mitochondrial outer membrane protein) functions as a receptor protein that interacts with Atg11 and Atg8 [87]. Recent studies show that a portion of the molecular mechanism involved in mitochondrial fission participates in mitophagy [88].

In *M. oryzae*, Atg24-assisted mitophagy in the foot cells is necessary for proper asexual differentiation and functions in redox homeostasis and nutrient modulation. MoAtg24, a sorting nexin related to yeast Snx4, is only required for mitophagy. The Δ*Moatg24* strain exhibited a decreased rate of conidiophore formation and reduced aerial hyphal growth. Subcellular localization of MoAtg24-GFP under ROS stress and starvation conditions found that MoAtg24-GFP was localized to the mitochondria and studies employing Mito-GFP clearly indicated signals in the foot cells. The Bin-Amphiphysin-Rvs (BAR) and Phox homology (PX) domains of MoAtg24 are essential for its mitochondrial localization [89].

4.3. Reticulophagy and ribophagy

The ribosome is a cellular ribonucleoprotein particle that is primarily composed of RNA (rRNA) and proteins, and its only function is to catalyze amino acids for proteins according to the instructions imparted by the mRNA. Ribophagy involves degradation of the 60S ribosomal subunit and is regulated by both ubiquitination and deubiquitination. Some proteins are synthesized in the endoplasmic reticulum (ER) shortly after synthesis begins. The ER also modulates the modification and processing of protein folding and assembly and transport of nascent peptides. Like other selective autophagy pathways, reticulophagy and nucleophagy should be highly controlled. ERphagy is differentially induced, depending on the intensity and type of the ER stress [90]. This shows that the Ypt/Rab GTPase module, formed by the Trs85 containing the Ypt1, the Atg11 effector and the TRAPPIII GEF, functions in reticulophagy. With the exception of Ypt/Rab GTPases [91], macroreticulophagy also depends on Atgs and their cargos. In conclusion, Atg9-dependent ERphagy involved autophagy of the endoplasmic reticulum, Ypt1- and core Atg-dependent phagy mediate the organization of PAS, and Ypt51-dependent phagy mediates the delivery of autophagosomes to the vacuole [92].

Recently, researchers identified two Atg8-binding proteins in *S. cerevisiae*, Atg39 and Atg40, both of which are receptors for reticulophagy. ER consists of 3 subdomains in *S. cerevisiae*, specifically, the cytoplasmic ER (cytoER), the cortical ER (cER), and the perinuclear ER (pnER), i.e., the nuclear envelope (NE). Atg39 specifically localizes to the pnER/NE and induces autophagic sequestration of double-membrane vesicles as well as some intranuclear components. Thus, the Atg39-dependent pathway should also be called nucleophagy. Atg40 is primarily responsible for cER/cytoERphagy. Atg39 mutant cells die earlier than wild-type cells under prolonged nitrogen starvation, while Atg40 mutant cells do not, suggesting the physiological significance of the pnER-autophagy pathway. It is still unknown whether the generation of nuclear envelope-derived vesicles and ER fragments are coupled with their sequestration into autophagosomes [93, 94]. The processes of reticulophagy and ribophagy still need to be studied in plant pathogens.

4.4. Nucleophagy

The removal of damaged/nonessential/entire (in some circumstances) nuclei under stress is crucial for cell survival. Nucleophagy is the selective degradation of nuclear material by autophagy. It has two modes: macronucleophagy and micronucleophagy. Two processes (piecemeal microautophagy of the nucleus, PMN, and late nucleophagy) exist in *S. cerevisiae* and are considered microautophagy. In the PMN, two proteins, Vac8 (located in the vacuolar membrane) and Nvj1 (in the nuclear envelope), recognize each other at nucleus-vacuole (NV) junctions. Then, nuclear membranes that are associated with NV junctions invaginate into the vacuolar lumen. Sequestration of the nuclear cargo occurs via the fission of nuclear membranes and the vacuolar membrane, and ultimately a triple-membrane PMN vesicle is released into the vacuolar lumen and is degraded. The cargo that is sequestered by PMN includes nonessential nuclear components; PMN requires the core *ATG* genes, and specific *ATG* genes are necessary at the step in which the vacuolar membrane fuses. Conversely, late nucleophagy can occur in the absence of Nvj1, Vac8, and Atg11 [95].

Micronucleophagy has been studied in *A. oryzae*. EGFP-tagged Atg8 has been utilized to track autophagosomal structures. Large autophagosomes (1–2 mm in diameter) are formed around the nuclei and then sequester whole targets. Subsequently, nuclear material is dispersed throughout neighboring vacuoles, suggesting that autophagosomes forming close to targeted nuclei in turn sequester whole nuclei [96]. Micronucleophagy has been studied in *M. oryzae*. A dome-shaped structure, called the appressorium, penetrates host tissues. To form appressoria, spores undergo a series of autophagy steps including nuclear degeneration. In contrast to piecemeal microautophagy of the nucleus, nuclear degradation in *M. oryzae* is dependent on core autophagy genes, and fungi, such as *A. oryzae*, possess large autophagosomes, while *M. oryzae*'s autophagosomes are smaller and more punctate [97].

In *M. oryzae*, core *ATG* genes such as *MoATG1*, *MoATG5*, and *MoATG8* were studied in the context of pathogenicity but given that these genes are essential for both selective and bulk autophagy, it is still unknown if selective autophagy has a critical function in the fungal pathogenicity of *M. oryzae* [98]. Kershaw et al. conducted a mass gene knockout in *M. oryzae* to analyze 22 *ATG* genes [51]. The results showed that *M. oryzae* becomes nonpathogenic on rice when it loses any one of the core *ATG* genes but is still pathogenic in the absence of *ATG* genes for selective autophagy (*ATG11, 24, 26–29*).

5. Conclusions

In this report, we have introduced a new concept to delineate the role of autophagy in homeostasis of plant pathogenic fungi and in their interaction with host cells, breach of host barrier, and in the mechanisms of plant fungal infection. More knowledge of the diverse modes of autophagy is likely to help us understand the mechanisms of fungal pathogen-plant host interactions. In addition, pathogenic fungi are multicellular organisms that undergo constant polar growth that is completely different from yeast. Due to the differences between yeast and multicellular organisms, greater divergence has been revealed in terms of the CVT pathway, SNARE proteins, and selective autophagy, as mentioned in this chapter. It is necessary to create a new system to analyze autophagy in filamentous fungi. Furthermore, with the development of metabolomics and proteomics, studies of autophagy in plant pathogenic fungi can be combined with new technologies based on the molecular mechanisms of autophagy.

Acknowledgements

This study was supported by grants (no. 31371890 and no. 31370171) from the National Natural Science Foundation of China and the Fundamental Research Funds for the Central Universities.

Author details

Huan-Bin Shi, Shuang Liang, Lu-Yao Huang, Xiao-Hong Liu, Xue-Ming Zhu, Ya-Hui Zhao and Fu-Cheng Lin*

*Address all correspondence to: fuchenglin@zju.edu.cn

State Key Laboratory for Rice Biology, Biotechnology Institute, Zhejiang University, Hangzhou, China

References

[1] Keen NT. Gene-for-gene complementarity in plant-pathogen interactions. Annu. Rev. Genet. 1990; 24: 447–463. Doi: 10.1146/annurev.ge.24.120190.002311

[2] Baker B, Zambryski P, Staskawicz B, Dinesh-Kumar SP. Signaling in plant-microbe interactions. Science. 1997; 276: 726–733. Doi: 10.1126/science.276.5313.726

[3] Greenberg JT. Programmed cell death in plant-pathogen interactions. Annu. Rev. Plant Physiol. Plant Mol. Biol. 1997; 48: 525–545. Doi: 10.1146/annurev.arplant.48.1.525

[4] Deising HB, Werner S, Wernitz M. The role of fungal appressoria in plant infection. Microbes Infect. 2000; 2: 1631–1641. Doi: 10.1016/S1286-4579(00)01319-8

[5] Liu XH, Lu JP, Zhang L, Dong B, Min H, et al. Involvement of a Magnaporthe grisea serine/threonine kinase gene, MgATG1, in appressorium turgor and pathogenesis. Eukaryot. Cell. 2007; 6: 997–1005. Doi: 10.1128/EC.00011-07

[6] Tanweer FA, Rafii MY, Sijam K, Rahim HA, Ahmed F, et al. Current advance methods for the identification of blast resistance genes in rice. C. R. Biol. 2015; 338: 321–334. Doi: 10.1016/j.crvi.2015.03.001

[7] Dean RA. Signal pathways and appressorium morphogenesis. Annu. Rev. Phytopathol. 1997; 35: 211–234. Doi: 10.1146/annurev.phyto.35.1.211

[8] Wilson RA, Talbot NJ. Under pressure: investigating the biology of plant infection by *Magnaporthe oryzae*. Nat. Rev. Microbiol. 2009; 7: 185–195. Doi: 10.1038/nrmicro2032

[9] Dean RA, Talbot NJ, Ebbole DJ, Farman ML, Mitchell TK, et al. The genome sequence of the rice blast fungus *Magnaporthe grisea*. Nature. 2005; 434: 980–986. Doi: 10.1038/nature03449

[10] O'Connell RJ, Thon MR, Hacquard S, Amyotte SG, Kleemann J, et al. Lifestyle transitions in plant pathogenic Colletotrichum fungi deciphered by genome and transcriptome analyses. Nat. Genet. 2012; 44: 1060–1065. Doi: 10.1038/ng.2372

[11] Dean R, Van Kan JA, Pretorius ZA, Hammond-Kosack KE, Di Pietro A, et al. The top 10 fungal pathogens in molecular plant pathology. Mol. Plant Pathol. 2012; 13: 414–430. Doi: 10.1111/j.1364-3703.2011.00783.x

[12] Brefort T, Doehlemann G, Mendoza-Mendoza A, Reissmann S, Djamei A, et al. Ustilago maydis as a pathogen. Annu. Rev. Phytopathol. 2009; 47: 423–445. Doi: 10.1146/annurev-phyto-080508-081923

[13] Cuervo AM. Autophagy: in sickness and in health. Trends Cell. Biol. 2004; 14: 70–77. Doi: 10.1016/j.tcb.2003.12.002

[14] Inoue Y, Klionsky DJ. Regulation of macroautophagy in *Saccharomyces cerevisiae*. Semin. Cell Dev. Biol. 2010; 21(7): 664–670. Doi: 10.1016/j.semcdb.2010.03.009

[15] Mijaljica D, Prescott M, Devenish RJ. Microautophagy in mammalian cells: revisiting a 40-year-old conundrum. Autophagy. 2011; 7(7): 673–682. Doi: 10.4161/auto.7.7.14733

[16] Parzych KR, Klionsky DJ. An overview of autophagy: morphology, mechanism, and regulation. Antioxid. Redox Signal. 2014; 20(3): 460–473. Doi: 10.1089/ars.2013.5371

[17] He C, Klionsky DJ. Regulation mechanisms and signaling pathways of autophagy. Annu. Rev. Genet. 2009; 43: 67–93. Doi: 10.1146/annurev-genet-102808-114910

[18] Reggiori F, Klionsky DJ. Autophagic processes in yeast: mechanism, machinery and regulation. Genetics. 2013; 194(2): 341–361. Doi: 10.1534/genetics.112.149013

[19] Mizushima N. Autophagy: process and function. Genes Dev. 2007; 21: 2861–2873. Doi: 10.1101/gad.1599207

[20] Klionsky DJ, Abdelmohsen K, Abe A, Abedin MJ, Abeliovich H, et al. Guidelines for the use and interpretation of assays for monitoring autophagy (3rd edition). Autophagy. 2016; 12(1): 1–222. Doi: 10.1080/15548627.2015.1100356

[21] Suzuki K, Kirisako T, Kamada Y, Mizushima N, Noda T, et al. The pre-autophagosomal structure organized by concerted functions of APG genes is essential for autophagosome formation. EMBO J. 2001; 20: 5971–5981. Doi: 10.1093/emboj/20.21.5971

[22] 22.Suzuki K, Ohsumi Y. Molecular machinery of autophagosome formation in yeast, *Saccharomyces cerevisiae*, FEBS Lett. 2007; 581: 2156–2161. Doi: 10.1016/j.febslet.2007.01.096

[23] Delorme-Axford E, Guimaraes RS, Reggiori F, Klionsky DJ. The yeast *Saccharomyces cerevisiae*: an overview of methods to study autophagy progression. Methods. 2015; 75: 3–12. Doi: 10.1016/j.ymeth.2014.12.008

[24] Dementhon K, Paoletti M, Pinan-Lucarré B, Loubradou-Bourges N, Sabourin M, et al. Rapamycin mimics the incompatibility reaction in the fungus *Podospora anserina*. Eukaryotic Cell. 2003; 2: 238–246. Doi: 10.1128/EC.2.2.238-246.2003

[25] Kim J, Kundu M, Viollet B, Guan KL. AMPK and mTOR regulate autophagy through direct phosphorylation of Ulk1. Nat. Cell Biol. 2011; 13: 132–141. Doi: 10.1038/ncb2152

[26] Chen YQ, Klionsky DJ. The regulation of autophagy—unanswered questions. J. Cell Sci. 2011; 124: 161–170. Doi: 10.1242/jcs.064576

[27] Tanida I. Autophagosome formation and molecular mechanism of autophagy. Antiox. Redox Signal. 2011; 14: 2201–2214. Doi: 10.1089/ars.2010.3482

[28] Pinan-Lucarré B, Balguerie A, Clavé C. Accelerated cell death in Podospora autophagy mutants. Eukaryotic Cell. 2005; 4: 1765–1774. Doi: 10.1128/EC.4.11.1765-1774.2005

[29] Teichert S, Wottawa M, Schönig B, Tudzynski B. Role of the Fusarium fujikuroi TOR kinase in nitrogen regulation and secondary metabolism. Eukaryotic Cell. 2006; 5: 1807–1819. Doi: 10.1128/EC.00039-06

[30] Yorimitsu T, Klionsky DJ. Autophagy: molecular machinery for self-eating. Cell Death Diff. 2005; 12: 1542–1552. Doi: 10.1038/sj.cdd.4401765

[31] Richie DL, Fuller KK, Fortwendel J, Miley MD, McCarthy JW, et al. Unexpected link between metal ion deficiency and autophagy in *Aspergillus fumigatus*. Eukaryotic Cell. 2007; 6: 2437–2447. Doi: 10.1128/Ec.00224-07

[32] Glass NL, Rasmussen C, Roca MG, Read ND. Hyphal homing, fusion and mycelial interconnectedness. Trends Microbiol. 2004; 12: 135–141. Doi: 10.1016/j.tim.2004.01.007

[33] Pollack JK, Zheng JL, Marten MR. Fungal mycelia show lag time before re-growth on endogenous carbon. Biotechnol. Bioeng. 2008; 100: 458–465. Doi: 10.1002/bit.21779

[34] Lu JP, Liu XH, Feng XX, Min H, Lin FC. An autophagy gene, MgATG5, is required for cell differentiation and pathogenesis in *Magnaporthe oryzae*. Curr. Genet. 2009; 55: 461–473. Doi: 10.1007/s00294-009-0259-5

[35] Dong B, Liu XH, Lu JP, Zhang FS, Gao HM, et al. MgAtg9 trafficking in *Magnaporthe oryzae*. Autophagy. 2009; 5: 946–953. Doi: 10.4161/auto.5.7.9161

[36] Liu TB, Liu XH, Lu JP, Zhang L, Min H, et al. The cysteine protease MoAtg4 interacts with MoAtg8 and is required for differentiation and pathogenesis in *Magnaporthe oryzae*. Autophagy. 2010; 6: 74–85. Doi: 10.4161/auto.6.1.10438

[37] Liu XH, Chen SM, Gao HM, Ning GA, Shi HB, et al. The small GTPase MoYpt7 is required for membrane fusion in autophagy and pathogenicity of *Magnaporthe oryzae*. Environ. Microbiol. 2015; 17: 4495–4510. Doi: 10.1111/1462-2920.12903

[38] Gao HM, Liu XG, Shi HB, Lu JP, Yang J, et al. MoMon1 is required for vacuolar assembly, conidiogenesis and pathogenicity in the rice blast fungus *Magnaporthe oryzae*. Res. Microbiol. 2013; 164: 300–309. Doi: 10.1016/j.resmic.2013.01.001

[39] Deng YZ, Ramospamplona M, Naqvi NI. Autophagy-assisted glycogen catabolism regulates asexual differentiation in *Magnaporthe oryzae*. Autophagy. 2009; 5: 33–43. Doi: 10.4161/auto.5.1.7175

[40] Long NN, Bormann J, Le GTT, Stärkel C, Olsson S, et al. Autophagy-related lipase FgATG15 of *Fusarium graminearum* is important for lipid turnover and plant infection. Fungal Genet. Biol. 2011; 48: 217–224. Doi: 10.1016/j.fgb.2010.11.004

[41] Pinan-Lucarre B, Paoletti M, Dementhon K, Coulary-Salin B, Clave C. Autophagy is induced during cell death by incompatibility and is essential for differentiation in the filamentous fungus *Podospora anserina*. Mol. Microbiol. 2003; 47: 321–333. Doi: 10.1046/j.1365-2958.2003.03208.x

[42] Palmer GE, Kelly MN, Sturtevant JE. Autophagy in the pathogen *Candida albicans*. Microbiology. 2007; 153: 51–58. Doi: 10.1046/j.1365-2958.2003.03208.x

[43] Noda NN, Inagaki F. Mechanisms of Autophagy. Ann. Rev. Biophys. 2015; 44: 101–122. Doi: 10.1146/annurev-biophys-060414-034248

[44] Bartoszewska M, Kiel JA. The role of macroautophagy in development of filamentous fungi. Antiox. Redox Signal. 2011; 14: 2271–2287. Doi: 10.1089/ars.2010.3528

[45] Feng Y, He D, Yao Z, Klionsky DJ. The machinery of macroautophagy. Cell Res. 2014; 24: 24–41. Doi: 10.1038/cr.2013.168

[46] Dufresne M, Bailey JA, Dron M, Langin T. Clk1, a serine/threonine protein kinase-encoding gene, is involved in pathogenicity of *Colletotrichum lindemuthianum* on common bean. Mol. Plant-Microbe. Inter. 1998; 11: 99–108. Doi: 10.1094/MPMI.1998.11.2.99

[47] Pinan-Lucarre B, Balguerie A, Clave C. Accelerated cell death in Podospora autophagy mutants. Eukaryotic cell. 2005; 4: 1765–1774. Doi: 10.1128/EC.4.11.1765-1774.2005

[48] AU: Please provide volume number and the page range for Refs. [48 and 83].Ren W, Zhang Z, Shao W, Yang Y, Zhou M, et al. The autophagy-related gene BcATG1 is involved in fungal development and pathogenesis in *Botrytis cinerea*. Mol. Plant Pathol. 2016. Doi: 10.1111/mpp.12396

[49] Yanagisawa S, Kikuma T, Kitamoto K. Functional analysis of Aoatg1 and detection of the Cvt pathway in *Aspergillus oryzae*. FEMS Microbiol. Lett. 2013; 338: 168–176. Doi: 10.1111/1574-6968.12047

[50] Ragusa MJ, Stanley RE, Hurley JH. Architecture of the Atg17 complex as a scaffold for autophagosome biogenesis. Cell. 2012; 151: 1501–1512. Doi: 10.1016/j.cell.2012.11.028

[51] Kershaw MJ, Talbot NJ. Genome-wide functional analysis reveals that infection-associated fungal autophagy is necessary for rice blast disease. Proc. Nat. Acad. Sci. U. S. A. 2009; 106: 15967–15972. Doi: 10.1073/pnas.0901477106

[52] Kikuma T, Kitamoto K. Analysis of autophagy in *Aspergillus oryzae* by disruption of Aoatg13, Aoatg4, and Aoatg15 genes. FEMS Microbiol. Lett. 2011; 316: 61–69. Doi: 10.1111/j.1574-6968.2010.02192.x

[53] Reggiori F, Tucker KA, Stromhaug PE, Klionsky DJ. The Atg1-Atg13 complex regulates Atg9 and Atg23 retrieval transport from the pre-autophagosomal structure. Dev. Cell. 2004; 6: 79–90. Doi: 10.1016/S1534-5807(03)00402-7

[54] Yen WL, Legakis JE, Nair U, Klionsky DJ. Atg27 is required for autophagy-dependent cycling of Atg9. Mol. Biol. Cell. 2007; 18: 581–593. Doi: 10.1091/mbc.E06-07-0612

[55] He C, Song H, Yorimitsu T, Monastyrska I, Yen WL, et al. Recruitment of Atg9 to the preautophagosomal structure by Atg11 is essential for selective autophagy in budding yeast. J. Cell Biol. 2006; 175: 925–935. Doi: 10.1083/jcb.200606084

[56] Papinski D, Schuschnig M, Reiter W, Wilhelm L, Barnes CA, et al. Early steps in autophagy depend on direct phosphorylation of Atg9 by the Atg1 kinase. Mol. Cell. 2014; 53: 471–483. Doi: 10.1016/j.molcel.2013.12.011

[57] Walczak M, Martens S. Dissecting the role of the Atg12-Atg5-Atg16 complex during autophagosome formation. Autophagy. 2013; 9: 424–425. Doi: 10.4161/auto.22931

[58] Zhang L, Wang J, Xie XQ, Keyhani NO, Feng MG, et al. The autophagy gene BbATG5, involved in the formation of the autophagosome, contributes to cell differentiation and growth but is dispensable for pathogenesis in the entomopathogenic fungus Beauveria bassiana. Microbiology. 2013; 159: 243–252. Doi: 10.1099/mic.0.062646-0

[59] Liu XH, Yang J, He RL, Lu JP, Zhang CL, et al. An autophagy gene, TrATG5, affects conidiospore differentiation in *Trichoderma reesei*. Res. Microbiol. 2011; 162: 756–763. Doi: 10.1016/j.resmic.2011.06.011

[60] Cheong H, Klionsky DJ. Biochemical methods to monitor autophagy-related processes in yeast. Methods Enzymol. 2008; 451: 1–26. Doi: 10.1016/S0076-6879(08)03201-1

[61] Nakatogawa H, Ichimura Y, Ohsumi Y. Atg8, a ubiquitin-like protein required for autophagosome formation, mediates membrane tethering and hemifusion. Cell. 2007; 130: 165–178. Doi: 10.1016/j.cell.2007.05.021

[62] Xie Z, Nair U, Klionsky DJ. Atg8 controls phagophore expansion during autophagosome formation. Mol. Biol. Cell. 2008; 19: 3290–3298. Doi: 10.1091/mbc.E07-12-1292

[63] Pollack JK, Harris SD, Marten MR. Autophagy in filamentous fungi. Fungal Genet. Biol. 2009; 46: 1–8. Doi: 10.1016/j.fgb.2008.10.010

[64] Josefsen L, Droce A, Sondergaard TE, Sorensen JL, Bormann J, et al. Autophagy provides nutrients for nonassimilating fungal structures and is necessary for plant colonization but not for infection in the necrotrophic plant pathogen *Fusarium graminearum*. Autophagy. 2012; 8: 326–337. Doi: 10.4161/auto.18705

[65] Corral-Ramos C, Roca MG, Di Pietro A, Roncero MI, Ruiz-Roldan C. Autophagy contributes to regulation of nuclear dynamics during vegetative growth and hyphal fusion in *Fusarium oxysporum*. Autophagy. 2015; 11: 131–144. Doi: 10.4161/15548627.2014.994413

[66] Voigt O, Poggeler S. Autophagy genes Smatg8 and Smatg4 are required for fruiting-body development, vegetative growth and ascospore germination in the filamentous ascomycete *Sordaria macrospora*. Autophagy. 2013; 9: 33–49. Doi: 10.4161/auto.22398

[67] Veneault-Fourrey C, Barooah M, Egan M, Wakley G, Talbot NJ. Autophagic fungal cell death is necessary for infection by the rice blast fungus. Science. 2006; 312: 580–583. Doi: 10.1126/science.1124550

[68] Asakura M, Ninomiya S, Sugimoto M, Oku M, Yamashita S, et al. Atg26-mediated pexophagy is required for host invasion by the plant pathogenic fungus *Colletotrichum orbiculare*. Plant Cell. 2009; 21: 1291–1304. Doi: 10.1105/tpc.108.060996

[69] Kikuma T, Ohneda M, Arioka M, Kitamoto K. Functional analysis of the ATG8 homologue Aoatg8 and role of autophagy in differentiation and germination in *Aspergillus oryzae*. Eukaryotic Cell. 2006; 5: 1328–1336. Doi: 10.1128/EC.00024-06

[70] Moreau K, Renna M, Rubinsztein DC. Connections between SNAREs and autophagy. Trends Biochem. Sci. 2013; 38: 57–63. Doi: 10.1016/j.tibs.2012.11.004

[71] Solinger JA, Spang A. Tethering complexes in the endocytic pathway: CORVET and HOPS. FEBS J. 2013; 280: 2743–2757. Doi: 10.1111/febs.12151

[72] AU: Please provide the page range for Ref. [72].Dou XY, Wang Q, Qi ZQ, Song WW, Wang W, et al. MoVam7, a conserved SNARE involved in vacuole assembly, is required for growth, endocytosis, ROS accumulation, and pathogenesis of *Magnaporthe oryzae*. PLoS ONE. 2011; 6. Doi: 10.1371/journal.pone.0016439

[73] Song WW, Dou XY, Qi ZQ, Wang Q, Zhang X, et al. R-SNARE homolog MoSec22 is required for conidiogenesis, cell wall integrity, and pathogenesis of *Magnaporthe oryzae*. PLoS ONE. 2010; 5. Doi: 10.1371/journal.pone.0013193

[74] Qi Z, Liu M, Dong Y, Zhu Q, Li L, et al. The syntaxin protein (MoSyn8) mediates intracellular trafficking to regulate conidiogenesis and pathogenicity of rice blast fungus. New Phytologist. 2016; 209: 1655–1667. Doi: 10.1111/nph.13710

[75] Zhang H, Li B, Fang Q, Li Y, Zheng X, et al. SNARE protein FgVam7 controls growth, asexual and sexual development, and plant infection in *Fusarium graminearum*. Mol. Plant Pathol. 2016; 17: 108–119. Doi: 10.1111/mpp.12267

[76] Ramanujam R, Calvert ME, Selvaraj P, Naqvi NI. The late endosomal HOPS complex anchors active G-protein signaling essential for pathogenesis in *Magnaporthe oryzae*. PLoS Pathog. 2013; 9: e1003527. Doi: 10.1371/journal.ppat.1003527

[77] Li Y, Li B, Liu L, Chen H, Zhang H, et al. FgMon1, a guanine nucleotide exchange factor of FgRab7, is important for vacuole fusion, autophagy and plant infection in *Fusarium graminearum*. Sci. Rep. 2015; 5: 18101. Doi: 10.1038/srep18101

[78] Zheng H, Zheng W, Wu C, Yang J, Xi Y, et al. Rab GTPases are essential for membrane trafficking-dependent growth and pathogenicity in *Fusarium graminearum*. Environ. Microbiol. 2015; 17: 4580–4599. Doi: 10.1111/1462-2920.12982

[79] Epple UD, Suriapranata I, Eskelinen EL, Thumm M. Aut5/Cvt17p, a putative lipase essential for disintegration of autophagic bodies inside the vacuole. J. Bacteriol. 2001; 183: 5942–5955. Doi: 10.1128/JB.183.20.5942-5955.2001

[80] Kiel JA, Van Der Klei IJ. Proteins involved in microbody biogenesis and degradation in Aspergillus nidulans. Fungal Genet. Biol. 2009; 46(Suppl 1): S62–S71. Doi: 10.1016/j.fgb.2008.07.009

[81] Nguyen LN, Bormann J, Le GT, Starkel C, Olsson S, et al. Autophagy-related lipase FgATG15 of *Fusarium graminearum* is important for lipid turnover and plant infection. Fungal Genet. Biol. 2011; 48: 217–224. Doi: 10.1016/j.fgb.2010.11.004

[82] Yang Z, Huang J, Geng J, Nair U, Klionsky DJ. Atg22 recycles amino acids to link the degradative and recycling functions of autophagy. Mol. Biol. Cell. 2006; 17: 5094–5104. Doi: 10.1091/mbc.E06-06-0479

[83] Zaffagnini G, Martens S. Mechanisms of selective autophagy. J. Mol. Biol. 2016. Doi: 10.1016/j.jmb.2016.02.004

[84] Bhambra GK, Wang ZY, Soanes DM, Wakley GE, Talbot NJ. Peroxisomal carnitine acetyl transferase is required for elaboration of penetration hyphae during plant infection by *Magnaporthe grisea*. Mol. Microbiol. 2006; 61: 46–60. Doi: 10.1111/j. 1365-2958.2006.05209.x

[85] Till A, Lakhani R, Burnett SF, Subramani S. Pexophagy: the selective degradation of peroxisomes. Int. J. Cell Biol. 2012; 2012: 512721. Doi: 10.1155/2012/512721

[86] Dunn JWA, Cregg JM, Kiel JAKW, Klei IJvd, Oku M, et al. Pexophagy: the selective autophagy of peroxisomes. Autophagy. 2005; 1: 75–83. Doi: 10.4161/auto.1.2.1737

[87] Kanki T, Wang K, Cao Y, Baba M, Klionsky DJ. Atg32 is a mitochondrial protein that confers selectivity during mitophagy. Dev. Cell. 2009; 17: 98–109. Doi: 10.1016/j.devcel. 2009.06.014

[88] Mao K, Klionsky DJ. Mitochondrial fission facilitates mitophagy in *Saccharomyces cerevisiae*. Autophagy. 2009; 9: 1900–1901. Doi: 10.4161/auto.25804

[89] He Y, Deng YZ, Naqvi NI. Atg24-assisted mitophagy in the foot cells is necessary for proper asexual differentiation in *Magnaporthe oryzae*. Autophagy. 2013; 9: 1818–1827. Doi: 10.4161/auto.26057

[90] Cebollero E, Reggiori F, Kraft C. Reticulophagy and ribophagy: regulated degradation of protein production factories. Int. J. Cell Biol. 2012; 2012: 182834. Doi: 10.1155/2012/182834

[91] Lipatova Z, Shah AH, Kim JJ, Mulholland JW, Segev N. Regulation of ER-phagy by a Ypt/Rab GTPase module. Mol. Biol. Cell. 2013; 24: 3133–3144. Doi: 10.1091/ mbc.E13-05-0269

[92] Lipatova Z, Segev N. A role for macro-ER-phagy in ER quality control. PLoS Genet. 2015; 11: e1005390. Doi: 10.1371/journal.pgen.1005390

[93] Mochida K, Oikawa Y, Kimura Y, Kirisako H, Hirano H, et al. Receptor-mediated selective autophagy degrades the endoplasmic reticulum and the nucleus. Nature. 2015; 522: 359–362. Doi: 10.1038/nature14506

[94] Nakatogawa H, Mochida K. Reticulophagy and nucleophagy: new findings and unsolved issues. Autophagy. 2015; 11: 2377–2378. Doi: 10.1080/15548627.2015.1106665

[95] Mijaljica D, Devenish RJ. Nucleophagy at a glance. J. Cell Sci. 2013; 126: 4325–4330. Doi: 10.1242/jcs.133090

[96] Shoji JY, Kikuma T, Arioka M, Kitamoto K. Macroautophagy-mediated degradation of whole nuclei in the filamentous fungus *Aspergillus oryzae*. PLoS ONE. 2010; 5: e15650. Doi: 10.1371/journal.pone.0015650

[97] He M, Kershaw MJ, Soanes DM, Xia Y, Talbot NJ. Infection-associated nuclear degeneration in the rice blast fungus *Magnaporthe oryzae* requires non-selective macroautophagy. PLoS ONE. 2012; 7: e33270. Doi: 10.1371/journal.pone.0033270

[98] Liu XH, Lu JP, Lin FC. Autophagy during conidiation, conidial germination and turgor generation in *Magnaporthe grisea*. Autophagy. 2010; 3: 472–473. Doi: 10.4161/auto.4339

Section 4

Autophagy in tissue morphogenesis, remodeling, regeneration

Chapter 13

Autophagy in Cystic Fibrosis Pathogenesis and Treatment

Estelle Cormet-Boyaka, Kyle Caution and
Amal O. Amer

Additional information is available at the end of the chapter

http://dx.doi.org/10.5772/64686

Abstract

Cystic fibrosis (CF) is a fatal, genetic disorder that critically affects the lungs and is directly caused by mutations in the *CF transmembrane conductance regulator* (*CFTR*) gene, resulting in defective CFTR function. In epithelial cells, the CFTR channel conducts anions and plays a critical role in regulating the volume and composition of airway surface liquid. This thin layer of aqueous fluid and mucus covering the airway surface facilitates mucociliary clearance, bacterial killing, and epithelial cell homeostasis. The importance of the CFTR channel in macrophages was revealed in recent work that demonstrated that defective CFTR function is accompanied by impaired innate immune responses to specific infections. Notably, most CF-associated infections are caused by microbes that are cleared by autophagy in healthy cells. Autophagy is a highly regulated biological process that provides energy during periods of stress and starvation. Autophagy clears pathogens, inflammatory molecules, and dysfunctional protein aggregates within macrophages. However, this process is impaired in CF patients and CF mice, as their cells exhibit limited autophagy activity. The mechanisms linking a malfunctioning ion channel function to the defective autophagy remains unclear. In this chapter, we describe and discuss the recent findings indicating the presence of several mechanisms leading to defective autophagy in CF cells. Thus, these novel data advance our understanding of mechanisms underlying the pathobiology of CF and provide a new therapeutic platform for restoring CFTR function and autophagy in patients with CF.

Keywords: cystic fibrosis (CF), autophagy, Rab GTPases (Rabs), CF-associated bacteria, autophagy therapeutics

1. Introduction

Cystic fibrosis (CF) is the most common life-threatening genetic disease in North America and Europe. The birth prevalence of CF is estimated to be one in 3500–4500, with 200–300 new cases each year in Europe. The typical form of CF is diagnosed during early childhood and is characterized by recurrent pulmonary infections, pancreatic insufficiency, and elevated chloride concentrations in sweat. CF is a multi-organ disorder, however infections often occur in the lungs accompanied with severe inflammation and tissue destruction [1]. CF is directly caused by mutations in the *CF transmembrane conductance regulator* (*CFTR*) gene, resulting in defective CFTR function. Over 1400 mutations have been identified in the CFTR gene, the most common mutation leading to CF is deletion of phenylalanine at position 508 (F508del) and is present in over 70% of the CF alleles [2, 3]. Most of the work discussed in this chapter refers to the F508del-CFTR.

The previous commonly accepted hypothesis for CF pathology is the excessive secretion of thick mucus that remains in the lungs and is accompanied by impaired mucociliary clearance. This viscous mucus layer predisposes CF patients to chronic pulmonary infections. An intriguing speculation arose from the specificity of organisms that tend to infect CF patients. We hypothesized that the susceptibility of CF patients to these infectious agents is due to weak autophagic activity since most of the organisms that tend to cause chronic infection in CF are controlled by autophagy in healthy cells [4–6]. Autophagy is a highly regulated biological process that provides energy during periods of stress and starvation [7] and is typically induced upon glucose or amino acid starvation. Autophagy clears pathogens, inflammatory molecules, and dysfunctional protein aggregates within macrophages. Autophagy proceeds through sequential steps that begin with the formation of the phagophore or isolation membrane at a pre-autophagosomal structure (discussed in chapter of this book) [8]. The nascent autophagic membrane elongates to form a double-membrane autophagosomes that captures regions of cytoplasm, damaged mitochondria, or aggregated proteins. Upon maturation, the autophagosome containing the isolated cargo then fuses with the lysosome to form a single membrane compartment called the autolysosome. The autophagosomal cargo is then degraded in this compartment by lysosomal acid hydrolases and other degradative enzymes. The resulting degradation products including free amino acids, fatty acids, and nucleotides are released to the cytoplasm by the action of lysosomal permeases, where they may be reutilized for anabolic pathways [9, 10]. However, this process is impaired in CF patients and CF mice, as their macrophages and epithelial cells exhibit limited autophagic activity. The mechanisms linking a malfunctioning ion channel to the defective autophagy remains unclear.

Until recently, it was believed that the production of thick mucus and the impairment of mucociliary clearance was the main underlying culprit, allowing the persistence of specific infections in the CF lung. The idea of the existence of an innate immune deficiency disorder in CF was not examined until lately. The discovery that the innate immune functions of macrophages and neutrophils are disrupted in CF was a turning point in the CF field and in understanding the pathobiology of CF. Several years were consumed to provide what is now undisputable data confirming that CF should be considered an innate immune disorder.

In this chapter, we describe and discuss the recent findings in the field, which demonstrate that CF is a newly recognized innate immune deficiency disorder. We will discuss several reports demonstrating that macrophage functions are disrupted in CF contributing to the pathobiology of the disease. This chapter, encompassing recent data in CF, suggests that targeting autophagy may be exploited as a novel strategy for treatment of CF.

2. Cystic fibrosis

The cystic fibrosis transmembrane conductance regulator (CFTR) protein is an integral membrane glycoprotein that functions as a cAMP-activated and phosphorylation-regulated Cl channel at the apical membrane of epithelial cells. CFTR is a member of the ATP-binding cassette transporter superfamily. It is a multi-domain glycoprotein whose biosynthesis, maturation, and functions involve multi-level posttranslational modifications and complex folding processes to reach its native, tertiary conformation. The topology of CFTR includes two transmembrane-spanning domains, two nucleotide-binding domains, and a regulatory domain, which is a unique feature among ATP-binding cassette transporters (**Figure 1**) [11]. The newly synthesized CFTR emerges out of the ribosome and is targeted through the signal recognition particle to the ER membrane translocon [12, 13]. The CFTR polypeptide chain emerges into the ER lumen, and its glycosylated helping stabilize the protein. CFTR follows the secretory pathway through the Golgi in order to reach the plasma membrane [14]. The recycling of internalized CFTR channels is important for maintaining a functional pool of CFTR at the plasma membrane.

Only 20–40% of the nascent chains achieve folded conformation, whereas the remaining molecules are targeted for degradation by endoplasmic reticulum, lysosomes, or autophagy. A large number of mutations impair processing of CFTR. Growing knowledge of CFTR biosynthesis has enabled understanding of the cellular basis of CF and has brought to light various potential targets for novel and promising therapies [15]. The most common mutation leading to CF is deletion of phenylalanine at position 508 (F508del) and is present in over 70% of the CF alleles [2, 3]. As mutant CFTR is targeted for degradation by the proteasome, the formation of protein aggregates occurs, hence provoking an unfolded protein response (UPR) [16]. Other common mutations such as the G551D exhibit defective CFTR function, however it does not aggregate or elicit a UPR. Comparing the phenotypes and immune functions of the F508del and G551D mutants will discern if defective bacterial clearance and uncontrolled inflammation are due to defective channel function, UPR, or both.

Hyper-inflammation and failure to clear infection is recognized as a leading cause of lung tissue destruction in CF [17] that can be explained, at least in part, by incompetent autophagy machinery in cells with a dysfunctional CFTR channel [18]. Bronchoalveolar lavages from CF patients contain high levels of the pro-inflammatory cytokine interleukin (IL)-1β [1, 19–26]. IL-1β is primarily expressed as a precursor inactive molecule that is later cleaved by caspase-1 to yield active 17-kDa IL-1β [27]. The biological activities of IL-1β include promoting inflammatory responses and leukocyte infiltration. Autophagy directly regulates the level of pro-

IL-1β in response to lipopolysaccharide (LPS) and infections [28, 29]. Interestingly, stimulation of autophagy by rapamycin dramatically reduced signs of inflammation in the murine CF lung [30, 31].

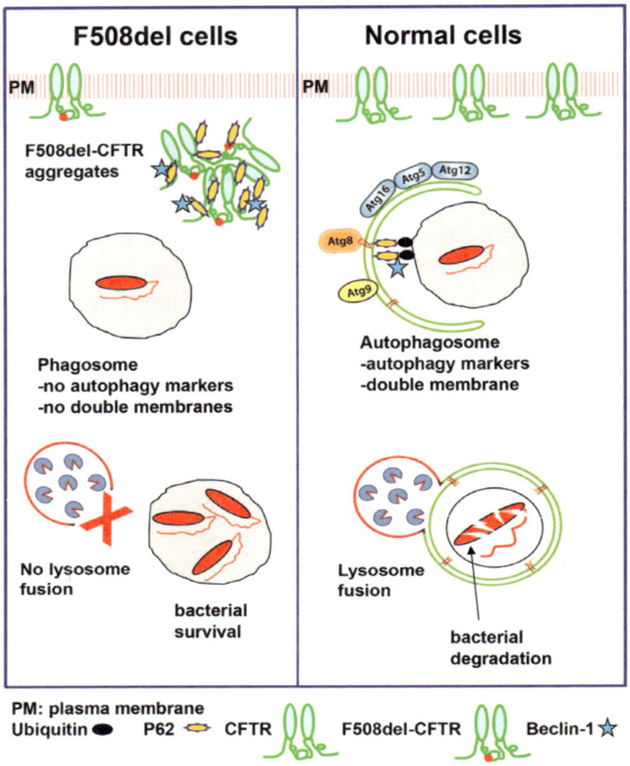

Figure 1. Autophagy process is impaired in CF cells preventing bacterial clearance.

Several reports demonstrated multiple malfunctions in adaptive and innate immune responses in CF. Lack of functional CFTR in CD3+ lymphocytes leads to aberrant cytokine secretion and hyper-inflammatory adaptive immune responses [32] while producing high levels of IL-1β [1, 19]. Naive cystic fibrosis T cells are intrinsically predisposed to differentiate toward a Th17 phenotype [33]. Therefore, CF is a multifaceted immune deficiency disease.

3. Rabs and the cytoskeleton: common modulators or innocent bystanders for autophagy and CFTR trafficking?

Rab (Ras-related proteins in brain) proteins are key regulators of both vesicular transport and trafficking of proteins [34]. Several Rab GTPases have been implicated in the regulation of the intracellular transport and the plasma membrane delivery of CFTR. The trafficking of CFTR from the plasma membrane to early endosomes is controlled by RAB5 [35, 36]. RAB7 regulates the movement of CFTR away from the recycling pathway and into late endosomes and also from late endosomes to lysosomes for degradation [37]. RAB9, however, can move CFTR away from lysosomal degradation by mediating its transport from late endosomes back to the trans-Golgi, from which CFTR may reenter the secretory pathway leading to plasma membrane insertion [37].

Growing knowledge of CFTR biosynthesis has enabled understanding of the cellular basis of CF and has brought to light various potential targets for novel and promising therapies [15]. Although some *in vitro* studies have shown that F508del-CFTR cell surface expression can be increased through the manipulation of key Rab GTPases, the mechanisms involved are still unclear. Rabs are also related to autophagy by regulating the transport and fusion of autophagosomes. However, it remains unclear how each cycle of Rab activation/inactivation is finely regulated. There is evidence indicating that RAB1, RAB5, and RAB7 participate in certain steps of autophagosome development and maturation [38], but the specific function of some of these Rab proteins remains poorly characterized. Conversely, RAB7, a low molecular weight GTPase found mainly on late endosomes, has been extensively studied [38]. By interacting with its partners (including upstream regulators and downstream effectors), RAB7 regulates mechanisms in endosomal sorting, biogenesis of lysosome, and phagocytosis [37]. Particularly, RAB7 governs early-to-late endosomal maturation, microtubule minus-end as well as plus-end directed endosomal migration, and endosome-lysosome transport through different protein–protein interaction cascades [34, 39]. In addition, RAB7 directs the maturation of autophagosomes, by guiding the trafficking of cargos along microtubules to participate in the fusion step with lysosomes [38]. Notably, activation of Rab7 is impaired by bacteria that tend to infect CF patients such as *Burkholderia cenocepacia*, accounting at least in part for the inability of the vacuole to merge with lysosomes [40]. *Staphylococcus aureus* also modulates Rabs to establish infection [41]. Whether specific nonfunctional Rabs promote aberrant CFTR trafficking and autophagy malfunction through common mechanisms in CF remains to be elucidated.

4. "Eat-me" signaling molecules: Are they trapped in CFTR aggregates?

The major molecular regulator of autophagy in response to starvation or energy depletion include the mammalian target of rapamycin complex 1 (mTORC1). Inhibition of mTORC1 by starvation or rapamycin results in the activation of autophagy and the start of autophagosome formation. The capacity of autophagy to clear intracellular pathogens such as bacteria, viruses, and parasites is collectively referred to as xenophagy [5], whereas the selective autophagic degradation of mitochondria is denoted as mitophagy and that of protein aggregates is termed aggrephagy. But how do autophagosomes find their targets? The modification of targets by ubiquitination represents a signal for selection of substrates to the autophagy pathway.

Mammalian cells ubiquitinate bacteria that erroneously enter the cytosol or their containing vacuole and target them for destruction by autophagy. Adaptors, including p62/SQSTM1 [42], optineurin (OPTN) [43], NBR1 (neighbor of BRCA1 gene 1) [44–46], and NDP52 (nuclear dot protein 52 kDa) [47], mediate "eat-me" functions by promoting autophagic sequestration of cargo. The adaptor molecule p62 is a ubiquitously expressed cellular protein and its quantity is critical for cell viability [48]. p62 has multiple protein-protein interaction domains, including the ubiquitin-associated domain for binding of ubiquitinated cargo and a LC3 interaction region for binding Atg8/LC3 [49]. p62 plays a role in amino acid sensing and the oxidative Stress response, in addition to its function as an autophagy receptor for ubiquitinated cargos [50]. Most p62 protein in the healthy cell is distributed in the cytoplasm. In response to various stressors though, it is translocated to autophagy substrates such as protein aggregates, damaged mitochondria, and intracellular bacteria [42]. Then, through its LC3-binding domain, p62 engages autophagosomes.

Autophagy is responsible for the degradation of p62. Therefore, impairment of autophagy is usually accompanied by massive accumulation of p62 followed by the formation of aggregate structures positive for p62 and ubiquitin [50]. This accrual is a defining characteristic of impaired autophagy. Aggregation occurs due to both the predilection for self-oligomerization and the ubiquitin-binding capabilities of p62 [51]. Notably, p62 accumulates in CF macrophages and promotes the sequestration of mutant CFTR (**Figure 1**). These aggregates, in turn, consume important autophagosome-needed proteins, such as BECN1 and LC3 [30, 31]. Recent reports demonstrate that the adapter protein NDP52 directly binds to ubiquitinated bacteria and facilitates the assembly of an autophagic membrane that surrounds these invaders [47]. Interestingly, NDP52 can also bind ubiquitinated bacteria-containing vacuole when p62 is drastically reduced [30, 31]. Optineurin can mediate the removal of protein aggregates through an ubiquitin-independent mechanism. In addition, this protein can induce autophagy upon overexpression or mutation [43]. NBR1 and p62 cooperate in the sequestration of misfolded and ubiquitinated proteins in p62 bodies and are both required for their degradation by autophagy. Recently, NBR1 was found to be necessary and sufficient for pexophagy [44–46]. Whether NBR1, optineurin, and NDP52 play important roles in CF-associated autophagy is still unknown.

5. Function of CFTR channel in epithelial cells

CFTR is an anion channel permeable to chloride and bicarbonate [2, 52]. Upon activation, CFTR transports chloride following its electrochemical gradient. In the lung, CFTR is expressed at the apical membrane of bronchial cells where it regulates chloride transport and fluid homeostasis [53]. The absence of functional CFTR in the lung results in abnormal surface hydration and decreased airway surface fluid. Thus, mutations in the CFTR protein results in the accumulation of thick mucus at the surface of epithelial cells, leading to impairment of pathogen clearance and dysregulated inflammatory responses that in turn results in chronic infection and inflammation [54]. In addition, epithelial cells expressing mutant CFTR exhibit weak autophagy activity.

6. Function of CFTR channel in macrophages

Macrophages are central innate immune cells that engulf invaders within a vacuole and target them to fuse with the lysosome for degradation. Therefore, lysosomes contribute to antimicrobial capacities by fusing with the pathogen-containing, intracellular vacuole [55]. Lysosomes are acidic compartments filled with various acid hydrolases, NADPH oxidases, and oxygen radicals that degrade and break down proteins, lipids, and polysaccharides.

The lung disease seen in CF was accordingly suggested to result in part from lysosomal dysfunction [56], yet the exact location and function of CFTR in macrophages are still debatable. A report showed impaired bacterial killing due to impaired function of antibacterial proteins at low pH in a CF pig model [57]. Another study demonstrated that acidification requires anion transport through CFTR [58]. More recently, defective lysosomal acidification was also invoked as the mechanism underlying CF lysosome malfunction [59]. Others failed to confirm the involvement of CFTR in acidification [60, 61]. Both these reports are plausible. When macrophages are infected with autophagosome-resident organisms such as *B. cenocepacia*, fusion with the lysosome is impaired and acidification is reduced [30, 31, 56]. Yet, when they are infected with phagosome-resident organism such as *Escherichia coli*, fusion with the lysosome moves swiftly and the bacterium is degraded within minutes in an acidified compartment [30, 31]. These results suggest that the defect in lysosomal degradation is only associated with autophagy. This explains the prevalence of several autophagy-related organisms in CF. Actually, an acceptable approach would be to examine organisms that tend to infect CF patient and test whether they are cleared by autophagosomes in healthy cells as was achieved for *B. cenocepacia* and *Pseudomonas aeruginosa*. However, on the other hand, not every organism that interacts with autophagosomes survives in CF macrophages. For example, in our hands, *Legionella pneumophila* that is cleared by autophagy in wild-type healthy macrophages and survives in autophagosomes in permissive macrophages fail to establish infection in CF macrophages (unpublished observation). How would a chloride channel alter autophagic activity is still unclear. Taken together, the role of CFTR in lysosomal function still awaits more conclusive reports.

7. CFTR in neutrophils

Polymorphonuclear neutrophils (PMNs, neutrophils) are responsible for the earliest innate immune response to infection and most of their antimicrobial activity against ingested microbes is confined within phagosomes. However, exuberant neutrophil activation can culminate in extracellular release of oxidants and granule contents that leads to local damage to healthy tissue. Neutrophils function to kill microbes through compartmentalization by use of membrane-bound phagosomes, where toxic oxidants such as hydrogen peroxide (H_2O_2) and hypochlorous acid (HOCl) are generated [62]. The azurophilic granule protein myeloperoxidase (MPO) catalyzes the oxidation of Cl^- to form HOCl. Neutrophils predominate in the CF patient's lung and is a major contributor to the inflammation and destruction of the lung [63].

Yet, despite a successful inflammatory response, neutrophils fail to eradicate invading microbes in the CF lung [64]. CFTR dysfunction results in impaired intraphagosomal HOCl production and neutrophil-mediated microbial killing. Notably, the *Cftr* $^{-/-}$ lungs are deficient in bacterial clearance, despite the sustained neutrophilic infiltration and accompanying inflammation [65]. These events demonstrate that neutrophils with nonfunctional CFTR have reduced capability to clear infection, yet remain capable of releasing inflammatory molecules that destroy the lung tissue leading to decline in lung function. Some studies suggested that CFTR channel expression in neutrophils and its dysfunction can affect neutrophil chlorination of phagocytosed bacteria [65] and that CFTR-dependent chloride anion transport contributes significantly to *P. aeruginosa* killing by normal neutrophils [63]. Others have reported that neutrophils from mice expressing the F508del-CFTR or mice lacking CFTR in myeloid cells have a pro-inflammatory phenotype after lipopolysaccharide or bacterial [64] challenge, thus suggesting again that CFTR expression in neutrophils might regulate their function [66].

In the CF lung, there is overproduction of the neutrophil chemotactic cytokine interleukin 8 (IL-8). This leads to excessive infiltration of neutrophils [67]. Lung infections cause significant morbidity and mortality in patients with CF, even in the presence of neutrophil infiltration into infected lungs. Thus, disturbance in innate neutrophil function in CF includes excessive recruitment [65], hyper-production of oxidants, and increased release of degradative enzymes [64]. Thus, it is evident that CF is indeed an innate immune disorder involving several malfunctioning immune cells.

8. Autophagy and cystic fibrosis

Several autophagy proteins are scarcely expressed in CF cells, yet the underlying mechanism is undefined [30, 31]. This strongly suggests the presence of an epigenic regulation that targets autophagy mRNA in cells bearing mutant F508del-CFTR. Considering the strong implications of microRNAs (miRs) in autophagy [28, 29, 68–70] and given the increasing evidence showing reduced expression of essential autophagy proteins in CF cells, we performed an *in silico* approach to recognize miRs that target autophagy and are highly expressed in CF cells [71]. miRs are evolutionarily conserved class of small (~21–24 nucleotides) noncoding RNAs that play key roles in the transcriptional and posttranscriptional regulation of gene expression [72]. We identified the *Mir17~92* cluster as being deregulated in CF [71]. This cluster generates a single polycistronic transcript that yields six mature *Mirs*: *Mir17*, *Mir18a*, *Mir19a*, *Mir20a*, *Mir19b*, and *Mir92* [73]. miRs can regulate individual stages of these processes [72]. The polycistronic *Mir17~92* cluster was initially linked to tumorigenesis. Whether the elevated levels in CF patients will promote cancer in aging CF population remains to be observed [73–80]. Several specific miRs comprising the *Mir17~92* cluster are overexpressed in CF human and murine macrophages. Their expression is indirectly proportional to the expression of their predicted autophagy-targeted genes. Notably, reducing the inherently elevated expression of *Mir17* and *Mir20a* improves ATG7 and ATG16 expression both *in vitro* and *in vivo* [71]. In addition, reducing *Mir17* and *Mir20a* expression improves CFTR function by restoring autophagy expression [71]. Whether other epigenetic regulatory elements contribute to low expression of autophagy proteins is still unclear.

In airway epithelial cells, absence of functional CFTR increases oxidative stress and transglutaminase 2 (TGM2), a calcium-dependent enzyme that creates intra- or intermolecular covalent bonds between proteins. TGM2-mediated cross-linking causes sequestration of BECN1 and its accumulation in histone deacetylase-6 (HDAC6), p62, and ubiquitin-containing cytoplasmic aggresomes. BECN1 sequestration in aggresomes results in the dislodgement of class III PtdIns3K complexes from the endoplasmic reticulum, thereby inhibiting autophagy [30, 81]. In addition, the sequestration of BECN1 within F508-CFTR protein aggregates deprives the cell for an essential factor needed for autophagosome formation.

High levels of p62 promote the aggregation of mutant F508del-CFTR sequestering several autophagy molecules such as BECN1. This p62 buildup could be due to reduced recycling in CF macrophages as a consequence of compromised autophagosome formation and maturation. Notably, p62 downregulation disassembles mutant CFTR and autophagy factors, thus improving autophagy activity and allowing the maturation and trafficking of CFTR to the cell surface in epithelial cells and bacterial clearance in macrophages [30, 31]. Similarly, genetic manipulation or autophagy stimulatory proteostasis regulators such as cystamine restore BECN1 availability and detangle SQSTM1/p62, which partially rescues F508del CFTR function in airway epithelial cells and reduces lung inflammation [82].

9. Cystic fibrosis and infection

There is evidence for autophagy dysregulation in a variety of disease states, including cancer, neurodegenerative diseases, chronic granulomatous diseases, infectious diseases, and autoimmune disorders [83–85]. For this reason, therapeutic modulation of autophagy is of great interest. Autophagy has emerged as a central component of the innate and adaptive immune responses where it plays roles in direct and indirect killing of intracellular and extracellular pathogens, the generation of bactericidal peptides, and antigen presentation [29]. Functions of autophagy that are compromised in CF include bacterial clearance, degradation of protein aggregates, and the elimination of dysfunctional mitochondria. Thus, the restoration of autophagy will have positive therapeutic effects in CF.

Patients with CF are susceptible to Nontypeable *Haemophilus influenzae* (NTHi), *Staphylococcus aureus* (*S. aureus*), *Burkholderia cenocepacia* (*B. cenocepacia*), *Pseudomonas aeruginosa* (*P. aeruginosa*), and nontuberculous *Mycobacterium* (NTM) [86–88]. The major cause of high morbidity and mortality in CF remains the chronic respiratory infections with *P. aeruginosa* [89].

In healthy cells, *P. aeruginosa* is targeted to the autophagy pathway through yet uncharacterized mechanisms [90, 91]. Increased susceptibility of CF cells to *P. aeruginosa* has been attributed to changes in airway liquid composition and enhanced bacterial binding to mucin and epithelial cell receptors such as asialo-GM1 [92]. In addition, disruption of lipid metabolism in CF cells increases innate inflammation in the presence of bacteria. The CFTR protein may also act as a receptor to *P. aeruginosa* and explains the high infection rate in CF patients [93]. Although largely considered an extracellular pathogen, *P. aeruginosa* can invade host airway epithelial cells where the bacteria can reside for extended periods of time. Pharmacological improvement in autophagy *in vivo* effectively promoted bacterial clearance of *P. aeruginosa* from the lungs.

S. aureus is one of the earliest bacteria detected in infants and children with CF. This pathogen is the single most common CF-associated opportunistic infection, colonizing between 50 and 68% of the population. The rise of methicillin-resistant *S. aureus* (MRSA) in the past 10 years has drawn attention to this organism [94, 95]. Unlike the other common CF-associated pathogens, *S. aureus* escapes from the phagosome upon entering the cell. In healthy cells, cytosolic *S. aureus*, or those contained within damaged phagosomes, are targeted to the autophagy pathway where they inhibit lysosomal fusion [96, 97]. The ability to escape degradation is increased in CF cells [98].

NTM strains infect between 5 and 22% of CF patients and are a growing concern among CF populations due to their increasing prevalence and multi-drug resistance. Infection is often associated with poor clinical outcomes [99]. Although autophagy contributes to clearance of *M. tuberculosis* in healthy cells, it is still unclear whether NTMs can be specifically targeted for degradation by the autophagy machinery [88].

NTHi chronically colonizes the airways of CF patients at a very young age. Recent reports suggest that autophagy may be actively subverted by NTHi by an unknown mechanism.

On the other hand, following phagocytosis, the degradation of *Aspergillus fumigatus* spores requires LC3-associated phagocytosis for effective lysosomal degradation [100]. Whether defective autophagy contributes to the prevalence of this infection in CF is still under investigation.

B. cenocepacia infections are not particularly common in CF patients, afflicting 3–5% of the population, yet they are very difficult to treat due to multi-drug resistance and are associated with a rapid decline in lung function. In healthy macrophages, *B. cenocepacia*-containing vacuoles are targeted to the lysosome for degradation via the autophagy pathway [30, 31]. However, in CF macrophages, *B. cenocepacia* persists in vacuoles that do not acquire LC3 or p62 like their counterparts in wild-type healthy cells. These *B. cenocepacia*-containing vacuoles do not fuse with lysosomes and the bacteria evade degradation (**Figure 1**) [30, 31]. Similar findings were reported in human macrophages derived from CF patients [101]. Together, these data show that correcting autophagy activity in CF will help prevent and eradicate infectious agents that are otherwise detrimental to the CF patients.

10. Targeting autophagy in CF improves CFTR function and bacterial clearance

Many of the opportunistic bacteria that infect the CF lung have employed mechanisms to target phagolysosomal maturation and/or autophagy, suggesting that these are major barriers that need to be overcome. Activation of autophagy through rapamycin treatment has shown efficacy in promoting clearance of certain bacteria *in vitro* and *in vivo* [30, 102–104]. Thus, the exploration of alternative autophagy inducing drugs is in the early stages.

Rapamycin was developed as an antifungal agent but its use was abandoned due to the potent immunosuppressive and antiproliferative properties. Recently, it was found that rapamycin inhibits mTOR and therefore induces autophagy [105]. Rapamycin promotes clearance of the CF-associated pathogens *P. aeruginosa* and *B. cenocepacia* both *in vitro* and *in vivo* in mice [30]. This could potentially be beneficial in the treatment of CF-associated lung infection since it is an important cause of decline in lung functions of CF patients [30]. However, rapamycin was most effective when administrated before *B. cenocepacia* infection but not after. The use of immunosuppressive drugs such as rapamycin to treat infections could negatively impact the ability of the patients to fight other opportunistic infections. In addition, rapamycin has been associated with significant lung toxicity in transplant recipients [106–108]. As a result, rapamycin proved the concept that targeting autophagy is needed in CF, but represents a poor candidate for autophagy-inducing therapy in the treatment of CF-associated lung infections, and other more specific candidates are urgently needed. Thus, the field requires a safe autophagy-enhancing approach for CF patients.

Among the most common autophagy-stimulating compounds were the antipsychotic drugs bromperidol, metergoline, thioridazine, and chlorpromazine. However, the psychoactive nature of these compounds and their potentially life-threatening side effects limit their utility, but nevertheless provide a strong theoretical basis for future drug development [109, 110].

Metformin is a drug that activates AMPK and therefore stimulates autophagy via TORC1-dependent and TORC-1-independent methods [111, 112]. Metformin probably has many other mechanisms of action that cannot be explained by the induction of autophagy. Metformin and resveratrol activate SIRT1 that in turn activates autophagy [113, 114].

Certain anticancer drugs have also been found to stimulate autophagy. For example, Perifosine inhibits mTOR signaling through a different mechanism than classical mTOR inhibitors such as rapamycin [115], whereas Tamoxifen, an antagonist of the estrogen receptor, is known to induce autophagy [116]. Tamoxifen stimulates autophagy by increasing the intracellular level of ceramide, which inhibits mTOR activation and/or stimulates expression of Atg genes.

The second generation of selective histamine H1-receptor antagonist astemizole is a potent inducer of autophagy at biologically achievable concentrations [117]. Astemizole exhibits antifungal activity and antimalarial properties making it attractive option for CF patients even though the mechanism by which it activates autophagy is still unclear [117, 118]. Safety and drug interaction profiles of astemizole are well characterized. However, due to the availability of superior next-generation histamine receptor agonists, it is not commonly used in Europe or North America.

The potential application of the TGM2 inhibitor cystamine in CF patients has recently been reviewed. Cystamine restores normal autophagy in CFTR-deficient cells and mouse models. Cystamine also restores normal trafficking of the F508del-CFTR and stabilizes the expression of the protein at the plasma membrane of airway epithelial cells [100]. In a pilot clinical trial involving 10 F508del-CFTR homozygous CF patients, the combination of cysteamine and epigallocatechin gallate (EGCG) restored the levels of the autophagy molecules BECN1 and p62 and improved CFTR function from nasal epithelial cells *in vivo*. These effects correlate with a decrease in chloride concentrations in sweat [119]. Although the mechanism of EGCG-mediated autophagy is unclear, it seems like a viable option for targeting autophagy in CF. Depletion of p62 in CF cells disintegrates the mutant CFTR aggregates and releases sequestered molecules such as BECN1, improving overall autophagic flux in the cell. This was accompanied by improved bacterial clearance [30, 31]. This indicates that targeting p62 in CF is a promising approach to improve bacterial clearance and reduce inflammation in CF [30, 31]. Since the expression of several autophagy proteins is low in CF cells, it is necessary to restore their levels to effectively improve autophagy activity. Identifying *Mir17* and *Mir20a* as new targets to improve the expression of autophagy proteins and CFTR function offers a viable target in the CF field. Further studies should explore whether other epigenetic regulatory elements contribute to low expression of autophagy proteins in CF.

11. Concluding remarks

There is now strong evidence that immune cells, such as macrophages and neutrophils, are intrinsically impaired in CF. We therefore, recommend that CF be added to the list of innate immune disorders. In fact, autophagy, an intracellular degradation process that contributes to bacterial clearance, protein degradation, and cell survival, is defective in CF. Therefore, we propose that altered autophagy in CF contributes to chronic lung infection and inflammation. The CF field is in desperate need for an approach to correct autophagy in CF patients. The autophagy-correcting agents should ameliorate the marginally positive effects of correctors (therapies that correct the trafficking defect of F508del CFTR) and activators (compounds that activate the F508del CFTR that reaches the cell membrane) typically used in CF. Thus, the development of safe novel autophagy stimulating agents will improve the clinical outcome of CF and promote the clearance of infectious agents.

Author details

Estelle Cormet-Boyaka[1], Kyle Caution[2] and Amal O. Amer[2*]

*Address all correspondence to: Amal.Amer@osumc.edu

1 Department of Veterinary Biosciences, The Ohio State University, Columbus, OH, USA

2 Department of Microbial Infection and Immunity, Center for Microbial Interface Biology, The Ohio State University, Columbus, OH, USA

References

[1] Berger M (2002) Inflammatory mediators in cystic fibrosis lung disease. Allergy Asthma Proc 23: 19–25.

[2] Rich DP, Gregory RJ, Cheng SH, Smith AE, Welsh MJ (1993) Effect of deletion mutations on the function of CFTR chloride channels. Receptors Channels 1: 221–232.

[3] Welsh MJ, Denning GM, Ostedgaard LS, Anderson MP (1993) Dysfunction of CFTR bearing the delta F508 mutation. J Cell Sci Suppl 17: 235–239.

[4] Campoy E, Colombo MI (2009) Autophagy in intracellular bacterial infection. Biochim Biophys Acta 1793: 1465–1477.

[5] Amer AO (2013) The many uses of autophagosomes. Autophagy 9: 633–634, in press.

[6] Abdelaziz HAD, Khalil H, Cormet-Boyaka E, Amer A (2015) The cooperation between the autophagy machinery and the inflammasome to implement an appropriate innate immune response: Do they regulate each other? Immunological Reviews 265: 194–204. Accepted.

[7] Klionsky DJ, Abdelmohsen K, Abe A, Abedin MJ, Abeliovich H, et al. (2016) Guidelines for the use and interpretation of assays for monitoring autophagy (3rd edition). Autophagy 12: 1–222.

[8] Mizushima N, Noda T, Yoshimori T, Tanaka Y, Ishii T, et al. (1998) A protein conjugation system essential for autophagy. Nature 395: 395–398.

[9] Zhang XD, Qi L, Wu JC, Qin ZH (2013) DRAM1 regulates autophagy flux through lysosomes. PLoS One 8: e63245.

[10] Kroemer G, Jaattela M (2005) Lysosomes and autophagy in cell death control. Nat Rev Cancer 5: 886–897.

[11] Naren AP, Cormet-Boyaka E, Fu J, Villain M, Blalock JE, et al. (1999) CFTR chloride channel regulation by an interdomain interaction. Science 286: 544–548.

[12] Pitonzo D, Yang Z, Matsumura Y, Johnson AE, Skach WR (2009) Sequence-specific retention and regulated integration of a nascent membrane protein by the endoplasmic reticulum Sec61 translocon. Mol Biol Cell 20: 685–698.

[13] Sadlish H, Skach WR (2004) Biogenesis of CFTR and other polytopic membrane proteins: new roles for the ribosome-translocon complex. J Membr Biol 202: 115–126.

[14] Cheng SH, Gregory RJ, Marshall J, Paul S, Souza DW, et al. (1990) Defective intracellular transport and processing of CFTR is the molecular basis of most cystic fibrosis. Cell 63: 827–834.

[15] Pranke IM, Sermet-Gaudelus I (2014) Biosynthesis of cystic fibrosis transmembrane conductance regulator. Int J Biochem Cell Biol 52: 26–38.

[16] Lukacs GL, Chang XB, Bear C, Kartner N, Mohamed A, et al. (1993) The delta F508 mutation decreases the stability of cystic fibrosis transmembrane conductance regulator in the plasma membrane. Determination of functional half-lives on transfected cells. J Biol Chem 268: 21592–21598.

[17] Bruscia EM, Bonfield TL (2016) Innate and adaptive immunity in cystic fibrosis. Clin Chest Med 37: 17–29.

[18] Hartl D, Gaggar A, Bruscia E, Hector A, Marcos V, et al. (2012) Innate immunity in cystic fibrosis lung disease. J Cyst Fibros 11: 363–382.

[19] Armstrong DS, Hook SM, Jamsen KM, Nixon GM, Carzino R, et al. (2005) Lower airway inflammation in infants with cystic fibrosis detected by newborn screening. Pediatr Pulmonol 40: 500–510.

[20] Rosenfeld M, Gibson RL, McNamara S, Emerson J, Burns JL, et al. (2001) Early pulmonary infection, inflammation, and clinical outcomes in infants with cystic fibrosis. Pediatr Pulmonol 32: 356–366.

[21] Reiniger N, Lee MM, Coleman FT, Ray C, Golan DE, et al. (2007) Resistance to *Pseudomonas aeruginosa* chronic lung infection requires cystic fibrosis transmembrane conductance regulator-modulated interleukin-1 (IL-1) release and signaling through the IL-1 receptor. Infect Immun 75: 1598–1608.

[22] Levy H, Murphy A, Zou F, Gerard C, Klanderman B, et al. (2009) IL1B polymorphisms modulate cystic fibrosis lung disease. Pediatr Pulmonol 44: 580–593.

[23] Corvol H, Fitting C, Chadelat K, Jacquot J, Tabary O, et al. (2003) Distinct cytokine production by lung and blood neutrophils from children with cystic fibrosis. Am J Physiol Lung Cell Mol Physiol 284: L997–L1003.

[24] Bonfield TL, Panuska JR, Konstan MW, Hilliard KA, Hilliard JB, et al. (1995) Inflammatory cytokines in cystic fibrosis lungs. Am J Respir Crit Care Med 152: 2111–2118.

[25] Greally P, Hussain MJ, Vergani D, Price JF (1993) Serum interleukin-1 alpha and soluble interleukin-2 receptor concentrations in cystic fibrosis. Arch Dis Child 68: 785–787.

[26] Kronborg G, Hansen MB, Svenson M, Fomsgaard A, Hoiby N, et al. (1993) Cytokines in sputum and serum from patients with cystic fibrosis and chronic *Pseudomonas aeruginosa* infection as markers of destructive inflammation in the lungs. Pediatr Pulmonol 15: 292–297.

[27] Abdelaziz DH, Gavrilin MA, Akhter A, Caution K, Kotrange S, et al. (2011) Apoptosis-associated speck-like protein (ASC) controls *Legionella pneumophila* infection in human monocytes. J Biol Chem 286: 3203–3208; PMCID:3030324.

[28] Fesus L, Demeny MA, Petrovski G (2011) Autophagy Shapes Inflammation. Antioxid Redox Signal. 11: 2233–2234.

[29] Levine B, Mizushima N, Virgin HW (2011) Autophagy in immunity and inflammation. Nature 469: 323–335.

[30] Abdulrahman BA, Khweek AA, Akhter A, Caution K, Kotrange S, et al. (2011) Autophagy stimulation by rapamycin suppresses lung inflammation and infection by *Burkholderia cenocepacia* in a model of cystic fibrosis. Autophagy 7: 1359–1370.

[31] Abdulrahman BA, Khweek AA, Akhter A, Caution K, Tazi M, et al. (2013) Depletion of the ubiquitin-binding adaptor molecule SQSTM1/p62 from macrophages harboring cftr DeltaF508 mutation improves the delivery of *Burkholderia cenocepacia* to the autophagic machinery. J Biol Chem 288: 2049–2058.

[32] Mueller C, Braag SA, Keeler A, Hodges C, Drumm M, et al. (2011) Lack of cystic fibrosis transmembrane conductance regulator in CD3+ lymphocytes leads to aberrant cytokine secretion and hyperinflammatory adaptive immune responses. Am J Respir Cell Mol Biol 44: 922–929.

[33] Kushwah R, Gagnon S, Sweezey NB (2013) Intrinsic predisposition of naive cystic fibrosis T cells to differentiate towards a Th17 phenotype. Respir Res 14: 138.

[34] Feng Y, Press B, Chen W, Zimmerman J, Wandinger-Ness A (2001) Expression and properties of Rab7 in endosome function. Methods Enzymol 329: 175–187.

[35] Jouret F, Devuyst O (2009) CFTR and defective endocytosis: new insights in the renal phenotype of cystic fibrosis. Pflugers Arch 457: 1227–1236.

[36] Villella VR, Esposito S, Bruscia EM, Vicinanza M, Cenci S, et al. (2013) Disease-relevant proteostasis regulation of cystic fibrosis transmembrane conductance regulator. Cell Death Differ 20: 1101–1115.

[37] Gentzsch M, Chang XB, Cui L, Wu Y, Ozols VV, et al. (2004) Endocytic trafficking routes of wild type and DeltaF508 cystic fibrosis transmembrane conductance regulator. Mol Biol Cell 15: 2684–2696.

[38] Saxena SK, Kaur S (2006) Regulation of epithelial ion channels by Rab GTPases. Biochem Biophys Res Commun 351: 582–587.

[39] Girard E, Chmiest D, Fournier N, Johannes L, Paul JL, et al. (2014) Rab7 is functionally required for selective cargo sorting at the early endosome. Traffic 15: 309–326.

[40] Huynh KK, Plumb JD, Downey GP, Valvano MA, Grinstein S (2010) Inactivation of macrophage Rab7 by *Burkholderia cenocepacia*. J Innate Immun 2: 522–533.

[41] Lopez de Armentia MM, Amaya C, Colombo MI (2016) Rab GTPases and the autophagy pathway: Bacterial targets for a suitable biogenesis and trafficking of their own vacuoles. Cells 5.

[42] Ichimura Y, Komatsu M (2010) Selective degradation of p62 by autophagy. Semin Immunopathol 32: 431–436.

[43] Ying H, Yue BY (2016) Optineurin: The autophagy connection. Exp Eye Res 144: 73–80.

[44] Deosaran E, Larsen KB, Hua R, Sargent G, Wang Y, et al. (2013) NBR1 acts as an autophagy receptor for peroxisomes. J Cell Sci 126: 939–952.

[45] Lamark T, Kirkin V, Dikic I, Johansen T (2009) NBR1 and p62 as cargo receptors for selective autophagy of ubiquitinated targets. Cell Cycle 8: 1986–1990.

[46] Kirkin V, Lamark T, Johansen T, Dikic I (2009) NBR1 cooperates with p62 in selective autophagy of ubiquitinated targets. Autophagy 5: 732–733.

[47] Ivanov S, Roy CR (2009) NDP52: the missing link between ubiquitinated bacteria and autophagy. Nat Immunol 10: 1137–1139.

[48] Komatsu M, Ichimura Y (2010) Physiological significance of selective degradation of p62 by autophagy. FEBS Lett 584: 1374–1378.

[49] Zheng YT, Shahnazari S, Brech A, Lamark T, Johansen T, et al. (2009) The adaptor protein p62/SQSTM1 targets invading bacteria to the autophagy pathway. J Immunol 183: 5909–5916.

[50] Bjorkoy G, Lamark T, Johansen T (2006) p62/SQSTM1: a missing link between protein aggregates and the autophagy machinery. Autophagy 2: 138–139.

[51] Ichimura Y, Kumanomidou T, Sou YS, Mizushima T, Ezaki J, et al. (2008) Structural basis for sorting mechanism of p62 in selective autophagy. J Biol Chem 283: 22847–22857.

[52] Hanrahan JW, Tabcharani JA, Becq F, Mathews CJ, Augustinas O, et al. (1995) Function and dysfunction of the CFTR chloride channel. Soc Gen Physiol Ser 50: 125–137.

[53] Sheppard DN, Welsh MJ (1999) Structure and function of the CFTR chloride channel. Physiol Rev 79: S23–S45.

[54] Sarkadi B, Bauzon D, Huckle WR, Earp HS, Berry A, et al. (1992) Biochemical characterization of the cystic fibrosis transmembrane conductance regulator in normal and cystic fibrosis epithelial cells. J Biol Chem 267: 2087–2095.

[55] Gong L, Devenish RJ, Prescott M (2012) Autophagy as a macrophage response to bacterial infection. IUBMB Life 64: 740–747.

[56] Lamothe J, Valvano MA (2008) *Burkholderia cenocepacia*-induced delay of acidification and phagolysosomal fusion in cystic fibrosis transmembrane conductance regulator (CFTR)-defective macrophages. Microbiology 154: 3825–3834.

[57] Pezzulo AA, Tang XX, Hoegger MJ, Alaiwa MH, Ramachandran S, et al. (2012) Reduced airway surface pH impairs bacterial killing in the porcine cystic fibrosis lung. Nature 487: 109–113.

[58] Di A, Brown ME, Deriy LV, Li C, Szeto FL, et al. (2006) CFTR regulates phagosome acidification in macrophages and alters bactericidal activity. Nat Cell Biol 8: 933–944.

[59] Teichgraber V, Ulrich M, Endlich N, Riethmuller J, Wilker B, et al. (2008) Ceramide accumulation mediates inflammation, cell death and infection susceptibility in cystic fibrosis. Nat Med 14: 382–391.

[60] Haggie PM, Verkman AS (2007) Cystic fibrosis transmembrane conductance regulator-independent phagosomal acidification in macrophages. J Biol Chem 282: 31422–31428.

[61] Barriere H, Bagdany M, Bossard F, Okiyoneda T, Wojewodka G, et al. (2009) Revisiting the role of cystic fibrosis transmembrane conductance regulator and counterion permeability in the pH regulation of endocytic organelles. Mol Biol Cell 20: 3125–3141.

[62] Qadri YJ, Cormet-Boyaka E, Benos DJ, Berdiev BK (2011) CFTR regulation of epithelial sodium channel. Methods Mol Biol 742: 35–50.

[63] Painter RG, Valentine VG, Lanson NA, Jr., Leidal K, Zhang Q, et al. (2006) CFTR Expression in human neutrophils and the phagolysosomal chlorination defect in cystic fibrosis. Biochemistry 45: 10260–10269.

[64] Witko-Sarsat V (2013) Neutrophils in the innate immunity conundrum of cystic fibrosis: A CFTR-related matter? J Innate Immun 5: 195–196.

[65] Ng HP, Zhou Y, Song K, Hodges CA, Drumm ML, et al. (2014) Neutrophil-mediated phagocytic host defense defect in myeloid Cftr-inactivated mice. PLoS One 9: e106813.

[66] Zhou J, Wang J, Cheng Y, Chi YJ, Fan B, et al. (2013) NBR1-mediated selective autophagy targets insoluble ubiquitinated protein aggregates in plant stress responses. PLoS Genet 9: e1003196.

[67] Downey DG, Martin SL, Dempster M, Moore JE, Keogan MT, et al. (2007) The relationship of clinical and inflammatory markers to outcome in stable patients with cystic fibrosis. Pediatr Pulmonol 42: 216–220.

[68] Yoshizaki T, Kusunoki C, Kondo M, Yasuda M, Kume S, et al. (2012) Autophagy regulates inflammation in adipocytes. Biochem Biophys Res Commun 417: 352–357.

[69] Jo EK, Shin DM, Choi AM (2012) Autophagy: Cellular defense to excessive inflammation. Microbes Infect 14: 119–125.

[70] Deretic V (2009) Links between autophagy, innate immunity, inflammation and Crohn's disease. Dig Dis 27: 246–251.

[71] Tazi MF, Dakhlallah DA, Caution K, Ahmed A, Khalil H, et al. (2016) Elevated miR-17~92 cluster expression negatively regulates autophagy and cystic fibrosis transmembrane conductance regulator (CFTR) function in CF macrophages. Autophagy, in press.

[72] Treiber T, Treiber N, Meister G (2012) Regulation of microRNA biogenesis and function. Thromb Haemost 107: 605–610.

[73] Mendell JT (2008) miRiad roles for the miR-17-92 cluster in development and disease. Cell 133: 217–222.

[74] Dakhlallah D, Batte K, Wang Y, Cantemir-Stone CZ, Yan P, et al. (2013) Epigenetic regulation of miR-17~92 contributes to the pathogenesis of pulmonary fibrosis. Am J Respir Crit Care Med 187: 397–405.

[75] Aguda BD, Kim Y, Piper-Hunter MG, Friedman A, Marsh CB (2008) MicroRNA regulation of a cancer network: Consequences of the feedback loops involving miR-17-92, E2F, and Myc. Proc Natl Acad Sci U S A 105: 19678–19683.

[76] Nana-Sinkam SP, Karsies T, Riscili B, Ezzie M, Piper M (2009) Lung microRNA: From development to disease. Expert Rev Respir Med 3: 373–385.

[77] Bonauer A, Dimmeler S (2009) The microRNA-17-92 cluster: still a miRacle? Cell Cycle 8: 3866–3873.

[78] Chen L, Li C, Zhang R, Gao X, Qu X, et al. (2011) miR-17-92 cluster microRNAs confers tumorigenicity in multiple myeloma. Cancer Lett 309: 62–70.

[79] Grillari J, Hackl M, Grillari-Voglauer R (2010) miR-17-92 cluster: ups and downs in cancer and aging. Biogerontology 11: 501–506.

[80] Hayashita Y, Osada H, Tatematsu Y, Yamada H, Yanagisawa K, et al. (2005) A polycistronic microRNA cluster, miR-17-92, is overexpressed in human lung cancers and enhances cell proliferation. Cancer Res 65: 9628–9632.

[81] Luciani A, Villella VR, Esposito S, Brunetti-Pierri N, Medina D, et al. (2010) Defective CFTR induces aggresome formation and lung inflammation in cystic fibrosis through ROS-mediated autophagy inhibition. Nat Cell Biol 12: 863–875.

[82] De Stefano D, Villella VR, Esposito S, Tosco A, Sepe A, et al. (2014) Restoration of CFTR function in patients with cystic fibrosis carrying the F508del-CFTR mutation. Autophagy 10: 2053–2074.

[83] Kroemer G, White E (2010) Autophagy for the avoidance of degenerative, inflammatory, infectious, and neoplastic disease. Curr Opin Cell Biol 22: 121–123.

[84] Shintani T, Klionsky DJ (2004) Autophagy in health and disease: a double-edged sword. Science 306: 990–995.

[85] Huang J, Klionsky DJ (2007) Autophagy and human disease. Cell Cycle 6: 1837–1849.

[86] Webb AK (1993) The difficulties of treating infection in adults with cystic fibrosis. Monaldi Arch Chest Dis 48: 657–661.

[87] Hoiby N (1988) Hemophilus influenzae, *Staphylococcus aureus*, *Pseudomonas cepacia*, and *Pseudomonas aeruginosa* in patients with cystic fibrosis. Chest 94: 97S-103S.

[88] Vergne I, Singh S, Roberts E, Kyei G, Master S, et al. (2006) Autophagy in immune defense against Mycobacterium tuberculosis. Autophagy 2: 175–178.

[89] Tatterson LE, Poschet JF, Firoved A, Skidmore J, Deretic V (2001) CFTR and pseudomonas infections in cystic fibrosis. Front Biosci 6: D890–D897.

[90] Yuan K, Huang C, Fox J, Laturnus D, Carlson E, et al. (2012) Autophagy plays an essential role in the clearance of *Pseudomonas aeruginosa* by alveolar macrophages. J Cell Sci. 125: 507–515.

[91] Junkins RD, Shen A, Rosen K, McCormick C, Lin TJ (2013) Autophagy enhances bacterial clearance during *P. aeruginosa* lung infection. PLoS One 8: e72263.

[92] Eidelman O, BarNoy S, Razin M, Zhang J, McPhie P, et al. (2002) Role for phospholipid interactions in the trafficking defect of Delta F508-CFTR. Biochemistry 41: 11161–11170.

[93] Goldberg JB, Pier GB (2000) The role of the CFTR in susceptibility to *Pseudomonas aeruginosa* infections in cystic fibrosis. Trends Microbiol 8: 514–520.

[94] Marks MI (1990) Clinical significance of *Staphylococcus aureus* in cystic fibrosis. Infection 18: 53–56.

[95] Goss CH, Muhlebach MS (2011) Review: *Staphylococcus aureus* and MRSA in cystic fibrosis. J Cyst Fibros 10: 298–306.

[96] Mestre MB, Colombo MI (2012) *Staphylococcus aureus* promotes autophagy by decreasing intracellular cAMP levels. Autophagy 8: 1865–1867.

[97] Schnaith A, Kashkar H, Leggio SA, Addicks K, Kronke M, et al. (2007) *Staphylococcus aureus* subvert autophagy for induction of caspase-independent host cell death. J Biol Chem 282: 2695–2706.

[98] Jarry TM, Cheung AL (2006) *Staphylococcus aureus* escapes more efficiently from the phagosome of a cystic fibrosis bronchial epithelial cell line than from its normal counterpart. Infect Immun 74: 2568–2577.

[99] Catherinot E, Roux AL, Vibet MA, Bellis G, Ravilly S, et al. (2013) *Mycobacterium avium* and *Mycobacterium abscessus* complex target distinct cystic fibrosis patient subpopulations. J Cyst Fibros 12: 74–80.

[100] Junkins RD, McCormick C, Lin TJ (2014) The emerging potential of autophagy-based therapies in the treatment of cystic fibrosis lung infections. Autophagy 10: 538–547.

[101] Assani K, Tazi MF, Amer AO, Kopp BT (2014) IFN-gamma stimulates autophagy-mediated clearance of *Burkholderia cenocepacia* in human cystic fibrosis macrophages. PLoS One 9: e96681.

[102] Zullo AJ, Jurcic Smith KL, Lee S (2014) Mammalian target of Rapamycin inhibition and mycobacterial survival are uncoupled in murine macrophages. BMC Biochem 15: 4.

[103] Zullo AJ, Lee S (2012) Mycobacterial induction of autophagy varies by species and occurs independently of mammalian target of rapamycin inhibition. J Biol Chem 287: 12668–12678.

[104] Foldenauer ME, McClellan SA, Berger EA, Hazlett LD (2013) Mammalian target of rapamycin regulates IL-10 and resistance to *Pseudomonas aeruginosa* corneal infection. J Immunol 190: 5649–5658.

[105] Abraham RT, Wiederrecht GJ (1996) Immunopharmacology of rapamycin. Annu Rev Immunol 14: 483–510.

[106] Berney T, Secchi A (2009) Rapamycin in islet transplantation: friend or foe? Transpl Int 22: 153–161.

[107] Ferrer IR, Araki K, Ford ML (2011) Paradoxical aspects of rapamycin immunobiology in transplantation. Am J Transplant 11: 654–659.

[108] Saunders RN, Metcalfe MS, Nicholson ML (2001) Rapamycin in transplantation: a review of the evidence. Kidney Int 59: 3–16.

[109] Cheng HW, Liang YH, Kuo YL, Chuu CP, Lin CY, et al. (2015) Identification of thioridazine, an antipsychotic drug, as an antiglioblastoma and anticancer stem cell agent using public gene expression data. Cell Death Dis 6: e1753.

[110] Shin SY, Lee KS, Choi YK, Lim HJ, Lee HG, et al. (2013) The antipsychotic agent chlorpromazine induces autophagic cell death by inhibiting the Akt/mTOR pathway in human U-87MG glioma cells. Carcinogenesis 34: 2080–2089.

[111] Feng Y, Ke C, Tang Q, Dong H, Zheng X, et al. (2014) Metformin promotes autophagy and apoptosis in esophageal squamous cell carcinoma by downregulating Stat3 signaling. Cell Death Dis 5: e1088.

[112] Tomic T, Botton T, Cerezo M, Robert G, Luciano F, et al. (2011) Metformin inhibits melanoma development through autophagy and apoptosis mechanisms. Cell Death Dis 2: e199.

[113] Delmas D, Solary E, Latruffe N (2011) Resveratrol, a phytochemical inducer of multiple cell death pathways: apoptosis, autophagy and mitotic catastrophe. Curr Med Chem 18: 1100–1121.

[114] Song YM, Lee YH, Kim JW, Ham DS, Kang ES, et al. (2015) Metformin alleviates hepatosteatosis by restoring SIRT1-mediated autophagy induction via an AMP-activated protein kinase-independent pathway. Autophagy 11: 46–59.

[115] Fu L, Kim YA, Wang X, Wu X, Yue P, et al. (2009) Perifosine inhibits mammalian target of rapamycin signaling through facilitating degradation of major components in the mTOR axis and induces autophagy. Cancer Res 69: 8967–8976.

[116] de Medina P, Silvente-Poirot S, Poirot M (2009) Tamoxifen and AEBS ligands induced apoptosis and autophagy in breast cancer cells through the stimulation of sterol accumulation. Autophagy 5: 1066–1067.

[117] Jakhar R, Paul S, Bhardwaj M, Kang SC (2016) Astemizole-Histamine induces Beclin-1-independent autophagy by targeting p53-dependent crosstalk between autophagy and apoptosis. Cancer Lett 372: 89–100.

[118] Chong CR, Chen X, Shi L, Liu JO, Sullivan DJ, Jr. (2006) A clinical drug library screen identifies astemizole as an antimalarial agent. Nat Chem Biol 2: 415–416.

[119] Tosco A, De Gregorio F, Esposito S, De Stefano D, Sana I, et al. (2016) A novel treatment of cystic fibrosis acting on-target: cysteamine plus epigallocatechin gallate for the autophagy-dependent rescue of class II-mutated CFTR. Cell Death Differ. 23: 1380–1393.

Chapter 14

The Role of Autophagy in Lung Disease

Michael P.A. Davies and Cornelia M. Wilson

Additional information is available at the end of the chapter

http://dx.doi.org/10.5772/64167

Abstract

The autophagy degradation pathway is a cellular pathway that sequesters cargo from the cytosol to autophagosomes that are transferred to lysosomes for degradation or recycled as precursor metabolites. The autophagy pathway allows the removal of damaged organelles/proteins and is emerging as an important aspect of multiple human pulmonary diseases. The autophagy process is important in both the function of the immune system and the control of inflammation. Xenophagy (autophagy of bacteria) is an example of selective autophagy which could play a role in host defense mechanisms in pulmonary diseases such as sepsis. Autophagy pathways involving the degradation of cytosolic cargo could play different roles in disease pathogenesis and progression. In the case of certain lung diseases, mitophagy is elevated and the cilia shorten (ciliophagy), which contribute to lung dysfunction in the pathogenesis of chronic obstructive pulmonary disease. In other types of lung diseases such as pulmonary vascular disease, autophagy may provide a protective role to allow cell proliferation, repair and control of cell death. Disruption of autophagy in cystic fibrosis and idiopathic pulmonary fibrosis could promote pathogenesis of the disease. In lung cancer, autophagy is a 'double-edged sword' it blocks progression, but at the same time promotes tumor growth. In this chapter, we will review the different types of autophagy, the role of autophagy and its significance to human lung diseases. In addition, we will discuss the potential of targeting autophagy with therapeutics for lung disease management.

Keywords: autophagy, cancer, COPD, infection, CF

1. Introduction

Autophagy is an evolutionarily-conserved cellular mechanism that allows the turnover of organelles and proteins, through a lysosome-dependent degradation pathway in the cell. The most common type of autophagy (also called macroautophagy) involves the sequestering of cytosolic molecules into double-membrane compartments called autophagosomes, which subsequently fuse to lysosomes where their contents are degraded and recycled into metabolic precursors (**Figure 1**). Autophagy is emerging as an important mechanism in the pathogenesis of several pulmonary diseases, including lung injury, sepsis, pulmonary vascular disease, idiopathic pulmonary fibrosis (IPF), cystic fibrosis and cancer. As autophagy can act as a modulator of pathogenesis, it is a therapeutic target. A number of studies indicate that autophagy may provide a protective function against disease and at the same time contribute to deleterious processes, usually when its activity is impeded or elevated. Autophagy is a known function in the clearance of subcellular material and the recycling of metabolites playing a role in cellular detoxification and adaptive or protective mechanisms. It is well accepted that autophagy plays a key role as a regulator of adaptive and innate immune responses, hence modulating autophagy could have a profound effect on disease pathogenesis. Xenophagy, which is the autophagic clearance of pathogens/bacteria, is one of the host's protective defense mechanisms against the pathogenesis of sepsis and inflammatory diseases. However, autophagy clearance of mitochondria may contribute to chronic obstructive pulmonary disease (COPD) through the activation of cell death pathways. In other types of pulmonary diseases, impairment of autophagy may aggravate and contribute to the pathogenesis process. In contrast, active autophagy in lung cancer could have many consequences including controlling carcinogenesis, modulating treatment efficacy and thus supporting tumor cell survival.

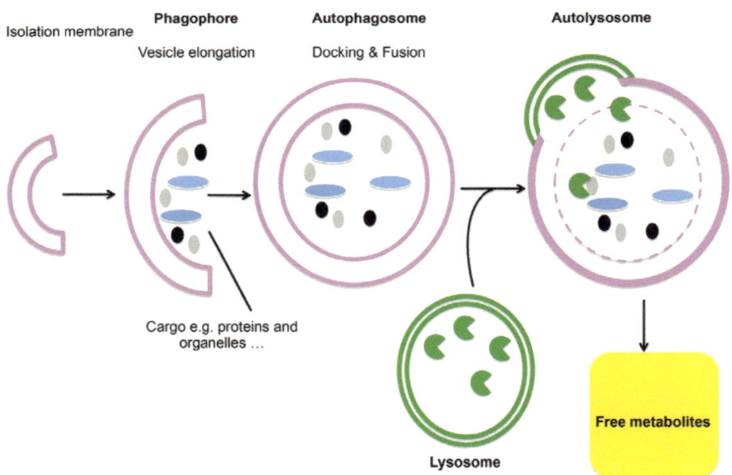

Figure 1. Schematic of the steps of autophagy. In the first step of autophagy the isolation membrane or phagophore is formed. In the next step of vesicle elongation, the proteins of the autophagy core machinery are believed to gather at the phagophore assembly site (PAS), and expand the phagophore into an autophagosome. In the next step, the autophagosome can engulf the cytosolic cargo including targeted cargo/organelles specifically. Following engulfment of cargo, the autophagosome can fuse with an endosome called an amphisome (not shown) or directly fuse with a lysosome to form an autophagolysosome. In the final step the cargo inside the autophagolysosome is degraded and free metabolites recycled.

2. Autophagy regulation

The autophagy process is comprised of the rearrangement of membrane components derived from the endoplasmic reticulum (ER), the endosome/Golgi, plasma membrane and mitochondrial membranes [1–9]. Autophagy proceeds through several sequential steps (**Figure 1**), starting with the formation of the phagophore or the isolation membrane at the preautophagosome site (PAS, see **Figure 1**) [10]. Next, the autophagic membrane elongates forming a double-membrane autophagosome engulfing a part of the cytoplasm. The maturation of the autophagosome and its cargo results in their fusion with the lysosome to form a single-membrane compartment, the autolysosome. This process is assisted by many proteins such as class C Vps proteins, small GTPases (e.g., Rab5 and 7), UVRAG and lysosome-associated membrane proteins (e.g., LAMP2) [11–14]. The cargo can then be degraded through the action of lysosomal degradative enzymes such as acid hydrolases that are activated at low pH (pH 5). This is accomplished by a proton pump located in the lysosomal membrane pumping H+ ions from the cytosol into the lysosome [15]. The degraded products such as free amino acids, nucleotides and fatty acids are then released into the cytoplasm by lysosomal permeases and then recycled through anabolic pathways [10].

A large number of autophagy-related (Atg) proteins regulate autophagy through various steps of initiation and execution [16, 17]. Autophagy is regulated by upstream cellular signals such as glucose or amino acid starvation and is an adaptive response in these conditions [10]. Autophagy contributes to cellular homeostasis along with the unfolded protein response (UPR) and endoplasmic reticulum stress (ERS). All these cellular responses are linked through crosstalk and can be activated by external stimuli such as hypoxia [18]. The mammalian target of rapamycin complex 1 (mTORC1) and the energy sensing, 5′-adenosine monophosphate-regulated kinase (AMPK) are the main regulators initiated under starvation conditions. The mammalian target of rapamycin (mTOR) pathway suppresses autophagy under starvation conditions can be activated by growth factors through the Class I phosphatidylinositol-3-kinase (PI3K)/Akt-pathway [19]. As well as mTOR protein, mTORC1 is comprised of the regulatory-associated protein of mTOR (Raptor), the mammalian lethal with SEC13 protein 8 (mLST8), and the 40 kDa proline-rich Akt/PKB substrate (PRAS40) [17]. mTORC1 can be inhibited by rapamycin or starvation leading to autophagy activation via derepression of the substrate complex, the uncoordinated 51-like kinase-1 (ULK1) complex, which is comprised of ULK1, FIP200/RB1CC1, Atg13 and Atg101 [20–24]. Upregulation of AMPK is directly related to an increase in AMP levels which inturn downregulates mTORC1 by Raptor phosphorylation and in addition activates ULK1 [25, 26]. A recent study has shown that the trafficking of mAtg9 which is involved in vesicle delivery to the PAS is regulated by AMPK-dependent ULK1 phosphorylation [27].

The Beclin1 complex is comprised of Beclin1, class III PI3K (PI3KC3/Vps34), UVRAG or p150 and Atg14L [28, 29]. The connection of ULK1 and Beclin 1 complex regulation via Vps34 phosphorylation has been demonstrated [30, 31]. Also, Beclin 1 complexes are known to associate with a number of inhibitory or activating proteins such as Ambra 1 [28]. Autophagosome formation requires the activation of PI3KC3 in the Beclin 1 complex leading to the formation of phosphatidylinositol-3-phosphate (PI3P). Various accessory protein factors are recruited by PI3P such as the double FYVE-containing protein-1 (DFCP1) and the WD-repeat protein in association with phosphoinositides (WIPI) proteins that are important in the assembly of autophagosomes [20, 32, 33]. In addition, the soluble N-ethylmaleimide-sensitive factor attachment protein receptor (SNARE) proteins are also implicated recruitment of membrane and the assembly of autophagosomes [34, 35].

There are two ubiquitin-like conjugation systems: the LC3/ATG8 conjugation system and the ATG5–ATG12 conjugation system [36]. Autophagosome assembly requires ATG5–ATG12 conjugates along with ATG16L1. The ubiquitin-like protein microtubule-associated protein-1 light chain 3 (LC3B) mediates autophagosome formation [17]. The pro-LC3 form is cleaved by the endopeptidase, ATG4B generating LC3B-I form. In eukaryotes, LC3B-I (unconjugated form) conversion to its conjugated phosphatidylethanolamine (PE) form, and the association of LC3B-II with the membrane is an important stage of autophagosome biogenesis [36, 37]. LC3B-II association with the autophagosomal membrane is prolonged until the final stages of autophagosome-lysosome fusion. At the final stages, the LC3B-II associated with the outer membrane can be recycled by ATG4B while LC3B-II found on the inner membrane will be degraded by lysosome degradative enzymes [17].

3. Autophagy

3.1. Lung injury

Acute lung injury (ALI) is characterized by an uncontrolled acute inflammation and dysfunction of both the endothelial and epithelial barriers in the lung, and disproportionate transepithelial leukocyte migration, affecting the integrity of the alveolar-capillary membrane and the increase of proinflammatory cytokines [38–40]. In addition, ALI is linked to hyperoxia, sepsis, trauma, xenobiotic exposure and mechanical ventilation.

The annual incidence of ALI in the USA in 2005 was around 78.9 per 100,000 [41]. However, current ALI therapy is fairly basic involving simple medical or surgical treatment allowing an improvement in ventilation of the patient, while remedial ALI therapy exists [39, 42, 43]. The ALI pathogenesis can be rationalized by the injury to both the alveolar epithelium and vascular endothelium. The innate immune system cells are both the target of damage and effectors of injury in ALI/acute respiratory distress syndrome (ARDS). Damage to type I alveolar epithelial cells promotes pulmonary edema and the degradation of the epithelial layer revealing the basement membrane increasing the chances of bacteremia and sepsis. Damage of type II alveolar cells results in perturbed synthesis of surfactant and metabolism giving rise to elevated alveolar surface tension and alveolar collapse [38]. A diverse range of biomarkers such as intracellular adhesion molecule-1 (ICAM-1), the receptor for advanced glycation end-products (RAGE), von Willebrand factor (vWf) and surfactant D (SP-D) have been discovered in the endothelium and the epithelium; these play a role in coagulation cascades and inflammation and may be used to predict outcome in patients with ALI [44–47]. In addition, aberrations in the coagulation cascade as a result of plasminogen activator inhibitor-1 and protein C also contribute to ALI [48]. Furthermore, the elevated levels of the inflammatory factors such as interleukins (IL-1, -6, -8) and tumor necrosis factor (TNF) are often associated with a response to cellular injury [49, 50].

Autophagy has been implicated in the pathogenesis of ALI [51] but could be due to long exposure to high levels of oxygen (hyperoxia) through mechanical ventilation treatment. Hyperoxia is sufficient to induce autophagy activation and LC3B knockdown effects cell survival [51]. Further studies have shown that in normal physiological conditions, the formation of a complex, p62/LC3B/truncated BH3-interacting domain death agonist (tBID) is necessary for cellular homeostasis [52]. In hyperoxia conditions, this complex dissociates and prevents the transport of tBID to the lysosome. The elevated levels of tBID result in the mitochondria releasing cytochrome c and caspase-dependent cell death [52]. The outcome of hyperoxia can lead to mitochondrial dysfunction in the lungs [53, 54]. Thus, mitophagy in hyperoxia has been implicated to play a role in ALI pathogenesis [55].

3.2. Infectious lung disease

Phagocytosis can act against infection in two ways: by ingesting bacteria (by macrophages and neutrophils) or by processing infectious agents for antigen presentation by dendritic cells. Similarly, autophagy is emerging as an essential process in innate and adaptive immune functions [56]. Autophagy plays an important role in protection of the host from various microbes such as viruses, bacteria and parasites [56–58]. The functions of antibacterial and antipathogenic autophagy are well characterized [59, 60]. It is known that phagocytosis of nonpathogenic mycobacteria by macrophages stimulates autophagy and apoptosis resulting in the removal of the pathogen. Conversely, the phagocytosis of pathogenic mycobacteria can perturb the autophagy pathway [61].

Tuberculosis is the result of a pathogen infection called *Mycobacterium tuberculosis* which is a serious infection affecting the lungs and is a burden worldwide [62]. The mycobacteria are resistant, remaining and replicating inside the immature phagosomes during tuberculosis infection. *M. tuberculosis* utilizes a protective survival strategy which allows them to interfere with a fusion event at the phagosomal compartments comprising *M. tuberculosis* and lysosomes [63]. Furthermore, the *M. tuberculosis* phagocytosis stimulates necrotic cell death allowing mycobacteria release to uninfected cells as opposed to stimulating macrophage apoptosis. Thus, this leads to a reduction in mycobacterial antigen presentation and promotion of *M. tuberculosis* infection [64]. The intracellular survival and replication rate of *M. tuberculosis* can be reduced through stimulation of autophagy [59, 60, 64–66]. Recent studies have shown that several compounds that stimulate autophagy through mTORC1 inhibition to alleviate *M. tuberculosis* infection [67, 68]. At the same time, autophagy inhibition promotes *M. tuberculosis* infection [60].

Autophagy can be important for antivirulence elements against *M. tuberculosis* generated substrate degradation [69, 70]. The host defense system against *M. tuberculosis* is protected by the production of interferon-gamma (IFN-γ) which stimulates macrophages, induces the autophagy process and thus protects the host against infection [60, 69]. In fact, stimulating with IFN-γ leads to bacteria degradation by p62-dependent selective autophagy [69]. Autophagy induced by IFN-γ requires p47 guanosine triphosphatase IRGM-1 [60, 71, 72] and increased sensitivity to *M. tuberculosis* infection is associated with single nucleotide polymorphisms in the IRGM-1 gene [73].

M. tuberculosis processing has been recently linked to selective autophagy [74]. It is known that the bacterial early secretory antigenic target (ESAT-6) system 1 (ESX-1) secretion system facilitates phagosomal permeabilisation allowing the ubiquitin-mediated autophagy entry to phagosomal *M. tuberculosis*. The extracellular bacterial DNA is recognized and tagged with ubiquitin by the stimulator of interferon gene (STING)-dependent cytosolic pathway. Then p62 and NDP52, the autophagy cargo adaptors, recognize ubiquitinated *M. tuberculosis* and target them to autophagosomes.

Autophagy is also described as a defense mechanism against other pathogens affecting the respiratory system. The growth of *Legionella pneumophila* which causes Legionnaire's disease is promoted in the absence of Atg9 suggesting that autophagy contributes to defense against this pathogen [75]. Influenza-A virus infection induces autophagy and autophagosome formation is necessary for virus replication [76], but it is believed that the influenza-A proteins such as M2 protein, can block autophagosome formation and subsequent fusion with the lysosome [77]. Recent studies on influenza infection have shown that autophagy is a requisite for maintaining memory B cells that are necessary in secondary antibody responses. Mice carrying a B cell-specific knockout of Atg7 showed perturbed secondary antibody responses and thus were more sensitive to influenza infection [78]. Therefore, future therapeutic developments should challenge the modulation of the autophagy pathway to eliminate infection and encourage adaptive immunity against infectious agents.

3.3. Emphysema and chronic obstructive pulmonary disease

Emphysema is characterized by the deterioration of the lung parenchyma as a result of an abnormal inflammatory response. In addition, there is degradation of the pulmonary matrix leading to an increase in the lung space and reduced respiratory function [79]. It is commonly observed that mice exposed to cigarette smoke give rise to an increase in lung space and can be used as a model of emphysema. Mice exposed to cigarette smoke over 3 months demonstrate raised autophagosomal numbers and elevated levels of LC3B-II in the resected lung tissue [80]. Interestingly, in the LC3B knockout mouse model demonstrated insensitivity to cigarette smoke leading to no change in the lung space. Depletion of the early growth response-1 (Egr-1) known to regulate LC3B transcription resulted in an increase in lung space with no changes induced with cigarette smoke. Conversely, knockout of caveolin-1 in mice led to sensitivity to cigarette smoke [80, 81]. Thus, LC3B is considered to be a crucial regulator of apoptosis and autophagic signaling.

Chronic obstructive pulmonary disease (COPD) is emerging as a global burden to society [82]. COPD is clinically defined into different types of lung disease, including emphysema and bronchitis leading to obstruction of the airway with mucus [83]. The main risk factor for COPD is cigarette smoke [84]. The mechanisms of COPD pathogenesis are still unknown. However, some theories indicate that abnormal cellular and inflammatory responses to cigarette smoke within the lung, the vessels and the airspace may be involved [84, 85]. The role of autophagy in COPD pathogenesis could be demonstrated by an excessive increase in autophagy or by mitophagy (selective degradation of mitochondria by autophagy) leading to cell death [80, 81, 86–88]. These studies indicate that the levels of the autophagy markers (such as LC3-II) and the mitophagy markers (such as PINK1) are elevated in COPD patient's lung tissue. Thus, an increase in autophagy or a defect in the autophagy flux could contribute to COPD pathogenesis [80, 87, 89]. In fact, a number of studies have shown that cigarette smoke can induce abnormal autophagy and mitophagy leading to bronchial cell death through apoptosis or necroptosis, respectively [81, 87]. Furthermore, there was an increase in histone deacetylase-6 levels as a result of oxidative stress, inducing hypomethylation leading to autophagy and cilia shortening which could contribute to mucociliary dysfunction [86]. Previous studies have shown that a block in autophagy promotes cilia growth [90]. Cigarette smoke has been shown to activate autophagy leading to elevated cell senescence along with a build-up of both p62 and ubiquitinated proteins [91]. As a result of autophagy blockade, cells display a 'senescence-associated secretory phenotype' secreting interleukin 8 (CXCL8) [91]. Autophagy was rescued with Torin 1, an mTOR inhibitor [91]. The genetic knockout Atg5−/− mouse model showed that p62 levels were elevated in the lung epithelial cells along with a shortening of cilia [92].

A recent study analyzing autophagy in macrophages in the alveoli from smokers' lung showed induction of LC3-II and the presence of autophagosomes [93]. The levels of autophagosomes in macrophages were elevated and autophagy activity was perturbed [93]. Altogether, these studies implicate that autophagy is an important cellular process in COPD pathogenesis. Future studies will need to address whether cigarette smoke promotes or blocks the autophagy flux in lung parenchyma, as well as the links between autophagy, apoptotic cell death and development of emphysema.

3.4. Pulmonary vascular disease

Hypoxia is an important contributing factor to pulmonary cardiovascular diseases such as pulmonary hypertension (PH), ischemia-reperfusion injury and atherosclerosis [94]. The hypoxia-inducible factors (HIFs), heterodimers of transcription family Per-Arnt-Sim/basic helix-loop-helix, regulate hypoxia. The heterodimer of HIF-1 is composed of two subunits, HIF-1β and HIF-1α, the O2-sensitive subunit. Under normoxia conditions, HIF-1α is degraded by the proteasome aided by HIF prolyl hydroxylase and the von Hippel-Lindau E3 ubiquitin ligase [94].

It is widely reported that the autophagy process can be regulated by hypoxia in cells and tissues. This process could be a crucial factor in cell injury induced by hypoxia in mitochondrial turnover [95–98]. Under *in vitro* conditions, autophagy activation requires Beclin 1 and an element that can be induced, encompassing HIF-1α stabilization and elevated ROS production [67, 96]. In addition, hypoxia-inducible autophagy is regulated by BNIP3 (Bcl-2/adenovirus E1B 19 kDa-interacting protein-3), a Bcl-2 family member [96].

The accumulation of abnormal, unfolded proteins during hypoxia or nutrient deprivation (endoplasmic reticulum stress, ERS) is the stimulus for the unfolded protein response. In its adaptive phase, this vital defense mechanism initially recruits autophagy to assist ER-associated degradation in the breakdown of abnormal proteins, while blocking cell death pathways. This prosurvival role for autophagy is partly responsible for resistance of hypoxic cells to therapies, such as chemotherapy [18]. Autophagy activation in tumor cells through hypoxia requires BNIP3 [97, 98]. In this case, autophagy could provide a mechanism of cell survival in hypoxia conditions via selective mitophagy [96]. It still remains unclear how BNIP3 acts during hypoxia either through cell survival or the cell death pathway.

Hypoxia conditions trigger autophagy in pulmonary vascular primary cells. In particular, hypoxia experiments performed *in vitro* show that LC3B-II accumulates and Beclin 1 expression increases both in pulmonary vascular endothelial cells and muscle cells. In addition, GFP-LC3 puncta are formed which is a common feature of autophagosome formation in cells transfected with GFP-LC3 [95]. In the same study, LC3-II and autophagosome formation were elevated in the presence of the autophagosomal inhibitor, bafilomycin A1, indicating elevated autophagic activity [95].

3.5. Idiopathic pulmonary fibrosis

Idiopathic pulmonary fibrosis (IPF) is characterized by the build-up of extracellular matrix proteins deposited in the interstitial tissue and basement membrane of damaged epithelium and an increase in active mesenchymal cells such as myofibroblasts [99]. The lung interstitium is primarily composed of fibroblasts, which play a role in maintaining the extracellular matrix and in wound healing. Altered autophagic function is implicated in the pathogenesis of IPF. Elevated cellular senescence and decreased autophagy activity as a result of decreased LC3B protein expression have been observed in the lung tissues from patients with IPF and in lung fibroblasts treated with TGF-β [91, 100]. TGF-β1 is known as an inhibitor of autophagy in human lung fibroblasts. In fibroblasts, genetic ablation of the autophagy proteins, LC3B or Beclin1, elevates the expression of fibronectin and α-smooth muscle actin (a myofibroblast marker) induced by TGF-β1 [100]. The mTORC1 inhibitor, rapamycin, was shown to protect against lung fibrosis when mice were treated [100]. Thus, reduced autophagy in IPF patients could increase the efficacy of TGF-β1 leading to a build-up of the extracellular matrix and conversion to a myofibroblast phenotype. Perturbation of lung IL-17A has been demonstrated to be protective against fibrosis, through autophagy recovery, in a mouse bleomycin model of pulmonary fibrosis [101]. Recent studies have shown that the multiple tyrosine kinase inhibitor nintedanib downregulated extracellular matrix production and promoted autophagy in IPF fibroblasts while inhibiting the TGF-β signaling pathway [102]. However, further investigations are required to better understand the role of autophagy and the molecular mechanism of fibrogenesis.

3.6. Cystic fibrosis

CF is a fatal genetic lung disease caused by a mutation in the gene encoding the cystic fibrosis transmembrane conductance regulator (CFTR); it is characterized by the accumulation of mucus in the airways. This accumulation can result in damage to the lungs that are linked to recurrent secondary infections. The CFTR mutation of a deletion of phenylalanine at position 508 (CFTRF508del) is the most common found in humans [103]. Recent studies are demonstrating defective autophagy with the CFTR mutation in airway epithelial cells from CF patients with the CFTR mutation. This is visualized by a reduction in autophagosome number along with the presence of LC3 puncta and p62 accumulation in starvation conditions [104]. Cells expressing CFTR F508del have been shown to be defective in autophagy and to accumulate polyubiquitinated proteins and aggresome-like structures [105]. Inflammatory response is enhanced in CF with defective autophagy [106]. In addition, defective CFTR can upregulate reactive oxygen species production and tissue transglutaminase [104]. A block in autophagy leads to cross-linking and inactivation of Beclin1 with PI3KC3 sequestering and p62 accumulation [104]. Protein trafficking of CFTR F508del to the cell surface could be restored with the overexpression of Beclin1 or the depletion of p62 [104]. Targeting p62 genetically rescued the functional expression of CFTR and improved the efficacy of CFTR channel activators [107]. Genetic targeting of p62 was also recently shown to improve the therapeutic effect of CFTR channel activators [107].

Xenophagy, the clearance of bacteria by autophagy, could play a role in defending against infections linked to CF. In a mouse CF model, rapamycin has been shown to alleviate *Burkholderia cenocepacia* infection and reduce lung inflammation [108]. *In vivo* studies have shown that pharmacological intervention of autophagy could enhance the clearance of *Pseudomonas aeruginosa* bacteria from the lung [109]. Therefore, dysfunctional autophagy leading to a compromise as a result secondary infection could promote CF pathogenesis. Hence, future CF therapy should include the rescue of autophagy [110].

3.7. Cancer

The autophagy pathway is a complex biological process that is believed to influence the induction, development and cancer therapy. Autophagy is the main cellular protection system with an anticarcinogenic effect through preserving mitochondria, recycling of precursors of metabolism, removal of cell debris/products and the control of inflammation that could promote gene instability [111]. Surprisingly, autophagy could provide a mechanism that contributes to tumor cell survival and cancer progression as a result of chemotherapy resistance. At the same time, autophagy could provide a means to treat the cancer through chemotherapy promoting autophagy-mediated cell death [111].

Early studies demonstrated that the monoallelic loss of Beclin1 from chromosome 17q21 occurs in many cancers such as prostate, breast and ovarian [112–114]. It is reported from *in vitro* studies that perturbing autophagy can promote cancer growth, and in addition heterozygous Beclin1 mice develop lung, liver and lymphomas [115, 116]. In addition, the aberrant expression of Beclin1 in the tumor is associated with poor outcome and aggressive disease [57, 58]. Currently, the role of autophagy in the lung and its effect on therapy is an unexplored area of research. Studies have shown that Beclin1 expression is reduced in NSCLC compared to healthy tissue and this reduction is further pronounced in higher stage, poorly differentiated tumors [117, 118]. Autophagy induction with mTOR inhibitors is linked to radiosensitization in NSCLC [119]. In addition, hydroxychloroquine, an autophagy inhibitor, has been tested in the clinic for treatment of NSCLC on the basis that inhibition of autophagy, which aids cell-survival, helps prevent resistance to chemotherapy [113]. Depletion of the autophagy protein, ATG5, both perturbs the progression of KRas (G12D) lung cancer and stimulates tumor survival in a mouse model. On the other hand, ATG5 depletion promotes the initiation of KRas (G12D) tumors [120]. These studies indicate the 'double-edged sword of autophagy' in cancer where autophagy prevents progression and promotes tumor growth.

4. Conclusion and future perspectives

It is emerging that autophagy can promote and perturb pathogenesis in human disease (see **Figure 2**). These effects are well illustrated in cancer, where the 'double-edged sword' of

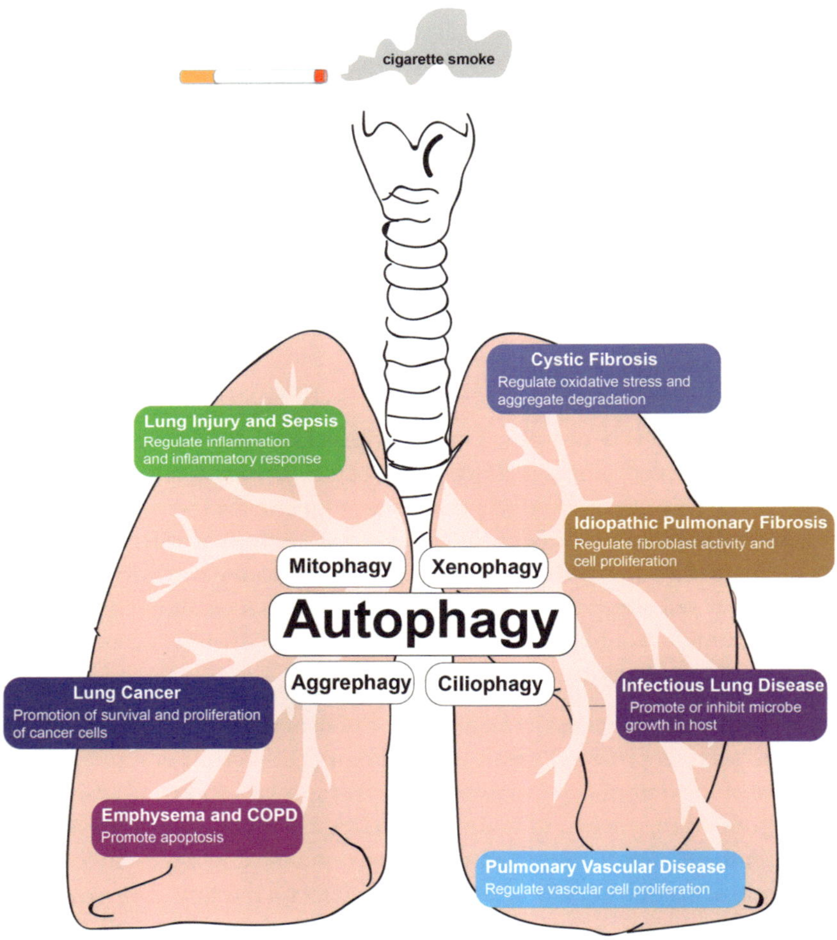

Figure 2. The effects of autophagy in human lung diseases. The role of different pathways of autophagy to remove bacteria (xenophagy), mitochondria (mitophagy), cilia (ciliophagy) and protein aggregates (aggrephagy). The autophagy process plays a role in the pathogenesis of many human lung diseases as shown in the schematic. In lung cancer, autophagy could perturb tumorigenesis, but at the same time promote cell proliferation and thus tumoral survival. In pulmonary infections, autophagy could reduce bacterial expansion and block survival of bacteria such as *M. tuberculosis*. The pathogenesis of human lung disease is promoted with cigarette smoke exposure which induces oxidative stress and triggers cilia protein damage.

autophagy provides protection in the early stages of disease, while in the later stages promotes tumor growth or resistance to therapy. In some lung diseases, such as sepsis, autophagy could provide protection against bacterial infection and allow an inflammatory response. In the case of exposure to cigarette smoke, this aggravates and promotes disease progression for many human lung diseases. Current therapeutic strategies are still in development, and a few molecules that regulate autophagy have already been tested in the clinic. For example, the autophagy inhibitors chloroquine and hydroxychloroquine and autophagy inducer, mTOR inhibitor rapamycin, have been analyzed in the clinic. Recent development of compounds that modulate autophagy has been the Tat-Beclin 1 peptide, an inhibitor of histone deacetylases [121], vitamin D and AMPK activators [58, 113, 122]. Due to the complexity of autophagy, further research is required to fully understand how modulating autophagy could be used as a therapeutic strategy in disease management. In addition, we will need to define the various pathways of selective autophagy and their relevance to disease pathogenesis. This will allow the design of novel compounds to be used in therapeutic strategies that could specifically target certain lung diseases.

Author details

Michael P.A. Davies[1] and Cornelia M. Wilson[1,2,3*]

*Address all correspondence to: cornelia.wilson@canterbury.ac.uk; Cornelia.Wilson@liverpool.ac.uk

1 Molecular and Clinical Cancer Medicine, Institute of Translational Medicine, University of Liverpool, Liverpool, UK

2 School of Human and Life Sciences, Canterbury Christ Church University, Discovery Park, Sandwich, UK

3 Biomolecular Research Group, Canterbury Christ Church University, School of Human and Life Sciences, Canterbury, UK

References

[1] Chan, S.N. and B.L. Tang. *Location and membrane sources for autophagosome formation—from ER-mitochondria contact sites to Golgi-endosome-derived carriers*. Mol Membr Biol, 2013. 30(8): p. 394–402.

[2] Cook, K.L., et al. *Mitochondria directly donate their membrane to form autophagosomes during a novel mechanism of parkin-associated mitophagy*. Cell Biosci, 2014. 4(1): p. 16.

[3] Hailey, D.W., et al. *Mitochondria supply membranes for autophagosome biogenesis during starvation*. Cell, 2010. 141(4): p. 656–667.

[4] Hamasaki, M., et al. *Autophagosomes form at ER-mitochondria contact sites*. Nature, 2013. 495(7441): p. 389–393.

[5] Hayashi-Nishino, M., et al. *A subdomain of the endoplasmic reticulum forms a cradle for autophagosome formation*. Nat Cell Biol, 2009. 11(12): p. 1433–1437.

[6] Ravikumar, B., et al. *Plasma membrane contributes to the formation of pre-autophagosomal structures*. Nat Cell Biol, 2010. 12(8): p. 747–757.

[7] Rubinsztein, D.C., T. Shpilka, and Z. Elazar. *Mechanisms of autophagosome biogenesis*. Curr Biol, 2012. 22(1): p. R29-R34.

[8] Tooze, S.A. and T. Yoshimori. *The origin of the autophagosomal membrane*. Nat Cell Biol, 2010. 12(9): p. 831–835.

[9] Yla-Anttila, P., et al. *3D tomography reveals connections between the phagophore and endoplasmic reticulum*. Autophagy, 2009. 5(8): p. 1180–1185.

[10] Mizushima, N. and M. Komatsu. *Autophagy: renovation of cells and tissues*. Cell, 2011. 147(4): p. 728–741.

[11] Gutierrez, M.G., et al. *Rab7 is required for the normal progression of the autophagic pathway in mammalian cells*. J Cell Sci, 2004. 117(Pt 13): p. 2687–2697.

[12] Jager, S., et al. *Role for Rab7 in maturation of late autophagic vacuoles*. J Cell Sci, 2004. 117(Pt 20): p. 4837–4848.

[13] Liang, C., et al. *Beclin1-binding UVRAG targets the class C Vps complex to coordinate autophagosome maturation and endocytic trafficking*. Nat Cell Biol, 2008. 10(7): p. 776–787.

[14] Tanida, I. *Autophagy basics*. Microbiol Immunol, 2011. 55(1): p. 1–11.

[15] Kinchen, J.M. and K.S. Ravichandran. *Phagosome maturation: going through the acid test*. Nat Rev Mol Cell Biol, 2008. 9(10): p. 781–795.

[16] Feng, Y., et al. *The machinery of macroautophagy*. Cell Res, 2014. 24(1): p. 24–41.

[17] Yang, Z. and D.J. Klionsky. *Mammalian autophagy: core molecular machinery and signaling regulation*. Curr Opin Cell Biol, 2010. 22(2): p. 124–131.

[18] Senft, D. and Z.A. Ronai. *UPR, autophagy, and mitochondria crosstalk underlies the ER stress response*. Trends Biochem Sci, 2015. 40(3): p. 141–148.

[19] Jung, C.H., et al. *mTOR regulation of autophagy*. FEBS Lett, 2010. 584(7): p. 1287–1295.

[20] Chan, E.Y. *Regulation and function of uncoordinated-51 like kinase proteins*. Antioxid Redox Signal, 2012. 17(5): p. 775–785.

[21] Ganley, I.G., et al. *ULK1.ATG13.FIP200 complex mediates mTOR signaling and is essential for autophagy*. J Biol Chem, 2009. 284(18): p. 12297–12305.

[22] Hosokawa, N., et al. *Nutrient-dependent mTORC1 association with the ULK1-Atg13-FIP200 complex required for autophagy*. Mol Biol Cell, 2009. 20(7): p. 1981–1991.

[23] Jung, C.H., et al. *ULK-Atg13-FIP200 complexes mediate mTOR signaling to the autophagy machinery*. Mol Biol Cell, 2009. 20(7): p. 1992–2003.

[24] Wong, P.M., et al. *The ULK1 complex: sensing nutrient signals for autophagy activation*. Autophagy, 2013. 9(2): p. 124–137.

[25] Kim, J., et al. *AMPK and mTOR regulate autophagy through direct phosphorylation of Ulk1*. Nat Cell Biol, 2011. 13(2): p. 132–141.

[26] Lee, J.W., et al. *The association of AMPK with ULK1 regulates autophagy*. PLoS One, 2010. 5(11): p. e15394.

[27] Mack, H.I., et al. *AMPK-dependent phosphorylation of ULK1 regulates ATG9 localization*. Autophagy, 2012. 8(8): p. 1197–1214.

[28] He, C. and B. Levine. *The Beclin 1 interactome*. Curr Opin Cell Biol, 2010. 22(2): p. 140–149.

[29] Itakura, E., et al. *Beclin 1 forms two distinct phosphatidylinositol 3-kinase complexes with mammalian Atg14 and UVRAG*. Mol Biol Cell, 2008. 19(12): p. 5360–5372.

[30] Nazarko, V.Y. and Q. Zhong. *ULK1 targets Beclin-1 in autophagy*. Nat Cell Biol, 2013. 15(7): p. 727–728.

[31] Russell, R.C., et al. *ULK1 induces autophagy by phosphorylating Beclin-1 and activating VPS34 lipid kinase*. Nat Cell Biol, 2013. 15(7): p. 741–750.

[32] Dooley, H.C., et al. *WIPI2 links LC3 conjugation with PI3P, autophagosome formation, and pathogen clearance by recruiting Atg12-5-16L1*. Mol Cell, 2014. 55(2): p. 238–252.

[33] Wirth, M., J. Joachim, and S.A. Tooze. *Autophagosome formation--the role of ULK1 and Beclin1-PI3KC3 complexes in setting the stage*. Semin Cancer Biol, 2013. 23(5): p. 301–309.

[34] Moreau, K., et al. *Autophagosome precursor maturation requires homotypic fusion*. Cell, 2011. 146(2): p. 303–317.

[35] Nair, U., et al. *SNARE proteins are required for macroautophagy*. Cell, 2011. 146(2): p. 290–302.

[36] Nakatogawa, H. *Two ubiquitin-like conjugation systems that mediate membrane formation during autophagy*. Essays Biochem, 2013. 55: p. 39–50.

[37] Kabeya, Y., et al. *LC3, GABARAP and GATE16 localize to autophagosomal membrane depending on form-II formation*. J Cell Sci, 2004. 117(Pt 13): p. 2805–2812.

[38] Matthay, M.A., L.B. Ware, and G.A. Zimmerman. *The acute respiratory distress syndrome*. J Clin Invest, 2012. 122(8): p. 2731–2740.

[39] Matthay, M.A. and R.L. Zemans. *The acute respiratory distress syndrome: pathogenesis and treatment*. Annu Rev Pathol, 2011. 6: p. 147–163.

[40] Johnson, E.R. and M.A. Matthay. *Acute lung injury: epidemiology, pathogenesis, and treatment*. J Aerosol Med Pulm Drug Deliv, 2010. 23(4): p. 243–252.

[41] Rubenfeld, G.D., et al. *Incidence and outcomes of acute lung injury*. N Engl J Med, 2005. 353(16): p. 1685–1693.

[42] Xu, F., et al. *Mesenchymal stem cells in acute lung injury: are they ready for translational medicine?* J Cell Mol Med, 2013. 17(8): p. 927–935.

[43] Crimi, E. and A.S. Slutsky. *Inflammation and the acute respiratory distress syndrome*. Best Pract Res Clin Anaesthesiol, 2004. 18(3): p. 477–492.

[44] Calfee, C.S., et al. *Plasma receptor for advanced glycation end products and clinical outcomes in acute lung injury*. Thorax, 2008. 63(12): p. 1083–1089.

[45] Flori, H.R., et al. *Early elevation of plasma von Willebrand factor antigen in pediatric acute lung injury is associated with an increased risk of death and prolonged mechanical ventilation*. Pediatr Crit Care Med, 2007. 8(2): p. 96–101.

[46] Ware, L.B., E.R. Conner, and M.A. Matthay. *von Willebrand factor antigen is an independent marker of poor outcome in patients with early acute lung injury*. Crit Care Med, 2001. 29(12): p. 2325–2331.

[47] Ware, L.B., et al. *Significance of von Willebrand factor in septic and nonseptic patients with acute lung injury*. Am J Respir Crit Care Med, 2004. 170(7): p. 766–772.

[48] Ware, L.B., et al. *Pathogenetic and prognostic significance of altered coagulation and fibrinolysis in acute lung injury/acute respiratory distress syndrome*. Crit Care Med, 2007. 35(8): p. 1821–1828.

[49] Meduri, G.U., et al. *Inflammatory cytokines in the BAL of patients with ARDS. Persistent elevation over time predicts poor outcome*. Chest, 1995. 108(5): p. 1303–1314.

[50] Parsons, P.E., et al. *Lower tidal volume ventilation and plasma cytokine markers of inflammation in patients with acute lung injury*. Crit Care Med, 2005. 33(1): p. 1–6; discussion 230–2.

[51] Tanaka, A., et al. *Hyperoxia-induced LC3B interacts with the Fas apoptotic pathway in epithelial cell death*. Am J Respir Cell Mol Biol, 2012. 46(4): p. 507–514.

[52] Liang, X., et al. *p62 sequestosome 1/light chain 3b complex confers cytoprotection on lung epithelial cells after hyperoxia*. Am J Respir Cell Mol Biol, 2013. 48(4): p. 489–496.

[53] Ratner, V., et al. *Mitochondrial dysfunction contributes to alveolar developmental arrest in hyperoxia-exposed mice*. Am J Respir Cell Mol Biol, 2009. 40(5): p. 511–518.

[54] Waxman, A.B. and N. Kolliputi. *IL-6 protects against hyperoxia-induced mitochondrial damage via Bcl-2-induced Bak interactions with mitofusins*. Am J Respir Cell Mol Biol, 2009. 41(4): p. 385–396.

[55] Youle, R.J. and D.P. Narendra. *Mechanisms of mitophagy*. Nat Rev Mol Cell Biol, 2011. 12(1): p. 9–14.

[56] Levine, B., N. Mizushima, and H.W. Virgin. *Autophagy in immunity and inflammation*. Nature, 2011. 469(7330): p. 323–335.

[57] Levine, B. and G. Kroemer. *Autophagy in the pathogenesis of disease*. Cell, 2008. 132(1): p. 27–42.

[58] Rubinsztein, D.C., P. Codogno, and B. Levine. *Autophagy modulation as a potential therapeutic target for diverse diseases*. Nat Rev Drug Discov, 2012. 11(9): p. 709–730.

[59] Nakagawa, I., et al. *Autophagy defends cells against invading group A Streptococcus*. Science, 2004. 306(5698): p. 1037–1040.

[60] Gutierrez, M.G., et al. *Autophagy is a defense mechanism inhibiting BCG and Mycobacterium tuberculosis survival in infected macrophages*. Cell, 2004. 119(6): p. 753–766.

[61] Flannagan, R.S., G. Cosio, and S. Grinstein. *Antimicrobial mechanisms of phagocytes and bacterial evasion strategies*. Nat Rev Microbiol, 2009. 7(5): p. 355–366.

[62] Jordao, L. and O.V. Vieira. *Tuberculosis: new aspects of an old disease*. Int J Cell Biol, 2011. 2011: p. 403623.

[63] Vergne, I., et al. *Cell biology of mycobacterium tuberculosis phagosome*. Annu Rev Cell Dev Biol, 2004. 20: p. 367–394.

[64] Gan, H., et al. *Mycobacterium tuberculosis blocks crosslinking of annexin-1 and apoptotic envelope formation on infected macrophages to maintain virulence*. Nat Immunol, 2008. 9(10): p. 1189–1197.

[65] Deretic, V., et al. *Autophagy in immunity against mycobacterium tuberculosis: a model system to dissect immunological roles of autophagy*. Curr Top Microbiol Immunol, 2009. 335: p. 169–188.

[66] Jagannath, C., et al. *Autophagy enhances the efficacy of BCG vaccine by increasing peptide presentation in mouse dendritic cells*. Nat Med, 2009. 15(3): p. 267–276.

[67] Lam, K.K., et al. *Nitazoxanide stimulates autophagy and inhibits mTORC1 signaling and intracellular proliferation of Mycobacterium tuberculosis*. PLoS Pathog, 2012. 8(5): p. e1002691.

[68] Floto, R.A., et al. *Small molecule enhancers of rapamycin-induced TOR inhibition promote autophagy, reduce toxicity in Huntington's disease models and enhance killing of mycobacteria by macrophages*. Autophagy, 2007. 3(6): p. 620–622.

[69] Ponpuak, M., et al. *Delivery of cytosolic components by autophagic adaptor protein p62 endows autophagosomes with unique antimicrobial properties*. Immunity, 2010. 32(3): p. 329–341.

[70] Alonso, S., et al. *Lysosomal killing of Mycobacterium mediated by ubiquitin-derived peptides is enhanced by autophagy*. Proc Natl Acad Sci U S A, 2007. 104(14): p. 6031–6036.

[71] Singh, S.B., et al. *Human IRGM induces autophagy to eliminate intracellular mycobacteria*. Science, 2006. 313(5792): p. 1438–1441.

[72] MacMicking, J.D., G.A. Taylor, and J.D. McKinney. *Immune control of tuberculosis by IFN-gamma-inducible LRG-47*. Science, 2003. 302(5645): p. 654–659.

[73] Che, N., et al. *Identification of a novel IRGM promoter single nucleotide polymorphism associated with tuberculosis*. Clin Chim Acta, 2010. 411(21–22): p. 1645–1649.

[74] Watson, R.O., P.S. Manzanillo, and J.S. Cox. *Extracellular M. tuberculosis DNA targets bacteria for autophagy by activating the host DNA-sensing pathway*. Cell, 2012. 150(4): p. 803–815.

[75] Tung, S.M., et al. *Loss of Dictyostelium ATG9 results in a pleiotropic phenotype affecting growth, development, phagocytosis and clearance and replication of Legionella pneumophila*. Cell Microbiol, 2010. 12(6): p. 765–780.

[76] Zhang, R., et al. *The regulation of autophagy by influenza A virus*. Biomed Res Int, 2014. 2014: p. 498083.

[77] Gannage, M., et al. *Matrix protein 2 of influenza A virus blocks autophagosome fusion with lysosomes*. Cell Host Microbe, 2009. 6(4): p. 367–380.

[78] Chen, M., et al. *Essential role for autophagy in the maintenance of immunological memory against influenza infection*. Nat Med, 2014. 20(5): p. 503–510.

[79] Celli, B.R. *Pathophysiology of chronic obstructive pulmonary disease*. Chest Surg Clin N Am, 1995. 5(4): p. 623–634.

[80] Chen, Z.H., et al. *Egr-1 regulates autophagy in cigarette smoke-induced chronic obstructive pulmonary disease*. PLoS One, 2008. 3(10): p. e3316.

[81] Chen, Z.H., et al. *Autophagy protein microtubule-associated protein 1 light chain-3B (LC3B) activates extrinsic apoptosis during cigarette smoke-induced emphysema*. Proc Natl Acad Sci U S A, 2010. 107(44): p. 18880–18885.

[82] Dal-Re, R. *Worldwide behavioral research on major global causes of mortality*. Health Educ Behav, 2011. 38(5): p. 433–440.

[83] Houghton, A.M. *Mechanistic links between COPD and lung cancer*. Nat Rev Cancer, 2013. 13(4): p. 233–245.

[84] Vestbo, J., et al. *Global strategy for the diagnosis, management, and prevention of chronic obstructive pulmonary disease: GOLD executive summary*. Am J Respir Crit Care Med, 2013. 187(4): p. 347–365.

[85] Barnes, P.J., S.D. Shapiro, and R.A. Pauwels. *Chronic obstructive pulmonary disease: molecular and cellular mechanisms*. Eur Respir J, 2003. 22(4): p. 672–688.

[86] Lam, H.C., et al. *Histone deacetylase 6-mediated selective autophagy regulates COPD-associated cilia dysfunction*. J Clin Invest, 2013. 123(12): p. 5212–5230.

[87] Mizumura, K., et al. *Mitophagy-dependent necroptosis contributes to the pathogenesis of COPD*. J Clin Invest, 2014. 124(9): p. 3987–4003.

[88] Fujii, S., et al. *Insufficient autophagy promotes bronchial epithelial cell senescence in chronic obstructive pulmonary disease*. Oncoimmunology, 2012. 1(5): p. 630–641.

[89] Hoffmann, R.F., et al. *Prolonged cigarette smoke exposure alters mitochondrial structure and function in airway epithelial cells*. Respir Res, 2013. 14: p. 97.

[90] Pampliega, O., et al. *Functional interaction between autophagy and ciliogenesis*. Nature, 2013. 502(7470): p. 194–200.

[91] Araya, J., et al. *Insufficient autophagy in idiopathic pulmonary fibrosis*. Am J Physiol Lung Cell Mol Physiol, 2013. 304(1): p. L56-L69.

[92] Hahn, D.R., C.L. Na, and T.E. Weaver. *Reserve autophagic capacity in alveolar epithelia provides a replicative niche for influenza A virus*. Am J Respir Cell Mol Biol, 2014. 51(3): p. 400–412.

[93] Monick, M.M., et al. *Identification of an autophagy defect in smokers' alveolar macrophages*. J Immunol, 2010. 185(9): p. 5425–5435.

[94] Semenza, G.L. *Involvement of hypoxia-inducible factor 1 in pulmonary pathophysiology*. Chest, 2005. 128(6 Suppl): p. 592S-594S.

[95] Lee, S.J., et al. *Autophagic protein LC3B confers resistance against hypoxia-induced pulmonary hypertension*. Am J Respir Crit Care Med, 2011. 183(5): p. 649–658.

[96] Zhang, H., et al. *Mitochondrial autophagy is an HIF-1-dependent adaptive metabolic response to hypoxia*. J Biol Chem, 2008. 283(16): p. 10892–10903.

[97] Bellot, G., et al. *Hypoxia-induced autophagy is mediated through hypoxia-inducible factor induction of BNIP3 and BNIP3L via their BH3 domains*. Mol Cell Biol, 2009. 29(10): p. 2570–2581.

[98] Azad, M.B., et al. *Hypoxia induces autophagic cell death in apoptosis-competent cells through a mechanism involving BNIP3*. Autophagy, 2008. 4(2): p. 195–204.

[99] Cheresh, P., et al. *Oxidative stress and pulmonary fibrosis*. Biochim Biophys Acta, 2013. 1832(7): p. 1028–1040.

[100] Patel, A.S., et al. *Autophagy in idiopathic pulmonary fibrosis*. PLoS One, 2012. 7(7): p. e41394.

[101] Mi, S., et al. *Blocking IL-17A promotes the resolution of pulmonary inflammation and fibrosis via TGF-beta1-dependent and -independent mechanisms*. J Immunol, 2011. 187(6): p. 3003–3014.

[102] Rangarajan, S., et al. *Novel mechanisms for the antifibrotic action of nintedanib*. Am J Respir Cell Mol Biol, 2016. 54(1): p. 51–59.

[103] Rommens, J.M., et al. *Identification of the cystic fibrosis gene: chromosome walking and jumping*. Science, 1989. 245(4922): p. 1059–1065.

[104] Luciani, A., et al. *Defective CFTR induces aggresome formation and lung inflammation in cystic fibrosis through ROS-mediated autophagy inhibition*. Nat Cell Biol, 2010. 12(9): p. 863–875.

[105] Bence, N.F., R.M. Sampat, and R.R. Kopito. *Impairment of the ubiquitin-proteasome system by protein aggregation*. Science, 2001. 292(5521): p. 1552–1555.

[106] Mayer, M.L., et al. *Rescue of dysfunctional autophagy attenuates hyperinflammatory responses from cystic fibrosis cells*. J Immunol, 2013. 190(3): p. 1227–1238.

[107] De Stefano, D., et al. *Restoration of CFTR function in patients with cystic fibrosis carrying the F508del-CFTR mutation*. Autophagy, 2014. 10(11): p. 2053–2074.

[108] Abdulrahman, B.A., et al. *Autophagy stimulation by rapamycin suppresses lung inflammation and infection by Burkholderia cenocepacia in a model of cystic fibrosis*. Autophagy, 2011. 7(11): p. 1359–1370.

[109] Junkins, R.D., et al. *Autophagy enhances bacterial clearance during P. aeruginosa lung infection*. PLoS One, 2013. 8(8): p. e72263.

[110] Luciani, A., et al. *Targeting autophagy as a novel strategy for facilitating the therapeutic action of potentiators on DeltaF508 cystic fibrosis transmembrane conductance regulator*. Autophagy, 2012. 8(11): p. 1657–1672.

[111] White, E. *Deconvoluting the context-dependent role for autophagy in cancer*. Nat Rev Cancer, 2012. 12(6): p. 401–410.

[112] Aita, V.M., et al. *Cloning and genomic organization of beclin 1, a candidate tumor suppressor gene on chromosome 17q21*. Genomics, 1999. 59(1): p. 59–65.

[113] Choi, A.M., S.W. Ryter, and B. Levine. *Autophagy in human health and disease*. N Engl J Med, 2013. 368(7): p. 651–662.

[114] Liang, X.H., et al. *Induction of autophagy and inhibition of tumorigenesis by beclin 1*. Nature, 1999. 402(6762): p. 672–676.

[115] Qu, X., et al. *Promotion of tumorigenesis by heterozygous disruption of the beclin 1 autophagy gene*. J Clin Invest, 2003. 112(12): p. 1809–1820.

[116] Yue, Z., et al. *Beclin 1, an autophagy gene essential for early embryonic development, is a haploinsufficient tumor suppressor*. Proc Natl Acad Sci U S A, 2003. 100(25): p. 15077–15082.

[117] Liu, J., et al. *The expression of p33(ING1), p53, and autophagy-related gene Beclin1 in patients with non-small cell lung cancer*. Tumour Biol, 2011. 32(6): p. 1113–1121.

[118] Won, K.Y., et al. *Decreased Beclin-1 expression is correlated with the growth of the primary tumor in patients with squamous cell carcinoma and adenocarcinoma of the lung*. Hum Pathol, 2012. 43(1): p. 62–68.

[119] Kim, E.J., et al. *mTOR inhibitors radiosensitize PTEN-deficient non-small-cell lung cancer cells harboring an EGFR activating mutation by inducing autophagy*. J Cell Biochem, 2013. 114(6): p. 1248–1256.

[120] Rao, S., et al. *A dual role for autophagy in a murine model of lung cancer*. Nat Commun, 2014. 5: p. 3056.

[121] Shoji-Kawata, S., et al. *Identification of a candidate therapeutic autophagy-inducing peptide*. Nature, 2013. 494(7436): p. 201–206.

[122] Campbell, G.R. and S.A. Spector. *Vitamin D inhibits human immunodeficiency virus type 1 and Mycobacterium tuberculosis infection in macrophages through the induction of autophagy*. PLoS Pathog, 2012. 8(5): p. e1002689.

Chapter 15

Autophagy in Ocular Pathophysiology

S.K. Mitter and M.E. Boulton

Additional information is available at the end of the chapter

http://dx.doi.org/10.5772/64667

Abstract

Autophagy is an evolutionarily conserved intracellular recycling pathway that is indispensable for cellular quality control. Dysfunctional autophagy has been associated with several neurodegenerative, myodegenerative, infectious, and cancerous disorders. Autophagic processes are not only important for cellular maintenance in the retina but also intimately involved with phagocytosis and the very core of retinal visual process. Additionally, excessively upregulated autophagy may culminate into a cell death modality, which may be detrimental to the non-dividing cells of various eye segments. Major advances have been made in understanding the role and fate of autophagy in different ocular tissue layers. In this chapter, we summarize the current understanding of autophagy in the eye in the context of development, aging, and disease. We also speculate on the putative therapeutic strategies where autophagy may be incorporated to treat oculopathies.

Keywords: retinal degeneration, RPE, photoreceptor, lens, cornea, retinal ganglion cells, phagocytosis, lipofuscin, autophagic flux, lysosomes, non-canonical autophagy, diurnal rhythm, circadian, cell death, aging, neovascularization, glaucoma, diabetic retinopathy, macular degeneration

1. Introduction

As a housekeeping cellular degradative and recycling process, autophagy is indispensable for the maintenance of ocular physiology. Since the early 2000s, research in understanding the mechanism and role of autophagy in development and disease has received a tremendous boost [1]. A rapidly growing wealth of data, focused on the diverse role of autophagy in ocular development, physiology, or disease, has enabled researchers to begin understanding this complex process with the hope of manipulating it as a therapeutic tool in treating a myriad of disorders that often lead to loss of visual acuity or complete blindness [2, 3].

The eye has a complex anatomy (**Figure 1A**), with a plethora of specialized cells working together to create the visual perception [4]. Almost all cells in the developed eye have some common characteristics; they have high metabolic rates; are highly differentiated and are either post-mitotic or slowly dividing [5–10]. In addition, owing to the high blood perfusion rates, the eye is an oxygen-rich organ that, along with stressors like UV radiation and visible light, provides a highly oxidative microenvironment leading to cellular damage [7, 11, 12]. In order to combat this onslaught of oxidative damage, cells require not only effective antioxidant defense mechanisms but also cellular repair both at the organellar and macromolecular levels. Autophagy (referring mainly to macroautophagy), along with proteasomal degradation and DNA repair mechanisms, provides this critical housekeeping service to almost every cell type in the eye from the cornea in the anterior part of the eye to the retina/choroid in the posterior [3].

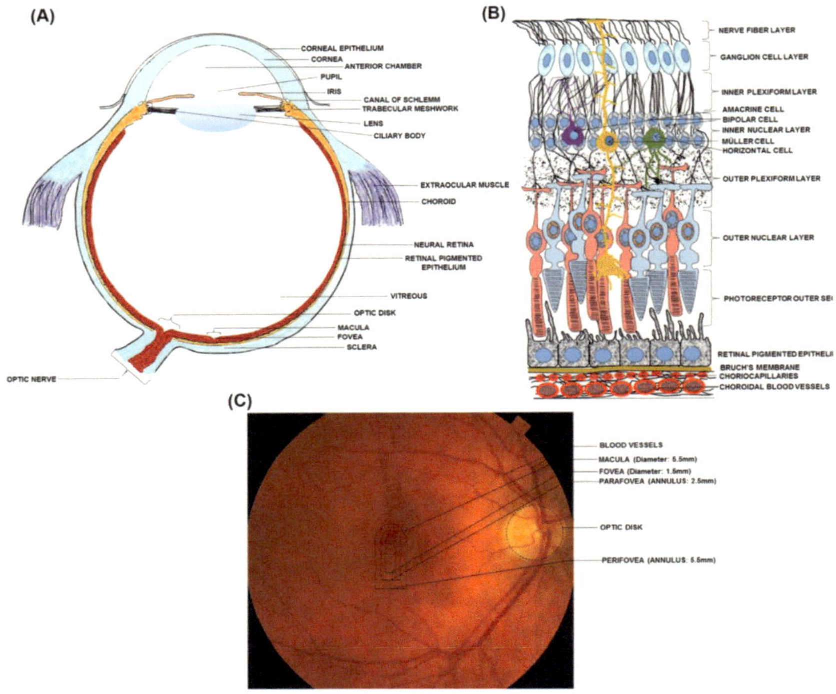

Figure 1. (A) Basic anatomy of the eye: principal ocular tissue components are shown. (B) Diagrammatic cross section of the retina. (C) Fundus image of a healthy adult eye (kindly provided by Dr. Yang Sun, MD, PhD, Department of Ophthalmology, IU School of Medicine, Indianapolis) with no ocular disease history is shown. The ganglion cell axons exit the eye at the optic disk, which is a 'blind spot' due to the complete absence of photoreceptors. The macula is tightly packed with photoreceptors and is critical for central vision. The fovea is a small pit-like area within the macula highly enriched with cone photoreceptors.

Perhaps, the most convincing evidence of the importance of autophagic activity in the eye is the preferential expression of autophagic proteins in the ocular cells and diurnal variation in expression of autophagic protein expression in the retina [13, 14]. As early as 1977, Reme et al. showed increased autophagosome formation in the inner segments of the photoreceptor cells

3 hours after maximum photoreceptor disk shedding in the rat retina [15, 16]. Diurnal variation in autophagosome formation rates was shown to be strongly dependent on light and amplitudes were severely dampened in animals kept under constant darkness [17, 18]. Recent reports have shown autophagic activity in the retinal pigmented epithelium (RPE) to be strongly correlated to the phagocytosis of photoreceptor outer segments (POS) underlining the importance of autophagy not just in being a housekeeping process but as an essential component of RPE function [19, 20].

Because of its dual role in cell survival and death, autophagy has often been referred to as a 'double-edged sword' [21, 22]. As a degradative and recycling pathway, autophagy is essential for sequestration and digestion of toxic waste that could otherwise lead to loss of cell function and eventually lead to cell death. Autophagy (specifically macroautophagy) remains the only known process by which damaged cellular organelles as large as the mitochondria can be digested and recycled [23]. Metabolites generated from autophagic digestion and recycling serve as essential components for new macromolecules and organelles. Because of the ability of this process to be upregulated when the cell is subjected to stress such as nutrient starvation, oxidative stress, hypoxia, and growth factor depletion, autophagy can be thought of as an adaptive process that can meet the energy demand under unfavorable conditions [24].

Due to the generally non-dividing nature of many of the constituent cells of the eye, most reports on autophagy in the eye have concluded it to be a necessary cytoprotective mechanism that prevents the accumulation of cellular damage and inflammation over the lifetime of an individual [25, 26]. However, reduced autophagic efficiency is implicated in a number of ocular pathologies such as age-related macular degeneration (AMD), glaucoma, diabetic retinopathy (DR), photoreceptor degeneration, and ocular infections. Because of its 'destructive self-eating' nature, when autophagic activity exceeds a certain threshold or duration, it may actually promote cellular demise. Moreover, autophagy can 'cross talk' with other cell death modalities like apoptosis to influence overall cell fate [26]. It is thus critical to understand the mechanism and functional role of autophagy in specific cells of the eye before autophagic modulation be incorporated in ocular therapeutic strategies.

In this chapter, we summarize the overall understanding of the role of autophagy in development and normal aging of the eye. We then describe the aspects of autophagy with respect to ocular diseases.

2. Autophagy in the healthy and aging eye

There is increasing evidence that autophagy plays a critical role in ocular development and homeostasis. Developmentally, the vertebrate eye derives from coordinated interactions between neuroepithelium, surface ectoderm, and extraocular mesenchyme (originating from neural crest and mesoderm) [27]. The major development of the eye occurs between the 3rd and 10th weeks of fetal development with the initial formation of the optic vesicles followed by the gradual formation of the lens and the optic nerve [27–30]. Development of the lens requires maturation of lens fibers by degrading the mitochondria, nuclei, Golgi apparatus, and endoplasmic reticulum to create the transparent organelle-free zone to allow passage of light into the ocular chamber (reviewed in [31]). During development, the hyaloid artery supplies the lens with much needed nutrition and eventually its distal end degrades in the inner vitreous of the eye bulb, while the proximal end becomes the central retinal artery [32]. While the pigmented layer and the retina form from the outer and inner layers of the posterior (4/5th) optic cup, respectively, iris and the ciliary body are formed from the anterior (1/5th) region. The sclera and choroid are formed from the mesenchyme on the outer side of the optic cup. The primary and secondary lens fibers form the lens. The vitreous humor is a gel-like substance formed from the mesenchymal cells of the neural crest (reviewed in [33]). In the normal human eye, the photoreceptors continue to mature after birth. Foveal cone photoreceptor cell size shrinks (from 7.5 to 2 μm diameter), while cell density increases (18–42/100 μm) until 3 years of age [34, 35].

Programmed cell death plays a crucial role in neuroretinal development [36]. Autophagic proteins AMBRA1 and Beclin1 are strongly expressed in chicken embryonic (E5) neural retina [37]. Autophagy supplies ATP to energize the externalization of phosphatidyl serine on the dying cell surface, an essential step in the clearance of cell corpses from the developing retinal neuroepithelium. Pharmacological inhibition (3-Methyladenine) of autophagy increases TUNEL-positive apoptotic cells [38]. It will be interesting to investigate the role of autophagy in the development of mammalian neuroretinal cells. Autophagy has also been shown to reduce cell size in other cell types [39–41]. Antagonistic mTOR and autophagic pathways control activity of YAP/TAZ transcription factors, thereby influencing cell size and proliferation [42]. Differential cellular signaling to modulate autophagy and mTOR in both dividing and differentiated photoreceptors during pre- and postnatal cone photoreceptor enrichment in the macular fovea is an area hitherto unexplored.

Autophagic vacuoles engulfing mitochondria were reported in the lens in 1984 [43, 44]. Autophagy was the expected pathway of choice for the developing lens' fiber cells to degrade cell organelles to create the organelle-free zone (reviewed in [45]). Autophagosomes were reported in both differentiating primary and secondary lens fiber cells [43, 46]. However, during the embryonic period, deletion of either Atg5 or Pik3c3 genes in mice did not affect lens organelle clearance [43, 46]. Costello et al. put forward an alternative hypothesis that since both Atg5- and Pik3c3-independent autophagy have been reported and that mutation in autophagy gene FYCO1 causes autosomal recessive congenital cataract, the role of autophagy (and mitophagy) cannot be ruled out in organelle clearance [47–50]. ATG5-independent non-canonical autophagy has also been implicated in the mitochondrial clearance required during metabolic reprogramming of induced pluripotent cells (iPSC) [51]. Perhaps a simpler approach, where overall lens autophagic flux is inhibited (possibly by inhibiting lysosomal fusion), needs to be adopted to confirm that organelle clearance in developing lens is not dependent on autophagy. Furthermore, it remains to be seen if the various proteolytic mechanisms active during lens fiber differentiation can compensate autophagic deficiency (for further reading, please refer [30]).

The adult eye is enclosed in the outer fibrous tunic, composed of the sclera (posterior 5/6th of the eye bulb) and cornea (transparent anterior part) (**Figure 1A**). The middle layer is known as the vascular tunic (or uvea) comprising of the choroid, ciliary body, and iris. The ciliary body supports the lens and controls the shape of the lens with the ciliary muscle. The innermost layer is the retina with ten distinct layers. Moving in a direction from the inside of the eye, these layers are arranged as (1) the inner limiting membrane; (2) the nerve fiber layer; (3) the ganglion cell layer; (4) the inner plexiform layer; (5) the inner nuclear layer; (6) the outer plexiform layer; (7) the outer nuclear layer; (8) the outer limiting membrane; (9) the photoreceptor (rods and cones) layer; and (10) the retinal pigmented epithelium (RPE) (**Figure 1B**) [52]. The innermost layers of the neural retina comprise of different classes of neuronal cells such as the ganglion cells, the Müller, horizontal, bipolar, amacrine cells, and the photoreceptor cells (rods and cones). Together, these cells constitute a complex network of visual sensory synapses that communicate the visual signals to the brain via the central nervous system.

The extraocular muscles (EOMs) control eye directional movement and eyelid movements and contain cells that are likely to accumulate mitochondrial damage during aging, resulting in slower eye muscle movements [53–55]. McMullen et al. reported that autophagy was severely impaired in 18- and 30-month-old compared with 6-month old Fisher 344-Brown Norway rat EOMs supported by their observation of decline in LC3, ATG5, and ATG7 [56].

The cornea, being a non-keratinized epithelial surface, requires to be kept moist by tear secretions from the lacrimal, meibomian (or tarsal) glands, and the conjunctival goblet cells [57, 58]. Basal autophagy-lysosomal activity in the constituent fibroblasts (also known as keratocytes) of the corneal stroma is critical for the clearance of transforming growth factor β-induced protein (TGF-βIp) (discussed in detail in next section along with other corneal abnormalities) [59].

The lacrimal glands produce the aqueous components of the tears (i.e., lacritin, lysozymes, lactoferrin, lipocalin, secretory immunoglobulin A (IgA) and complements), which protect the cornea against a large number of infectious agents (reviewed in [60]). The meibomian glands produce the lipid components of tears called meibum consisting of a variety of esters and fatty acids that prevent evaporation of the tears from the conjunctiva (reviewed in [58]). The mucous secretion from the conjunctival goblet cells allows for even distribution of the tears over the conjunctival surface (reviewed in [61]).

Earliest data on autophagic activity in the acinar cells of the lacrimal glands showed dramatic buildup of autophagosomes upon treatment with vinblastine (microtubule inhibitor that blocks autophagosome–lysosome fusion), strongly suggesting the existence of basal autophagy [62, 63]. Autophagy (along with apoptosis) is upregulated in response to inflammation induced in BALB/c mice lacrimal glands, resulting in acinar cell death. It remains to be investigated whether this phenomenon is critical for tissue repair and remodeling post-inflammation injury [64]. The tear component glycoprotein, lacritin, has been reported to protect *in vitro*-cultured corneal epithelial cells under inflammatory cytokine-induced stress via upregulation of autophagic flux [65–67]. Lacritin-stimulated acetylation of transcription factor FOXO3a, followed by acetylated FOXO3a-ATG101 coupling and coupling of stress-acetylated FOXO1 with ATG7, is critical for this autophagic response [68].

The earliest publication on autophagy in the conjunctiva described autophagic structures in guinea pigs [69].

The trabecular meshwork (TM) is located in the iridocorneal junction of the eye and is responsible for draining the aqueous humor from the eye via the anterior chamber (**Figure 1B**) [70]. Porcine TM cells under hypoxic conditions show increased autophagy, perhaps as a response to increase reactive oxygen species (ROS) [71]. TM cells when chronically exposed to oxidative stress tend to develop lysosomal basification and membrane permeabilization owing to the increased lysosomal iron content. Although autophagic activity is elevated in oxidatively stressed TM cells, the levels of ATG5, ATG7, and ATG12 are significantly reduced [72, 73]. This paradoxical observation hints at the possibility of the existence of active and potentially novel non-canonical autophagic pathways that may use another enzymatic network to modify LC3. *Ex vivo* human trabecular meshwork, cells collected from healthy donor eyes, displays an increase in both oxidative DNA damage (8-hydroxy-2-deoxyguanosine) and autophagic activity markers (increased LC3II/I ratio and reduced p62/SQSTM1) in older donors [74]. However, *in vitro* data have failed to establish autophagy as either an inducer of senescence or an inhibitor. Trehalose-induced autophagy and oxidative stress-induced senescence-associated-β-galactosidase (SA-β-gal) activity, while rapamycin treatment did not show the same effect [73].

The iris modulates the amount of light entering the retina by controlling the size and diameter of the pupil. The iris contains pigmented epithelial cells that have the same origin as RPE and contain melanosomes [75]. *In vitro*-cultured newt iris epithelial cells dedifferentiated to lens cells and this process was accompanied by the sequestration of the melanosomes, ribosomes, and multivesicular bodies (MVB) in autophagic vesicles [76]. Autophagy has not been widely studied in the iris. However, since the iris-pigmented epithelial cells and RPE have some common features like melanosome content as well as phagocytic ability, it is expected that they would have high metabolic rates making them susceptible to accumulating cellular damage [77, 78]. Therefore, autophagy, among other housekeeping processes, must be at an efficient level in order to exacerbate this cellular damage. Consistent with this hypothesis, Petrovski et al. have reported an increase in basal LC3 levels in iris sections from aging human cadaver eyes [79]. Interestingly, the authors observed LC3 expression in mouse iris sections. They also reported an increase in basal LC3 levels in the non-pigmented aqueous humor producing epithelium of the ciliary body (connection between iris and choroid) of aging human cadaver eyes and hypothesized that autophagy might play a key role in the maintenance of intraocular pressure (IOP) in the aging eye. Similar LC3 expression was observed in mouse ciliary body [79]. One must however exercise caution before interpreting autophagic protein expression in any ocular tissue. We and others have observed significant oscillations of autophagic protein expression in the retinal layers of rodents [19]. Recent reports show oscillatory expression of circadian clock genes Bmal1, Clock, Cry1, Cry2, Per1, and Per2 in the irisciliary body of C57BL/6 mice with significant correlation with diurnal IOP variations over a 12-h/12-h light/dark cycle [80]. Per1-deficient mice show increased susceptibility to neuronal injury after cerebral ischemia and one of the key reasons behind this has been hypothesized to be a dramatic attenuation of autophagic activity [81]. Therefore, before conclusions are made about the absence of expression of autophagic proteins in the iris of mice, diurnal/circadian studies on autophagic protein expression must be conducted.

During normal aging, basal LC3 levels are elevated in the lens of human cadaver eyes [79]. Analysis of autophagic gene expression by a combined approach of microarray, qRT-PCR, and Western blotting revealed as many as 42 autophagy-related genes in microdissected human lens epithelium and fiber cells (age range: 47–69 years) providing convincing evidence of autophagy in the lens [82, 83]. Two independent reports suggest the critical role of autophagy in maintaining lens homeostasis in mutant models of αβ-crystallin (R120G) and αA-crystallin (R120A) (discussed in detail in the next section) [84, 85].

The mean retinal thickness of the human eye reduces by about 0.53 μm annually with concurrent loss of macular thickness, implying that there is a significant cell loss in the retina even with no pathology [86, 87]. It remains a challenge to researchers to determine whether changes in cell biology of the retinal cells that trigger onset of disease are different from the normal aging process. The nerve fiber layer shows substantial thinning over time (nearly 150/mm^2 during an average lifetime) due to the significant loss of retinal ganglion cells (RGCs) that make up the ganglion cell layer of the retina and convey visual signals from the photoreceptor layer to the optic nerve that constitutes of RGC axons and glial cells [88–91]. The optic nerve also suffers some detrimental changes during aging due to the loss of ganglial axons [92]. The neuroretinal rim area reduces at a rate of 0.28–0.39% annually, while the optic cup area and the vertical cup diameter start to increase especially after the third decade of life [93]. Apoptosis is considered the primary cell death mode for RGC loss [94–96]. Reports suggest that autophagy may promote RGC survival after optic nerve axotomy in mice [97]. Atg5, 7 and 12, LC3, and Beclin-1 expression is elevated in the mouse RGCs up to 7 days after optic nerve injury [98]. It has been suggested that autophagic flux impairment in the RGC axons may lead to age-associated changes in the optic nerve [99]. We will elaborate the role of autophagy in terms of RGC and disease in Section 3.

Photoreceptor density decreases at a rate of 0.2–0.4% annually with a greater degree of rod cell loss than cones causing reduced dark adaptation in aged individuals [100, 101]. This loss is mostly in the peripheral and the parafoveal retina rather than at the fovea (**Figure 1C**) [102]. We have previously shown strong expression of Atg9 and LC3 in the ganglion cell layer, retinal vessels, a subpopulation of the inner nuclear layer, the outer nuclear layer of rods and cones, and the RPE [103]. Deletion of Beclin1 or Atg7 or mitophagy-specific Parkin gene in mice causes severe retinal degeneration along with accumulation of abnormal mitochondria [104].

The RPE monolayer consists of perhaps the most multifunctional cells of the eye. The RPE has a plethora of functions such as phagocytosis of photoreceptor outer segments, renewal of chromophores in visual transduction cycle, supplying nutrients from the choroidal side to the photoreceptors, ion, and metabolite exchange and light absorption (reviewed in [7]). Like the entire retina, the RPE is also prone to age-associated decline in function and vitality and accumulate massive cell damage even during normal aging [104, 105]. Additionally, the RPE has to combat light and reactive oxygen species induced damage not just to its own cellular components but also to those of the photoreceptors. Aging changes in the RPE layer is not uniform across the retina [105, 106]. It appears that the peripheral cell area increases while that at the central retina declines. Cell density decreases with increasing distance from the fovea, but the foveal RPE cell density is relatively very stable. Surprisingly, the aged non-diseased macula shows a population of apoptotic RPE [107]. Both the RPE and photoreceptors are highly metabolic and a healthy pool of mitochondria is required to meet this energy demand. There is a significant reduction in the number of healthy mitochondria and extensive damages to mitochondrial cristae, and matrices are observed [108]. A number of publications together have shown RPE and photoreceptors expressing autophagy proteins (p62, LC3, ATG7, ATG9, and Beclin1) in both human cadaver and rodent model retina sections [13, 14, 19, 103, 109]. Furthermore, as mentioned earlier, diurnal oscillations of autophagic proteins and autophagosomes in the RPE/photoreceptor layers confirm a functional role of autophagy that is integrally linked to POS phagocytosis [19, 110, 111]. Autophagic digestion of rhodopsin light pigment in rod photoreceptors is also necessary for adaptation to changes in light intensity (3–200-lx) [110]. Lentiviral shRNA-mediated silencing of autophagic genes (Beclin1 and ATG7) or 3-Methyadenine (3-MA)-mediated autophagy inhibition in human RPE in cell culture increases susceptibility to oxidative stress with compromised mitochondria, increased lipofuscin, and reduced cell viability [13]. Deletion of RB1CC1 in rodent RPE caused severe retinal degeneration underlining the importance of basal autophagy in the RPE [109]. Levels of autophagic protein such as ATG7, ATG9, and LC3 increase with aging in the retinal layers including RPE and photoreceptors in both human cadaver donor and c57Bl/6 mouse retina sections [13].

Non-canonical LC3-associated phagocytosis (LAP), dependent on Atg5 and Beclin1 but independent of the autophagy pre-initiation complex consisting of Ulk1/Atg13/Fip200, was reported to be critical for degradation of POS and renewal of retinoids required for chromophore synthesis for optimal visual function (**Figure 2**) [111, 112]. Melanoregulin, a 28 KDa membrane-associated protein is critical for lysosomal hydrolase activity in the RPE as well as for RILP-p150Glued complex-mediated retrograde melanosome transport via actin filaments in melanocytes [113, 114]. Frost et al. demonstrated a diurnal variation in melanoregulin expression in the RPE and its distinct association with the ATG5-dependent LAP [111]. Loss of melanoregulin causes accumulation of phagosomes and lipofuscin in the RPE with elevated cathepsin-D secretion that could injure not only the RPE but also the adjacent ocular layers [111, 114]. Furthermore, ROS generated from NADPH oxidase activity resulting from the delayed clearance of all-trans retinal (activated visual chromophore of the visual transduction cycle) shows severe RPE cytotoxicity [104]. LC3 association with phagosomes is signaled by elevated NAPDH oxidase activity in other 'professionally' phagocytic cells like macrophages [115]. Park2 (mitophagy receptor protein) and LC3 activity are indispensable for RPE defense against all-trans retinal induced cytotoxicity [104]. It is now evident that while basal rate canonical autophagy is critical for quality control and stress adaptation, non-canonical forms of autophagy where some but not all components and mechanisms of the canonical form participate, supports the very core of retinal visual function.

The ocular vasculature has several indispensable functions including supply of oxygen and nutrients to the ocular components; transportation of ions and metabolites; circulation of immune-surveillant cells; and exclusion of pro-inflammatory cytokines and molecular toxins

http://dx.doi.org/10.5772/64667

[116]. The study of autophagic flux and its role in ocular vascular endothelial physiology is still at rudimentary stages. However, recently it has been reported that conditional deletion of endothelial Atg7 in $ApoE^{-/-}$ mice results in accumulation of oxidized LDL within the RPE and choroidal vascular endothelium of the eye underscoring the importance of autophagy in vascular lipid homeostasis [117]. Conflicting opinions exist regarding the role of autophagy in angiogenesis most likely due to the different tissue source of the endothelial cells under study. Lee et al. have recently reported that Beclin1 deficiency leads to increased hypoxia-induced angiogenesis in human pulmonary artery endothelial cells [118]. On the other hand, in bovine aortic endothelial cells, Du et al. suggest that autophagy promotes angiogenesis and elevated ROS levels [119]. Autocrine vascular endothelial growth factor (VEGF) suppresses autophagy in human umbilical vascular endothelial cells (HUVECs) to maintain cell viability. The role of $\alpha v\beta 3$ and $\alpha v\beta 5$ integrins has long been implicated in retinal neovascularization [120, 121]. $\alpha v\beta 5$ integrins act downstream of VEGF activating focal adhesion kinases (FAKs) that are critical for cell migration [122]. Recent reports suggest a critical role of autophagy in restricting integrin activity and thus inhibiting cell migration [123]. Autophagy receptor NBR1 has been shown to be a specific cargo receptor for targeting focal adhesion components to the lysosome for degradation [124].

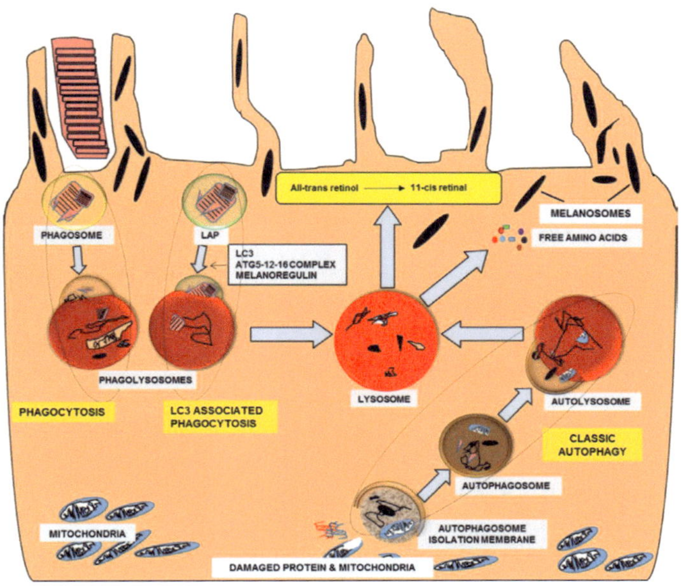

Figure 2. Classic and non-canonical LC3-associated phagocytosis (LAP) in the RPE: basal autophagy is essential in the RPE to maintain organelle and protein quality. Phagocytosis of ingested outer segments may be mediated by autophagy components LC3, ATG5-12-16 complex, and delivery of the phagosome to the lysosome is dependent on these proteins underlining the existence of a non-canonical autophagic pathway in the eye that supports RPE phagocytic function. Furthermore, melanoregulin (MREG) facilitates LC3 recruitment to the phagosomes. Phagocytosis is essential for renewal of all-trans retinol to 11-cis retinal visually active chromophore that is sent to the photoreceptors for enabling the visual cycle. Hence, while basal canonical autophagy is essential for basic housekeeping of the RPE, non-canonical autophagy supports at least in part the visual cycle and photoreceptor disk processing.

These mechanisms need to be reinvestigated in retinal endothelial cells in order to elucidate the role of autophagy in maintenance of retinal vasculature. Inhibiting autophagy in the RPE *in vitro* elevates the levels of pro-angiogenic intercellular adhesion molecule (ICAM), stromal cell-derived factor (SDF-1) and VEGF A in response to challenge by lipofuscin component A2E [125]. Therefore, retinal vascular stability may not only be influenced by autophagy in the vascular endothelial cells but also by the cross talk from adjacent cell layers.

One of the inevitable consequences of oxidative damage in the aging retina is the accompanying inflammatory response and elevated levels of damage-associated molecular patterns (DAMP) [126]. Although the eye was considered immune privileged for a long time, immunocompetent cells like the monocyte-derived cells, microglia, dendritic cells, and perivascular macrophages have been detected in the retina [127–130]. Ample evidence suggests that the inflammation observed during normal tissue aging is an adaptive response and the word coined for this inflammation is 'para-inflammation' [129, 131–133]. Para-inflammation is required for retinal tissue homeostasis plays a crucial role in tissue repair/remodeling, but when para-inflammation becomes chronic or progresses to destructive inflammation, retinal damage and pathology may ensue (reviewed in [134]). As mentioned earlier, the aging retina shows an increase in apoptotic cells. However, several reports have recently indicated that other cell death modalities like autophagy and necrosis may also exist in the eye that may become particularly active in retinal degenerative conditions [135, 136]. While apoptosis restricts the release of inflammatory danger signals, late-stage apoptosis and necrosis can initiate DAMP-mediated inflammation. Autophagy, at least in the early stages, has been considered a protective response that suppresses inflammatory signals [137]. Shi et al. showed the activation of autophagy by sterile inflammation (NLRP3- and AIM2-mediated inflammation) limited caspase-1-mediated maturation of IL-1β and IL-18 [138]. Impairing autophagy in RPE leads to not only inflammasome activation but also macrophage-mediated angiogenesis [139–141]. The age pigment, lipofuscin, is a common feature of many post-mitotic cells throughout the body and is largely derived from autophagic removal of damaged organelles [142]. Lipofuscin accumulation occurs in an age-dependent manner in both photoreceptor cells and the RPE. In both cell types, lipofuscin is derived at least in part via autophagy of damaged organelles (e.g., mitochondria) [142], but the situation is more complicated in the RPE where (a) lipofuscin is also an inevitable consequence of phagocytosis of spent photoreceptor outer segments [143] and (b) phagocytosis is linked to a non-canonical autophagy pathway [111, 112]. Lipofuscin is both a cause and consequence of oxidative stress and oxidative stress-mediated accumulation of lipofuscin increases dramatically in the RPE when autophagy is pharmacologically inhibited [13].

Considerable cross talk exists between apoptosis and autophagy. p53 and Bc1-2 family proteins and calpain have been classically considered as apoptotic proteins but can also modulate autophagy [144–146]. For example, Beclin1 is cleaved by caspase upon depletion of IL-3 in Ba/F3 cells leading to inactivation of autophagy and release of proapoptotic cytochrome c from the mitochondria [147]. Direct cleavage of ATG3 (a ubiquitin-like-conjugating enzyme involved in autophagosome biogenesis) by activated caspase-8 can lead to inhibition of autophagy and cell death [148]. Yet other reports show autophagy prevents necrosis by reducing metabolic stress [149]. Although cross talk between cell death pathways needs to be confirmed in ocular cells, it is safe to assume that autophagy in the aging eye plays a critical role in maintaining balance between the cell death modalities to avoid a pathological scenario. Arrested autophagic flux by lysosomal disruption enhances buildup of ubiquitinated protein aggregates and cell death under oxidative stress that cannot be prevented by apoptotic caspase inhibitor (zVAD-FMK) [150].

3. Autophagy in ocular disease

3.1. Autophagy in congenital ocular abnormalities

Several congenital deformities of the eye occur such as coloboma, congenital glaucoma, congenital cataracts, congenital detached retina, partially persistent iridopupillary membrane, persistent hyaloid artery, microphthalmia and Peter's anomaly, Leber's Congenital Ameurosis [151–156]. While the cause of most of these diseases is rooted deep in mutation of genes such as PAX2, PAX6, CYP1B1, GLC3A, GLC3B, GLC3C, FOXC1, CEP290, CRB1, GUCY2D, RPE65, several reports suggest autophagy may be compromised in some of these diseases [157]. Persistent hyaloid artery and persistent hyperplastic primary vitreous (PHPV) could result from incomplete involution of the hyaloid vessel [32, 158]. Studies have shown that hypoxic conditions, as seen in the developing eye, increase autophagic activity in vascular endothelial cells. Hypoxia plays a major role in triggering the hyaloid vessel regression and activation of autophagy seems to enhance hyaloid regression in the developing eye [159]. Recessive mutations in EPG5 cause a rare inherited congenital multisystem disorder called Vici syndrome with defective systemic autophagy [160]. EPG5, the human homolog of the *Caenorhabditis elegans* autophagy gene epg-5, encodes a key protein (ectopic P-granules autophagy protein 5) that regulates the formation of autolysosomes [160]. Retinal hypopigmentation and bilateral cataracts are among the chief manifestations of this disorder, once again suggesting the importance of autophagy in retinal development (reviewed in [157]). Mutations in WDR45 (also known as WIPI14) cause a rare biphasic X-chromosome-linked disorder called beta-propeller protein-associated neurodegeneration (BPAN) [161, 162]. WDR45 interacts with ATG2 and ATG9 that regulate lipid and membrane supplies for autophagosome formation and elongation [163–165]. A subset of BPAN patients has optic nerve atrophy suggesting a possibility that defective autophagy may be part of the disease etiology [166]. Congenital eye disorders typically emerge from underlying genetic mutations, which may often prove difficult to manage therapeutically. Understanding the effect of these genes on cell biology leading to the disease may, in some cases, provide a better therapeutic avenue.

Autophagy is an integral part of developmental cell biology that is coordinated by a vast network of genes [167]. Although still at rudimentary stages, research on the role and fate of autophagy in ocular development must be intensified in the search of more promising therapies in debilitating congenital eye disorders.

3.2. Conjunctiva

Several topical eye ointments contain benzalkonium chloride (BAC) as a preservative that has been shown to induce caspase-independent cell death in a conjunctival cell line, was reversible by autophagy induction [168].

3.3. Cornea

The consequences of corneal infection can be devastating with corneal scars that would require corneal transplant [20]. *Toxoplasma gondii* and herpes simplex virus-1 (HSV-1) are two most common pathogens that can directly infect the cornea. Alternately, HSV-1 can infect the cornea indirectly via first infecting the oral mucosa [169]. In some extremely infectious cases, HSV-1 infections may cause stromal keratitis leading to blindness and is the leading cause of corneal blindness globally [170]. A key virulence mechanism of HSV-1 is to hijack and inhibit autophagy in the host cell via the binding of Beclin1 with the viral protein ICP34.5 [171, 172]. Additionally, IPC34.5 can inhibit autophagy induction by inhibiting the antiviral eIF2alpha kinase-signaling pathway (including PKR and eIF2alpha) [173]. Contradicting opinions exist as to whether autophagy promotes HSV-1 infection or inhibits it. Petrovski et al. recently showed that autophagy was induced in HSV-1-infected rabbit corneal cell survival against apoptosis [174]. Yakoub et al. showed that HSV-1 does not increase autophagic activity in the cell but basal autophagy is required for it to infect the host cells [175] and pharmacological induction of autophagy in host cells suppresses HSV-1 infection [176]. Yet in another independent study, Alexander et al. reported no effect on HSV-1 replication in autophagy-deficient host mouse fibroblast cells [177]. It seems that the experimental model of choice from one research team to another may affect the final outcome of autophagy on HSV-1 in the infected host cell.

T. gondii infection in the cornea is a classic example where an invading pathogen disrupts cellular endosomal-lysosomal fusion and therefore prevents itself from degradation by lysosomal enzymes [178]. Defects in the CD40 pathway that activates macrophages to eliminate *T. gondii* cause ocular toxoplasmosis, suggesting autophagy may prevent *T. gondii*-mediated infection [179, 180].

Granular Corneal Dystrophy 2 (GCD2), an autosomal dominant disorder caused by mutation R124H in the transforming growth factor β-induced gene (TGFBI) on chromosome 5q31, shows dramatic accumulation of mutant transforming growth factor β-induced protein (TGF-βIp) in autophagosomes and/or lysosomes of corneal fibroblasts [59]. Autophagy is activated, but the rate of autophagic degradation is not sufficient to inhibit the accumulation of the aberrant protein or polyubiquitinated proteins that are also digested in part by autophagy [59, 181].

3.4. Trabecular meshwork

Defects in fluid drainage by the TM can lead to elevated IOPs and eventually cause irreversible damage to the optic nerve leading to glaucoma. Glaucoma is manifested with loss of peripheral vision leading eventually to complete blindness [182–184]. Both elevated IOP and biaxial TM stretching have been independently shown to promote autophagosome formation [185, 186]. Additionally, aging TM is subjected to both hypoxic and highly oxidative conditions that cause increased ROS production and accumulation of non-degradable material along with lipofuscin in lamina cribrosa as well as TM cells [187]. In both cell types, autophagic flux is severely impaired contributing to glaucoma pathogenesis [73, 188]. Autophagy seems to be protective from apoptotic caspase signals in TM cells [189].

3.5. Optic nerve

Optic nerve damage is a commonality in all glaucoma subtypes [190]. Other retinal neuropathies such as optic neuritis, hereditary optic atrophy, and traumatic injury may also lead to degeneration of retinal ganglion cell axons in the optic nerve [191]. Optineurin overexpression in retinal ganglion cells (RGC-5) *in vitro* was found to be beneficial via its stimulatory effect on autophagy [192]. However, some reports contradict this hypothesis suggesting that autophagy may, in chronic hypertensive glaucoma models, be neuropathic to the optic nerve [99, 186]. This disagreement may arise from variations in disease stage and models under investigation (reviewed in [191]).

3.6. Lens

As mentioned in the earlier section, there still seems to be considerable debate over the role of autophagy in digestion of organelles of differentiating lens fiber cells to create the organelle-free zone. However, even the reports that argue against the role of autophagy in organelle clearance suggest that it is indispensable for lens quality control. Morishita et al. showed that Pik3kc3/VPS34 deletion in mouse caused congenital lens defects including cataract and that in ATG5 deletion mice lens, the lens developed age-related cataracts although not congenital cataracts [46]. Mutations in FYCO1 (facilitates microtubule-dependent directional transport of autophagosome vesicles) show severe autosomal-recessive congenital cataracts in patients [48, 193]. αβ-Crystallin mutation (R120G) in hereditary cataract mouse model causes concurrent increase in autophagosome fractional volumes and p62-positive aggregates in lens suggesting impaired autophagic flux that leads to increased lens opacity [194]. Similar results were also observed in a hereditary mutant double knock-in (R49C$^{+/+}$) mouse model where autophagic flux also seemed impaired [85]. Recently, an ESCRT-III subunit CHMP4B has been proposed to be involved in autophagosomal clearance of extranuclear DNA and chromatin [195]. CHMP4B mutation is associated with autosomal dominant posterior polar cataract formation [196].

3.7. Retina

Many studies have investigated the implications of autophagy in retinal degenerative diseases [12–14, 103, 109, 125, 150, 194, 197, 198]. Age-related macular degeneration (AMD) is an aging-associated neuropathy that affects primarily the photoreceptors and RPE in the macula resulting in loss of peripheral vision and eventual legal blindness [199]. Early in AMD pathology, sub-RPE deposits known as drusen are observed on Bruch's membrane (BM) by fundoscopy. There are two types of AMD: 'dry AMD' characterized by geographic atrophy (GA) and 'wet AMD' characterized by neovascularization. Although a heterogenous disease, the key reason behind the pathology is the increased susceptibility of the RPE to oxidative stress [200, 201]. Diseased retina shows a significantly greater extent of damaged organelles (mitochondria, peroxisome, melanosomes, etc.) and protein aggregates compared to age-matched healthy retina [108]. The overall accumulation of damaged organelles and macromolecules suggests a collapse of overall antioxidant and proteolytic capacity of the RPE that sets up the stage for disease [202]. Not surprisingly, autophagy has been shown to be severely impaired in AMD retinas of both human cadaver eyes compared with age-matched donors as well as in AMD mouse models [13, 14, 198]. *In vivo* deletion of autophagic gene RB1CC1 in mouse retina results in retinal degeneration that shares many features with AMD disease [109]. RPE cells *in vitro* accumulate lipofuscin and greater loss of mitochondrial activity and membrane integrity under oxidative stress when autophagy is inhibited [13, 125, 203]. Autofluorescent lipofuscin in the RPE destabilizes the lysosome and is a hallmark of RPE senescence that has been widely implicated in AMD [204, 205]. Lysosomal destabilization leads to a severely impeded autophagic flux that has been recently suggested in dry AMD where a higher accumulation of p62 was observed in the foveomacular regions of AMD patient retinas compared with age-matched donors [14]. Interestingly, as mentioned earlier, autophagy inhibition in lipofuscin chromophore A2E-laden RPE results in elevated levels of several pro-inflammatory and pro-angiogenic factors that suggest a possible role of autophagy in both dry and wet forms of AMD [125].

Diabetic retinopathy (DR) is a retinal complication characterized by pericyte loss, microvascular instability, blood retinal barrier (BRB) leakage, and abnormalities in the retinal vasculature [206, 207]. Since pericyte loss is a key feature of DR, the effect of autophagy was investigated in a combination mouse model of diabetes and hypercholesterolaemia. The authors showed that autophagy promoted pericyte survival under mild stress but under chronic stress conditions resulted in pericyte death [208]. This may be considered as a perfect example of the dual role of autophagy both as a protector and as a destructive pathway. Extravascular oxidized low-density lipoprotein (LDL) has been reported to be damaging to the BRB and to cause apoptotic pericyte loss [209]. Du et al. suggested that oxidized LDL may cause RPE injury by excessive oxidative stress, ER stress, autophagy, and apoptosis [210]. High glucose (30 mM) conditions in the RPE result in higher levels of p62 and LC3 accompanied by an increase in the number of autophagosomes [211]. This increase in autophagosomes is possibly to accommodate for the increased ROS damage sustained by the mitochondria, but it needs to be determined whether autophagic flux is reduced as lysosomal pH is reported to be elevated under high glucose conditions [212]. As described earlier, circadian rhythmicity and diurnal variations in expression amplitudes of autophagic proteins is a prominent feature of the retina. Disruption of the peripheral clock has been reported in DR pathology affecting cellular processes such as regulation of inflammation and lipid metabolism [213–216]. Our unpublished data show dramatic phase-shift and amplitude dampening of key autophagic proteins in the retina of rodent models of diabetes (manuscript under preparation). It remains to be elucidated how disruption of diurnal rhythm dysregulates the normal balance between retinal cell metabolism and autophagy, which contributes to DR pathology.

Photoreceptor degeneration is widely observed both in AMD as well as in retinitis pigmentosa (RP). The latter is a highly hetergenous disease with hereditary mutations in multiple gene loci

[217]. Both caspase-dependent and caspase-independent pathways are involved in photoreceptor cell death in RP [218–220]. rd/rd mouse, the rds/rds mouse, and the light-damage model in albino mice show several elements of the autophagic pathway to be upregulated. This induction seemed secondary to an increase in oxidative stress markers, suggesting that autophagy may be upregulated specifically to remove damaged photoreceptors [221].

Inherited lysosomal storage disease Niemann–Pick type C (NPC) disease is caused by mutations in genes NPC 1 and 2 [222]. LC3 and autophagosmal numbers are elevated in the ganglion cell layer of Balb/cNctr-Npc1m1N/J mouse model possibly because of disruption of autophagic flux and reduced degradation of autophagosomes in the lysosome [223].

3.8. Retinal detachment

Retinal detachment has a number of causes and could be rhegmatogenous or may due to other causes such as traumatic brain injury, severe myopia, retinal tear, or vascular abnormalities frequently encountered in diabetic retina and hypertension [224–227]. In rodent models, retinal detachment induced by subretinal injection of 1% hyaluronic acid resulted in a rapid increase in autophagic activity 3 days after insult. However, 7 days post-injury the autophagic response declined with a simultaneous rise in calpain activity resulting in photoreceptor cell death. Calpain inhibition resulted in increased autophagy and prolonged the survival of photoreceptors [228]. Furthermore, activating autophagy in the same model in Fas-dependent manner inhibited apoptotic death of photoreceptors [229]. Unpublished results suggest hypoxia (increased Hif1α and Hif2α protein levels) induced by the retina-RPE separation is a key inducer of autophagy *in vivo*. In an independent study, Dong et al. confirmed the increase in autophagy 3 days after retinal detachment induction. They also showed induction of necroptotic cell death as seen by increased RIP kinase activation [230]. Although they concluded that autophagic and necroptotic cell death can be blocked by the use of necrostatin, it must be considered that all procedures were done at 3 days post-injury when autophagy is at its peak. It would be interesting to see whether necroptosis is still active at 5 days post-injury when autophagy has subsided.

3.9. Uvea

The uvea (consisting of the choroid and the ciliary body) may be affected in some disease conditions such as uveitis and uveal melanoma. Autoantigen-induced experimental autoimmune uveitis (EAU) in Lewis rats shows an increased autophagic activity in infiltrating T lymphocytes that was required for disease recurrence [231].

Uveal melanoma results from malignant tumors arising from melanocytes in the uvea and is the most common intraocular cancer [232]. Mutations in GNAQ and GNA11 genes contribute to a majority of uveal melanoma cases [233, 234]. Ambrosini et al. showed that mutant GNAQ promoted AKT activation via phosphorylation and deletion of mutant GNAQ upregulated AMP kinase-dependent autophagic cell death in primary choroidal uveal melanoma cell lines [235].

3.10. Autophagy in ocular inflammation

Inflammation is an unavoidable phenomenon of aging. Elevated inflammation in the eye contributes to disease pathologies including uveitis, diabetic retinopathy, or maculopathies [236, 237]. As discussed earlier, chronic inflammation in diseased eye is destructive and detrimental to ocular health compared with para-inflammatory immune surveillance that responds to, and repairs, localized tissue injuries. In AMD, drusen deposits play a major role in eliciting inflammation via both the inflammasomes and the complement pathway [238]. While mild upregulation of NLRP3 inflammasome has been shown to be protective, accumulation of lipofuscin, drusen, and damaged mitochondrial DNA have all been implicated in pathological upregulation of inflammasome activity [239]. Furthermore, complement pathway element C5a has been shown to prime the RPE cell for upregulating NLRP3 inflammasome activity in response to light-induced damage [240].

Ocular tissue	**Autophagic role**	**Disease implications**	**References**
Lens	Critical for proteolytic digestion and lens quality control. Role in lens organelle clearance during eye development controversial.	Disrupted autophagy due to *FYCO1* mutations leads to and autosomal recessive congenital cataract. *CHMP4B* mutation leads to autosomal dominant posterior polar cataract. Hereditary cataract mouse models show disruption of autophagic flux.	Costello et al. [47] Chen et al. [48] Wignes et al. [194] Shiels et al. [196]
Cornea	Cellular housekeeping and defense against infectious pathogens	Insufficient autophagic degradation leads to accumulation of TGF-βIp in autolysosomes. *T. gondii* disrupts cellular endosomal-lysosomal fusion machinery as a part of the infectious process.	Choi et al. [251, 252, 254]
Optic nerve	Promotes cell survival in RGC post-optic nerve axotomy. Optineurin mediated RGC *in vitro* survival dependent on autophagy stimulation. Alternative opinions exist of autophagy effecting cell death in chronic hypertensive model of glaucoma.	Some BPAN patients have optic nerve atrophy possibly due to defective autophagy.	Park et al, [186] Gregory et al. [166] Sternberg et al. [189]
Trabecular meshwork	Autophagy upregulated under hypoxia, elevated IOP and biaxial TM stretching. Oxidative stress increases autophagy but ATG5, ATG7, and ATG12 reduced.	Disruption of endoplasmic reticulum autophagy may be a key feature of myocilin accumulation seen in a majority of glaucoma cases.	Pulliero et al. [74] Porter et al. [185] McElnea et al. [187]

Ocular tissue	Autophagic role	Disease implications	References
	Autophagic markers increase during aging in donor eyes.		
Ocular vasculature	Critical for proteolytic and lipid homeostasis. Potential regulator of vascular stability by prevention of angiogenesis. Autophagy essential for hyaloid regression and clearance during development.	Diabetic Retinopathy: Accumulation of extravascular oxidized LDLs due to disrupted autophagy leads to blood-retina barrier damage and causes pericyte loss.	Torisu et al. [117] Lee et al. [118] Kim et al. [159] Fu et al. [208] Wu et al. [209]
Retina	Diurnal modulation of Autophagic proteins. Non-canonical LC3 associated Autophagy is essential for phagocytosis is critical for degradation of POS. Basal autophagy indispensable for maintenance of RPE and photoreceptor homeostasis. Restricts sterile inflammation in the retina	AMD: Autophagy initiation and flux are disrupted in human donors and mouse models for AMD. Diabetic Retinopathy: Disrupted autophagy may cause pericyte loss.	Kim et al. [112] Yao et al. [19, 109]; Mitter et al. [13] Viiri et al. [14] Fu et al. [208] Liu et al. [139]
Uvea	Autophagy promotes cellular survival	Uveal Melanoma: Inhibition of autophagy may be effective in Uveal melanoma therapy	Ambrosini et al. [232]

Table 1. Description of role of autophagy along with disease implications in various ocular tissues.

Autophagy plays a critical role in controlling NLRP3 inflammasome activity in the retina. Several *in vitro* experiments show dramatic upregulation of inflammasome activity in RPE and secretion of IL-1β after autophagic flux inhibition [139–141] [203]. Wortmannin inhibits autophagy by inhibiting class III PI3Kinase [241]. Intravitreal injection of wortmannin inhibits autophagy *in vivo* in mouse retina-induced inflammasome activity and CCR2+ monocyte-derived macrophage augmentation that promotes angiogenesis [139]. Although the role of autophagy in activating and augmenting the retinal complement pathway needs to be deeply investigated, evidence does exist of C3- and CD63-positive deposits on Bruch's membrane in aged mice possibly as a result of increased autophagy and exocytosis [197] (see **Table 1**).

4. Autophagy as a therapeutic target in ocular pathology

Since autophagy is a pathway with a dual role in cell maintenance as well as cell death; multiple stages such as initiation, maturation, and lysosomal fusion; and cross talk with multiple cellular pathways, its manipulation is a challenging therapeutic option in diseases. Nevertheless, autophagy has received special attention in cancer, metabolic, neurodegenerative, and infectious diseases [242–245]. Since overall proteolytic capacity is attenuated in a majority of ocular diseases, autophagy modulation must be incorporated in current therapeutic regimens to achieve a better outcome.

Treatment strategies for immature cataracts have been sought after for more than a century [245, 246]. Topical solutions with antioxidants glutathione, cysteine ascorbate, L-taurine, riboflavin, and 2% N-acetyl carnosine showed some promise in reducing immature cataracts [247]. Including autophagy stimulation may improve this treatment strategy. Posterior capsule opacification, a common post-surgical complication of mature cataracts, results from the remnant lens fibers and epithelial cells that proliferate and damage the new lens implant [248]. Laser capsulotomy surgery although usually successful may in rare occasions give rise to retinal detachment and is also extremely challenging to execute when treating congenital cataracts in younger patients [249]. It is not known whether autophagy (as well as other mechanisms) supports cell survival of the remnant lens epithelial cells. Pharmacologically stimulating cell death may involve autophagy that either promotes or inhibits survival of these cells.

As mentioned earlier, removal of TGF-βIp deposits is a focus in granular corneal dystrophy 2 (GCD2) research. Lithium, which has shown some success in removing these deposits from *in vitro*-cultured corneal fibroblasts from GCD2 patients, has also been shown to induce autophagy as a part of its cytoprotective mechanism [250, 251]. Since TGF-βIp accumulates in autolysosomes, autophagy stimulation by lithium or by rapamycin and melatonin as suggested in another study must be explored as a treatment strategy in this disease (please refer to Choi et al. for more details) [59, 252–254].

Non-infectious uveitis treatment with subconjunctival injections of rapamycin as an immunosuppressive agent has shown promise in clinical studies with patients showing improved visual acuity and reduced vitreous haze with no noticeable adverse effects [255]. Mechanistic studies may reveal that at least a part of this immune suppressive ability of rapamycin may be credited to autophagy stimulation.

Antitumor activity is seen in combinatorial therapy involving mTOR inhibition and autophagy inhibition with hydroxychloroquine has been shown to restrict melanoma and these treatments are currently under phase-1 trial [256, 257]. Such treatment strategies may be adopted in treatment of uveal melanoma although the fact that chloroquine may induce cataract formation, demands that careful dose-response studies be conducted to ensure no adverse effects [258, 259].

Autophagic degradation is attenuated in AMD. Lipofuscin accumulation in the disease has been shown to perturb the lysosomes that have serious implications on RPE health [142, 260–262]. Lysosomal activity disruption affects both autophagic flux and phagocytosis [263]. Hence, putative therapies should first focus on restoring lysosomal activity to improve degradation of existing autophagosomes. Rapamycin administration to senescence-accelerated OXYS rats improved the RPE morphology in the retina [264]. Clinical trials using rapamycin to treat GA in advanced stages of dry AMD showed 'no positive anatomic or functional effects' [265]. The treatment failure may partly be attributed partly to the fact that the intervention may have been attempted at a time when the disease was well underway with well-developed AMD lesions. An earlier intervention in addition to stimulating lysosomal activity may produce better results.

To include autophagy in ocular therapeutic strategy for better treatment outcomes, the following aspects must be considered. (1) Stimulating autophagy initiation: Several pharmacological activators of autophagy have been identified for possible therapeutic treatments. Rapamycin and its analogs (CCI-779, RAD001 and AP23573) act via inhibiting the mTOR pathway. Metformin mediates AMP kinase activity which stimulates autophagy initiation. Yet other drugs such as lithium and valproic acid have been identified that stimulate autophagy induction. Studies using rapamycin or resveratrol have shown promising results in treatment of cardiac hypertrophy [266, 267]. Clearance of α-synuclein and polyQ mutant Huntingtin aggregates has also resulted from using rapamycin in Parkison's and Huntingtin disease, respectively [268–270]. Also, small molecule enhancers of rapamycin have also been reported that show positive results in neuroprotection. However, whether stimulation of autophagy would be at all beneficial in retinal diseases perspective depends significantly on the status of lysosomal machinery at the stage of the disease when intervention is attempted. Stimulating autophagosome biogenesis when lysosomes are destabilized will not alleviate the cytotoxic burden resulting from damaged protein aggregates. (2) Stimulating lysosomal activity: An effective strategy to clear aggregate proteins may be attempted by improving lysosomal activity and thereby increasing (or restoring) the autophagic flux. Transcription Factor EB (TFEB) is considered a 'master regulator of autophagy' and drives the expression of several autophagy and lysosomal genes including p62, Atg9b, LC3B, Wipi1, and Lamp1 [271]. Gene therapy with TFEB in mouse model of hepatic disease improved clearance of protein aggregation and rescued alpha-1-anti-trypsin deficiency [272]. High efficiency gene transfers have been achieved to specific retinal layers previously with different adeno-associated virus (AAV) serotypes. TFEB gene transfer may dramatically improve lysosomal biogenesis and overall autophagic flux in the RPE and may be of particular importance in AMD therapeutic strategies.

5. Summary and future directions

The autophagic machinery consists of a fine-tuned complex network of genes whose mysteries are still being unraveled by researchers. Autophagy research in the eye so far has established it as an essential housekeeping pathway indispensable for ocular homeostasis. While therapeutic strategies to regulate autophagy in ocular diseases are still in rudimentary stages, promising results from initial trials have raised hope of autophagic modulation moving gradually from bench to clinic. The challenge lies in modulation of autophagy to the levels required in the particular disease scenario, that is do we want cell death in malignant conditions or just to restore autophagy to levels where it can not only clear cellular waste but also effectively reverse inflammation and contain cell death signals.

Author details

S.K. Mitter and M.E. Boulton*

*Address all correspondence to: mboulton@iupui.edu

Department of Ophthalmology, Indiana University, Indianapolis, IN, USA

References

[1] Klionsky DJ. Autophagy: from phenomenology to molecular understanding in less than a decade. Nat Rev Mol Cell Biol. 2007;8(11):931–937. doi:10.1038/nrm2245.

[2] Li YJ, Jiang Q, Cao GF, Yao J, Yan B. Repertoires of autophagy in the pathogenesis of ocular diseases. Cell Physiol Biochem. 2015;35(5):1663–1676. doi:10.1159/000373980.

[3] Frost LS, Mitchell CH, Boesze-Battaglia K. Autophagy in the eye: implications for ocular cell health. Exp Eye Res. 2014;124:56–66. doi:10.1016/j.exer.2014.04.010.

[4] McCaa CS. The eye and visual nervous system: anatomy, physiology and toxicology. Environ Health Perspect. 1982;44:1–8.

[5] Rehen SK, Neves DD, Fragel-Madeira L, Britto LR, Linden R. Selective sensitivity of early postmitotic retinal cells to apoptosis induced by inhibition of protein synthesis. Eur J Neurosci. 1999;11(12):4349–4356.

[6] Strauss O. The Retinal Pigment Epithelium. In: Kolb H, Fernandez E, Nelson R, editors. Webvision: The Organization of the Retina and Visual System. Salt Lake City (UT)1995.

[7] Strauss O. [The role of retinal pigment epithelium in visual functions]. Ophthalmologe. 2009;106(4):299–304. doi:10.1007/s00347-008-1869-x.

[8] Slingsby C, Wistow GJ. Functions of crystallins in and out of lens: roles in elongated and post-mitotic cells. Prog Biophys Mol Biol. 2014;115(1):52–67. doi:10.1016/j.pbiomolbio.2014.02.006.

[9] Espana EM, Sun M, Birk DE. Existence of Corneal Endothelial Slow-Cycling Cells. Invest Ophthalmol Vis Sci. 2015;56(6):3827–3837. doi:10.1167/iovs.14-16030.

[10] Chen Q, Hung FC, Fromm L, Overbeek PA. Induction of cell cycle entry and cell death in postmitotic lens fiber cells by overexpression of E2F1 or E2F2. Invest Ophthalmol Vis Sci. 2000;41(13):4223–4231.

[11] Williamson TH, Harris A. Ocular blood flow measurement. Br J Ophthalmol. 1994;78(12):939–945.

[12] Kaarniranta K, Sinha D, Blasiak J, Kauppinen A, Vereb Z, Salminen A, et al. Autophagy and heterophagy dysregulation leads to retinal pigment epithelium dysfunction and

development of age-related macular degeneration. Autophagy. 2013;9(7):973–984. doi: 10.4161/auto.24546.

[13] Mitter SK, Song C, Qi X, Mao H, Rao H, Akin D, et al. Dysregulated autophagy in the RPE is associated with increased susceptibility to oxidative stress and AMD. Autophagy. 2014;10(11):1989–2005. doi:10.4161/auto.36184.

[14] Viiri J, Amadio M, Marchesi N, Hyttinen JM, Kivinen N, Sironen R, et al. Autophagy activation clears ELAVL1/HuR-mediated accumulation of SQSTM1/p62 during proteasomal inhibition in human retinal pigment epithelial cells. PLoS One. 2013;8(7):e69563. doi:10.1371/journal.pone.0069563.

[15] Reme CE. Autography in visual cells and pigment epithelium. Invest Ophthalmol Vis Sci. 1977;16(9):807–814.

[16] Reme CE, Sulser M. Diurnal variation of autophagy in rod visual cells in the rat. Albrecht Von Graefes Arch Klin Exp Ophthalmol. 1977;203(3–4):261–270.

[17] Reme C, Wirz-Justice A. [Circadian rhythm, the retina and light]. Klin Monbl Augenheilkd. 1985;186(3):175–179.

[18] Reme C, Wirz-Justice A, Rhyner A, Hofmann S. Circadian rhythm in the light response of rat retinal disk-shedding and autophagy. Brain Res. 1986;369(1–2):356–360.

[19] Yao J, Jia L, Shelby SJ, Ganios AM, Feathers K, Thompson DA, et al. Circadian and noncircadian modulation of autophagy in photoreceptors and retinal pigment epithelium. Invest Ophthalmol Vis Sci. 2014;55(5):3237–3246. doi:10.1167/iovs.13–13336.

[20] Trachoma. World's leading preventable cause of blindness. Glob Child Health New Rev. 1995;3(1):23.

[21] Shintani T, Klionsky DJ. Autophagy in health and disease: a double-edged sword. Science. 2004;306(5698):990–995. doi:10.1126/science.1099993.

[22] Thorburn A. Autophagy and its effects: making sense of double-edged swords. PLoS Biol. 2014;12(10):e1001967. doi:10.1371/journal.pbio.1001967.

[23] Feng Y, He D, Yao Z, Klionsky DJ. The machinery of macroautophagy. Cell Res. 2014;24(1):24–41. doi:10.1038/cr.2013.168.

[24] Takagi A, Kume S, Kondo M, Nakazawa J, Chin-Kanasaki M, Araki H, et al. Mammalian autophagy is essential for hepatic and renal ketogenesis during starvation. Sci Rep. 2016;6:18944. doi:10.1038/srep18944.

[25] Rodriguez-Muela N, Koga H, Garcia-Ledo L, de la Villa P, de la Rosa EJ, Cuervo AM, et al. Balance between autophagic pathways preserves retinal homeostasis. Aging Cell. 2013;12(3):478–488. doi:10.1111/acel.12072.

[26] Maiuri MC, Zalckvar E, Kimchi A, Kroemer G. Self-eating and self-killing: crosstalk between autophagy and apoptosis. Nat Rev Mol Cell Biol. 2007;8(9):741–752. doi: 10.1038/nrm2239.

[27] Fuhrmann S. Eye morphogenesis and patterning of the optic vesicle. Curr Top Dev Biol. 2010;93:61–84. doi:10.1016/B978-0-12-385044-7.00003-5.

[28] Svoboda KK, O'Shea KS. An analysis of cell shape and the neuroepithelial basal lamina during optic vesicle formation in the mouse embryo. Development. 1987;100(2):185–200.

[29] Garcia CM, Huang J, Madakashira BP, Liu Y, Rajagopal R, Dattilo L, et al. The function of FGF signaling in the lens placode. Dev Biol. 2011;351(1):176–185. doi:10.1016/j.ydbio.2011.01.001.

[30] Wride MA. Lens fibre cell differentiation and organelle loss: many paths lead to clarity. Philos Trans R Soc Lond B Biol Sci. 2011;366(1568):1219–1233. doi:10.1098/rstb.2010.0324.

[31] Bassnett S. On the mechanism of organelle degradation in the vertebrate lens. Exp Eye Res. 2009;88(2):133–139. doi:10.1016/j.exer.2008.08.017.

[32] Saint-Geniez M, D'Amore PA. Development and pathology of the hyaloid, choroidal and retinal vasculature. Int J Dev Biol. 2004;48(8–9):1045–1058. doi:10.1387/ijdb.041895ms.

[33] Graw J. Eye development. Curr Top Dev Biol. 2010;90:343–386. doi:10.1016/S0070-2153(10)90010-0.

[34] Banks MS, Bennett PJ. Optical and photoreceptor immaturities limit the spatial and chromatic vision of human neonates. J Opt Soc Am A. 1988;5(12):2059–2079.

[35] Yuodelis C, Hendrickson A. A qualitative and quantitative analysis of the human fovea during development. Vision Res. 1986;26(6):847–855.

[36] Braunger BM, Demmer C, Tamm ER. Programmed cell death during retinal development of the mouse eye. Adv Exp Med Biol. 2014;801:9–13. doi: 10.1007/978-1-4614-3209-8_2.

[37] Mellen MA, de la Rosa EJ, Boya P. Autophagy is not universally required for phosphatidyl-serine exposure and apoptotic cell engulfment during neural development. Autophagy. 2009;5(7):964–972.

[38] Mellen MA, de la Rosa EJ, Boya P. The autophagic machinery is necessary for removal of cell corpses from the developing retinal neuroepithelium. Cell Death Differ. 2008;15(8):1279–1290. doi:10.1038/cdd.2008.40.

[39] Chang TK, Shravage BV, Hayes SD, Powers CM, Simin RT, Wade Harper J, et al. Uba1 functions in Atg7- and Atg3-independent autophagy. Nat Cell Biol. 2013;15(9):1067–1078. doi:10.1038/ncb2804.

[40] Hosokawa N, Hara Y, Mizushima N. Generation of cell lines with tetracycline-regulated autophagy and a role for autophagy in controlling cell size. FEBS Lett. 2007;581(15): 2623–2629. doi:10.1016/j.febslet.2007.05.061.

[41] Wang RC, Levine B. Autophagy in cellular growth control. FEBS Lett. 2010;584(7):1417–1426. doi:10.1016/j.febslet.2010.01.009.

[42] Liang N, Zhang C, Dill P, Panasyuk G, Pion D, Koka V, et al. Regulation of YAP by mTOR and autophagy reveals a therapeutic target of tuberous sclerosis complex. J Exp Med. 2014;211(11):2249–2263. doi:10.1084/jem.20140341.

[43] Matsui M, Yamamoto A, Kuma A, Ohsumi Y, Mizushima N. Organelle degradation during the lens and erythroid differentiation is independent of autophagy. Biochem Biophys Res Commun. 2006;339(2):485–489. doi:10.1016/j.bbrc.2005.11.044.

[44] Garcia-Porrero JA, Colvee E, Ojeda JL. The mechanisms of cell death and phagocytosis in the early chick lens morphogenesis: a scanning electron microscopy and cytochemical approach. Anat Rec. 1984;208(1):123–136. doi:10.1002/ar.1092080113.

[45] Morishita H, Mizushima N. Autophagy in the lens. Exp Eye Res. 2016;144:22–28. doi:10.1016/j.exer.2015.08.019.

[46] Morishita H, Eguchi S, Kimura H, Sasaki J, Sakamaki Y, Robinson ML, et al. Deletion of autophagy-related 5 (Atg5) and Pik3c3 genes in the lens causes cataract independent of programmed organelle degradation. J Biol Chem. 2013;288(16):11436–11447. doi:10.1074/jbc.M112.437103.

[47] Costello MJ, Brennan LA, Basu S, Chauss D, Mohamed A, Gilliland KO, et al. Autophagy and mitophagy participate in ocular lens organelle degradation. Exp Eye Res. 2013;116:141–150. doi:10.1016/j.exer.2013.08.017.

[48] Chen J, Ma Z, Jiao X, Fariss R, Kantorow WL, Kantorow M, et al. Mutations in FYCO1 cause autosomal-recessive congenital cataracts. Am J Hum Genet. 2011;88(6):827–838. doi:10.1016/j.ajhg.2011.05.008.

[49] Nishida Y, Arakawa S, Fujitani K, Yamaguchi H, Mizuta T, Kanaseki T, et al. Discovery of Atg5/Atg7-independent alternative macroautophagy. Nature. 2009;461(7264):654–658. doi:10.1038/nature08455.

[50] Zhou X, Wang L, Hasegawa H, Amin P, Han BX, Kaneko S, et al. Deletion of PIK3C3/Vps34 in sensory neurons causes rapid neurodegeneration by disrupting the endosomal but not the autophagic pathway. Proc Natl Acad Sci U S A. 2010;107(20):9424–9429. doi:10.1073/pnas.0914725107.

[51] Ma T, Li J, Xu Y, Yu C, Xu T, Wang H, et al. Atg5-independent autophagy regulates mitochondrial clearance and is essential for iPSC reprogramming. Nat Cell Biol. 2015;17(11):1379–1387. doi:10.1038/ncb3256.

[52] Perlman I. The Electroretinogram: ERG. In: Kolb H, Fernandez E, Nelson R, editors. Webvision: The Organization of the Retina and Visual System. Salt Lake City (UT), 1995.

[53] Tian JR, Crane BT, Wiest G, Demer JL. Impaired linear vestibulo-ocular reflex initiation and vestibular catch-up saccades in older persons. Ann N Y Acad Sci. 2002;956:574–578.

[54] McKelvie P, Friling R, Davey K, Kowal L. Changes as the result of ageing in extraocular muscles: a post-mortem study. Aust N Z J Ophthalmol. 1999;27(6):420–425.

[55] Muller-Hocker J, Seibel P, Schneiderbanger K, Kadenbach B. Different in situ hybridization patterns of mitochondrial DNA in cytochrome c oxidase-deficient extraocular muscle fibres in the elderly. Virchows Arch A Pathol Anat Histopathol. 1993;422(1):7–15.

[56] McMullen CA, Ferry AL, Gamboa JL, Andrade FH, Dupont-Versteegden EE. Age-related changes of cell death pathways in rat extraocular muscle. Exp Gerontol. 2009;44(6–7):420–425. doi:10.1016/j.exger.2009.03.006.

[57] Tiffany JM, Bron AJ. Role of tears in maintaining corneal integrity. Trans Ophthalmol Soc U K. 1978;98(3):335–338.

[58] Butovich IA. Tear film lipids. Exp Eye Res. 2013;117:4–27. doi:10.1016/j.exer.2013.05.010.

[59] Choi SI, Kim EK. Autophagy in granular corneal dystrophy type 2. Exp Eye Res. 2016;144:14–21. doi:10.1016/j.exer.2015.09.008.

[60] McDermott AM. Antimicrobial compounds in tears. Exp Eye Res. 2013;117:53–61. doi: 10.1016/j.exer.2013.07.014.

[61] Johansson ME, Hansson GC. Mucus and the goblet cell. Dig Dis. 2013;31(3–4):305–309. doi:10.1159/000354683.

[62] Jordan MA, Thrower D, Wilson L. Effects of vinblastine, podophyllotoxin and nocodazole on mitotic spindles. Implications for the role of microtubule dynamics in mitosis. J Cell Sci. 1992;102(Pt 3):401–416.

[63] Kelly RB, Oliver C, Hand AR. The effects of vinblastine on acinar cells of the exorbital lacrimal gland of the rat. Cell Tissue Res. 1978;195(2):227–237.

[64] Zoukhri D, Fix A, Alroy J, Kublin CL. Mechanisms of murine lacrimal gland repair after experimentally induced inflammation. Invest Ophthalmol Vis Sci. 2008;49(10):4399–4406. doi:10.1167/iovs.08-1730.

[65] Padeletti L. Radioimmunoassay of digoxin: aspecific interferences of human plasma. J Nucl Biol Med. 1975;19(1):22–28.

[66] Fujii A, Morimoto-Tochigi A, Walkup RD, Shearer TR, Azuma M. Lacritin-induced secretion of tear proteins from cultured monkey lacrimal acinar cells. Invest Ophthalmol Vis Sci. 2013;54(4):2533–2540. doi:10.1167/iovs.12-10394.

[67] Karnati R, Talla V, Peterson K, Laurie GW. Lacritin and other autophagy associated proteins in ocular surface health. Exp Eye Res. 2016;144:4–13. doi:10.1016/j.exer.2015.08.015.

[68] Wang N, Zimmerman K, Raab RW, McKown RL, Hutnik CM, Talla V, et al. Lacritin rescues stressed epithelia via rapid forkhead box O3 (FOXO3)-associated autophagy that restores metabolism. J Biol Chem. 2013;288(25):18146–18161. doi:10.1074/jbc.M112.436584.

[69] Latkovic S, Nilsson SE. The ultrastructure of the normal conjunctival epithelium of the guinea pig. I. The basal and intermediate layers of the perilimbal zone. Acta Ophthalmol (Copenh). 1979;57(1):106–122.

[70] Llobet A, Gasull X, Gual A. Understanding trabecular meshwork physiology: a key to the control of intraocular pressure? News Physiol Sci. 2003;18:205–209.

[71] Liton PB, Lin Y, Luna C, Li G, Gonzalez P, Epstein DL. Cultured porcine trabecular meshwork cells display altered lysosomal function when subjected to chronic oxidative stress. Invest Ophthalmol Vis Sci. 2008;49(9):3961–3969. doi:10.1167/iovs.08-1915.

[72] Lin Y, Epstein DL, Liton PB. Intralysosomal iron induces lysosomal membrane permeabilization and cathepsin D-mediated cell death in trabecular meshwork cells exposed to oxidative stress. Invest Ophthalmol Vis Sci. 2010;51(12):6483–6495. doi:10.1167/iovs.10-5410.

[73] Porter K, Nallathambi J, Lin Y, Liton PB. Lysosomal basification and decreased autophagic flux in oxidatively stressed trabecular meshwork cells: implications for glaucoma pathogenesis. Autophagy. 2013;9(4):581–594. doi:10.4161/auto.23568.

[74] Pulliero A, Seydel A, Camoirano A, Sacca SC, Sandri M, Izzotti A. Oxidative damage and autophagy in the human trabecular meshwork as related with ageing. PLoS One. 2014;9(6):e98106. doi:10.1371/journal.pone.0098106.

[75] Rezai KA, Lai WW, Farrokh-Siar L, Pearlman J, Shu J, Patel SC, et al. A new method of culturing and transferring iris pigment epithelium. Invest Ophthalmol Vis Sci. 1997;38(11):2255–2260.

[76] Yamada T, Dumont JN, Moret R, Brun JP. Autophagy in dedifferentiating newt iris epithelial cells in vitro. Differentiation. 1978;11(3):133–147.

[77] Rezai KA, Lappas A, Farrokh-siar L, Kohen L, Wiedemann P, Heimann K. Iris pigment epithelial cells of long evans rats demonstrate phagocytic activity. Exp Eye Res. 1997;65(1):23–29. doi:10.1006/exer.1997.0307.

[78] Khalil AK, Kubota T, Tawara A, Inomata H. Ultrastructural age-related changes on the posterior iris surface. A possible relationship to the pathogenesis of exfoliation. Arch Ophthalmol. 1996;114(6):721–725.

[79] Petrovsky G, Albert R, Kaarniranta K, Morte M, et al. Autophagy in the eye: a double-edged sword. Autophagy: Principles, Regulation and Roles in Disease. Gorbunov, Nikolai (Editor) Hauppauge, N.Y.: Nova Science Publishers; 2012.41 157-180.

[80] Dalvin LA, Fautsch MP. Analysis of circadian rhythm gene expression with reference to diurnal pattern of intraocular pressure in mice. Invest Ophthalmol Vis Sci. 2015;56(4): 2657–2663. doi:10.1167/iovs.15-16449.

[81] Wiebking N, Maronde E, Rami A. Increased neuronal injury in clock gene Per-1 deficient-mice after cerebral ischemia. Curr Neurovasc Res. 2013;10(2):112–125.

[82] Brennan LA, Kantorow WL, Chauss D, McGreal R, He S, Mattucci L, et al. Spatial expression patterns of autophagy genes in the eye lens and induction of autophagy in lens cells. Mol Vis. 2012;18:1773–1786.

[83] Hawse JR, DeAmicis-Tress C, Cowell TL, Kantorow M. Identification of global gene expression differences between human lens epithelial and cortical fiber cells reveals specific genes and their associated pathways important for specialized lens cell functions. Mol Vis. 2005;11:274–283.

[84] Andley UP, Goldman JW. Autophagy and UPR in alpha-crystallin mutant knock-in mouse models of hereditary cataracts. Biochim Biophys Acta. 2016;1860(1 Pt B):234–239. doi:10.1016/j.bbagen.2015.06.001.

[85] Andley UP, Hamilton PD, Ravi N, Weihl CC. A knock-in mouse model for the R120G mutation of alphaB-crystallin recapitulates human hereditary myopathy and cataracts. PLoS One. 2011;6(3):e17671. doi:10.1371/journal.pone.0017671.

[86] Alamouti B, Funk J. Retinal thickness decreases with age: an OCT study. Br J Ophthalmol. 2003;87(7):899–901.

[87] Eriksson U, Alm A. Macular thickness decreases with age in normal eyes: a study on the macular thickness map protocol in the Stratus OCT. Br J Ophthalmol. 2009;93(11): 1448–1452. doi:10.1136/bjo.2007.131094.

[88] Sehi M, Grewal DS, Feuer WJ, Greenfield DS. The impact of intraocular pressure reduction on retinal ganglion cell function measured using pattern electroretinogram in eyes receiving latanoprost 0.005% versus placebo. Vis Res. 2011;51(2):235–242. doi: 10.1016/j.visres.2010.08.036.

[89] Harman A, Abrahams B, Moore S, Hoskins R. Neuronal density in the human retinal ganglion cell layer from 16–77 years. Anat Rec. 2000;260(2):124–131.

[90] Cavallotti C, Artico M, Pescosolido N, Leali FM, Feher J. Age-related changes in the human retina. Can J Ophthalmol. 2004;39(1):61–68.

[91] Parikh RS, Parikh SR, Sekhar GC, Prabakaran S, Babu JG, Thomas R. Normal age-related decay of retinal nerve fiber layer thickness. Ophthalmology. 2007;114(5):921–926. doi: 10.1016/j.ophtha.2007.01.023.

[92] Dolman CL, McCormick AQ, Drance SM. Aging of the optic nerve. Arch Ophthalmol. 1980;98(11):2053–2058.

[93] Garway-Heath DF, Wollstein G, Hitchings RA. Aging changes of the optic nerve head in relation to open angle glaucoma. Br J Ophthalmol. 1997;81(10):840–845.

[94] Rakic P, Riley KP. Overproduction and elimination of retinal axons in the fetal rhesus monkey. Science. 1983;219(4591):1441–1444.

[95] Insausti R, Blakemore C, Cowan WM. Ganglion cell death during development of ipsilateral retino-collicular projection in golden hamster. Nature. 1984;308(5957):362–365.

[96] Provis JM, Penfold PL. Cell death and the elimination of retinal axons during development. Prog Neurobiol. 1988;31(4):331–347.

[97] Rodriguez-Muela N, Germain F, Marino G, Fitze PS, Boya P. Autophagy promotes survival of retinal ganglion cells after optic nerve axotomy in mice. Cell Death Differ. 2012;19(1):162–169. doi:10.1038/cdd.2011.88.

[98] Kim SH, Munemasa Y, Kwong JM, Ahn JH, Mareninov S, Gordon LK, et al. Activation of autophagy in retinal ganglion cells. J Neurosci Res. 2008;86(13):2943–2951. doi: 10.1002/jnr.21738.

[99] Munemasa Y, Kitaoka Y. Autophagy in axonal degeneration in glaucomatous optic neuropathy. Prog Retin Eye Res. 2015;47:1–18. doi:10.1016/j.preteyeres.2015.03.002.

[100] Panda-Jonas S, Jonas JB, Jakobczyk-Zmija M. Retinal photoreceptor density decreases with age. Ophthalmology. 1995;102(12):1853–1859.

[101] Gao H, Hollyfield JG. Aging of the human retina. Differential loss of neurons and retinal pigment epithelial cells. Invest Ophthalmol Vis Sci. 1992;33(1):1–17.

[102] Curcio CA, Medeiros NE, Millican CL. Photoreceptor loss in age-related macular degeneration. Invest Ophthalmol Vis Sci. 1996;37(7):1236–1249.

[103] Mitter SK, Rao HV, Qi X, Cai J, Sugrue A, Dunn WA, Jr., et al. Autophagy in the retina: a potential role in age-related macular degeneration. Adv Exp Med Biol. 2012;723:83–90. doi:10.1007/978-1-4614-0631-0_12.

[104] Chen Y, Sawada O, Kohno H, Le YZ, Subauste C, Maeda T, et al. Autophagy protects the retina from light-induced degeneration. J Biol Chem. 2013;288(11):7506–7518. doi: 10.1074/jbc.M112.439935.

[105] Boulton M, Dayhaw-Barker P. The role of the retinal pigment epithelium: topographical variation and ageing changes. Eye (Lond). 2001;15(Pt 3):384–389. doi:10.1038/eye.2001.141.

[106] Bonilha VL. Age and disease-related structural changes in the retinal pigment epithelium. Clin Ophthalmol. 2008;2(2):413–424.

[107] Del Priore LV, Kuo YH, Tezel TH. Age-related changes in human RPE cell density and apoptosis proportion in situ. Invest Ophthalmol Vis Sci. 2002;43(10):3312–3318.

[108] Feher J, Kovacs I, Artico M, Cavallotti C, Papale A, Balacco Gabrieli C. Mitochondrial alterations of retinal pigment epithelium in age-related macular degeneration. Neurobiol Aging. 2006;27(7):983–993. doi:10.1016/j.neurobiolaging.2005.05.012.

[109] Yao J, Jia L, Khan N, Lin C, Mitter SK, Boulton ME, et al. Deletion of autophagy inducer RB1CC1 results in degeneration of the retinal pigment epithelium. Autophagy. 2015;11(6):939–953. doi:10.1080/15548627.2015.1041699.

[110] Reme CE, Wolfrum U, Imsand C, Hafezi F, Williams TP. Photoreceptor autophagy: effects of light history on number and opsin content of degradative vacuoles. Invest Ophthalmol Vis Sci. 1999;40(10):2398–2404.

[111] Frost LS, Lopes VS, Bragin A, Reyes-Reveles J, Brancato J, Cohen A, et al. The contribution of melanoregulin to microtubule-associated protein 1 light chain 3 (LC3) associated phagocytosis in retinal pigment epithelium. Mol Neurobiol. 2015;52(3): 1135–1151. doi:10.1007/s12035-014-8920-5.

[112] Kim JY, Zhao H, Martinez J, Doggett TA, Kolesnikov AV, Tang PH, et al. Noncanonical autophagy promotes the visual cycle. Cell. 2013;154(2):365–376. doi:10.1016/j.cell. 2013.06.012.

[113] Ohbayashi N, Maruta Y, Ishida M, Fukuda M. Melanoregulin regulates retrograde melanosome transport through interaction with the RILP–p150Glued complex in melanocytes. J Cell Sci 2012 125: 1508–1518; doi:10.1242/jcs.094185

[114] Damek-Poprawa M, Diemer T, Lopes V, Lillo C, Harper D, Marks M, Wu Y, Sparrow J, Rachel R, Williams D, Boesze-Battaglia K. Melanoregulin (MREG) Modulates lysosome function in pigment epithelial cells. J. Biol. Chem. 2009 284: 10877-. doi:10.1074/ jbc.M808857200

[115] Martinez J, Malireddi S, Qun L, Cunha LD, Pelletier S, Gingras S, Orchard R el al. Molecular characterization of LC3-associated phagocytosis reveals distinct roles for Rubicon, NOX2 and autophagy proteins. Nat Cell Biol 2015 17, 893–906 doi:10.1038/ ncb3192.

[116] Bharadwaj AS, Appukuttan B, Wilmarth PA, Pan Y, Stempel AJ, Chipps TJ, et al. Role of the retinal vascular endothelial cell in ocular disease. Prog Retin Eye Res. 2013;32:102–180. doi:10.1016/j.preteyeres.2012.08.004.

[117] Torisu K, Singh KK, Torisu T, Lovren F, Liu J, Pan Y, et al. Intact endothelial autophagy is required to maintain vascular lipid homeostasis. Aging Cell. 2016;15(1):187–191. doi: 10.1111/acel.12423.

[118] Lee SJ, Kim HP, Jin Y, Choi AM, Ryter SW. Beclin 1 deficiency is associated with increased hypoxia-induced angiogenesis. Autophagy. 2011;7(8):829–839.

[119] Du J, Teng RJ, Guan T, Eis A, Kaul S, Konduri GG, et al. Role of autophagy in angiogenesis in aortic endothelial cells. Am J Physiol Cell Physiol. 2012;302(2):C383–391. doi: 10.1152/ajpcell.00164.2011.

[120] Friedlander M, Theesfeld CL, Sugita M, Fruttiger M, Thomas MA, Chang S, et al. Involvement of integrins alpha v beta 3 and alpha v beta 5 in ocular neovascular diseases. Proc Natl Acad Sci U S A. 1996;93(18):9764–9769.

[121] Brooks PC, Clark RA, Cheresh DA. Requirement of vascular integrin alpha v beta 3 for angiogenesis. Science. 1994;264(5158):569–571.

[122] Hood JD, Frausto R, Kiosses WB, Schwartz MA, Cheresh DA. Differential alphav integrin-mediated Ras-ERK signaling during two pathways of angiogenesis. J Cell Biol. 2003;162(5):933–943. doi:10.1083/jcb.200304105.

[123] Tuloup-Minguez V, Hamai A, Greffard A, Nicolas V, Codogno P, Botti J. Autophagy modulates cell migration and beta1 integrin membrane recycling. Cell Cycle. 2013;12(20):3317–3328. doi:10.4161/cc.26298.

[124] Kenific CM, Stehbens SJ, Goldsmith J, Leidal AM, Faure N, Ye J, et al. NBR1 enables autophagy-dependent focal adhesion turnover. J Cell Biol. 2016;212(5):577–590. doi: 10.1083/jcb.201503075.

[125] Zhang J, Bai Y, Huang L, Qi Y, Zhang Q, Li S, et al. Protective effect of autophagy on human retinal pigment epithelial cells against lipofuscin fluorophore A2E: implications for age-related macular degeneration. Cell Death Dis. 2015;6:e1972. doi:10.1038/cddis.2015.330.

[126] Huang J, Xie Y, Sun X, Zeh HJ, 3rd, Kang R, Lotze MT, et al. DAMPs, ageing, and cancer: the 'DAMP Hypothesis'. Ageing Res Rev. 2015;24(Pt A):3–16. doi:10.1016/j.arr.2014.10.004.

[127] Mendes-Jorge L, Ramos D, Luppo M, Llombart C, Alexandre-Pires G, Nacher V, et al. Scavenger function of resident autofluorescent perivascular macrophages and their contribution to the maintenance of the blood-retinal barrier. Invest Ophthalmol Vis Sci. 2009;50(12):5997–6005. doi:10.1167/iovs.09-3515.

[128] Xu H, Dawson R, Forrester JV, Liversidge J. Identification of novel dendritic cell populations in normal mouse retina. Invest Ophthalmol Vis Sci. 2007;48(4):1701–1710. doi:10.1167/iovs.06-0697.

[129] Fukuda S, Nagano M, Yamashita T, Kimura K, Tsuboi I, Salazar G, et al. Functional endothelial progenitor cells selectively recruit neurovascular protective monocyte-derived F4/80(+) /Ly6c(+) macrophages in a mouse model of retinal degeneration. Stem Cells. 2013;31(10):2149–2161. doi:10.1002/stem.1469.

[130] Kezic J, McMenamin PG. Differential turnover rates of monocyte-derived cells in varied ocular tissue microenvironments. J Leukoc Biol. 2008;84(3):721–729. doi:10.1189/jlb.0308166.

[131] Medzhitov R. Origin and physiological roles of inflammation. Nature. 2008;454(7203): 428–435. doi:10.1038/nature07201.

[132] Juel HB, Faber C, Svendsen SG, Vallejo AN, Nissen MH. Inflammatory cytokines protect retinal pigment epithelial cells from oxidative stress-induced death. PLoS One. 2013;8(5):e64619. doi:10.1371/journal.pone.0064619.

[133] Chen M, Forrester JV, Xu H. Dysregulation in retinal para-inflammation and age-related retinal degeneration in CCL2 or CCR2 deficient mice. PLoS One. 2011;6(8):e22818. doi:10.1371/journal.pone.0022818.

[134] Xu H, Chen M, Forrester JV. Para-inflammation in the aging retina. Prog Retin Eye Res. 2009;28(5):348–368. doi:10.1016/j.preteyeres.2009.06.001.

[135] Ryan SJ. Retina. 5th ed. London: Saunders/Elsevier; 2013.

[136] Murakami Y, Matsumoto H, Roh M, Giani A, Kataoka K, Morizane Y, et al. Programmed necrosis, not apoptosis, is a key mediator of cell loss and DAMP-mediated inflammation in dsRNA-induced retinal degeneration. Cell Death Differ. 2014;21(2):270–277. doi: 10.1038/cdd.2013.109.

[137] Fesus L, Demeny MA, Petrovski G. Autophagy shapes inflammation. Antioxid Redox Signal. 2011;14(11):2233–2243. doi:10.1089/ars.2010.3485.

[138] Shi CS, Shenderov K, Huang NN, Kabat J, Abu-Asab M, Fitzgerald KA, et al. Activation of autophagy by inflammatory signals limits IL-1beta production by targeting ubiquitinated inflammasomes for destruction. Nat Immunol. 2012;13(3):255–263. doi:10.1038/ni.2215.

[139] Liu J, Copland DA, Theodoropoulou S, Chiu HA, Barba MD, Mak KW, et al. Impairing autophagy in retinal pigment epithelium leads to inflammasome activation and enhanced macrophage-mediated angiogenesis. Sci Rep. 2016;6:20639. doi:10.1038/srep20639.

[140] Kauppinen A, Niskanen H, Suuronen T, Kinnunen K, Salminen A, Kaarniranta K. Oxidative stress activates NLRP3 inflammasomes in ARPE-19 cells--implications for age-related macular degeneration (AMD). Immunol Lett. 2012;147(1–2):29–33. doi: 10.1016/j.imlet.2012.05.005.

[141] Piippo N, Korkmaz A, Hytti M, Kinnunen K, Salminen A, Atalay M, et al. Decline in cellular clearance systems induces inflammasome signaling in human ARPE-19 cells. Biochim Biophys Acta. 2014;1843(12):3038–3046. doi:10.1016/j.bbamcr.2014.09.015.

[142] Qian Q, Mitter SK, Pay SL, Qi X, Rickman CB, Grant MB, et al. A non-canonical role for beta-secretase in the retina. Adv Exp Med Biol. 2016;854:333–339. doi: 10.1007/978-3-319-17121-0_44.

[143] Sparrow JR, Boulton M. RPE lipofuscin and its role in retinal pathobiology. Exp Eye Res. 2005;80(5):595–606. doi:10.1016/j.exer.2005.01.007.

[144] Tasdemir E, Chiara Maiuri M, Morselli E, Criollo A, D'Amelio M, Djavaheri-Mergny M, et al. A dual role of p53 in the control of autophagy. Autophagy. 2008;4(6):810–814.

[145] Colunga A, Bollino D, Schech A, Aurelian L. Calpain-dependent clearance of the autophagy protein p62/SQSTM1 is a contributor to DeltaPK oncolytic activity in melanoma. Gene Ther. 2014;21(4):371–378. doi:10.1038/gt.2014.6.

[146] Norman JM, Cohen GM, Bampton ET. The in vitro cleavage of the hAtg proteins by cell death proteases. Autophagy. 2010;6(8):1042–1056.

[147] Wirawan E, Vande Walle L, Kersse K, Cornelis S, Claerhout S, Vanoverberghe I, et al. Caspase-mediated cleavage of Beclin-1 inactivates Beclin-1-induced autophagy and enhances apoptosis by promoting the release of proapoptotic factors from mitochondria. Cell Death Dis. 2010;1:e18. doi:10.1038/cddis.2009.16.

[148] Oral O, Oz-Arslan D, Itah Z, Naghavi A, Deveci R, Karacali S, et al. Cleavage of Atg3 protein by caspase-8 regulates autophagy during receptor-activated cell death. Apoptosis. 2012;17(8):810–820. doi:10.1007/s10495-012-0735-0.

[149] Degenhardt K, Mathew R, Beaudoin B, Bray K, Anderson D, Chen G, et al. Autophagy promotes tumor cell survival and restricts necrosis, inflammation, and tumorigenesis. Cancer Cell. 2006;10(1):51–64. doi:10.1016/j.ccr.2006.06.001.

[150] Yoon YH, Cho KS, Hwang JJ, Lee SJ, Choi JA, Koh JY. Induction of lysosomal dilatation, arrested autophagy, and cell death by chloroquine in cultured ARPE-19 cells. Invest Ophthalmol Vis Sci. 2010;51(11):6030–6037. doi:10.1167/iovs.10-5278.

[151] Gregory-Evans CY, Williams MJ, Halford S, Gregory-Evans K. Ocular coloboma: a reassessment in the age of molecular neuroscience. J Med Genet. 2004;41(12):881–891. doi:10.1136/jmg.2004.025494.

[152] Bhandari R, Ferri S, Whittaker B, Liu M, Lazzaro DR. Peters anomaly: review of the literature. Cornea. 2011;30(8):939–944. doi:10.1097/ICO.0b013e31820156a9.

[153] Moore DB, Tomkins O, Ben-Zion I. A review of primary congenital glaucoma in the developing world. Surv Ophthalmol. 2013;58(3):278–285. doi:10.1016/j.survophthal.2012.11.003.

[154] Yorston D, Yang YF, Sullivan PM. Retinal detachment following surgery for congenital cataract: presentation and outcomes. Eye (Lond). 2005;19(3):317–321. doi:10.1038/sj.eye.6701463.

[155] Sheth JU, Sharma A, Chakraborty S. Persistent hyaloid artery with an aberrant peripheral retinal attachment: A unique presentation. Oman J Ophthalmol. 2013;6(1):58–60. doi:10.4103/0974-620X.111924.

[156] Weleber RG, Francis PJ, Trzupek KM, Beattie C. Leber Congenital Amaurosis. In: Pagon RA, Adam MP, Ardinger HH, Wallace SE, Amemiya A, Bean LJH, et al., editors. GeneReviews(R). Seattle (WA), 1993.

[157] Ebrahimi-Fakhari D, Saffari A, Wahlster L, Lu J, Byrne S, Hoffmann GF, et al. Congenital disorders of autophagy: an emerging novel class of inborn errors of neuro-metabolism. Brain. 2016;139(Pt 2):317–337. doi:10.1093/brain/awv371.

[158] Goldberg MF. Persistent fetal vasculature (PFV): an integrated interpretation of signs and symptoms associated with persistent hyperplastic primary vitreous (PHPV). LIV Edward Jackson Memorial Lecture. Am J Ophthalmol. 1997;124(5):587–626.

[159] Kim JH, Kim JH, Yu YS, Mun JY, Kim KW. Autophagy-induced regression of hyaloid vessels in early ocular development. Autophagy. 2010;6(7):922–928. doi:10.4161/auto.6.8.13306.

[160] Cullup T, Kho AL, Dionisi-Vici C, Brandmeier B, Smith F, Urry Z, et al. Recessive mutations in EPG5 cause Vici syndrome, a multisystem disorder with defective autophagy. Nat Genet. 2013;45(1):83–87. doi:10.1038/ng.2497.

[161] Rathore GS, Schaaf CP, Stocco AJ. Novel mutation of the WDR45 gene causing beta-propeller protein-associated neurodegeneration. Mov Disord. 2014;29(4):574–575. doi:10.1002/mds.25868.

[162] Haack TB, Hogarth P, Kruer MC, Gregory A, Wieland T, Schwarzmayr T, et al. Exome sequencing reveals de novo WDR45 mutations causing a phenotypically distinct, X-linked dominant form of NBIA. Am J Hum Genet. 2012;91(6):1144–1149. doi:10.1016/j.ajhg.2012.10.019.

[163] Velikkakath AK, Nishimura T, Oita E, Ishihara N, Mizushima N. Mammalian Atg2 proteins are essential for autophagosome formation and important for regulation of size and distribution of lipid droplets. Mol Biol Cell. 2012;23(5):896–909. doi:10.1091/mbc.E11-09-0785.

[164] Suzuki K, Akioka M, Kondo-Kakuta C, Yamamoto H, Ohsumi Y. Fine mapping of autophagy-related proteins during autophagosome formation in *Saccharomyces cerevisiae*. J Cell Sci. 2013;126(Pt 11):2534–2544. doi:10.1242/jcs.122960.

[165] Lu Q, Yang P, Huang X, Hu W, Guo B, Wu F, et al. The WD40 repeat PtdIns(3)P-binding protein EPG-6 regulates progression of omegasomes to autophagosomes. Dev Cell. 2011;21(2):343–357. doi:10.1016/j.devcel.2011.06.024.

[166] Gregory A, Hayflick S. Neurodegeneration with brain iron accumulation disorders overview. In: Pagon RA, Adam MP, Ardinger HH, Wallace SE, Amemiya A, Bean LJH, et al., editors. GeneReviews(R). Seattle (WA), 1993.

[167] Hale AN, Ledbetter DJ, Gawriluk TR, Rucker EB, 3rd. Autophagy: regulation and role in development. Autophagy. 2013;9(7):951–972. doi:10.4161/auto.24273.

[168] Buron N, Micheau O, Cathelin S, Lafontaine PO, Creuzot-Garcher C, Solary E. Differential mechanisms of conjunctival cell death induction by ultraviolet irradiation and benzalkonium chloride. Invest Ophthalmol Vis Sci. 2006;47(10):4221–4230. doi:10.1167/iovs.05-1460.

[169] McGowin CL, Pyles RB. Mucosal treatments for herpes simplex virus: insights on targeted immunoprophylaxis and therapy. Future Microbiol. 2010;5(1):15–22. doi: 10.2217/fmb.09.111.

[170] Liesegang TJ. Herpes simplex virus epidemiology and ocular importance. Cornea. 2001;20(1):1–13.

[171] Leib DA, Alexander DE, Cox D, Yin J, Ferguson TA. Interaction of ICP34.5 with Beclin 1 modulates herpes simplex virus type 1 pathogenesis through control of CD4+ T-cell responses. J Virol. 2009;83(23):12164–12171. doi:10.1128/JVI.01676-09.

[172] Orvedahl A, Alexander D, Talloczy Z, Sun Q, Wei Y, Zhang W, et al. HSV-1 ICP34.5 confers neurovirulence by targeting the Beclin 1 autophagy protein. Cell Host Microbe. 2007;1(1):23–35. doi:10.1016/j.chom.2006.12.001.

[173] Talloczy Z, Virgin HWt, Levine B. PKR-dependent autophagic degradation of herpes simplex virus type 1. Autophagy. 2006;2(1):24–29.

[174] Petrovski G, Pasztor K, Orosz L, Albert R, Mencel E, Moe MC, et al. Herpes simplex virus types 1 and 2 modulate autophagy in SIRC corneal cells. J Biosci. 2014;39(4):683–692.

[175] Yakoub AM, Shukla D. Basal Autophagy Is Required for Herpes simplex Virus-2 Infection. Sci Rep. 2015;5:12985. doi:10.1038/srep12985.

[176] Yakoub AM, Shukla D. Autophagy stimulation abrogates herpes simplex virus-1 infection. Sci Rep. 2015;5:9730. doi:10.1038/srep09730.

[177] Alexander DE, Ward SL, Mizushima N, Levine B, Leib DA. Analysis of the role of autophagy in replication of herpes simplex virus in cell culture. J Virol. 2007;81(22): 12128–12134. doi:10.1128/JVI.01356-07.

[178] Subauste CS. Autophagy in immunity against *Toxoplasma gondii*. Curr Top Microbiol Immunol. 2009;335:251–265. doi:10.1007/978-3-642-00302-8_12.

[179] Andrade RM, Wessendarp M, Gubbels MJ, Striepen B, Subauste CS. CD40 induces macrophage anti-*Toxoplasma gondii* activity by triggering autophagy-dependent fusion of pathogen-containing vacuoles and lysosomes. J Clin Invest. 2006;116(9):2366–2377. doi:10.1172/JCI28796.

[180] Subauste CS, Andrade RM, Wessendarp M. CD40-TRAF6 and autophagy-dependent anti-microbial activity in macrophages. Autophagy. 2007;3(3):245–248.

[181] Settembre C, Fraldi A, Jahreiss L, Spampanato C, Venturi C, Medina D, et al. A block of autophagy in lysosomal storage disorders. Hum Mol Genet. 2008;17(1):119–129. doi: 10.1093/hmg/ddm289.

[182] Fan BJ, Wiggs JL. Glaucoma: genes, phenotypes, and new directions for therapy. J Clin Invest. 2010;120(9):3064–3072. doi:10.1172/JCI43085.

[183] Stamer WD, Acott TS. Current understanding of conventional outflow dysfunction in glaucoma. Curr Opin Ophthalmol. 2012;23(2):135–143. doi:10.1097/ICU.0b013e32834ff23e.

[184] Ray K, Mukhopadhyay A, Acharya M. Recent advances in molecular genetics of glaucoma. Mol Cell Biochem. 2003;253(1–2):223–231.

[185] Porter KM, Jeyabalan N, Liton PB. MTOR-independent induction of autophagy in trabecular meshwork cells subjected to biaxial stretch. Biochim Biophys Acta. 2014;1843(6):1054–1062. doi:10.1016/j.bbamcr.2014.02.010.

[186] Park HY, Kim JH, Park CK. Activation of autophagy induces retinal ganglion cell death in a chronic hypertensive glaucoma model. Cell Death Dis. 2012;3:e290. doi:10.1038/cddis.2012.26.

[187] McElnea EM, Hughes E, McGoldrick A, McCann A, Quill B, Docherty N, et al. Lipofuscin accumulation and autophagy in glaucomatous human lamina cribrosa cells. BMC Ophthalmol. 2014;14:153. doi:10.1186/1471-2415-14-153.

[188] Porter K, Hirt J, Stamer WD, Liton PB. Autophagic dysregulation in glaucomatous trabecular meshwork cells. Biochim Biophys Acta. 2015;1852(3):379–385. doi:10.1016/j.bbadis.2014.11.021.

[189] Sternberg C, Benchimol M, Linden R. Caspase dependence of the death of neonatal retinal ganglion cells induced by axon damage and induction of autophagy as a survival mechanism. Braz J Med Biol Res. 2010;43(10):950–956.

[190] Bourne RR. The optic nerve head in glaucoma. Community Eye Health. 2006;19(59):44–45.

[191] Koch JC, Lingor P. The role of autophagy in axonal degeneration of the optic nerve. Exp Eye Res. 2016;144:81–89. doi:10.1016/j.exer.2015.08.016.

[192] Ying H, Turturro S, Nguyen T, Shen X, Zelkha R, Johnson EC, et al. Induction of autophagy in rats upon overexpression of wild-type and mutant optineurin gene. BMC Cell Biol. 2015;16:14. doi:10.1186/s12860-015-0060-x.

[193] Ma J, Becker C, Reyes C, Underhill DM. Cutting edge: FYCO1 recruitment to dectin-1 phagosomes is accelerated by light chain 3 protein and regulates phagosome maturation and reactive oxygen production. J Immunol. 2014;192(4):1356–1360. doi:10.4049/jimmunol.1302835.

[194] Wignes JA, Goldman JW, Weihl CC, Bartley MG, Andley UP. p62 expression and autophagy in alphaB-crystallin R120G mutant knock-in mouse model of hereditary cataract. Exp Eye Res. 2013;115:263–273. doi:10.1016/j.exer.2013.06.026.

[195] Sagona AP, Nezis IP, Stenmark H. Association of CHMP4B and autophagy with micronuclei: implications for cataract formation. Biomed Res Int. 2014;2014:974393. doi:10.1155/2014/974393.

[196] Shiels A, Bennett TM, Knopf HL, Yamada K, Yoshiura K, Niikawa N, et al. CHMP4B, a novel gene for autosomal dominant cataracts linked to chromosome 20q. Am J Hum Genet. 2007;81(3):596–606. doi:10.1086/519980.

[197] Wang AL, Lukas TJ, Yuan M, Du N, Tso MO, Neufeld AH. Autophagy and exosomes in the aged retinal pigment epithelium: possible relevance to drusen formation and age-related macular degeneration. PLoS One. 2009;4(1):e4160. doi:10.1371/journal.pone.0004160.

[198] Zhao Z, Chen Y, Wang J, Sternberg P, Freeman ML, Grossniklaus HE, et al. Age-related retinopathy in NRF2-deficient mice. PLoS One. 2011;6(4):e19456. doi:10.1371/journal.pone.0019456.

[199] Swaroop A, Chew EY, Rickman CB, Abecasis GR. Unraveling a multifactorial late-onset disease: from genetic susceptibility to disease mechanisms for age-related macular degeneration. Annu Rev Genomics Hum Genet. 2009;10:19–43. doi:10.1146/annurev.genom.9.081307.164350.

[200] Winkler BS, Boulton ME, Gottsch JD, Sternberg P. Oxidative damage and age-related macular degeneration. Mol Vis. 1999;5:32.

[201] Beatty S, Koh H, Phil M, Henson D, Boulton M. The role of oxidative stress in the pathogenesis of age-related macular degeneration. Surv Ophthalmol. 2000;45(2):115–134.

[202] Ferrington DA, Sinha D, Kaarniranta K. Defects in retinal pigment epithelial cell proteolysis and the pathology associated with age-related macular degeneration. Prog Retin Eye Res. 2016;51:69–89. doi:10.1016/j.preteyeres.2015.09.002.

[203] Saadat KA, Murakami Y, Tan X, Nomura Y, Yasukawa T, Okada E, et al. Inhibition of autophagy induces retinal pigment epithelial cell damage by the lipofuscin fluorophore A2E. FEBS Open Bio. 2014;4:1007–1014. doi:10.1016/j.fob.2014.11.003.

[204] Bergmann M, Schutt F, Holz FG, Kopitz J. Inhibition of the ATP-driven proton pump in RPE lysosomes by the major lipofuscin fluorophore A2-E may contribute to the pathogenesis of age-related macular degeneration. FASEB J. 2004;18(3):562–564. doi:10.1096/fj.03-0289fje.

[205] Holz FG, Schutt F, Kopitz J, Eldred GE, Kruse FE, Volcker HE, et al. Inhibition of lysosomal degradative functions in RPE cells by a retinoid component of lipofuscin. Invest Ophthalmol Vis Sci. 1999;40(3):737–743.

[206] Shin ES, Sorenson CM, Sheibani N. Diabetes and retinal vascular dysfunction. J Ophthalmic Vis Res. 2014;9(3):362–373. doi:10.4103/2008-322X.143378.

[207] Nguyen TT, Wong TY. Retinal vascular changes and diabetic retinopathy. Curr Diab Rep. 2009;9(4):277–283.

[208] Fu D, Wu M, Zhang J, Du M, Yang S, Hammad SM, et al. Mechanisms of modified LDL-induced pericyte loss and retinal injury in diabetic retinopathy. Diabetologia. 2012;55(11):3128–3140. doi:10.1007/s00125-012-2692-0.

[209] Wu M, Chen Y, Wilson K, Chirindel A, Ihnat MA, Yu Y, et al. Intraretinal leakage and oxidation of LDL in diabetic retinopathy. Invest Ophthalmol Vis Sci. 2008;49(6):2679–2685. doi:10.1167/iovs.07-1440.

[210] Du M, Wu M, Fu D, Yang S, Chen J, Wilson K, et al. Effects of modified LDL and HDL on retinal pigment epithelial cells: a role in diabetic retinopathy? Diabetologia. 2013;56(10):2318–2328. doi:10.1007/s00125-013-2986-x.

[211] Yao J, Tao ZF, Li CP, Li XM, Cao GF, Jiang Q, et al. Regulation of autophagy by high glucose in human retinal pigment epithelium. Cell Physiol Biochem. 2014;33(1):107–116. doi:10.1159/000356654.

[212] Liu J, Lu W, Reigada D, Nguyen J, Laties AM, Mitchell CH. Restoration of lysosomal pH in RPE cells from cultured human and ABCA4(−/−) mice: pharmacologic approaches and functional recovery. Invest Ophthalmol Vis Sci. 2008;49(2):772–780. doi:10.1167/iovs.07-0675.

[213] Busik JV, Tikhonenko M, Bhatwadekar A, Opreanu M, Yakubova N, Caballero S, et al. Diabetic retinopathy is associated with bone marrow neuropathy and a depressed peripheral clock. J Exp Med. 2009;206(13):2897–2906. doi:10.1084/jem.20090889.

[214] Wang Q, Tikhonenko M, Bozack SN, Lydic TA, Yan L, Panchy NL, et al. Changes in the daily rhythm of lipid metabolism in the diabetic retina. PLoS One. 2014;9(4):e95028. doi:10.1371/journal.pone.0095028.

[215] Wang Q, Bozack SN, Yan Y, Boulton ME, Grant MB, Busik JV. Regulation of retinal inflammation by rhythmic expression of MiR-146a in diabetic retina. Invest Ophthalmol Vis Sci. 2014;55(6):3986–3994. doi:10.1167/iovs.13-13076.

[216] O'Brien IA, Lewin IG, O'Hare JP, Arendt J, Corrall RJ. Abnormal circadian rhythm of melatonin in diabetic autonomic neuropathy. Clin Endocrinol (Oxf). 1986;24(4):359–364.

[217] Ferrari S, Di Iorio E, Barbaro V, Ponzin D, Sorrentino FS, Parmeggiani F. Retinitis pigmentosa: genes and disease mechanisms. Curr Genomics. 2011;12(4):238–249. doi:10.2174/138920211795860107.

[218] Cottet S, Schorderet DF. Mechanisms of apoptosis in retinitis pigmentosa. Curr Mol Med. 2009;9(3):375–383.

[219] Liu C, Li Y, Peng M, Laties AM, Wen R. Activation of caspase-3 in the retina of transgenic rats with the rhodopsin mutation s334ter during photoreceptor degeneration. J Neurosci. 1999;19(12):4778–4785.

[220] Chang GQ, Hao Y, Wong F. Apoptosis: final common pathway of photoreceptor death in rd, rds, and rhodopsin mutant mice. Neuron. 1993;11(4):595–605.

[221] Lohr HR, Kuntchithapautham K, Sharma AK, Rohrer B. Multiple, parallel cellular suicide mechanisms participate in photoreceptor cell death. Exp Eye Res. 2006;83(2): 380–389. doi:10.1016/j.exer.2006.01.014.

[222] Chang TY, Reid PC, Sugii S, Ohgami N, Cruz JC, Chang CC. Niemann-Pick type C disease and intracellular cholesterol trafficking. J Biol Chem. 2005;280(22):20917–20920. doi:10.1074/jbc.R400040200.

[223] Claudepierre T, Paques M, Simonutti M, Buard I, Sahel J, Maue RA, et al. Lack of Niemann-Pick type C1 induces age-related degeneration in the mouse retina. Mol Cell Neurosci. 2010;43(1):164–176. doi:10.1016/j.mcn.2009.10.007.

[224] Bedgood A, Rand SE, Major J, Jr. Occult retinal detachment after mild traumatic brain injury: case report and literature review. Clin J Sport Med. 2015;25(1):e26–27. doi: 10.1097/JSM.0000000000000114.

[225] Shukla D, Ramchandani B, Vignesh TP, Rajendran A, Neelakantan N. Localized serous retinal detachment of macula as a marker of malignant hypertension. Ophthalmic Surg Lasers Imaging. 2010:1–7. doi:10.3928/15428877-20100215-74.

[226] Villalba-Pinto L, Hernandez-Ortega MA, de Los Mozos FJ, Pascual-Camps I, Dolz-Marco R, Arevalo JF, et al. Massive bilateral serous retinal detachment in a case of hypertensive chorioretinopathy. Case Rep Ophthalmol. 2014;5(2):190–194. doi: 10.1159/000364942.

[227] Feltgen N, Walter P. Rhegmatogenous retinal detachment--an ophthalmologic emergency. Dtsch Arztebl Int. 2014;111(1–2):12–21; quiz 22. doi:10.3238/arztebl.2014.0012.

[228] Chinskey ND, Zheng QD, Zacks DN. Control of photoreceptor autophagy after retinal detachment: the switch from survival to death. Invest Ophthalmol Vis Sci. 2014;55(2): 688–695. doi:10.1167/iovs.13-12951.

[229] Besirli CG, Chinskey ND, Zheng QD, Zacks DN. Autophagy activation in the injured photoreceptor inhibits fas-mediated apoptosis. Invest Ophthalmol Vis Sci. 2011;52(7): 4193–4199. doi:10.1167/iovs.10-7090.

[230] Dong K, Zhu ZC, Wang FH, Ke GJ, Yu Z, Xu X. Activation of autophagy in photoreceptor necroptosis after experimental retinal detachment. Int J Ophthalmol. 2014;7(5): 745–752. doi:10.3980/j.issn.2222-3959.2014.05.01.

[231] Jia X, Li J, Shi D, Zhao Y, Dong Y, Ju H, et al. Grouping annotations on the subcellular layered interactome demonstrates enhanced autophagy activity in a recurrent experimental autoimmune uveitis T cell line. PLoS One. 2014;9(8):e104404. doi:10.1371/journal.pone.0104404.

[232] Singh AD, Topham A. Survival rates with uveal melanoma in the United States: 1973–1997. Ophthalmology. 2003;110(5):962–965. doi:10.1016/S0161-6420(03)00077-0.

[233] Van Raamsdonk CD, Griewank KG, Crosby MB, Garrido MC, Vemula S, Wiesner T, et al. Mutations in GNA11 in uveal melanoma. N Engl J Med. 2010;363(23):2191–2199. doi: 10.1056/NEJMoa1000584.

[234] Van Raamsdonk CD, Bezrookove V, Green G, Bauer J, Gaugler L, O'Brien JM, et al. Frequent somatic mutations of GNAQ in uveal melanoma and blue naevi. Nature. 2009;457(7229):599–602. doi:10.1038/nature07586.

[235] Ambrosini G, Musi E, Ho AL, de Stanchina E, Schwartz GK. Inhibition of mutant GNAQ signaling in uveal melanoma induces AMPK-dependent autophagic cell death. Mol Cancer Ther. 2013;12(5):768–776. doi:10.1158/1535-7163.MCT-12-1020.

[236] Buschini E, Piras A, Nuzzi R, Vercelli A. Age related macular degeneration and drusen: neuroinflammation in the retina. Prog Neurobiol. 2011;95(1):14–25. doi:10.1016/j.pneurobio.2011.05.011.

[237] Prete M, Dammacco R, Fatone MC, Racanelli V. Autoimmune uveitis: clinical, pathogenetic, and therapeutic features. Clin Exp Med. 2015 doi:10.1007/s10238-015-0345-6.

[238] Nozaki M, Raisler BJ, Sakurai E, Sarma JV, Barnum SR, Lambris JD, et al. Drusen complement components C3a and C5a promote choroidal neovascularization. Proc Natl Acad Sci U S A. 2006;103(7):2328–2333. doi:10.1073/pnas.0408835103.

[239] Brandstetter C, Mohr LK, Latz E, Holz FG, Krohne TU. Light induces NLRP3 inflammasome activation in retinal pigment epithelial cells via lipofuscin-mediated photooxidative damage. J Mol Med (Berl). 2015;93(8):905–916. doi:10.1007/s00109-015-1275-1.

[240] Brandstetter C, Holz FG, Krohne TU. Complement component C5a primes retinal pigment epithelial cells for inflammasome activation by lipofuscin-mediated photooxidative damage. J Biol Chem. 2015;290(52):31189–31198. doi:10.1074/jbc.M115.671180.

[241] Blommaart EF, Krause U, Schellens JP, Vreeling-Sindelarova H, Meijer AJ. The phosphatidylinositol 3-kinase inhibitors wortmannin and LY294002 inhibit autophagy in isolated rat hepatocytes. Eur J Biochem. 1997;243(1–2):240–246.

[242] Li X, Xu HL, Liu YX, An N, Zhao S, Bao JK. Autophagy modulation as a target for anticancer drug discovery. Acta Pharmacol Sin. 2013;34(5):612–624. doi:10.1038/aps.2013.23.

[243] Puri P, Chandra A. Autophagy modulation as a potential therapeutic target for liver diseases. J Clin Exp Hepatol. 2014;4(1):51–59. doi:10.1016/j.jceh.2014.04.001.

[244] Rubinsztein DC, Codogno P, Levine B. Autophagy modulation as a potential therapeutic target for diverse diseases. Nat Rev Drug Discov. 2012;11(9):709–730. doi: 10.1038/nrd3802.

[245] Smith MH. The treatment of immature cataract. Trans Am Ophthalmol Soc. 1908;11(Pt 3):603–617.

[246] Wang LQ. Surgical treatment of senile cataract in the immature stage. Zhonghua Yan Ke Za Zhi. 1991;27(6):354–355.

[247] Williams DL, Munday P. The effect of a topical antioxidant formulation including N-acetyl carnosine on canine cataract: a preliminary study. Vet Ophthalmol. 2006;9(5):311–316. doi:10.1111/j.1463-5224.2006.00492.x.

[248] Shah GD, Vasavada AR, Praveen MR, Shah AR, Trivedi RH. Incidence and influence of posterior capsule striae on the development of posterior capsule opacification after 1-piece hydrophobic acrylic intraocular lens implantation. J Cataract Refract Surg. 2012;38(2):202–207. doi:10.1016/j.jcrs.2011.07.038.

[249] Medsinge A, Nischal KK. Pediatric cataract: challenges and future directions. Clin Ophthalmol. 2015;9:77–90. doi:10.2147/OPTH.S59009.

[250] Motoi Y, Shimada K, Ishiguro K, Hattori N. Lithium and autophagy. ACS Chem Neurosci. 2014;5(6):434–442. doi:10.1021/cn500056q.

[251] Choi SI, Kim BY, Dadakhujaev S, Jester JV, Ryu H, Kim TI, et al. Inhibition of TGFBIp expression by lithium: implications for TGFBI-linked corneal dystrophy therapy. Invest Ophthalmol Vis Sci. 2011;52(6):3293–3300. doi:10.1167/iovs.10-6405.

[252] Choi SI, Kim KS, Oh JY, Jin JY, Lee GH, Kim EK. Melatonin induces autophagy via an mTOR-dependent pathway and enhances clearance of mutant-TGFBIp. J Pineal Res. 2013;54(4):361–372. doi:10.1111/jpi.12039.

[253] Choi SI, Maeng YS, Kim KS, Kim TI, Kim EK. Autophagy is induced by raptor degradation via the ubiquitin/proteasome system in granular corneal dystrophy type 2. Biochem Biophys Res Commun. 2014;450(4):1505–1511. doi:10.1016/j.bbrc.2014.07.035.

[254] Choi SI, Kim BY, Dadakhujaev S, Oh JY, Kim TI, Kim JY, et al. Impaired autophagy and delayed autophagic clearance of transforming growth factor beta-induced protein (TGFBI) in granular corneal dystrophy type 2. Autophagy. 2012;8(12):1782–1797. doi:10.4161/auto.22067.

[255] Nguyen QD, Ibrahim MA, Watters A, Bittencourt M, Yohannan J, Sepah YJ, et al. Ocular tolerability and efficacy of intravitreal and subconjunctival injections of sirolimus in patients with non-infectious uveitis: primary 6-month results of the SAVE Study. J Ophthalmic Inflamm Infect. 2013;3(1):32. doi:10.1186/1869-5760-3-32.

[256] Xie X, White EP, Mehnert JM. Coordinate autophagy and mTOR pathway inhibition enhances cell death in melanoma. PLoS One. 2013;8(1):e55096. doi:10.1371/journal.pone.0055096.

[257] Rangwala R, Chang YC, Hu J, Algazy KM, Evans TL, Fecher LA, et al. Combined MTOR and autophagy inhibition: phase I trial of hydroxychloroquine and temsirolimus in

patients with advanced solid tumors and melanoma. Autophagy. 2014;10(8):1391–1402. doi:10.4161/auto.29119.

[258] Drenckhahn D. Anterior polar cataract and lysosomal alterations in the lens of rats treated with the amphiphilic lipidosis-inducing drugs chloroquine and chlorphentermine. Virchows Arch B Cell Pathol. 1978;27(3):255–266.

[259] Bernstein HN. Chloroquine ocular toxicity. Surv Ophthalmol. 1967;12(5):415–447.

[260] Cai J, Qi X, Kociok N, Skosyrski S, Emilio A, Ruan Q, et al. beta-Secretase (BACE1) inhibition causes retinal pathology by vascular dysregulation and accumulation of age pigment. EMBO Mol Med. 2012;4(9):980–991. doi:10.1002/emmm.201101084.

[261] Shamsi FA, Boulton M. Inhibition of RPE lysosomal and antioxidant activity by the age pigment lipofuscin. Invest Ophthalmol Vis Sci. 2001;42(12):3041–3046.

[262] Wassell J, Davies S, Bardsley W, Boulton M. The photoreactivity of the retinal age pigment lipofuscin. J Biol Chem. 1999;274(34):23828–23832.

[263] Deretic V. Autophagosome and phagosome. Methods Mol Biol. 2008;445:1–10. doi: 10.1007/978-1-59745-157-4_1.

[264] Kolosova NG, Muraleva NA, Zhdankina AA, Stefanova NA, Fursova AZ, Blagosklonny MV. Prevention of age-related macular degeneration-like retinopathy by rapamycin in rats. Am J Pathol. 2012;181(2):472–477. doi:10.1016/j.ajpath.2012.04.018.

[265] Wong WT, Dresner S, Forooghian F, Glaser T, Doss L, Zhou M, et al. Treatment of geographic atrophy with subconjunctival sirolimus: results of a phase I/II clinical trial. Invest Ophthalmol Vis Sci. 2013;54(4):2941–2950. doi:10.1167/iovs.13-11650.

[266] Nishida K, Kyoi S, Yamaguchi O, Sadoshima J, Otsu K. The role of autophagy in the heart. Cell Death Differ. 2009;16(1):31–38. doi:10.1038/cdd.2008.163.

[267] Petrovski G, Gurusamy N, Das DK. Resveratrol in cardiovascular health and disease. Ann N Y Acad Sci. 2011;1215:22–33. doi:10.1111/j.1749-6632.2010.05843.x.

[268] Lynch-Day MA, Mao K, Wang K, Zhao M, Klionsky DJ. The role of autophagy in Parkinson's disease. Cold Spring Harb Perspect Med. 2012;2(4):a009357. doi:10.1101/cshperspect.a009357.

[269] Rui YN, Xu Z, Patel B, Cuervo AM, Zhang S. HTT/Huntingtin in selective autophagy and Huntington disease: A foe or a friend within? Autophagy. 2015;11(5):858–860. doi: 10.1080/15548627.2015.1039219.

[270] Martin DD, Ladha S, Ehrnhoefer DE, Hayden MR. Autophagy in Huntington disease and huntingtin in autophagy. Trends Neurosci. 2015;38(1):26–35. doi:10.1016/j.tins.2014.09.003.

[271] Settembre C, Di Malta C, Polito VA, Garcia Arencibia M, Vetrini F, Erdin S, et al. TFEB links autophagy to lysosomal biogenesis. Science. 2011;332(6036):1429–1433. doi: 10.1126/science.1204592.

[272] Pastore N, Blomenkamp K, Annunziata F, Piccolo P, Mithbaokar P, Maria Sepe R, et al. Gene transfer of master autophagy regulator TFEB results in clearance of toxic protein and correction of hepatic disease in alpha-1-anti-trypsin deficiency. EMBO Mol Med. 2013;5(3):397–412. doi:10.1002/emmm.201202046.

Chapter 16

The Role of Autophagy in Maintaining Pregnancy

Akitoshi Nakashima, Aiko Aoki and Shigeru Saito

Additional information is available at the end of the chapter

http://dx.doi.org/10.5772/63475

Abstract

Autophagy is an evolutionarily conserved process in eukaryotes by which cytoplasmic cargo sequestered inside double-membrane vesicles is delivered to the lysosome for degradation. In early pregnancy, trophoblasts and the fetus experience hypoxic and low-nutrient conditions; nevertheless, extravillous trophoblasts (EVTs) invade the uterine myometrium up to one-third of its depth and migrate along the lumina of spiral arterioles, replacing the maternal endothelial lining. An enhancement of autophagy induced by physiological hypoxia takes part in the invasion and vascular remodeling in EVTs. On the other hand, soluble endoglin, which increased in sera in preeclamptic cases, suppresses EVT-invasion or -vascular remodeling by inhibiting autophagy *In vitro*. In addition, a substance selectively degraded by autophagy, p62/SQSTM1, accumulates in EVT cells in preeclamptic placental biopsy samples showing impaired autophagy *in vivo*. Thus, alternation of autophagy could affect fates of mothers and babies. Recently increasing evidence of modulating autophagy has accumulated during pregnancy. In this chapter, we introduce the role of autophagy in embryogenesis, implantation, and maintaining pregnancy.

Keywords: autophagy, extravillous trophoblast, hypoxia, invasion, preeclampsia, p62/SQSTM1, soluble endoglin

1. Introduction

The placenta acts to mediate the exchange of materials between mother and fetus under hypoxic and low-nutrient conditions during the early gestation period [1, 2]. It has been reported that hypoxia and low nutrient, which are generally harmful for cells, are preferable for trophoblasts during the early gestation period [3–5], indicating that trophoblasts possess evolutionary mechanisms allowing them to adjust to stress. In other words, disruption of these adaptive mechanisms may contribute to placental dysfunction, which induces preeclampsia and fetal growth restriction (FGR). Preeclampsia is one of the important diseases for life threatening a baby as well as a mother. It also causes FGR. Ten million women develop preeclampsia each year, and 76,000 mothers among them die each year all over the world. A recent interest for autophagy researchers is how autophagy contributes to the human diseases. It is getting clarified the role of autophagy for pathophysiology of preeclampsia or pregnancy-related diseases. In this chapter, we focus on the role of autophagy as a cellular cytoprotective mechanism, especially for embryogenesis, implantation, and maintaining pregnancy.

2. Autophagy in embryogenesis and implantation

Macroautophagy (herein referred to as autophagy) has long been considered a nonselective process for bulk degradation of either long-lived proteins, or cytoplasmic components during nutrient deprivation. Autophagy works for not only cellular energy metabolism, but also quality control for cellular protein (by eliminating protein aggregates, damaged organelles, lipid droplets, and intracellular pathogens) [6]. Though lysosomal degradation is served as a final step of autophagic machinery, this machinery can be deployed in some cellular processes: phagocytosis, exocytosis, secretion, antigen presentation, and regulation of inflammatory signaling [7]. Consequently, the autophagy pathway is mediated with human diseases, such as protection against aging, suppression or development of cancers, infections, neurodegenerative disorders, metabolic diseases, inflammatory diseases, and muscle diseases [8–12].

Autophagy has a variety of functions during embryogenesis (**Figure 1**). Autophagy is highly induced in fertilized oocytes, but not unfertilized oocytes, within 4 hours after fertilization [13]. As oocytes lacking Atg5 are fertilized normally *in vivo*, autophagy is not important for folliculogenesis and oogenesis. Similarly, ovulation, fertilization, implantation, and ovary size are not affected by the ablation of Beclin1 (BECN1) in luteal cells of ovary [14]. During oocyte-to-embryo transition, many of maternal proteins stored in oocytes are provided to zygotes (fertilized embryos), the stock of these proteins is largely degraded, and newly synthesized proteins encoded by the zygotic genome are translated. In mice, it is known that zygotic transcripts are detected at the late one-cell stage, and most of the maternal RNAs are eliminated by the two-cell stage [15–17]. During this period, autophagy is transiently suppressed from the late one-cell to middle two-cell stages, and reactivated after the late two-cell stage, suggesting that autophagy is involved in the degradation of maternal proteins in oocytes. In addition, complete autophagy-deficient embryos, which are derived from oocytes lacking Atg5 fertilized with Atg5-null sperm, arrest at four- to eight-cell stages, but not embryos derived from oocytes with Atg5-plus sperm [13]. Protein synthesis rates are reduced in complete autophagy-deficient embryos, suggesting that the degradation of maternal factors by autophagy is essential for preimplantation embryo development in mammals.

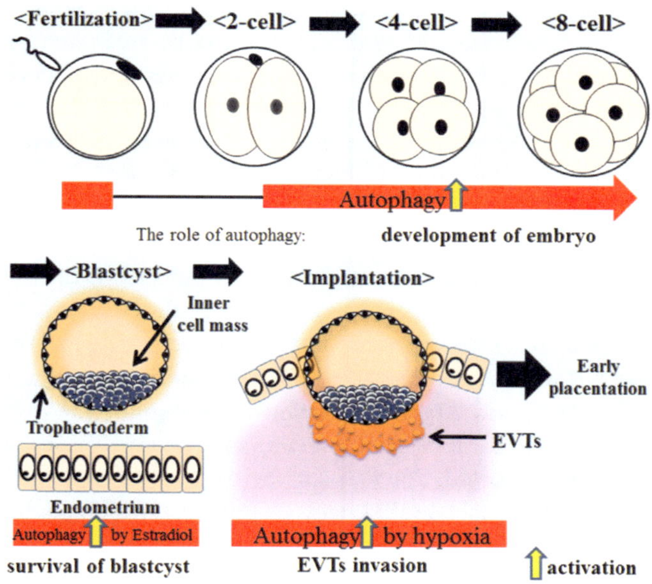

Figure 1. The role of autophagy during embryogenesis: Activation of autophagy occurs from the late two-cell stage to the eight-cell stage. Autophagy helps survival of blastocysts treated with estradiol (E2). Subsequently, EVTs' invasion, which is necessary for normal placentation, is supported by hypoxia-activated autophagy.

A mouse model of experimentally delayed implantation, established by ovariectomy before blastocyst implantation, showed that Atg7 and LC3 expression is upregulated in blastocysts made dormant by the elimination of 17β-estradiol (E2), in comparison with E2-activated blastocysts, suggesting that autophagy is sustained during the prolonged survival of dormant blastocysts [18]. Activation of autophagy is also observed in the inner cell mass, known as the embryoblast or pluriblast, which will evolutionally give rise to the definitive structures of the fetus. On the other hand, E2 or progesterone activates autophagy simultaneously with a decrease in activation/phosphorylation of mTOR (mammalian target of rapamycin) in bovine mammary epithelial cells [19].

Mice lacking BECN1, which is necessary for autophagy [20], died for early embryonic period (E7.5 or earlier) with defects in proamniotic canal closure [21]. Ablation of FIP200 (focal adhesion kinase family interacting protein of 200 kD) in mice also leads to embryonic lethality at mid/late gestation, which is caused by defective heart and liver development [22]. Most mouse embryos with functional deficiency of Ambra1 (activating molecule in beclin1-regulated autophagy) exhibit neural tube defects in midbrain/hindbrain exencephaly and/or spina bifida during E10–E14.5 [23]. Therefore, these types of autophagy deficiency are mainly involved in the impairment of embryonal development. However, it is unknown whether autophagy deficiency affects placental development or physiological status in dams.

Mice lacking Atg5 are born normally, but die within the first day after birth [24]. Atg7, which is essential for Atg5-Atg12 conjugation, knockout mice are also born at expected ratios, with large healthy major organs. Similar to Atg5 knockout mice, Atg7 knockout mice also die earlier than wild-type controls under nonsuckling conditions after cesarean delivery, and have lower plasma amino acid levels [25]. Atg3, Atg9, and Atg16L1 conventional knockout mice demonstrate similar phenotypes [26–28]. The exact reason for neonatal death in those knockout mice is still controversial, and observation of a suckling defect may indicate a neuronal defect. However, the above studies demonstrate that amino acid supply through autophagy is important for mouse neonatal survival, preventing sudden starvation at birth. In addition, autophagy knockout mice that are born normally exhibit slightly inhibited fetal growth. It is still unknown whether inhibited growth is a result of defects within the placenta, or by mechanisms within the pups themselves. This question could be elucidated with the use of placenta-specific autophagy-deficient mouse. The BECN1-containing PI(3)K and ULK1-FIP200 complexes function early in the autophagy process at the autophagosome nucleation step [9, 29], while Atg3, Atg5, Atg7, and Atg16L1 function later at autophagosome elongation. As shown in **Table 1**, genetic knockout upstream factors may produce more severe phenotypes, or alternatively, as recently reported, downstream factors may be responsible for a particular type of macroautophagy [30]. In fact, LC3B knockout pups survive longer than their wild-type counterparts without a compensatory increase in LC3A, demonstrating opposite results to the other Atg-knockout mice [31]. Atg4C, a family of cysteine proteinases for processing and delipidation of Atg8, knockout mice also indicate that Atg4C is dispensable for embryonic and adult mouse development, as well as for normal growth and fertility [32].

Genes	Phenotypes
$Atg3^{-/-}$, $Atg5^{-/-}$, $Atg7^{-/-}$, $Atg9^{-/-}$, $Atg16L1^{-/-}$	Neonatal lethal with reduced amino acid levels, suckling defect (Atg5: the failure of the four- and eight-cell stages in embryogenesis)
Beclin $1^{-/-}$	Early embryonic lethal (E7.5 or earlier) with defects in proamniotic canal closure (heterozygous mice show increased susceptibility to spontaneous tumor)
$FIP200^{-/-}$	Embryonic lethal (E13.5–E16.5) dueto defective heart and liver development
$Ambra1^{gt/gt}$	Embryonic lethal (-E14) with defects in neural tube development, and hyperproliferation of neural tissues
$ULK1^{-/-}$	Increased reticulocyte number with delayed mitochondrial clearance
$Atg4C^{-/-}$	Viable, fertile, increased susceptibility to carcinogen-induced fibrosarcoma
$LC3B^{-/-}$	Normal phenotype, production higher levels of IL-1β and IL-18 in response to LPS
$GABARAP^{-/-}$	Normal phenotype

Table 1. Phenotypes of systemic knockout mice of Atg-related genes.

3. The role of autophagy in trophoblasts for normal development of the placenta

In humans, trophoblast stem cells differentiate into two cell types: villous trophoblasts and extravillous trophoblasts (EVTs). Invading trophoblasts called interstitial EVTs migrate into the decidualized endometrium, and endovascular EVTs migrate along the lumina of spiral arterioles. The invasion of spiral arteries by EVTs starts early in pregnancy, and the endovascular trophoblastic cells aggregate in the lumen of the vessel forming the "trophoblastic plug" to allow the growth of the embryo and the placenta in a low-oxygen environment (**Figure 2**). As the EVTs migrate away from the villi and invade the maternal decidua, they progressively develop an invasive phenotype without proliferation [33], and stop at one-third of the depth of the myometrium in the uterus. EVTs invade the maternal decidua under harsh conditions, including low oxygen (2–5% O_2) and low-glucose concentrations (1 mM), until 11 weeks of gestation [33, 34]. During the invasion, the hypoxia inducible factor (HIF) system plays a critical role in their functions. After 12 weeks of gestation, endovascular EVTs invade the uterine spiral arteries, replace their endothelial cells, and participate in the degradation of tunica media smooth muscle cells under moderate hypoxia (approximately 8%). This remodeling of the spiral arteries is essential to allow a proper placental perfusion to sustain fetal growth (**Figure 2**). In other words, impairment of trophoblast invasion or remodeling may contribute to fetal loss during early pregnancy, or poor placentation during the middle of the pregnancy period.

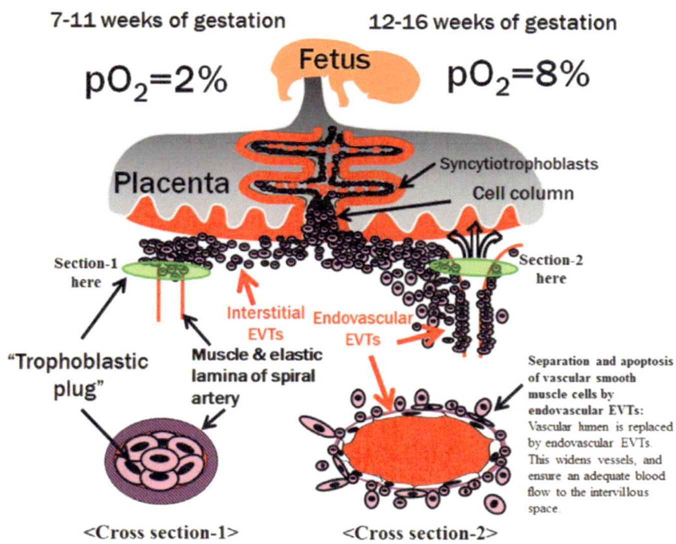

Figure 2. Autophagy supports EVT functions under physiological hypoxia: Interstitial EVT invades during 7–11 weeks of gestation, and vascular remodeling by EVTs occurs during 12–16 weeks of gestation. Cross section-1 indicates the "trophoplastic plug" to maintain hypoxia in the conceptus during 7–11 weeks of gestation. Cross section-2 indicates separation and apoptosis of vascular smooth muscle cells by endovascular EVTs.

Activation of autophagy is observed in EVTs invading into the maternal decidua at 7 weeks of gestation, under physiological hypoxic conditions [35]. Hypoxia induces autophagy in primary trophoblasts [35, 36], and inhibition of autophagy induced by silencing Atg7 in primary trophoblasts decreases apoptosis under 1% oxygen conditions in the presence of bafilomycin A1, an inhibitor of lysosome, indicating that autophagy mediates apoptosis in trophoblasts under specific conditions. Not only hypoxia, but also starvation induces autophagy in trophoblasts or some choriocarcinoma cell lines, and LC3-II/actin levels, a marker of autophagic activation, vary depending on the cell lines (unpublished data). Though many independent studies have shown that hypoxia enhances invasion of EVTs, the role of autophagy in trophoblast functions is still unclear. To elucidate the specific role of autophagy in trophoblast functions, we constructed autophagy-suppressed cells by stably transfecting ATG4B^{C74A}, an inactive mutant of ATG4B, which inhibits autophagic degradation and lipidation of LC3B paralogs [37]. In these cell lines, the conversion of LC3-I to LC3-II was abolished by starvation, a commonly used autophagy stimulator. Hypoxia enhances the invasive capacity of EVT cell lines, such as HTR-8/SVneo, the most widely used EVT cell line, or HchEpC1b, immortalized by infection with retroviral expression vectors containing the type 16 human papillomaviruses E6 and E7 in combination with a human telomerase reverse transcriptase, with a normal chromosomal number and no tumorigenic activity [38]. The invasion was significantly reduced in autophagy-suppressed EVT cell lines, compared with autophagy-normal EVT cell lines, under 2% oxygen tension, which matches the placental condition until 11 weeks of gestation.

The other function of EVTs, vascular remodeling, is also necessary to precisely develop placentation. To clarify the role of autophagy in vascular remodeling, tube formation assays with EVT cells and human umbilical vascular endothelial cells (HUVECs), an in vitro model of vascular remodeling by EVT cells, were performed under 8% oxygen tension, simulating physiological levels of oxygen tension after 12 weeks of gestation [39]. Both the autophagy-normal EVT cells and the autophagy-suppressed cells formed a tube structure with HUVECs at 12 h, but did not form a tube structure in the absence of HUVECs. In the culture with the autophagy-normal EVT cells and HUVECs, the tubes were mostly occupied by the autophagy-normal EVT cells at 12 h or later; whereas the tubes were not occupied by EVT cells when the autophagy-suppressed cells were cocultured with HUVECs, suggesting that replacement of endothelial cells by EVT cells requires autophagy. Thus, autophagy plays an important role in the endovascular interaction between EVT and endothelial cells [35].

No difference in HIF1α expression was observed between autophagy-normal and autophagy-suppressed EVT cell lines [35], suggesting that HIF1α is not affected by autophagy status. A number of studies have investigated the role of the HIF1 pathway in EVT invasion. A decrease in HIF1α expression induced by siRNA markedly reduced the invasiveness of HTR8/SVneo cells under hypoxia and normoxia [40]. Hypoxia-induced autophagy is modulated by the inactivation of mTOR via AMPK (5′-AMP-activated protein kinase) [41]. Additionally, Atg5 knockout, but not wild type, mouse embryo fibroblasts (MEFs) showed no activation of autophagy under hypoxia [42]. Thus, HIF1α may activate EVT functions, at least partially, by manipulating autophagy status. Rapamycin or siRNA-mediated mTOR knockdown, an activator of autophagy, reduced the invasiveness of HTR8/SVneo cells under normoxia [43]. However, our study revealed no effect of rapamycin on invasion of HTR8/SVneo cells [44]. This result might be explained by a rapamycin-induced cell cycle arrest in the G1 phase [45].

Hypoxia activates macroautophagy via the HIF1 pathway, and HIF1α seems to be controlled by autophagy. Chaperone-mediated autophagy (CMA) has recently shown to degrade

HIF1α, which is mostly controlled by the oxygen-dependent proteasome, through LAMP2A, a lysosomal transporter protein. Interestingly, this new pathway for degradation of HIF1α does not depend on the presence of oxygen and is activated in response to nutrient deprivation in rat livers [46]. CMA-mediated excessive degradation of HIF1α compromises cells' ability to respond to and survive under hypoxia, suggesting that the impairment of this pathway might be of pathophysiological importance in conditions that combine hypoxia with starvation, including the early pregnancy period, during which EVTs invade.

Placenta expresses LC3B in both cytotrophoblasts and syncytiotrophoblasts. LC3-II/actin levels, which indicate activation of autophagy, are higher in placentas from cesarean sections than from vaginal deliveries [47], suggesting that uterine contraction might inhibit autophagy in placentas. In other words, autophagy might be activated in normal placenta before delivery. Autophagic cell death is also reported in amniotic epithelium following the rupture of membranes in term placentas [48]. Atg9L2 (Atg9b) is specifically expressed in placenta (trophoblast cells) and pituitary gland, while another homolog Atg9L1 (Atg9a) is ubiquitously expressed in adult human tissues [49]. Atg9L1 and Atg9L2 are involved in autophagosome formation [26]. Mouse Atg9L2 is found to be more widely expressed at embryonic stages than in adulthood. In humans, the expression of Atg9L2 is significantly higher than that of Atg9L1 in human primary cytotrophoblasts, suggesting that Atg9L2 contributes to tissue-specific and developmental activation of autophagy. Autophagy-deficient mice, similar to Atg7 or Atg5 knockout mice, are born at expected ratios, and fetal weight is only slightly lower than that of wild-type fetuses. Kojima et al. [50] recently reported the role of autophagy in preeclampsia in $Atg9a^{-/-}$ mice mated with heterozygous $p57^{Kip2}$ mice, which develop hypertension and proteinuria in dams. Fetal death was increased in $Atg9a^{-/-}$ and $Atg9a^{+/-}$ pups, compared with wild-type controls [51]. In addition, the body weight of fetuses in $Atg9a^{-/-}$ pups was significantly lower than those of $Atg9a^{+/-}$ or wild-type pups, suggesting that autophagy sustains fetal spare ability under stress. A future task is to clarify the role of Atg9b, which is highly expressed in the placenta.

4. The role of autophagy in the pathophysiology of preeclampsia

The current hypothesis regarding the etiology of preeclampsia is focused on shallow trophoblast invasion and poor placentation [52, 53] (**Figure 3**). We recently reported that autophagy was enhanced in EVTs in early gestation placental tissues, which are under physiological hypoxia [35]. As mentioned previously, the impairment of the invasion and vascular remodeling under hypoxia, which are thought to be a cause of preeclampsia, are observed in autophagy-deficient EVT cell lines. Furthermore, soluble endoglin (sENG), levels of which increase in sera before the onset of preeclampsia, suppresses invasion in EVT cell lines by inhibiting autophagy. The sENG-inhibited EVT invasion is recovered by TGF-β treatment in a dose-dependent manner. A low dose of sENG also inhibits the replacement of HUVECs by EVT cell lines in the *in vitro* model of vascular remodeling. This is the first report to show the role of autophagy in poor vascular remodeling during preeclampsia. Several studies have linked TGF-β to inhibition of EVT invasion [54–56]. Conversely, one study reported that TGF-β augments EVT invasion [57]. It is well known that sENG binds TGF-β, thus neutralizing its effects. Paradoxically, our data showed that TGF-β neutralized the effect of sENG, resulting in recovery of HTR-8/SVneo cell invasion under hypoxic conditions (2% oxygen tension) mimicking the physiological hypoxia during the early pregnancy period, but TGF-β showed no effect on HTR-8/SVneo cell invasion under 20% oxygen tension. Thus, physiological hypoxia, in which EVT cells invade into maternal side, was the key to understand the mechanism by which sENG inhibits autophagy in EVT cells. Though TGF-β has been shown to induce autophagy accompanied with transcriptional increase of BECN1, Atg5, and Atg7 mRNA in human hepatoma cells, and increases autophagy in mammary carcinoma cells [58, 59], neutralizing effects to sENG by TGF-β might be more important than autophagy activation in EVT cells.

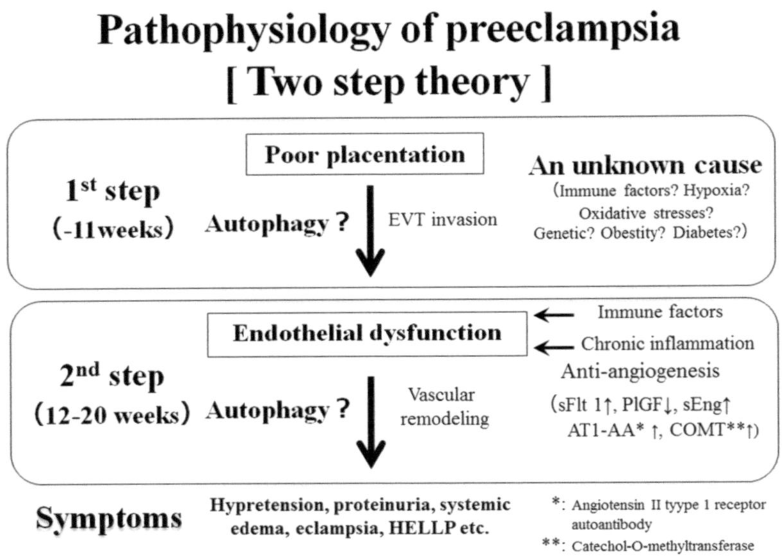

Figure 3. Two-step model of preeclampsia: Poor placentation is happened before 11 weeks of gestation, and endothelial dysfunction is happened after 12 weeks of gestation. Autophagy might be involved in each step of the pathophysiology of preeclampsia.

In regard to the role of autophagy in placenta, it remains to be elucidated whether autophagy is activated or inhibited in preeclamptic placenta. Increased numbers of LC3B dots, a marker of autophagy activation, in villous trophoblasts were observed in cases of preeclampsia with IUGR or idiopathic IUGR placentas, compared with normal human pregnancy [60]. Furthermore, sphingolipids may be involved in autophagic activation in trophoblasts in preeclamptic placentas [61]. These support the activation of autophagy in villous trophoblasts during preeclampsia and IUGR. On the other hand, p62/SQSTM1, a protein specifically digested by autophagy, accumulated in autophagy-suppressed human cell lines, suggesting inhibition of autophagy. The expression of p62/SQSTM1 is significantly higher in EVTs in preeclamptic placentas, demonstrating the inhibition of autophagy in EVTs during preeclampsia. However, p62/SQSTM1expression in syncytiotrophoblasts is not observed in either preeclamptic placentas or normal placentas, indicating the activation of autophagy in syncytiotrophoblasts [35]. Together, these results suggest that autophagy inhibition occurs specifically in EVTs, and that there appears to be a difference in autophagic activity between syncytiotrophoblasts and EVTs in preeclamptic placentas. Sera from preeclamptic patients induce hypertension, proteinuria, and FGR in pregnant IL-10$^{-/-}$ mice, indicating that a variety of factors, including sENG and soluble Flt-1, contribute to the occurrence of preeclampsia [39]. Additionally, rapamycin-induced autophagy in peripheral blood mononuclear cells is suppressed in the presence of sera from women with preeclampsia, but not women with normotension [62]. Thus, sera from women with preeclampsia modulate the status of autophagy in the placenta.

BECN1 acts as an initiator of autophagy in mammals, and upregulation of BECN1 expression represses cellular proliferation under hypoxia. The expression of BECN1 mRNA or protein is significantly higher in IUGR without preeclampsia, than in normal pregnancy [60], and no significant difference in BECN1 expression is reported between syncytiotrophoblasts of preeclampsia or normal pregnancy patients [63]. A recent report demonstrated the importance of BECN1 for maintaining pregnancy in mice. Pregnant dams lacking BECN1 in the ovarian granulosa cell population showed impaired progesterone production during preterm labor. Luteal cells in this mouse model exhibit p62 accumulation, which indicates deficiency of autophagy, and a failure of neutral lipid storage, which is needed for steroidogenesis [14]. Progesterone in humans is produced in syncytiotrophoblasts of the placenta after 8 weeks of gestation. Despite the difference between mouse and humans, a BECN1 deficiency can be able to affect preterm birth in humans by different ways. This is the case in the hormonally-induced preterm labor model, while inflammation-induced preterm labor is also enhanced by alternation of autophagy flux, resulting in NF-kB mediated hyperinflammation in the placenta [64]. Finally, an unexpected function of autophagy in the placenta has been revealed: micro RNAs delivered from human placental trophoblasts to nonplacental recipient cells confer resistance to infection with different types of viruses, such as human cytomegalovirus, herpes simplex virus-1, vaccinia virus, poliovirus, or coxsackievirus B3. This involves exosome-mediated transfer of a unique set of placental-specific effector micro RNAs, indicating that the placenta is involved in regulating systemic immunity in humans [65].

5. Attention for estimating autophagy

Mizushima et al. [66] stated that there is no single "gold standard" for methods to monitor or modulate autophagic activity or flux. Rather, one should consider the use of several different concurrent methods (with nonoverlapping limitations) to accurately assess the status and functions of autophagic activity in any given biological setting. Of note, a common misconception for estimating autophagic activity often occurs in human tissue samples. Increased numbers of autophagosomes, LC3 dots, in cells does not invariably correspond to increased cellular autophagic activity. Tissue samples reflect only one time point when they are fixed, but "autophagy flux" could have occurred before fixation. Thus, autophagosome accumulation may represent either autophagy induction, or alternatively, suppression of steps in the autophagy pathway downstream of autophagosome formation. The blockade of any step "downstream" of autophagosome formation increases the number of autophagosomes. In contrast, the blockade of any step "upstream" of autophagosome formation decreases the numbers of all autophagic structures. Therefore, simply determining the quantity of autophagosomes is insufficient for an overall estimation of autophagic activity. Indeed, in tissue samples, the quantities not only of autophagosomes, but also of autolysosomes, are available for estimating autophagic activity. In addition, we found that p62/SQSTM1, which becomes incorporated into the completed autophagosome and is degraded in autolysosomes, was accumulated in some trophoblast cell lines in which autophagy is suppressed by Atg4B^{C74A} mutant [35]. The accumulation of p62/SQSTM1 could assist in the estimation of the level of impairment of autophagy in the placenta. Autophagy researchers anticipate that better assays will be developed to monitor autophagy, and that more specific agents will be developed to modulate autophagy. Indeed, more advances are necessary to accurately assess the status of autophagy in human tissues in order to improve clinical therapies involving modulation of autophagy.

6. Future directions

A growing body of evidence indicates that autophagy plays a key role in placentation and contributes to differences observed between normal pregnancy, preeclampsia, and IUGR. The placenta contributes to systemic immunity by activating autophagy in extraplacental cells. We believe that autophagy affects numerous functions, including protection from stress, energy regulation, immune regulation, differentiation, proliferation, and cell death in the placenta. Better molecular characterization of the autophagic pathways, as well as the possibility of genetically manipulating these cellular processes, will further elucidate the link between autophagic abnormalities and disease. The first and most important issue is how we determine the status of autophagy in the placenta. The status of a preeclamptic placenta is affected by severity, infarction, infection, degree of cell damage, genetic background, immunological status, and age. As mentioned previously, the elevation of autophagosomes observed in many disease conditions, initially interpreted as an increase in macroautophagy, is now more cautiously interpreted, because blockage of further downstream in this pathway can also produce a similar morphological signature. Thus, conditions initially labeled as having 'too much autophagy' are being currently reinterpreted as having 'a blockage in autophagic clearance'. Precise estimation of autophagy status will likely help to elucidate autophagic mechanisms implicated in placental disorders.

Acknowledgements

We thank all the staff at the Department of Obstetrics, University of Toyama for help with patient identification and recruitment. We thank Dr. Tamotsu Yoshimori for giving some materials and providing advice, and Akemi Ushijima and Kaori Nomoto for technical assistance with some assays. This research was supported by grants from Tamura Science and Technology Foundation, Yamaguchi Endocrine Research Foundation, Kanzawa Medical Research Foundation, and the Ministry of Education, Culture, Sports, Science and Technology, Japan [Grant-in-Aid for Scientific Research (B) – 15H04980 and Grant-in-Aid for Challenging Exploratory Research – 26670717].

Author details

Akitoshi Nakashima, Aiko Aoki and Shigeru Saito*

*Address all correspondence to: s30saito@med.u-toyama.ac.jp

Department of Obstetrics and Gynecology, Faculty of Medicine, University of Toyama, Toyama, Japan

References

[1] J.L. James, P.R. Stone, L.W. Chamley: The regulation of trophoblast differentiation by oxygen in the first trimester of pregnancy, Hum Reprod Update 2006; 12: 137–144. Pii: dmi043, doi: 10.1093/humupd/dmi043.

[2] K.G. Pringle, K.L. Kind, A.N. Sferruzzi-Perri, J.G. Thompson, C.T. Roberts: Beyond oxygen: complex regulation and activity of hypoxia inducible factors in pregnancy, Hum Reprod Update 2010; 16: 415–431. Pii: dmp046, doi: 10.1093/humupd/dmp046.

[3] T.G. Graeber, C. Osmanian, T. Jacks, D.E. Housman, C.J. Koch, S.W. Lowe, A.J. Giaccia: Hypoxia-mediated selection of cells with diminished apoptotic potential in solid tumours, Nature 1996; 379: 88–91. Doi: 10.1038/379088a0.

[4] B. Levine, J. Yuan: Autophagy in cell death: an innocent convict, J Clin Invest 2005; 115: 2679–2688. Doi: 10.1172/JCI26390.

[5] J. Pouyssegur, F. Dayan, N.M. Mazure: Hypoxia signalling in cancer and approaches to enforce tumour regression, Nature 2006; 441: 437–443. Pii: nature04871, doi: 10.1038/nature04871.

[6] N. Mizushima, M. Komatsu: Autophagy: renovation of cells and tissues, Cell 2011; 147: 728–741. Pii: S0092-8674(11)01276-1, doi: 10.1016/j.cell.2011.10.026.

[7] P. Boya, F. Reggiori, P. Codogno: Emerging regulation and functions of autophagy, Nat Cell Biol 2013; 15: 713–720. Pii: ncb2788, doi: 10.1038/ncb2788.

[8] I. Beau, M. Mehrpour, P. Codogno: Autophagosomes and human diseases, Int J Biochem Cell Biol 2011; 43: 460–464. Pii: S1357-2725(11)00008-2, doi: 10.1016/j.biocel. 2011.01.006.

[9] B. Levine, G. Kroemer: Autophagy in the pathogenesis of disease, Cell 2008; 132: 27–42. Pii: S0092-8674(07)01685-6, doi: 10.1016/j.cell.2007.12.018.

[10] N. Mizushima, B. Levine, A.M. Cuervo, D.J. Klionsky: Autophagy fights disease through cellular self-digestion, Nature 2008; 451: 1069–1075. Pii: nature06639, doi: 10.1038/nature06639.

[11] S. Sarkar: Chemical screening platforms for autophagy drug discovery to identify therapeutic candidates for Huntington's disease and other neurodegenerative disorders, Drug Discov Today Technol 2013; 10: e137-e144. Pii: S1740-6749(12)00070-4, doi: 10.1016/j.ddtec.2012.09.010.

[12] J.L. Schneider, A.M. Cuervo: Autophagy and human disease: emerging themes, Curr Opin Genet Dev 2014; 26: 16–23. Pii: S0959-437X(14)00035-5, doi: 10.1016/j.gde. 2014.04.003.

[13] S. Tsukamoto, A. Kuma, M. Murakami, C. Kishi, A. Yamamoto, N. Mizushima: Autophagy is essential for preimplantation development of mouse embryos, Science 2008; 321: 117–120. Pii: 321/5885/117, doi: 10.1126/science.1154822.

[14] T.R. Gawriluk, C. Ko, X. Hong, L.K. Christenson, E.B. Rucker, 3rd: Beclin-1 deficiency in the murine ovary results in the reduction of progesterone production to promote preterm labor, Proc Natl Acad Sci U S A 2014; 111: E4194-E4203. Pii: 1409323111, doi: 10.1073/pnas.1409323111.

[15] F. Aoki, D.M. Worrad, R.M. Schultz: Regulation of transcriptional activity during the first and second cell cycles in the preimplantation mouse embryo, Dev Biol 1997; 181: 296–307. Pii: S0012-1606(96)98466-3, doi: 10.1006/dbio.1996.8466.

[16] L. Piko, K.B. Clegg: Quantitative changes in total RNA, total poly(A), and ribosomes in early mouse embryos, Dev Biol 1982; 89: 362–378. Pii: 0012-1606(82)90325-6.

[17] R.M. Schultz: Regulation of zygotic gene activation in the mouse, Bioessays 1993; 15: 531–538. Doi: 10.1002/bies.950150806.

[18] J.E. Lee, H.A. Oh, H. Song, J.H. Jun, C.R. Roh, H. Xie, S.K. Dey, H.J. Lim: Autophagy regulates embryonic survival during delayed implantation, Endocrinology 2011; 152: 2067–2075. Pii: en.2010-1456, doi: 10.1210/en.2010-1456.

[19] A. Sobolewska, M. Gajewska, J. Zarzynska, B. Gajkowska, T. Motyl: IGF-I, EGF, and sex steroids regulate autophagy in bovine mammary epithelial cells via the mTOR pathway, Eur J Cell Biol 2009; 88: 117–130. Pii: S0171-9335(08)00135-0, doi: 10.1016/j.ejcb. 2008.09.004.

[20] C. He, B. Levine: The Beclin 1 interactome, Curr Opin Cell Biol 2010; 22: 140–149. Pii: S0955-0674(10)00003-7, doi: 10.1016/j.ceb.2010.01.001.

[21] Z. Yue, S. Jin, C. Yang, A.J. Levine, N. Heintz: Beclin 1, an autophagy gene essential for early embryonic development, is a haploinsufficient tumor suppressor, Proc Natl Acad Sci U S A 2003; 100: 15077–15082. Pii: 2436255100, doi: 10.1073/pnas.2436255100.

[22] B. Gan, X. Peng, T. Nagy, A. Alcaraz, H. Gu, J.L. Guan: Role of FIP200 in cardiac and liver development and its regulation of TNFalpha and TSC-mTOR signaling pathways. J Cell Biol 2006; 175: 121–133. Pii: jcb.200604129, doi: 10.1083/jcb.200604129.

[23] G.M. Fimia, A. Stoykova, A. Romagnoli, L. Giunta, S. Di Bartolomeo, R. Nardacci, M. Corazzari, C. Fuoco, A. Ucar, P. Schwartz, P. Gruss, M. Piacentini, K. Chowdhury, F. Cecconi: Ambra1 regulates autophagy and development of the nervous system, Nature 2007; 447: 1121–1125. Pii: nature05925, doi: 10.1038/nature05925.

[24] A. Kuma, M. Hatano, M. Matsui, A. Yamamoto, H. Nakaya, T. Yoshimori, Y. Ohsumi, T. Tokuhisa, N. Mizushima: The role of autophagy during the early neonatal starvation period, Nature 2004; 432: 1032–1036. Pii: nature03029, doi: 10.1038/nature03029.

[25] M. Komatsu, S. Waguri, T. Ueno, J. Iwata, S. Murata, I. Tanida, J. Ezaki, N. Mizushima, Y. Ohsumi, Y. Uchiyama, E. Kominami, K. Tanaka, T. Chiba: Impairment of starvation-induced and constitutive autophagy in Atg7-deficient mice, J Cell Biol 2005; 169: 425–434. Pii: jcb.200412022, doi: 10.1083/jcb.200412022.

[26] T. Saitoh, N. Fujita, T. Hayashi, K. Takahara, T. Satoh, H. Lee, K. Matsunaga, S. Kageyama, H. Omori, T. Noda, N. Yamamoto, T. Kawai, K. Ishii, O. Takeuchi, T. Yoshimori, S. Akira: Atg9a controls dsDNA-driven dynamic translocation of STING and the innate immune response, Proc Natl Acad Sci U S A 2009; 106: 20842–20846. Pii: 0911267106, doi: 10.1073/pnas.0911267106.

[27] T. Saitoh, N. Fujita, M.H. Jang, S. Uematsu, B.G. Yang, T. Satoh, H. Omori, T. Noda, N. Yamamoto, M. Komatsu, K. Tanaka, T. Kawai, T. Tsujimura, O. Takeuchi, T. Yoshimori, S. Akira: Loss of the autophagy protein Atg16L1 enhances endotoxin-induced IL-1beta production, Nature 2008; 456: 264–268. Pii: nature07383, doi: 10.1038/nature07383.

[28] Y.S. Sou, S. Waguri, J. Iwata, T. Ueno, T. Fujimura, T. Hara, N. Sawada, A. Yamada, N. Mizushima, Y. Uchiyama, E. Kominami, K. Tanaka, M. Komatsu: The Atg8 conjugation system is indispensable for proper development of autophagic isolation membranes in mice, Mol Biol Cell 2008; 19: 4762–4775. Pii: E08-03-0309, doi: 10.1091/mbc.E08-03-0309.

[29] A. Longatti, S.A. Tooze: Vesicular trafficking and autophagosome formation, Cell Death Differ 2009; 16: 956–965. Pii: cdd200939, doi: 10.1038/cdd.2009.39.

[30] Y. Nishida, S. Arakawa, K. Fujitani, H. Yamaguchi, T. Mizuta, T. Kanaseki, M. Komatsu, K. Otsu, Y. Tsujimoto, S. Shimizu: Discovery of Atg5/Atg7-independent

alternative macroautophagy, Nature 2009; 461: 654–658. Pii: nature08455, doi: 10.1038/nature08455.

[31] G.M. Cann, C. Guignabert, L. Ying, N. Deshpande, J.M. Bekker, L. Wang, B. Zhou, M. Rabinovitch: Developmental expression of LC3alpha and beta: absence of fibronectin or autophagy phenotype in LC3beta knockout mice, Dev Dyn 2008; 237: 187–195. Doi: 10.1002/dvdy.21392.

[32] G. Marino, N. Salvador-Montoliu, A. Fueyo, E. Knecht, N. Mizushima, C. Lopez-Otin: Tissue-specific autophagy alterations and increased tumorigenesis in mice deficient in Atg4C/autophagin-3, J Biol Chem 2007; 282: 18573–18583. Pii: M701194200, doi: 10.1074/jbc.M701194200.

[33] O. Genbacev, Y. Zhou, J.W. Ludlow, S.J. Fisher: Regulation of human placental development by oxygen tension, Science 1997; 277: 1669–1672.

[34] E. Jauniaux, J. Hempstock, C. Teng, F.C. Battaglia, G.J. Burton: Polyol concentrations in the fluid compartments of the human conceptus during the first trimester of pregnancy: maintenance of redox potential in a low oxygen environment, J Clin Endocrinol Metab 2005; 90: 1171–1175. Pii: jc.2004-1513, doi: 10.1210/jc.2004-1513.

[35] A. Nakashima, M. Yamanaka-Tatematsu, N. Fujita, K. Koizumi, T. Shima, T. Yoshida, T. Nikaido, A. Okamoto, T. Yoshimori, S. Saito: Impaired autophagy by soluble endoglin, under physiological hypoxia in early pregnant period, is involved in poor placentation in preeclampsia, Autophagy 2013; 9: 303–316. Pii: 22927, doi: 10.4161/auto.22927.

[36] N. Fujita, M. Hayashi-Nishino, H. Fukumoto, H. Omori, A. Yamamoto, T. Noda, T. Yoshimori: An Atg4B mutant hampers the lipidation of LC3 paralogues and causes defects in autophagosome closure, Mol Biol Cell 2008; 19: 4651–4659. Pii: E08-03-0312, doi: 10.1091/mbc.E08-03-0312.

[37] N. Fujita, T. Noda, T. Yoshimori: Atg4B(C74A) hampers autophagosome closure: a useful protein for inhibiting autophagy, Autophagy 2009; 5: 88–89. Pii: 7183.

[38] H. Omi, A. Okamoto, T. Nikaido, M. Urashima, R. Kawaguchi, N. Umehara, K. Sugiura, M. Saito, T. Kiyono, T. Tanaka: Establishment of an immortalized human extravillous trophoblast cell line by retroviral infection of E6/E7/hTERT and its transcriptional profile during hypoxia and reoxygenation, Int J Mol Med 2009; 23: 229–236.

[39] S. Kalkunte, R. Boij, W. Norris, J. Friedman, Z. Lai, J. Kurtis, K.H. Lim, J.F. Padbury, L. Matthiesen, S. Sharma: Sera from preeclampsia patients elicit symptoms of human disease in mice and provide a basis for an in vitro predictive assay, Am J Pathol 2010; 177: 2387–2398. Pii: S0002-9440(10)60291-X, doi: 10.2353/ajpath.2010.100475.

[40] J.H. Choi, H.J. Lee, T.H. Yang, G.J. Kim: Effects of hypoxia inducible factors-1alpha on autophagy and invasion of trophoblasts, Clin Exp Reprod Med 2012; 39: 73–80. Doi: 10.5653/cerm.2012.39.2.73.

[41] I. Papandreou, A.L. Lim, K. Laderoute, N.C. Denko: Hypoxia signals autophagy in tumor cells via AMPK activity, independent of HIF-1, BNIP3, and BNIP3L, Cell Death Differ 2008; 15: 1572–1581. Pii: cdd200884, doi: 10.1038/cdd.2008.84.

[42] H. Zhang, M. Bosch-Marce, L.A. Shimoda, Y.S. Tan, J.H. Baek, J.B. Wesley, F.J. Gonzalez, G.L. Semenza: Mitochondrial autophagy is an HIF-1-dependent adaptive metabolic response to hypoxia, J Biol Chem 2008; 283: 10892–10903. Pii: M800102200, doi: 10.1074/jbc.M800102200.

[43] S. Busch, S.J. Renaud, E. Schleussner, C.H. Graham, U.R. Markert: mTOR mediates human trophoblast invasion through regulation of matrix-remodeling enzymes and is associated with serine phosphorylation of STAT3, Exp Cell Res 2009; 315: 1724–1733. Pii: S0014-4827(09)00051-2, doi: 10.1016/j.yexcr.2009.01.026.

[44] M. Yamanaka-Tatematsu, A. Nakashima, N. Fujita, T. Shima, T. Yoshimori, S. Saito: Autophagy induced by HIF1alpha overexpression supports trophoblast invasion by supplying cellular energy, PLoS One 2013; 8: e76605. Pii: PONE-D-13-18219, doi: 10.1371/journal.pone.0076605.

[45] S. Chan: Targeting the mammalian target of rapamycin (mTOR): a new approach to treating cancer, Br J Cancer 2004; 91: 1420–1424. Pii: 6602162, doi: 10.1038/sj.bjc.6602162.

[46] J.V. Ferreira, H. Fofo, E. Bejarano, C.F. Bento, J.S. Ramalho, H. Girao, P. Pereira: STUB1/CHIP is required for HIF1A degradation by chaperone-mediated autophagy, Autophagy 2013; 9: 1349–1366. Pii: 25190, doi: 10.4161/auto.25190.

[47] P. Signorelli, L. Avagliano, E. Virgili, V. Gagliostro, P. Doi, P. Braidotti, G.P. Bulfamante, R. Ghidoni, A.M. Marconi: Autophagy in term normal human placentas, Placenta 2011; 32: 482–485. Pii: S0143-4004(11)00108-1, doi: 10.1016/j.placenta.2011.03.005.

[48] Z.Y. Shen, E.M. Li, S.Q. Lu, J. Shen, Y.M. Cai, Y.E. Wu, R.M. Zheng, L.J. Tan, L.Y. Xu: Autophagic and apoptotic cell death in amniotic epithelial cells, Placenta 2008; 29: 956–961. Pii: S0143-4004(08)00302-0, doi: 10.1016/j.placenta.2008.09.001.

[49] T. Yamada, A.R. Carson, I. Caniggia, K. Umebayashi, T. Yoshimori, K. Nakabayashi, S.W. Scherer: Endothelial nitric-oxide synthase antisense (NOS3AS) gene encodes an autophagy-related protein (APG9-like2) highly expressed in trophoblast, J Biol Chem 2005; 280: 18283–18290. Pii: M413957200, doi: 10.1074/jbc.M413957200.

[50] N. Kanayama, K. Takahashi, T. Matsuura, M. Sugimura, T. Kobayashi, N. Moniwa, M. Tomita, K. Nakayama: Deficiency in p57Kip2 expression induces preeclampsia-like symptoms in mice, Mol Hum Reprod 2002; 8: 1129–1135.

[51] T. Kojima, T. Yamada, R. Akaishi, I. Furuta, T. Saitoh, K. Nakabayashi, K.I. Nakayama, K. Nakayama, S. Akira, H. Minakami: Role of the Atg9a gene in intrauterine growth and survival of fetal mice, Reprod Biol 2015; 15: 131–138. Pii: S1642-431X(15)00043-1, doi: 10.1016/j.repbio.2015.05.001.

[52] S. Saito, A. Nakashima: A review of the mechanism for poor placentation in early-onset preeclampsia: the role of autophagy in trophoblast invasion and vascular remodeling,

J Reprod Immunol 2014; 101–102: 80–88. Pii: S0165-0378(13)00078-8, doi: 10.1016/j.jri.2013.06.002.

[53] J.M. Roberts, C.A. Hubel: The two stage model of preeclampsia: variations on the theme, Placenta 2009; 30 Suppl A: S32-S37. Pii: S0143-4004(08)00384-6, doi: 10.1016/j.placenta.2008.11.009.

[54] I. Caniggia, S. Grisaru-Gravnosky, M. Kuliszewsky, M. Post, S.J. Lye: Inhibition of TGF-beta 3 restores the invasive capability of extravillous trophoblasts in preeclamptic pregnancies, J Clin Invest 1999; 103: 1641–1650. Doi: 10.1172/JCI6380.

[55] C.H. Graham, P.K. Lala: Mechanism of control of trophoblast invasion in situ, J Cell Physiol 1991; 148: 228–234. Doi: 10.1002/jcp.1041480207.

[56] G.E. Lash, H.A. Otun, B.A. Innes, J.N. Bulmer, R.F. Searle, S.C. Robson: Inhibition of trophoblast cell invasion by TGFB1, 2, and 3 is associated with a decrease in active proteases, Biol Reprod 2005; 73: 374–381. Pii: biolreprod.105.040337, doi: 10.1095/biolreprod.105.040337.

[57] N. Wolf, W. Yang, C.E. Dunk, I. Gashaw, S.J. Lye, T. Ring, M. Schmidt, E. Winterhager, A. Gellhaus: Regulation of the matricellular proteins CYR61 (CCN1) and NOV (CCN3) by hypoxia-inducible factor-1{alpha} and transforming-growth factor-{beta}3 in the human trophoblast, Endocrinology 2010; 151: 2835–2845. Pii: en.2009-1195, doi: 10.1210/en.2009-1195.

[58] K. Kiyono, H.I. Suzuki, H. Matsuyama, Y. Morishita, A. Komuro, M.R. Kano, K. Sugimoto, K. Miyazono: Autophagy is activated by TGF-beta and potentiates TGF-beta-mediated growth inhibition in human hepatocellular carcinoma cells, Cancer Res 2009; 69: 8844–8852. Pii: 0008–5472.CAN-08-4401, doi: 10.1158/0008-5472.CAN-08-4401.

[59] H.I. Suzuki, K. Kiyono, K. Miyazono: Regulation of autophagy by transforming growth factor-beta (TGF-beta) signaling, Autophagy 2010; 6: 645–647. Pii: 12046, doi: 10.4161/auto.6.5.12046.

[60] T.H. Hung, S.F. Chen, L.M. Lo, M.J. Li, Y.L. Yeh, T.T. Hsieh: Increased autophagy in placentas of intrauterine growth-restricted pregnancies, PLoS One 2012; 7: e40957. Pii: PONE-D-12-09139, doi: 10.1371/journal.pone.0040957.

[61] M. Melland-Smith, L. Ermini, S. Chauvin, H. Craig-Barnes, A. Tagliaferro, T. Todros, M. Post, I. Caniggia: Disruption of sphingolipid metabolism augments ceramide-induced autophagy in preeclampsia, Autophagy 2015; 11: 653–669. Doi: 10.1080/15548627.2015.1034414.

[62] T.T. Kanninen, A. Jayaram, S. Jaffe Lifshitz, S.S. Witkin: Altered autophagy induction by sera from pregnant women with pre-eclampsia: a case–control study, BJOG 2014; 121: 958–964. Doi: 10.1111/1471-0528.12755.

[63] S.Y. Oh, S.J. Choi, K.H. Kim, E.Y. Cho, J.H. Kim, C.R. Roh: Autophagy-related proteins, LC3 and Beclin-1, in placentas from pregnancies complicated by preeclampsia, Reprod Sci 2008; 15: 912–920. Pii: 15/9/912, doi: 10.1177/1933719108319159.

[64] V. Agrawal, M.K. Jaiswal, T. Mallers, G.K. Katara, A. Gilman-Sachs, K.D. Beaman, E. Hirsch: Altered autophagic flux enhances inflammatory responses during inflammation-induced preterm labor, Sci Rep 2015; 5: 9410. Pii: srep09410, doi: 10.1038/srep09410.

[65] E. Delorme-Axford, R.B. Donker, J.F. Mouillet, T. Chu, A. Bayer, Y. Ouyang, T. Wang, D.B. Stolz, S.N. Sarkar, A.E. Morelli, Y. Sadovsky, C.B. Coyne: Human placental trophoblasts confer viral resistance to recipient cells, Proc Natl Acad Sci U S A 2013; 110: 12048–12053. Pii: 1304718110, doi: 10.1073/pnas.1304718110.

[66] N. Mizushima, T. Yoshimori, B. Levine: Methods in mammalian autophagy research, Cell 2010; 140: 313–326. Pii: S0092-8674(10)00063-2, doi: 10.1016/j.cell.2010.01.028.

Section 5

Selective Autophagy in Health, Disease and Therapy

Chapter 17

Autophagy Modulation for Organelle-Targeting Therapy

Waleska K. Martins and Mauricio S. Baptista

Additional information is available at the end of the chapter

http://dx.doi.org/10.5772/63976

Abstract

Autophagy is a crucial metabolic pathway that sustains cellular homeostasis in health and that can also play either a protective or a destructive role in disease. During the last decade, progress made in understanding of the molecular basis of autophagy has uncovered an exciting opportunity to target it for the treatment of several human illnesses. In fact, there is emerging interest in autophagy-modulating and autophagy-targeted therapy with a variety of pharmacologic agents. However, to develop effective autophagy-targeted therapy, it is essential to identify the pharmacologic key targets in the autophagy pathway. In this chapter, we reviewed the cases of success and pitfalls of activating or inhibiting autophagy attempting therapeutic intervention of diseases, including cancer, neurologic disorders, and infectious diseases. In all these histopathologic states, autophagy is considered as the principal cellular mechanisms of defense and immunochemical homeostasis. In the last section of this chapter, we discuss main directions that may be of particular use in the future investigations, including a promissory avenue for autophagy modulation for organelle-targeting therapy through a promotion of parallel damage in lysosomal and mitochondrial membranes.

Keywords: autophagy-targeted therapy, activation/inhibition of autophagy, triterpenoids, lysosomal and mitochondrial membranes

1. Introduction

The impact of autophagy in human pathogenesis comprises its critical function for the degradation and recycling of long-lived proteins, lipid droplets, protein aggregates, mature ribosomes, glycogen, and even entire organelles such as the endoplasmic reticulum, mitochondria, and Golgi apparatus [1, 2]. For example, when the efficiency of mitophagy (a type of autophagy that specifically targets dysfunctional mitochondria [2–7]) is reduced, the maintenance of cellular homeostasis decreases, leading to cell aging, genomic instability, and senescence [4, 8, 9]. However, the molecular mechanism by which deficient mitophagy jeopardizes genomic stability are unclear [10].

To develop effective autophagy-based therapy, it has been essential to identify the pharmacologic key targets in the autophagy pathway for the development of new therapeutic agents. As discussed earlier in this book, autophagy can play either a protective or a destructive role in disease states and thus for therapeutic purposes is valuable to identify and develop pharmacologic agents that might activate or inhibit this cellular process [11].

We will evolve this part of the chapter to review the cases of success and pitfalls of activating or inhibiting autophagy attempting therapeutic intervention of diseases, including cancer, neurologic disorders, and infectious diseases. In the following sections, we discuss a few directions that may be of particular use in future investigations. As recently proposed by our group, the promotion of parallel damage in lysosomes and mitochondria represents a promissory avenue for therapeutic autophagy targeting and aiding controlled cell death and senescence [12, 13].

2. Autophagy-modulating drugs

During the last decade, progress made in understanding of the molecular basis of autophagy has uncovered an exciting opportunity to target it for the treatment of human illnesses [14]. In principle, understanding the role of autophagy in diseases has helped identify new avenues of pharmacologic modulation of autophagy as novel therapeutic intervention. Thus, knowing the process of autophagy targeting might facilitate the search of new drugs or concepts for the treatment of several types of diseases whose etiology or progression is associated with autophagy including cancer [8, 15–17], degenerative diseases 25 [18–23], and lysosomal storage disorders [24–26].

2.1. Autophagy inhibitors

The pharmacologic inhibition targeting the early or later autophagy process has been demonstrated to play pivotal roles in cellular outcome and may affect disease processes. Because inhibition of autophagy by pharmacologic agents also may have some off-target effects on cellular functions, the question of whether, for example, cell death can truly occur due to autophagy alone remains to be clarified [11, 27]. **Table 1** lists the compounds identified as inhibitors of autophagy.

The inhibitors that target the early stage of autophagy include 3-MA, Wortmannin, and LYS294002, all of which inhibit the class III PI3K (VSP34) and disable the formation of autophagosome. The inhibitors act on the later stage of autophagy, including compounds (see **Table 1**) that are capable of preventing lysosomal degradation or blocking the fusion of autophagosomes with lysosomes. For example, the neutralization of intralysosomal pH by lysosomotropic agents such as Bafilomycin A_1, Chloroquine (CQ), Hydroxychloroquine (HCQ), or NH_4Cl prevents the digestive activity of hydrolases, leading to inhibition of degradative activity of autolysosomes [33, 34, 37, 44]. In their unprotonated form, CQ and HCQ can diffuse across cell membranes to become protonated and accumulated only in acidic organelles. Once trapped within lysosomes, they interfere with prosurvival autophagy, resulting in controlled cell death [57–60]. This unique property has established CQ as the most widely used drug to inhibit autophagy in vitro and in vivo. Bafilomycin A_1 is a selective vacuolar-type H^+-ATPase [V-ATPase] inhibitor responsible for acidifying lysosomes and endosomes [33, 34]. Of note Bafilomycin A_1 blocks the fusion of autophagosomes with lysosomes, which results from inhibition of ATP2A/SERCA activity independently of its effect on intralysosomal pH [35, 61].

Compounds	Autophagy signaling pathway
3-methyladenine	An inhibitor of autophagic/lysosomal protein degradation [28], but not a specific autophagy inhibitor [29] may also inhibit the activity of Phosphatidylinositol 3-kinase [30], effectively blocking the early stage of autophagy. 3-MA does not inhibit BECN1-independent autophagy [29].
ARN5187	4-[[[1-(2-fluorophenyl)cyclopentyl]amino]methyl]-2-[(4-methylpiperazin-1 -yl)methyl]phenol, 1 is a lysosomotropic compound with a dual inhibitory activity against the circadian regulator NR1D2/REV-ERBβ and autophagy [31, 32].
Bafilomycin A_1	A V-ATPase inhibitor that causes an increase in lysosomal/vacuolar pH and, ultimately, blocks fusion of autophagosomes with the vacuole; the latter may result from inhibition of ATP2A/SERCA [33–35].
Betulinic acid	A pentacyclic triterpenoid that promotes parallel damage in mitochondrial and lysosomal compartments and, ultimately, triggers autophagy associated cell death in human keratinocytes [12].
CA074	*N*-(L-3-trans-propylcarbamoyloxirane-2-carbonyl)-L-isoleucyl-L-proline is a potent and specific inhibitor of cathepsin B in vitro [36].
Chloroquine	Chloroquine and its analog Hydroxychloroquine are lysosomotropic compounds that elevate/neutralize the lysosomal/vacuolar pH [37].
Colchicine	A microtubule depolarizing agent that may block autophagosome maturation to autolysosomes and increased LC3II protein levels [38].
Desmethyl clomipramine	3-(2-chloro-5,6-dihydrobenzo[b][1]benzazepin-11-yl)-N-methylpropan-1 -Amine, an active metabolite of clomipramine inhibits late autophagy through a significant blockage of the degradation of autophagic cargo [39]; may induce an increase in the steady-state levels of p62/SQSTM1 by inhibiting the autophagic flux as opposed to an activation of the autophagic pathway [39].
E64d	Inhibits papain-like cathepsin cysteine proteases and calpain-activated neutral proteases [40]; should be used in combination with pepstatin A to inhibit lysosomal protein degradation [29].
Eflornithine	2,5-diamino-2-(difluoromethyl)pentanoic acid, an irreversible inhibitor of ODC1 (ornithine decarboxylase 1) that blocks spermidine synthesis and *ATG5* gene expression acting as a novel autophagy inhibitor [41].

Compounds	Autophagy signaling pathway
Leupeptin	An inhibitor of cysteine, serine, and threonine proteases that causes significant inhibition of the intracellular maturation of cathepsin B, L, and H [37, 42, 43]; decreases the degradation of short- and long-lived proteins [44]; should be used in combination with pepstatin A and/or E-64d to block lysosomal protein degradation [29].
Lucanthone	Interferes with lysosomal function and leads to the accumulation of undegraded proteins and induces a cathepsin D-mediated apoptosis [45].
LYS294002	2-(4-morpholi-nyl)-8-phenylchromone may prevent autophagic sequestration by inhibiting phosphatidylinositol 3-kinase activity [30].
Monensin	An inhibitor of protein transport, acts as proton exchange for potassium or sodium and inhibits autophagy by preventing the fusion of the autophagosome with the lysosome [46].
NH4Cl	Lysosomotropic compound that elevate/neutralize the lysosomal/vacuolar pH, inhibiting the lysosomal pathway of protein degradation [37].
Nocodazole	A depolymerizer of nonacetylated microtubules and impairs tubulin acetylation but does not affect polymerized acetylated microtubules; may impair the conversion of LC3I to LC3II but does not block the degradation of LC3II-associated autophagosomes [47].
Pepstatin A	An aspartyl protease inhibitor that can be used to partially block lysosomal degradation [44]; should be used in combination with other inhibitors such as E-64d [29].
PES	2-Phenyl-ethynesulfonamide , a small molecule inhibitor of heat shock protein 70 (HSP70), impairs autophagy through its inhibitory effects on lysosomal functions showing an accumulation of the precursor procathepsin L and a markedly reduced abundance of the smaller, mature form of the enzyme [48].
Propofol	May exert protective effects on neuronal cells and cardiomyocytes, in part through the inhibition of early autophagy [49–51].
Spautin	A specific and potent autophagy inhibitor 1 that promotes degradation of the Vps34 (a phosphoinositide 3-kinase class III isoform) via inhibiting ubiquitin-specific processing protease 10 (USP10) and USP13, two ubiquitin-specific peptidases that target the deubiquitination of Beclin1 [52].
Thapsigargin	A sarco/endoplasmic reticulum Ca (2+)-ATPase inhibitor that inhibits autophagic sequestration of cytosolic material through the depletion of intracellular Ca2+ stores [53, 54]; also may lead to the accumulation of mature autophagosomes by blocking autophagosome fusion with the endocytic system by interfering with the recruitment of RAB7 [55].
Vacuolin-1	2-N-[(3-iodophenyl)methylideneamino]-6-morpholin-4-yl-4-N,4-N-diphenyl-1,3,5-triazine-2,4-diamine, an activator of RAB5A GTPase activity that potently and reversibly inhibits autophagosome-lysosome fusion; also may alkalinize lysosomal pH and decrease lysosomal Ca^{2+} content [56].
Vinblastine	A depolymerizer of both nonacetylated and acetylated microtubules that interferes with both LC3I-LC3II conversion and LC3II-associated autophagosome fusion with lysosomes [47].
Wortmannin	An inhibitor of PI3K and PtdIns3K that blocks autophagy, but not a specific inhibitor that prevents autophagic sequestration such as 3-methyladenine [30].

Table 1. Compounds known to inhibit autophagy.

Other inhibitors of autophagy that impair the autolysosome formation include the antidepressant drug Desmethylclomipramine, the anti-schistome agent Lucanthone, Eflornithine, Monensin, PES, Spautin, Thapsigargin, Vacuolin-1, and Vinblastine (see **Table 1**).

The digestive phase of autophagy may also be blocked by lessening lysosome-mediated proteolysis such as the cysteine protease inhibitor E-64d; the aspartic protease inhibitor Pepstatin A; the active inhibitor of cathepsin B CA074; and the cysteine, serine, and threonine protease inhibitor Leupeptin [37, 44, 62]. Autophagosomes and lysosomes move along the microtubules to fuse, so microtubule-disrupting agents, including taxanes, Nocodazole, Colchicine, and Vinca alkaloids, may inhibit the fusion of autophagosomes with lysosomes [11, 63, 64].

2.2. Autophagy activators

It is now generally believed that modulating the activity of autophagy through targeting specific regulatory molecules in the autophagy machinery may improve clinical outcome for diverse diseases [11, 14]. In this context, mTOR inhibitors has been considered as the most potent activators of autophagy by playing pivotal key negative regulatory role [29]. **Table 2** lists the compounds that have been identified as activators of autophagy.

Compounds	Autophagy signaling pathway
10-NCP	10-(4'-N-diethylamino)butyl)-2-chlorophenoxazine that may promote potential and safe upregulation of autophagy in neurons in an AKT- and mTOR-independent fashion [65].
17-AAG	17-Allylamino-17-Demethoxygeldanamycin that may inhibit the HSP90 CDC37 chaperone complex activating autophagy in certain systems (e.g., neurons), but impairs starvation-induced autophagy and mitophagy in others by promoting the turnover of ULK1 [66].
Akti-1/2	Akt inhibitor VIII isozyme-selective Akti-1/2 can promote allosteric inhibition of AKT1 and AKT2 and activates autophagy in B-cell lymphoma [67].
AUTEN-67	An inhibitor of MTMR14, a myotubularin-related phosphatase that may antagonize the formation of autophagic membrane [68].
AZD8055	5-[2,4-bis[(3S)-3-methylmorpholin-4-yl]pyrido[2,3- d]pyrimidin-7-yl]-2-methoxyphenyl methanol that may inhibit both mTORC1 and mTORC2 [69].
Everolimus	An inhibitor of mTORC1 that induces both autophagy and apoptosis in B-cell lymphoma primary cultures [67].
KU-0063794	A specific mTOR inhibitor that may bind the catalytic site and activates autophagy [70, 71].
MLN4924	A small molecule inhibitor of NEDD8 activating enzyme (NAE) [72] that triggers autophagy through the blockage of mTOR signals via DEPTOR as well as the HIF1A-DDIT4/REDD1-TSC1/2 axis [73] as a result of inactivation of cullin-RING ligases [74].

Compounds	Autophagy signaling pathway
Oleanolic acid	A pentacyclic triterpenoid that promotes damage in mitochondrial compartments, and ultimately, activates prosurvival autophagy in human keratinocytes [12].
NAADP-AM	Nicotinic acid adenine dinucleotide phosphate (NAADP) can mobilize Ca^{2+} from acidic Ca^{2+} stores through lysosomal two-pore channels (TPCs) in primary cultured rat astrocytes and present evidence that NAADP-evoked Ca^{2+} signals regulate autophagy [75].
NVP-BEZ235	NVP-BEZ235 is an imidazo[4,5-c]quinoline derivative that can inhibit the activity of target proteins in the PI3K/AKT/mTOR cascade and activates autophagy in human gliomas [76, 77].
PMI	Is a pharmacological P62-mediated mitophagy inducer (PMI) that activates mitophagy without recruiting Parkin or collapsing the mitochondrial membrane potential [78].
PP242	2-(4-amino-1-isopropyl-1H- pyrazolo[3,4-d]pyrimidin-3-yl)-1H-indol-5-ol is a ATP-competitive inhibitor of mTORC1 and mTORC2 [79, 80]; should be more effective mTORC1 inhibitor than rapamycin [81].
PP30	3-(4-amino-1-isopropyl-1H-pyrazolo[3,4-d]pyrimidin-3-yl)-N-(4,5-dihydrothiazol-2-yl) benzamide is a ATP-competitive inhibitor of mTORC1 and mTORC2 [81].
Rapamycin	Binds to FKBP1A/FKBP12 and inhibits mTORC1; the complex binds to the FRB domain of mTOR and limits its interaction with RPTOR, thus inducing autophagy, but only providing partial mTORC1 inhibition [82].
Resveratrol	A natural polyphenol that affects many proteins [83] via both AMPK activation and JNK-mediated p62/SQSTM1 expression activates autophagy [84, 85].
Ridaforolimus	Binds to and inhibits the mammalian target of Rapamycin (mTOR), which may result in cell cycle arrest and, consequently, the inhibition of tumor cell growth and proliferation.
RSVAs	Synthetic small-molecule analogs of resveratrol that potently activate AMPK and induce autophagy [86].
Saikosaponin-d	A natural small-molecule inhibitor of ATP2A/SERCA that induces autophagy via direct inhibition of sarcoplasmic/endoplasmic reticulum Ca^{2+} ATPase (SERCA), leading to the increase of intracellular calcium ion levels and activating the Ca^{2+}/calmodulin-dependent kinase kinase-b (CaMKKβ)/AMPK signaling cascade [87].
Tat-Beclin 1	A cell penetrating peptide that potently induces autophagy [88, 89].
Temsirolimus	Is Rapamycin ester analog CCI-779 with better stability and pharmacological properties compared to rapamycin that activates autophagy in neurons in Alzheimer's disease [90–92].
TMS	Trans-3,5,4-trimethoxystilbene upregulates the expression of the transient receptor potential canonical channel 4 (TRPC4), resulting in mTOR inhibition and autophagy activation [93].

Compounds	Autophagy signaling pathway
Torin1	1-[4-(4-propanoylpiperazin-1-yl)- 3-(trifluoromethyl)phenyl]-9-quinolin-3-ylbenzo [h][1, 6]naphthyridin-2-one, a catalytic mTORC1 and mTORC2 inhibitor that induces autophagy [94].
Trehalose	mTOR-independent, autophagic enhancer that may be relevant for the treatment of different neurodegenerative diseases [20, 95, 96].
Tunicamycin	A glycosylation inhibitor that induces autophagy due to endoplasmic reticulum stress [97].
WYE-125132	1-[4-[1-(1,4-dioxaspiro[4.5]decan- 8-yl)-4-(8-oxa-3- azabicyclo[3.2.1]octan-3-yl)pyrazolo [3,4- d]pyrimidin-6-yl]phenyl]-3-methylurea is an ATP-competitive and specific inhibitor of mTORC1 and mTORC2 [98].

Table 2. Compounds known to activate autophagy.

There are several other agents that negatively regulate autophagy, such as inositol 1,4,5-trisphosphate (IP3), epidermal growth factor receptor (EGFR), Bcl-2, and Bcl-xl. The mTORC1 inhibitor Rapamycin and its analogs Temsirolimus (CCI-79, Torisel), Everolimus (RAD001, Afinitor), and Ridaforolimus (AP-23573, Deferolimus, MK-8669) are strong inducers of autophagy [14], as are the ATP-competitive inhibitors of mTOR such as Torin 1 [94], PP242 [79, 80], PP30 [81] and AZD8055 [69], and WYE-125132 [98], but their autophagy-inducing efficacy has not been well documented [11]. Other autophagic enhancers that induces autophagy via a mTOR-independent pathway, include AUTEN-67 [68], 10-NCP [65], PMI [78], Resveratrol [84, 85], Trehalose [20, 95, 96], and Tunicamycin [97].

3. Modulation of autophagy as a cancer therapy

The human cancer represents a significant worldwide public health problem considered as the main cause of death [99]. Its worldwide incidence is expected to show more than 21 thousand million new cases in 2030 [100]. To deal with such increased incidence, 47,608 clinical 15 trials have been currently carried out according to *Clinical Trials.Gov* [101]. Therapeutic targeting of the autophagy pathway as a new anticancer strategy has been under extensive investigation [11, 64, 102, 103]. Several data indicate that prosurvival autophagy confers a tumor growth advantage through the supplementation of required nutrition of growth, and thus it represents a novel therapeutic target [46, 104]. Actually, autophagy may represent a major impediment to successful cancer therapy by radiation, drugs (e.g., Doxorubicin, Temozolomide, and Etoposide), histone deaceltylase inhibitors, Arsenic trioxide, TNF-alpha, IFN-gamma, Imatinib, and Rapamycin and the anti-estrogen hormonal therapy Tamoxifen as reviewed [46, 104, 105]. Dalby and colleagues propose that inhibitors of autophagy may either enhance the efficacy of anti-tumor therapy or promote cell death not only in primary cancer types but also in advanced-stage cancers and metastatic tumors that are considered drug resistant or apoptosis resistant, such as chemotherapy-resistant cancer [105]. However, depending on the context, such as tumor type or stage, the autophagy-enhancing agents believed to induce a type II programmed cell death mechanism through an extensive autophagic degradation of intracellular content may also elicit beneficial effects in the treatment of cancer [105]. Both the approaches, inhibition of either the prosurvival or induction of prodeath mechanism of autophagy, will be discussed further.

3.1. Use of autophagy activators in cancer treatment

Several evidence have suggested that increased autophagy may kill cells. However, the weakness of many studies has been that the demonstration of autophagy after a cytotoxic treatment does not prove that autophagy contributed to cell death, only that it was associated with it [11]. It is equally plausible that increased autophagy in these settings was more a failed effort to maintain cell survival than triggering per se cell death. If autophagy would act more definitively as a prodeath cell role than a prosurvival one, its inhibition would have to increase survival. In fact, most studies have showed that a cell death had been counterbalanced by an autophagic salvage response rather than demonstrate a causative role for autophagy in the promotion of cell death [106, 107]. Nonetheless, induction of autophagy-associated cell death has been suggested as a potential strategy to eradicate human cancers [108]. In fact, several anticancer drugs have been reported to kill tumor cells through autophagy-mediated mechanisms, include Photodynamic Therapy, Cisplatin, 5-Fluoroacil, Etoposide, Imatinib, and Paclitaxel, as reviewed [109].

Rapamycin and its more soluble analog Temsirolimus trigger autophagy, as does KU-63794, whose selective mTOR inhibition has been attributed to its antitumor mechanism regardless apoptosis induction [110–112] in which the disruption of the PI3K/Akt signaling pathway might greatly enhance the effectiveness of mTOR inhibitors [110]. Likewise, the inhibitory effect of the mTOR inhibitor Everolimus on acute lymphoblastic leukemia was associated with autophagy activation [113]. The combination of Temsirolimus, an mTOR inhibitor, and HCQ, an autophagy inhibitor, augments cell death in preclinical models [114].

The ATP-competitive inhibitors of mTORC1/mTORC2, WYE-125132 [98], and AZD8055 [69] have demonstrable anticancer activity by growth inhibition, and potentially autophagy both in vitro and in vivo. In case of rapamycin-resistant T37/46 phosphorylation sites on 4E-BP1, AZD8055 may fully inhibit mTOR [69]. AZD8055 is currently in Phase I clinical trials as an antitumor agent (NCT00973076, NCT01316809, NCT00999882 and NCT00731263). The tyrosine kinase inhibitor Dasatinib (BMS-354825) has been reported to enhance the antiglioma effect of Temozolomide through triggering significant decrease in cell proliferation while simultaneously increasing autophagy, and this action can be antagonized by the autophagy inhibitor 3-MA [115].

Other types of drugs possessing an autophagy-inducing effect have also found their potential application in cancer treatment (see **Table 3**). For instance, autophagy-associated cell death may contribute to the anticancer actions of the histone deacetylase (HDAC) inhibitor suberoylanilide hydroxamic acid (Vorinostat) [116–118]. Vorinostat may induce autophagy through downregulation of Akt/mTOR signaling and induction of ER stress response, whose biological effects might be antagonized by the autophagy inhibitor 3-MA [117]. Coadministration of Vorinostat and a poly (ADP-ribose) polymerase (PARP) inhibitor Olaparib synergistically

http://dx.doi.org/10.5772/63976

inhibits the growth of triple-negative breast cancer cells through increased apoptotic and autophagy-associated cell death [119]. In Tamoxifen-resistant MCF-7 breast cancer cells the HDAC inhibitor MHY218 induces apoptosis or autophagy-related cell death [120]. The estrogen receptor antagonist Raloxifene induces autophagy via the activation of AMPK by sensing decreases in ATP, leading to a nonapoptotic autophagy-associated cell death in breast cancer [121]. However, it has been proposed that autophagy is sterol-dependent and is associated with cell survival rather than cytotoxicity [122]. The natural products Resveratrol [84], triterpenoids Ursolic acid [123–125], and Saponin [126] promote cancer cell death associated with activation of autophagy. It is conceivable that some autophagy-inducing agents may also be useful in cancer therapies because of their ability to trigger autophagy-associated cell death [11, 113, 115]. The same attention given to inhibitors of autophagy should be given to autophagy-inducing or autophagy-enhancing agents [114, 127–131].

Cancer Type	Identifier	Study	Phase	Status	
Renal cell cancer	NCT00830895	Everolimus for nonclear cell renal cell carcinoma (RCC)	II	1	[132]
	NCT01090466	Gemcitabine Hydrochloride, Cisplatin, and Temsirolimus as first-line therapy to treat patients with locally advanced and/or metastatic transitional cell cancer of the urothelium	I/II	1	
Prostate cancer	NCT01313559	Pasireotide (SOM230) with or without Everolimus to treat patients with hormone-resistant chemotherapy naive prostate cancer	II	1	
	NCT00574769	Docetaxel with Everolimus and Bevacizumab in men with advanced prostate cancer	I/II	1	
	NCT02339168	Enzalutamide and Metformin Hydrochloride to treat patients with hormone-resistant prostate cancer	I	2	
	NCT01748500	Pantoprazole and Docetaxel for men with metastatic castration-resistant prostate cancer	II	2	
	NCT01497925	ADIPEG 20 and Docetaxel in solid tumors with emphasis on prostate cancer and nonsmall cell lung cancer	I	2	
Breast cancer	NCT00411788	Rapamycin and Trastuzumab for patients with HER-2 receptor positive metastatic breast cancer	II	3	
	NCT01111825	Temsirolimus and Neratinib for the treatment of patients with metastatic HER2-amplified or triple negative breast cancer	I/II	2	
	NCT00736970	Ridaforolimus in combination with Trastuzuma in patients with metastatic, HER2-positive breast cancer who have developed resistance to Trastuzumab.	III	1	[133]
	NCT01605396	Ridaforolimus and Exemestane, compared with Ridaforolimus, Dalotuzumab and Exemestane to	II	2	

Cancer Type	Identifier	Study	Phase	Status	
		treat breast cancer			
	NCT01234857	Ridaforolimus in combination with Dalotuzumab compared to the standard of care treatment in estrogen receptor positive breast cancer patients	II	1	
Nonsmall cell lung cancer	NCT00079235	Temsirolimus to treat patients with stage III-B (with pleural effusion) or stage IV nonsmall cell lung cancer	II	1	
	NCT00923273	Sirolimus and Pemetrexed to treat nonsmall cell lung cancer	I/II	1	
Small cell lung cancer	NCT00374140	Everolimus in previously treated small cell lung cancer	II	1	
	NCT01079481	Combination anticancer therapy of Everolimus and Paclitaxel for relapsed or refractory small cell lung cancer	I/II	1	
Pancreatic cancer	NCT01648465	Everolimus to treat newly diagnosed patients with advanced gastrointestinal neuroendocrine tumors	II	4	
	NCT01537107	Sirolimus and Vismodegib to treat patients with solid tumors or pancreatic cancer that is metastatic or cannot be removed by surgery	I	5	
Glioblastoma	NCT00329719	Temsirolimus and Sorafenib to treat patients with recurrent glioblastoma	I/II	1	
	NCT01062399	Everolimus, Temozolomide, and Radiation therapy to treat patients with newly diagnosed glioblastoma multiforme	I/II	2	
	NCT01956734	Virus DNX2401 and Temozolomide to treat recurrent glioblastoma	I	2	
Colorectal cancer	NCT00522665	Second-line therapy with Irinotecan, Cetuximab, and Everolimus to treat colorectal cancer	I/II	1	
	NCT01154335	Everolimus and Linsitinib to treat patients with refractory metastatic colorectal cancer	I	1	[134]
Chronic myeloid leukemia	NCT01188889	Everolimus to treat chronic phase chronic myeloid leukemia with persistent molecular disease.	I/II	6	
Chronic lymphocytic leukemia	NCT00935792	Everolimus and Alemtuzumab to treat patients with recurrent chronic lymphocytic leukemia or small lymphocytic lymphoma	I/II	1	
Advanced solid tumor	NCT00849550	Everolimus in combination with current standard treatment of XELOX-A (Bevacizumab, Oxaliplatin, Capecitabine) to treat advanced solid tumors	I	1	
	NCT01020305	Temsirolimus to reverse androgen insensitivity for	I/II	1	

http://dx.doi.org/10.5772/63976

Cancer Type	Identifier	Study	Phase	Status
		castration-resistant prostate cancer		
	NCT00657982	Everolimus in a neoadjuvant setting in men with intermediate or high risk prostate cancer	II	4
	NCT01155258	Temsirolimus and Vinorelbine Ditartrate to treat patients with unresectable or metastatic solid tumors	I	2
	NCT01295632	Ridaforolimus with MK-2206 or MK-0752 for participants with advanced cancer	I	1 [135]
	NCT01169532	Ridaforolimus and the HDAC inhibitor Vorinostat to treat patients with advanced cancer	I	1 [136]
	NCT00781846	Ridaforolimus in combination with Bevacizumab for patients with advanced cancers	I	1 [137]
Endometrial carcinoma	NCT00739830	Ridaforolimus in advanced endometrial carcinoma	II	1 [138]
Multiple myeloma	NCT00693433	Temsirolimus and Dexamethasone to treat patients with recurrent or refractory multiple myeloma	I	1
	NCT00398515	Temsirolimus and Lenalidomide to treat patients with previously treated multiple myeloma	I	1
	NCT00918333	Everolimus and Panobinostat to treat patients with recurrent multiple myeloma, non-Hodgkin lymphoma, or Hodgkin lymphoma	I/II	2
	NCT00474929	Everolimus and Sorafenib to treat patients with relapsed or refractory lymphoma or multiple myeloma	I/II	2
Ovarian cancer	NCT01460979	Temsirolimus to treat ovarian cancer of women who progressed during previous platinum chemotherapy or within 6 months after therapy or advanced endometrial carcinoma	II	1
	NCT00982631	Temsirolimus and Pegylated Liposomal Doxorubicin to treat advanced or recurrent breast, endometrial and ovarian cancer	I	3
	NCT01196429	Temsirolimus, Carboplatin, and Paclitaxel as first-line therapy to treat patients with newly diagnosed stage III–IV clear cell ovarian cancer	II	1
	NCT01010126	Temsirolimus and Bevacizumab to treat advanced endometrial, ovarian, liver, carcinoid, or islet cell cancer	II	2
	NCT01031381	Everolimus and Bevacizumab to treat recurrent ovarian, peritoneal, and fallopian tube cancer	II	1
	NCT01281514	Everolimus and Carboplatin, Pegylated Liposomal Doxorubicin Hydrochloride to treat patients with relapsed ovarian	I	4

Cancer Type	Identifier	Study	Phase	Status	
		epithelial, fallopian tube, or peritoneal cavity cancer			
Melanoma	NCT01166126	Temsirolimus and AZD 6244 to treat naive with BRAF mutant unresectable stage IV	II	1	
	NCT01014351	Everolimus with Paclitaxel and Carboplatin to treat metastatic melanoma	II	1	
	NCT01092728	Dasatinib to treat acral lentiginous, mucosal, or chronic sun-damaged melanoma	II	1	
Sarcoma	NCT00112372	Ridaforolimus to treat patients with refractory or advanced malignancies and sarcomas	I/II	1	[139]
	NCT00093080	Ridaforolimus to treat patients with advanced sarcoma	II	1	

1 Completed or terminated; 2 Active, not recruiting; 3 Unknown; 4 Recruiting; 5 Suspended; 6 Withdrawn.

Table 3. Clinical trials of the effects of autophagy activators on human cancers.

3.2. Use of autophagy inhibitors in cancer treatment

Autophagy confers stress tolerance that enables tumor cells to maintain metabolic homeostasis and the adaptation to hypoxic, nutrient-limiting, and metabolically stressed environments as well as resistance to therapy-induced stress, such as chemotherapy or radiotherapy [11, 64, 105]. Since autophagy activation confers an advantage to tumor growth, it would be one of the hallmarks of tumor progression [140]. For example, K-Ras^{V12} transforming malignant cells are capable of evading metabolic stress and cell death through activation of autophagy cascades. In an attempt to overcome this advantage of tumor behavior, the treatment with autophagy inhibitors Bafilomycin A_1 or 3-MA successfully decreases the growth of human breast epithelial cells in vitro [141]. Also, targeting autophagy inhibition using CQ suppressed growth and tumorigenicity of K-Ras mutation tumor cells leading to prolonged survival in pancreatic cancer xenografts and genetic mouse models [142]. These preclinical results suggest that autophagy might be exploited as a new therapeutic target in the setting of tumors driven by oncogenic RAS, which may improve clinical outcome of the patients with RAS-driven tumors, such as pancreatic cancer and malignant melanoma; however, recently reported KRAS-driven tumor lines may not require autophagy for growth [143]. By profiling 47 cell lines with pharmacological and genetic loss-of-function tools, Eng and colleagues suggested that KRAS mutation status would not predict the sensitivity of cancer cells to autophagy inhibition with CQ [143]. Accordingly, oncogenic B-RAF signaling in melanoma impairs the therapeutic advantage of autophagy inhibition [144].

Despite this controversial relation regarding the activation of MAPK pathway and the prediction of the efficacy of autophagy inhibition, the pharmacologic inhibition targeting the early or late autophagy process may increase controlled cell death of several other human tumors during chemotherapy aiding improved clinical outcomes [59, 102, 109]. The therapeutic modulation of autophagy for cancer treatment has been supported by preclinical models in which inhibition of autophagy restored chemosensitivity and enhanced tumor cell death [64]. For example, CQ and its analog HCQ given in combination with chemotherapy suppressed tumor growth and triggered cell death to a greater extent than did chemotherapy alone, both in vitro and in vivo as reviewed [64].

Moreover, suppression of autophagy via use of chemical inhibitors of autophagy such as 3-MA can sensitize tumor cells to the effects of chemotherapeutic drugs [11, 14], 5-Fluorouracil [128, 145], TNF-a [146], proteasome inhibitors [147], and Src family kinase (SFK) inhibitor Saracatinib [148].

The sensitizing effects of inhibiting autophagy on the antitumor efficacy of chemotherapeutic agents have been recapitulated in preclinical models of Myc-induced lymphoma [67, 149], colon cancer [45, 59, 127, 128, 131, 145, 150–153], ovarian cancer [154], breast cancer [31, 32, 155, 156] hepatocellular cancer [157], prostate cancer [148, 156], bladder cancer [156], melanoma [114], and glioma [152, 158]. Preclinical evidence reveals the efficacy of CQ to inhibit the genesis and self-renewal of cancer stem cells (CSC) and underlines the impact of this "old drug" as repurposing strategy to open a new CSC-targeted chemoprevention era [153].

Several clinical trials that have been conducted or are in progress have shown favorable effects of CQ as a novel antitumor drug as reviewed [159, 160]. Autophagy inhibition may contribute to the anticancer actions of the histone deacetylase (HDAC) inhibitor Vorinostat [127, 130, 136]. **Table 4** compiles recent clinical trials therapeutic targeting autophagy inhibition.

Cancer Type	Identifier	Study	Phase	Status
Renal cell cancer	NCT01144169	Hydroxychloroquine before surgery in patients with primary renal cell carcinoma	I	1
	NCT01480154	Akt Inhibitor MK2206 and Hydroxychloroquine to treat advanced solid tumors, melanoma, prostate, or kidney cancer	I	2
	NCT01510119	Everolimus and Hydroxychloroquine to treat renal cell carcinoma	I/II	4
Prostate cancer	NCT00726596	Hydroxychloroquine to treat patients with rising PSA levels after local therapy for prostate cancer	II	2
	NCT00786682	Docetaxel and Hydroxychloroquine to treat metastatic prostate cancer	II	1
	NCT01480154	Akt Inhibitor MK2206 and Hydroxychloroquine to treat advanced solid tumors, melanoma, prostate, or kidney cancer	I	2
	NCT01828476	Navitoclax and Abiraterone with or without Hydroxychloroquine to treat progressive metastatic castrate refractory prostate cancer	II	1
Breast cancer	NCT01292408	Hydrochloroquine to treat breast cancer patients	II	3
	NCT00765765	Hydroxychloroquine and Ixabepilone to treat metastatic breast cancer	I/II	1

Cancer Type	Identifier	Study	Phase	Status	
	NCT01023477	Chloroquine to treat ductal carcinoma *in situ*	I/II	4	
	NCT02333890	Chloroquine to treat breast cancer	II	4	
Non–small cell lung cancer	NCT00977470	Erlotinib with or without Hydroxychloroquine in chemo-naive advanced NSCLC and (EGFR) mutations	II	2	
	NCT00809237	Hydroxychloroquine and Gefitinib to treat lung cancer	I/II	3	
	NCT00933803	Hydroxychloroquine, Carboplatin, Paclitaxel, and Bevacizumab to treat recurrent advanced non-small cell lung cancer	I/II	1	
	NCT01649947	Hydroxychloroquine, Carboplatin, Paclitaxel, and Bevacizumab to treat advanced/recurrent nonsmall cell lung cancer	II	4	
	NCT00728845	Hydroxychloroquine, Carboplatin, Paclitaxel, and Bevacizumab to treat recurrent advanced non–small cell lung cancer	I/II	1	
Small cell lung cancer	NCT00969306	Chloroquine to treat stage IV small cell lung cancer	I	4	
Pancreatic cancer	NCT01273805	Hydroxychloroquine to treat patients with metastatic pancreatic cancer	II	2	
	NCT01128296	Study of presurgery Gemcitabine and Hydroxychloroquine to treat stage IIB or III adenocarcinoma of the pancreas	I/II	2	
	NCT01506973	Hydroxychloroquine in combination with Gemcitabine/ Abraxane to inhibit autophagy in pancreatic cancer	I/II	4	
	NCT01978184	Gemcitabine and Abraxane with or without Hydroxychloroquine	II	4	
Glioblastoma	NCT00486603	Hydroxychloroquine, radiation therapy, and Temozolomide to treat patients with newly diagnosed glioblastoma multiforme	I/II	1	[161]
	NCT00224978	Chloroquine to treat glioblastoma multiforme	III	1	[162]
	NCT02432417	Chloroquine and chemoradiation to treat glioblastoma	II	5	
	NCT02378532	Chloroquine and Chemoradiation to treat glioblastoma	I	5	
Colorectal cancer	NCT01206530	FOLFOX, Bevacizumab and Hydroxychloroquine to treat colorectal cancer	I/II	4	
	NCT01006369	Hydroxychloroquine, Capecitabine, Oxaliplatin, and Bevacizumab to treat metastatic colorectal cancer	II	6	
	NCT02316340	Vorinostat and Hydroxychloroquine versus Regorafenib to treat colorectal cancer	II	4	
Chronic myeloid leukemia	NCT01227135	Imatinib Mesylate with or without Hydroxychloroquine to treat patients with chronic myeloid leukemia	II	3	
	NCT00771056	Hydroxychloroquine in untreated B-CLL Patients	II	6	
Advanced solid tumor	NCT00813423	Sunitinib Malate and Hydroxychloroquine to treat patients with advanced solid tumors	I	2	

Cancer Type	Identifier	Study	Phase	Status
		that have not responded to chemotherapy		
	NCT00714181	Hydroxychloroquine and Temozolomide to treat patients with metastatic or unresectable solid tumors	I	1
	NCT00909831	Hydroxychloroquine and Temsirolimus to treat patients with metastatic solid tumors that have not responded to treatment	I	2
	NCT01023737	Hydroxychloroquine and with histone deacetylase (HDAC) inhibitor Vorinostat in patients with advanced solid tumors	I	4
	NCT01266057	Sirolimus or Vorinostat and Hydroxychloroquine in advanced solid tumors	I	4
Multiple myeloma	NCT00568880	Hydroxychloroquine and Bortezomib to treat patients with relapsed or refractory multiple myeloma	I/II	3
	NCT01689987	Hydroxychloroquine, Cyclophosphamide, Dexamethasone, and Sirolimus to treat patients with relapsed or refractory multiple myeloma	I	2
	NCT01438177	Chloroquine and VELCADE and Cyclophosphamide to treat relapsed and refractory multiple myeloma	II	1
Melanoma	NCT00962845	Hydroxychloroquine to treat patients with stage III or stage IV melanoma that can be removed by surgery	I	2
	NCT01480154	Akt Inhibitor MK2206 and Hydroxychloroquine to treat patients with advanced solid tumors, melanoma, prostate, or kidney cancer	I	2
	NCT02257424	Dabrafenib, Trametinib and Hydroxychloroquine in patients with advanced BRAF mutant melanoma	I/II	4

1 Completed or terminated; 2 Active, not recruiting; 3 Unknown; 4 Recruiting; 5 Not yet recruiting; 6 suspended.

Table 4. Clinical trials of the effects of autophagy inhibitors on human cancers.

4. Therapeutic effects of autophagy modulators on cardiovascular diseases

Autophagy plays a dichotomous role on many cardiac pathologic states in which it may exert both protective and detrimental effects through context-dependent mechanisms. As a protective mechanism, autophagy closely protects the heart from myocardial ischemia-reperfusion (I/R) and attenuates cardiac remodeling after myocardial infarction [49, 163]. In fact, as demonstrated in ischemia-reperfusion-induced heart injury, Parkin-mediated mitophagy showed a protective role against the cell death of cardiomyocytes [164]. Moreover, recent evidence indicates that basal levels of autophagy are required for the maintenance of normal cardiovascular function and morphology [165]. However, by contrast, excessive levels of autophagy—or perhaps distinct forms of autophagic flux—contribute to several types of cardiomyopathy by functioning as a controlled cell death pathway [165, 166]. In line with these findings, the selection of activators or inhibitors of autophagy for prevention or treatment of cardiovascular diseases will be complicated. Nevertheless, in practice successful therapeutic approaches that regulate autophagy have been reported recently, suggesting that the autophagic machinery may be properly manipulated to treat heart failure or to prevent rupture of atherosclerotic plaques and sudden death [165, 166]. Whereas there have been no clinical data reporting the efficacy of pharmacologic modulation of autophagy in cardiac diseases, as reviewed in 2011, nine patents have disclosed the pharmacologic modulation of autophagy as a new therapeutic strategy against cardiovascular diseases [166]. In this section, we will review the fundamental use of autophagy modulators on heart diseases, whose biological effects have been identified through both in vitro and in vivo models.

5. Use of autophagy inhibitors in treatment of heart disease

Pharmacologic suppression of autophagy pathway comprises potential new targets for treating cardiac disorders [11]. In vitro and in vivo studies demonstrated that the inhibitor of histone deacetylases Trichostatin A may attenuate both load- and agonist-induced hypertrophic growth and abolish the associated activation of autophagy, reducing pathologic cardiac remodeling during severe pressure overload [167]. Through negative modulation of early stage of autophagy process by inhibiting the expression of *Beclin-1* induced by myocardial I/R injury parallel to phosphorylation of mTOR, Propofol reduces autophagy-associated cell death induced by the myocardial I/R injury [49]. Although these strategies for suppressing the excessive activation of autophagy for treating cardiac disorders are in theory promising, a comprehensive view of myocardial autophagy will be obligatory to avoid disrupting homeostatic mechanisms [166]. Thus, although the major challenges remain, patients with heart disease are likely to benefit from these efforts.

In other situations, such as in response to stress, activation rather than suppressing autophagy might be beneficial, since it increases the clearance of misfolded and other harmful proteins. In fact, recent reports had established a requirement for autophagy for cardioprotection in rodent models mediated by a variety of agents including the adenosine A1 receptor agonist Chlorocyclopentyladenosine, Sulfaphenazole, and ischemic preconditioning [168]. On the fate of ischemic-reperfused cardiomyocytes, autophagy plays a protective role [169, 170].

The development of a pharmacological agent to salvage myocardium after an ischemic insult has been explored. For example, attempting to enhance the heart's tolerance to ischemia-reperfusion through inducing autophagy, the antimicrobial agent Sulfaphenazole might be used [163]. Likewise, chloramphenicol succinate has been shown to activate autophagy and reduce myocardial damage during I/R [169, 170]. In the case of regression of established increase in myocyte cell size (i.e., cardiac hypertrophy) induced by ascending aortic constriction (i.e., pressure overload), the administration of Rapamycin, a mTOR inhibitor, may improve cardiac function [171, 172]. In line with this finding, animal studies suggest that mTOR inhibition attenuates cardiac allograft remodeling secondary to downregulation of mTOR downstream targets and increased autophagy. To increase the paucity of data regarding effect of Sirolimus, a mTOR inhibitor, on human heart remodeling, a current clinical trial Phase 1 has been conducted (NCT01889992).

Macrophages play a central role in atherosclerotic plaque destabilization, leading to acute coronary syndromes and sudden death, and therefore their clearance from atherosclerotic plaques through autophagy has been suggested as an attractive therapeutic strategy for atherosclerosis [173]. In line with these findings, the stent-based delivery of Everolimus was shown to selectively clear macrophages from atherosclerotic plaques in rabbits by activating autophagy without altering smooth muscle cells [174]. As suggested recently, mTOR inhibition represents a promising strategy for stabilization of atherosclerotic plaques [175], as this also prevents adverse left ventricular remodeling and limits infarct size following myocardial infarction [176]. Though the benefits afforded by autophagic activation depend on cardiac pathologic states, vigilance for extra-cardiac effects may be critical [166].

6. Use of autophagy modulators for neurologic disorders

In contrast to other cell types, neurons for being nondividing cells are particularly sensitive to changes in autophagic degradation [177]. As most neurons must survive for the lifetime of the organism, maintenance of organelle function and clearance of aberrant, misfolded, and aggregate proteins are critical processes regulated by autophagy [19]. In fact, many aggregate-prone forms of such proteins, including tau [178], α-synuclein [179, 180], mutant huntingtin [181], and mutant ataxin 3 [178] have a higher dependency on autophagy for their clearance. While autophagy clears these aggregate-prone proteins, upregulation of autophagy may also contribute to amyloid-β pathology [182], as autophagic vacuoles may represent one site of amyloid-β generation. The intracellular accumulation of these aggregate proteins are features of many late-onset neurodegenerative diseases, including Alzheimer's disease (AD), Parkinson's disease (PD), tauopathies, and polyglutamine expansion diseases—such as Huntington's disease (HD) and various spinocerebellar ataxias (SCAs) [183]. Currently, there are no effective therapeutic strategies capable of attenuating or preventing the neurodegeneration resulting from these diseases in humans.

Autophagy has been considered as a potentially novel approach for treating neurodegenerative disorders [160, 183], although its role in neurodegenerative disorders remains unclear [11]. Even though autophagy may be initially induced as a neuroprotective response, due to excessive, imbalanced induction or defects in completing degradation, it may also contribute to neuronal atrophy, neurite degeneration, and cell death [19, 177]. It is noteworthy that the failed attempt of autophagy at neuron survival has been closely associated with age-related autophagy insufficiency and lysosomal aging [19]. In fact, several evidences recently suggest a possible role for autophagic dysfunction in the pathogenesis of neurodegenerative diseases [184]. Conversely, autophagy also has the ability to decrease the accumulation of toxic, aggregate-prone proteins that cause neurodegeneration [11, 183]. As summarized in the following paragraph, multiple studies provide proof of principle for the activation of autophagy as a therapy for neurodegenerative disease. To date, there are still very few reported clinical results demonstrating that modulation of autophagy indeed represents an effective therapeutic intervention for these devastating diseases [11].

HD disease is caused by a polyglutamine expansion mutation in the huntingtin protein (polyQ-expanded Htt) that confers a toxic gain-of-function and causes the protein to become aggregate-prone proteins, which are cleared by autophagy. It is noteworthy that upregulating this process by Rapamycin attenuates their toxicity in various HD models [181, 185]. The autophagy inducer Rapamycin or its analog CCI-779 has been reported to promote autophagic clearance of polyQ-expanded Htt protein [178, 181, 186]. Likewise, Rapamycin increases the clearance of α-synuclein and lessens the formation of aggregates (Lewy bodies) in neurons [179]. Rapamycin in combination with lithium showed a greater protection against neurodegeneration in an HD fly model [185]. Interestingly, the disaccharide bilayer membrane-protector Trehalose [187] accelerates the autophagic clearance of mutant Huntingtin and α-synuclein through an mTOR-independent pathway [95]. However, in a combination of mTOR inhibitor, Rapamycin Trehalose's effect on autophagic activity increases, resulting in an additive effect on the clearance of the above proteins [95]. Together these studies demonstrate that autophagy upregulation and promotion of aggregation-prone protein degradation ameliorate neurodegenerative pathology, but conversely autophagy inhibition enhances the toxicity of these proteins [184].

7. Autophagy modulators for treatment of other diseases

Similar to the details outlined above for neurodegenerative disorders, autophagy upregulation may enhance the clearance of a range of infectious agents, including multidrug-resistant (MDR) strains of *Mycobacterium tuberculosis*. In some cases, mouse models and preclinical data have strengthened the protective role of autophagy against microbial infections, as summarized in **Table 5**.

AR-12 induces autophagic clearance of *Francisella tularensis* from the human leukemic cell line THP-1 macrophages [188] and *Salmonella enterica serovar Typhimurium* in murine macrophages, both in vitro and in vivo [189]. Additionally, experimental findings underscore the importance of host autophagy in orchestrating successful antimicrobial responses to *Mycobacterium tuberculosis* during chemotherapy with Isoniazid and Pyrazinamide [190]. Likewise, the most active form of vitamin D 1,25D3 may inhibit replication of human immunodeficiency virus (HIV) in human macrophages through autophagy activation [191].

Based on these preclinical findings, researchers have raised the possibility that some antimycobacterial chemotherapies already used in clinical for the treatment of infectious diseases anti-infection effects, at least partially, via inducing autophagy. However, it is still unclear whether those findings can be translated into the clinical treatment of certain infections [11]. Stimulation of autophagy with Rapamycin reduces intracellular survival of mycobacteria in macrophages [192].

Drugs	Effects
1,25D3	1α,25-dihydroxycholecalciferol inhibits HIV replication and mycobacterial growth [191].
AR-12	2-amino-*N*-4-5-(2 phenanthrenyl)-3-(trifluoromethyl)-1*H*-pyrazol-1-yl phenyl-acetamide inhibits activity of phosphoinositide-dependent kinase-1 and promotes autophagic clearance of bacteria in human and murine macrophages [188, 189].
Carbamazepine	Induces antimicrobial autophagy through a mTOR-independent pathway controlled by cellular depletion of myo-inositol [193].
Fluoxetine	A selective serotonin reuptake inhibitor, enhances secretion of proinflammatory cytokine TNF-α and induces autophagy in *Mycobacterium tuberculosis*–infected macrophages [194].
Gefitinib	An inhibitor of the Epidermal Growth Factor Receptor (EGFR), activates autophagy, and restricts growth of *Mycobacterium tuberculosis* in the lungs of infected mice [194].
Isoniazid or pyrazinamide	Reduces *Mycobacterium tuberculosis (Mtb)*–induced proinflammatory responses by promoting autophagy activation and phagosomal maturation in *Mtb*-infected host cells [190].
Nitozoxanide	Nitazoxanide and its active metabolite Tizoxanide strongly stimulate autophagy and inhibit mTORC1 signaling and intracellular proliferation of *Mycobacterium tuberculosis* [195].
Nortriptyline	Induces the formation of autophagosomes that progressively acidify over time and become competent for *Mycobacterium tuberculosis* degradation in infected macrophages [196].
Prochlorperazine edisylate	Modulates autophagy that correlates with delivery of *Mycobacterium tuberculosis* to lysosomes leading to mycobacterial degradation [196].
Rapamycin	Induces autophagy and suppresses intracellular survival of *M. tuberculosis* [192].
Statins	Enhances autophagy and phagosome maturation leading to reduction the *Mycobacterium tuberculosis* burden in human macrophages and in mice [197].
Valproic acid	Stimulates autophagic killing of intracellular *Mycobacterium tuberculosis* within primary human macrophages [193].

Table 5. Preclinical studies of the effects of autophagy activators on infectious diseases.

Remarkably, several FDA-approved drugs counter *M. tuberculosis* infection, possibly through autophagy, which disrupts the host-pathogen equilibrium in favor of the host (see **Table 5**). The antidepressants Fluoxetine [194] and Nortriptyline [196], the anticonvulsants Carbamazepine and Valproic acid, and the antipsychotic Prochlorperazine edisylate reveal relevant antimycobacterial properties by targeting autophagy in the host (i.e., infected macrophages). Notably, in mice infected with a highly virulent MDR-strain of *Mycobacterium tuberculosis*, Carbamazepine reduces bacterial burden, improve lung pathology, and stimulate adaptive immunity [193]. Furthermore, the tyrosine kinase inhibitor (Gefitinib) also activates autophagy and suppress *Mycobacterium tuberculosis* in macrophages and, to some extent, in infected mice [194]. Other autophagy-inducing candidate drugs attempting to Antituberculosis Host-

Directed Therapy (HDT) include antiprotozoal drug Nitozoxanide [195] and cholesterol-lowering drugs, i.e., Statin [197]. Together these findings support that autophagy enhancement by repurposed drugs provides an easily implementable potential therapy for the treatment of multidrug-resistant mycobacterial infection.

8. Promising future therapeutic strategies

8.1. The case of the pentacyclic triterpenoids, the working hypothesis

After decades of scientific discoveries and discussions [10, 198] the general agreement is that autophagy associated cell-death is commonly linked to failure in either the fusion of autophagosomes with lysosomes or in the digestion activity of autolysosomes [27, 198]. Whereas the understanding of this process at the molecular level needs a deeper knowledge of the competition between its activation and inhibition pathways, autophagy has been explored as a potential therapeutically target for treating several diseases [11, 14]. Consequently, the impact of activating mitophagy on the condition of autophagy impairment is a noteworthy subject to explore. A recent work has proposed that by modulating parallel damage in membranes of mitochondria and lysosome, autophagy turns into a destructive process [12]. Comparative analysis of the biological effects of two chemical isomers, i.e., pentacyclic triterpenoids Betulinic and Oleanolic acids (BA and OA, respectively), Martins and colleagues showed that the main differences between the activity of BA and OA is due to their efficiency in interacting and damaging membranes [12] (see **Figure 1**). These triterpenoids are new promising drugs with various pharmacological actions (anti-inflammatory, antiviral, antifungal, antimalarial, among others), being easily extracted from plants [199]. So far, about 2167 patents for AB and 1018 for OA have been deposited.

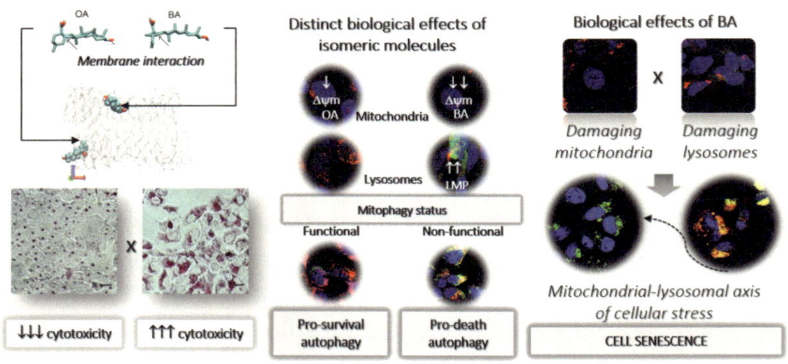

Figure 1. Modulation of membrane damage provided insights into biological effects of triterpenoids in vitro (Figure kindly supplied by Martins WK).

By comparing these triterpenoids, it was realized that the fate of autophagy may depend on the extent of lysosomal and mitochondrial membrane damage. In case of OA, there is marked

http://dx.doi.org/10.5772/63976

cytoplasmic vacuolization and mitochondria shrinkage with remarkable cellular recovery that was intrinsically associated with autophagy activation. However, cell recovery failed upon concomitant lysosome inhibition with CQ or Bafilomycin A_1 [12]. Of note, the lysosomal damage BA-mediated is per se capable of compromising autophagy, without any incremental damage when lysosomal function was deeply altered by lysosomal inhibitors, such as CQ and Bafilomycin A_1.

BA and OA differ significantly on their ability to penetrate membranes, which appears to be mainly related to the twisted backbone structure of OA, in contrast to the fully planar structure of BA. Interestingly, this stronger efficiency in interacting and damaging membrane mimics ascribed to BA correlates with a higher ability of disturbing mitochondrial and lysosomal membranes of human keratinocytes [12]. The ability of BA to disturb the mitochondrial membrane is in agreement with other published results [106, 200, 201]. For example, by inhibiting the activity of steroyl-CoA-desaturase (SCD-1) BA may also directly and rapidly impact on the saturation level of cardiolipin (CL), a specific mitochondrial phospholipid lipid that has important structural and metabolic functions, and at the same time regulates mitochondria-dependent cell death [202]. Interestingly, thermodynamic analyses of Langmuir monolayers and AFM study of Langmuir-Blodgett monolayers provide insights into the ability of BA interacting with CL-enriched membranes. BA may orient nearly perpendicularly with hydroxyl group toward water, which causes phase separation and changes the permeability of CL film [203]. BA was also shown to disrupt membranes of human red blood cells (RBC) in vitro, with release of calcein from the RBC ghosts in a way similar to Digitonin in membrane permeabilization experiments [204].

Of note, the damage in lysosomal function caused by BA may not be explained by traditional justifications (lack of lysosome acidification or neutralization of its internal pH). Otherwise, BA disturbs lysosome's membrane integrity that dramatically jeopardizes the lysosomal function, leading to a lysosomal-mitochondrial axis of cellular stress that causes autophagy-associated cell death [12]. Remarkably, in the survival of BA-challenged cells occurs sustained formation of reactive oxygen species (ROS) inside nonfunctional lysosomes, which in the long-term response leads to lipofuscinogenesis, genomic instability, and cell senescence [13]. Thus, promotion of concomitant damage in mitochondrial and lysosomal membranes seems to be an efficient strategy for inducing autophagy-associated cell death and cell aging.

The AB's ability to promote parallel damage in lysosomes and mitochondria could be the explication for the positive synergistic action of BA in different antitumor protocols including radiation [205, 206], chemotherapy drugs, such as Cisplatin [207] and Vincristine [208]. Therefore, the possible increase of cell death potentially relates to the AB ability of suppressing prosurvival autophagy. The knowledge of this premise at molecular level may contribute to the development of new autophagy modulators.

Since 1995, BA has been considered as a highly promising anticancer drug showing remarkable antitumor effects against several human tumors [106, 208–222]. In addition, in the last decade, many studies have shown further effects that justify the expectation that triterpenes and synthetic analogs are useful to treat cancer by several modes of action [223]. For example, BA acid derivatives are under evaluation as chemotherapeutic agent against several types of human tumors in vitro and in vivo [152, 199, 223–235]. The synthetic analog of OA [2-cyano-3,12-dioxooleana-1,9-dien-28-oic acid (CDDO)] is currently under clinical Phase I study (NCT00322140) for treating solid tumors and lymphoma [101]. Introduced in 2009, the new semisynthetic candidate drug designated NVX-207 (3-acetyl-betulinic 2-amino-3-hydroxy-2-hydroxymethyl-hi-ethyl propanoate) enhanced apoptosis-inducing activity and dramatically enhanced solubility over BA [230]. However, limited solubility and often ultimately modest efficacy have hampered the development of this class of compounds [230]. Thus, scientific efforts focused on the elucidation of molecular mechanisms triggered by these triterpenoids, attending the interests of the scientific community as well as of the pharmaceutical industry.

9. Perspectives for drug development

During the last decade, important progress was made in understanding the molecular basis of autophagy uncovering its potential in anticancer therapies [105, 236, 237]. Because the abrogation of autophagy via knockdown of autophagy-related molecules increases the sensibility of therapy-resistant cancer cells to conventional cancer therapies, there has been great interest in developing clinically relevant autophagy inhibitors [238]. As reviewed by Yang and colleagues, multiple studies have shown that genetic knockdown of autophagy-related genes (Atgs) or pharmacological inhibition of autophagy can effectively enhance tumor cell death induced by diverse anticancer drugs in preclinical models [64].

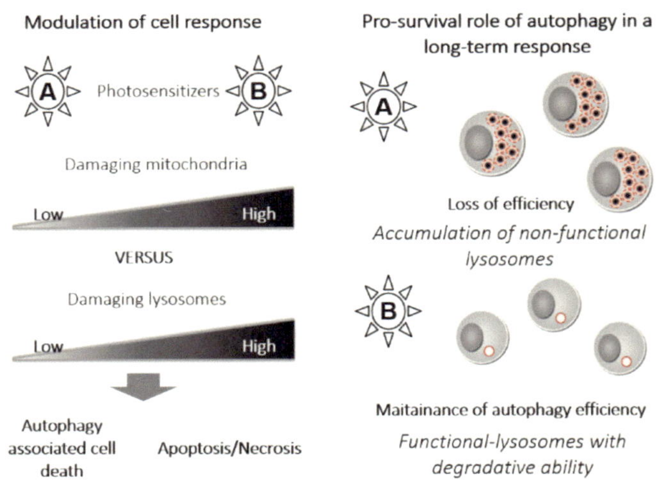

Figure 2. Modulation of membrane damage provided insights into biological effects after PDT (Figure kindly supplied by Martins WK).

http://dx.doi.org/10.5772/63976

Photodynamic therapy (PDT) is a procedure that has applications in the selective eradication of cancer where sites of tumor lesions are clearly delineated. It is a two- step process whereby cells are first incubated with photosensitizers and then photoirradiated. This results in the formation of singlet molecular oxygen and other reactive oxygen species (ROS) that can cause photodamage at sites where the photosensitizing agent has localized [239]. Photosensitizers found to be clinically useful, showing affinity for the endoplasmic reticulum, mitochondria, lysosomes, or combinations of these sites [240]. The induction of cell death triggered by apoptosis and/or autophagy in photosensitized cells is a common outcome of PDT [239–242]. Therefore, the photosensitizers are drugs that are used to treat a series of different diseases. Our group has addressed the concept of parallel photodamage in mitochondria and lysosome with the consequent induction of cell death and senescence after PDT (**Figure 2**). In near future, we will exploit the possible use of this concept in the development of new photosensitizers targeting autophagy as cell death mechanism.

Author details

Waleska K. Martins and Mauricio S. Baptista*

*Address all correspondence to: baptista@iq.usp.br

Santo Amaro University, São Paulo, Brazil

São Paulo University, São Paulo, Brazil

References

[1] Glick D, Barth S, Macleod KF. Autophagy: cellular and molecular mechanisms. J Pathol. 2010 May;221(1):3–12.

[2] Reggiori F, Komatsu M, Finley K, Simonsen A. Autophagy: More Than a Nonselective Pathway. Int J Cell Biol. 2012 Jan;2012 (Figure 1):1–18.

[3] Ding W-X, Yin X-M. Mitophagy: mechanisms, pathophysiological roles, and analysis. Biol Chem. 2012 Jan 1;393(7):547–64.

[4] May AI, Devenish RJ, Prescott M. The many faces of mitochondrial autophagy: making sense of contrasting observations in recent research. Int J Cell Biol. 2012;2012:431684.

[5] Lemasters JJ. Selective mitochondrial autophagy, or mitophagy, as a targeted defense against oxidative stress, mitochondrial dysfunction, and aging. Rejuvenation Res. 2005 Mar;8(1):3–5.

[6] Reggiori F, Komatsu M, Finley K, Simonsen A. Selective types of autophagy. Int J Cell Biol. 2012 Jan;2012:156272.

[7] Youle RJ, Narendra DP. Mechanisms of mitophagy. Nat Rev Mol Cell Biol. Nature Publishing Group; 2011 Jan;12(1):9–14.

[8] Vessoni AT, Filippi-Chiela EC, Menck CF, Lenz G. Autophagy and genomic integrity. Cell Death Differ. 2013 Nov;20(11):1444–54.

[9] Lu H, Li G, Liu L, Feng L, Wang X, Jin H. Regulation and function of mitophagy in development and cancer. Autophagy. 2013 Nov 1;9(11):1720–36.

[10] Levine B, Kroemer G. Autophagy in the pathogenesis of disease. Cell. Elsevier; 2008 Jan 11;132(1):27–42.

[11] Cheng Y, Ren X, Hait WN, Yang J-M. Therapeutic targeting of autophagy in disease: biology and pharmacology. Pharmacol Rev. 2013;65(4):1162–97.

[12] Martins WK, Costa ÉT, Cruz MC, Stolf BS, Miotto R, Cordeiro RM, et al. Parallel damage in mitochondrial and lysosomal compartments promotes efficient cell death with autophagy: The case of the pentacyclic triterpenoids. Sci Rep. 2015 Jul 27;5:12425.

[13] Martins WK, Gomide AB, Junqueira H, Stolf BS, Itri R, Baptista MS. Membrane damage by Betulinic Acid provides insights into cellular aging. Biochimica et Biophysica Acta (BBA) - General Subjects, in press, 10.1016/j.bbagen.2016.10.018

[14] Rubinsztein DC, Levine PC and B. Autophagy modulation as a potential therapeutic target for diverse diseases. Nat Rev Drug Discov. 2012;29(6):709–30.

[15] Pellegrini P, Strambi A, Zipoli C, Hägg-Olofsson M, Buoncervello M, Linder S, et al. Acidic extracellular pH neutralizes the autophagy-inhibiting activity of chloroquine: Implications for cancer therapies. Autophagy. 2014;10(4):1–10.

[16] Mathew R, Karantza-Wadsworth V, White E. Role of autophagy in cancer. Nat Rev Cancer. 2007;7(12):961–7.

[17] Choi KS. Autophagy and cancer. Exp Mol Med. 2012;44(2):109.

[18] Johri A, Beal MF. Mitochondrial dysfunction in neurodegenerative diseases. J Pharmacol Exp Ther. 2012 Sep;342(3):619–30.

[19] Cherra SJ, Dagda RK, Chu CT. Review: Autophagy and neurodegeneration: Survival at a cost? Vol. 36, Neuropath Appl Neuro. 2010. p. 125–32.

[20] Casarejos MJ, Solano RM, Gómez a., Perucho J, De Yébenes JG, Mena M a. The accumulation of neurotoxic proteins, induced by proteasome inhibition, is reverted by trehalose, an enhancer of autophagy, in human neuroblastoma cells. Neurochem Int. 2011;58(4):512–20.

[21] Nixon R a. The role of autophagy in neurodegenerative disease. Nat Med. Nature Publishing Group; 2013;19(8):983–97.

[22] Batlevi Y, La Spada AR. Mitochondrial autophagy in neural function, neurodegenerative disease, neuron cell death, and aging. Vol. 43, Neurobiol Dis. 2011. p. 46–51.

[23] Sarkar S, Ravikumar B, Rubinsztein DC. Chapter 5 Autophagic clearance of aggregate-prone proteins associated with neurodegeneration. 1st ed. Vol. 453, Methods in Enzymology. Elsevier Inc.; 2009. 83–110 p.

[24] Kiselyov K, Jennigs JJ, Rbaibi Y, Chu CT. Autophagy, mitochondria and cell death in lysosomal storage diseases. Autophagy. 2007;3(3):259–62.

[25] Appelqvist H, Wäster P, Kågedal K, Öllinger K. The lysosome: from waste bag to potential therapeutic target. J Mol Cell Biol. 2013 Aug;5(4):214–26.

[26] Ordonez MP. Defective mitophagy in human Niemann-Pick type C1 neurons is due to abnormal autophagy activation. Autophagy. 2012;8(7):1157–8.

[27] Kroemer G, Levine B. Autophagic cell death: the story of a misnomer. Nat Rev Mol Cell Biol. 2008 Dec;9(12):1004–10.

[28] Seglen PO, Gordon PB. 3-Methyladenine: specific inhibitor of autophagic/lysosomal protein degradation in isolated rat hepatocytes. Proc Natl Acad Sci U S A. 1982;79(6): 1889–92.

[29] Klionsky DJ, Abdelmohsen K, Abe A, Abedin MJ, Abeliovich H, Acevedo Arozena A, et al. Guidelines for the use and interpretation of assays for monitoring autophagy (3rd edition). Autophagy. 2016;12(1):1–222.

[30] Blommaart EF, Krause U, Schellens JP, Vreeling-Sindelárová H, Meijer a J. The phosphatidylinositol 3-kinase inhibitors wortmannin and LY294002 inhibit autophagy in isolated rat hepatocytes. Eur J Biochem. 1997;243:240–6.

[31] De Mei C, Ercolani L, Parodi C, Veronesi M, Lo Vecchio C, Bottegoni G, et al. Dual inhibition of REV-ERBβ and autophagy as a novel pharmacological approach to induce cytotoxicity in cancer cells. Oncogene. 2015 May 14;34(20):2597–608.

[32] Torrente E, Parodi C, Ercolani L, De Mei C, Ferrari A, Scarpelli R, et al. Synthesis and in vitro anticancer activity of the first class of dual inhibitors of REV-ERBβ and autophagy. J Med Chem. 2015 Aug 13;58(15):5900–15.

[33] Yoshimori T, Yamamoto A, Moriyama Y, Futai M, Tashiro Y. Bafilomycin A1, a specific inhibitor of vacuolar-type H + –ATPase, inhibits acidification and protein degradation in lysosomes of cultured cells. J Biol Chem. 1991;266(26):17707–12.

[34] Yamamoto a, Tagawa Y, Yoshimori T, Moriyama Y, Masaki R, Tashiro Y. Bafilomycin A1 prevents maturation of autophagic vacuoles by inhibiting fusion between autophagosomes and lysosomes in rat hepatoma cell line, H-4-II-E cells. Cell Struct Funct. 1998;23(1):33–42.

[35] Klionsky DJ, Elazar Z, Seglen PO, Rubinsztein DC. Does bafilomycin A1 block the fusion of autophagosomes with lysosomes? Vol. 4, Autophagy. 2008. p. 849–50.

[36] Murata M, Miyashita S, Yokoo C, Tamai M, Hanada K, Hatayama K, et al. Novel epoxysuccinyl peptides. Selective inhibitors of cathepsin B, in vitro. FEBS Lett. 1991 Mar 25;280(2):307–10.

[37] Seglen PO, Grinde B, Solheim a E. Inhibition of the lysosomal pathway of protein degradation in isolated rat hepatocytes by ammonia, methylamine, chloroquine and leupeptin. Eur J Biochem. 1979;95(2):215–25.

[38] Ju JS, Varadhachary AS, Miller SE, Weihl CC. Quantitation of "autophagic flux" in mature skeletal muscle. Autophagy. 2010;6(7):929–35.

[39] Rossi M, Munarriz ER, Bartesaghi S, Milanese M, Dinsdale D, Guerra-Martin MA, et al. Desmethylclomipramine induces the accumulation of autophagy markers by blocking autophagic flux. J Cell Sci. 2009;122(18):3330–9.

[40] Tamai M, Matsumoto K, Omura S, Koyama I, Ozawa Y, Hanada K. In vitro and in vivo inhibition of cysteine proteinases by EST, a new analog of E-64. J Pharmacobiodyn. 1986 Aug;9(8):672–7.

[41] Vanrell MC, Cueto J a, Barclay JJ, Carrillo C, Colombo MI, Gottlieb R a, et al. Polyamine depletion inhibits the autophagic response modulating Trypanosoma cruzi infectivity. Autophagy. 2013;9(7):1080–93.

[42] Nishimura Y, Kato K, Furuno K, Himeno M. Inhibitory effect of leupeptin on the intracellular maturation of lysosomal cathepsin L in primary cultures of rat hepatocytes. Biol Pharm Bull. 1995 Jul;18(7):945–50.

[43] Kirschke H, Langner J, Wiederanders B, Ansorge S, Bohley P, Broghammer U. Intracellular protein breakdown. VII. Cathepsin L and H; two new proteinases from rat liver lysosomes. Acta Biol Med Ger. 1976;35(3–4):285–99.

[44] Ahlberg J, Berkenstam A, Henell F, Glaumann H. Degradation of short and long lived proteins in isolated rat liver lysosomes. Effects of pH, temperature, and proteolytic inhibitors. J Biol Chem. 1985;260(9):5847–54.

[45] Carew JS, Espitia CM, Esquivel J a., Mahalingam D, Kelly KR, Reddy G, et al. Lucanthone is a novel inhibitor of autophagy that induces cathepsin d-mediated apoptosis. J Biol Chem. 2011;286(8):6602–13.

[46] Kondo Y, Kanzawa T, Sawaya R, Kondo S. The role of autophagy in cancer development and response to therapy. Nat Rev Cancer. 2005;5(9):726–34.

[47] Xie R, Nguyen S, McKeehan WL, Liu L. Acetylated microtubules are required for fusion of autophagosomes with lysosomes. BMC Cell Biol. 2010;11(1):89.

[48] Leu JI-J, Pimkina J, Frank A, Murphy ME, George DL. A small molecule inhibitor of inducible heat shock protein 70. Mol Cell. Elsevier Ltd; 2009;36(1):15–27.

[49] Noh HS, Shin IW, Ha JH, Hah Y-S, Baek SM, Kim DR. Propofol protects the autophagic cell death induced by the ischemia/reperfusion injury in rats. Mol Cells. 2010 Nov;30(5): 455–60.

[50] Cui D, Wang L, Qi A, Zhou Q, Zhang X, Jiang W. Propofol prevents autophagic cell death following oxygen and glucose deprivation in PC12 cells and cerebral ischemia-reperfusion injury in rats. PLoS One. 2012;7(4).

[51] Cui DR, Wang L, Jiang W, Qi AH, Zhou QH, Zhang XL. Propofol prevents cerebral ischemia-triggered autophagy activation and cell death in the rat hippocampus through the NF-kB/p53 signaling pathway. Neuroscience. 2013 Aug 29;246:117–32.

[52] Liu J, Xia H, Kim M, Xu L, Li Y, Zhang L, et al. Beclin1 controls the levels of p53 by regulating the deubiquitination activity of USP10 and USP13. Cell. Elsevier Inc.; 2011;147(1):223–34.

[53] Gordon PB, Holen I, Fosse M, Røtnes JS, Seglen PO. Dependence of hepatocytic autophagy on intracellularly sequestered calcium. J Biol Chem. 1993;268(35):26107–12.

[54] Engedal N, Torgersen ML, Guldvik IJ, Barfeld SJ, Bakula D, Saetre F, et al. Modulation of intracellular calcium homeostasis blocks autophagosome formation. Autophagy. 2013;9(10):1475–90.

[55] Ganley IG, Wong P-M, Gammoh N, Jiang X. Distinct autophagosomal-lysosomal fusion mechanism revealed by thapsigargin-induced autophagy arrest. Mol Cell. 2011 Jun; 42(6):731–43.

[56] Lu Y, Dong S, Hao B, Li C, Zhu K, Guo W, et al. Vacuolin-1 potently and reversibly inhibits autophagosome-lysosome fusion by activating RAB5A. Autophagy. 2014;10(11):1895–905.

[57] Yoon YH, Cho KS, Hwang JJ, Lee SJ, Choi J a., Koh JY. Induction of lysosomal dilatation, arrested autophagy, and cell death by chloroquine in cultured ARPE-19 cells. Investig Ophthalmol Vis Sci. 2010;51(11):6030–7.

[58] Maycotte P, Aryal S, Cummings CT, Thorburn J, Morgan MJ, Thorburn A. Chloroquine sensitizes breast cancer cells to chemotherapy independent of autophagy. Autophagy. 2012;8(2):200–12.

[59] Schonewolf C a, Mehta M, Schiff D, Wu H, Haffty BG, Karantza V, et al. Autophagy inhibition by chloroquine sensitizes HT-29 colorectal cancer cells to concurrent chemoradiation. World J Gastrointest Oncol. 2014;6(3):74–82.

[60] Geng Y, Kohli L, Klocke BJ, Roth K a. Chloroquine-induced autophagic vacuole accumulation and cell death in glioma cells is p53 independent. Neuro Oncol. 2010 May;12(5):473–81.

[61] Mauvezin C, Nagy P, Juhász G, Neufeld TP. Autophagosome—lysosome fusion is independent of V-ATPase-mediated acidification. Nat Commun. 2015;6(May): 7007.

[62] Montaser M, Lalmanach G, Mach L. CA-074, but not its methyl ester CA-074Me, is a selective inhibitor of cathepsin B within living cells. Biol Chem. 2002;383(7–8):1305–8.

[63] Amaravadi RK, Lippincott-Schwartz J, Yin XM, Weiss W a., Takebe N, Timmer W, et al. Principles and current strategies for targeting autophagy for cancer treatment. Clin Cancer Res. 2011;17(4):654–66.

[64] Yang ZJ, Chee CE, Huang S, Sinicrope F a. The role of autophagy in cancer: therapeutic implications. Mol Cancer Ther. 2011;10(9):1533–41.

[65] Tsvetkov AS, Miller J, Arrasate M, Wong JS, Pleiss M a, Finkbeiner S. A small-molecule scaffold induces autophagy in primary neurons and protects against toxicity in a Huntington disease model. Proc Natl Acad Sci U S A. 2010;107(39):16982–7.

[66] Joo JH, Dorsey FC, Joshi A, Hennessy-Walters KM, Rose KL, McCastlain K, et al. Hsp90-Cdc37 chaperone complex regulates Ulk1- and Atg13-mediated mitophagy. Mol Cell. 2011 Aug;43(4):572–85.

[67] Colomer D, Rosich L. Counteracting autophagy overcomes resistance to everolimus in mantle cell lymphoma. Clin Cancer Res. 2012;18(19):5278–89.

[68] Papp D, Kovács T, Billes V, Varga M, Tarnóci A, Hackler L, et al. AUTEN-67, an autophagy-enhancing drug candidate with potent antiaging and neuroprotective effects. Autophagy. 2015 Aug 27;

[69] Chresta CM, Davies BR, Hickson I, Harding T, Cosulich S, Critchlow SE, et al. AZD8055 is a potent, selective, and orally bioavailable ATP-competitive mammalian target of rapamycin kinase inhibitor with in vitro and in vivo antitumor activity. Cancer Res. 2010;70(1):288–98.

[70] García-Martínez JM, Moran J, Clarke RG, Gray A, Cosulich SC, Chresta CM, et al. Ku-0063794 is a specific inhibitor of the mammalian target of rapamycin (mTOR). Biochem J. 2009;421(1):29–42.

[71] Nyfeler B, Bergman P, Triantafellow E, Wilson CJ, Zhu Y, Radetich B, et al. Relieving autophagy and 4EBP1 from rapamycin resistance. Mol Cell Biol. 2011;31(14):2867–76.

[72] Soucy TA, Smith PG, Milhollen MA, Berger AJ, Gavin JM, Adhikari S, et al. An inhibitor of NEDD8-activating enzyme as a new approach to treat cancer. Nature. 2009 Apr 9;458(7239):732–6.

[73] Yang D, Zhao Y, Liu J, Sun Y, Jia L. Protective autophagy induced by RBX1/ROC1 knockdown or CRL inactivation via modulating the DEPTOR-MTOR axis. Autophagy. 2012 Dec 27;8(12):1856–8.

[74] Zhao Y, Xiong X, Jia L, Sun Y. Targeting Cullin-RING ligases by MLN4924 induces autophagy via modulating the HIF1-REDD1-TSC1-mTORC1-DEPTOR axis. Cell Death Dis. Nature Publishing Group; 2012;3(9):e386.

[75] Pereira GJS, Hirata H, Fimia GM, do Carmo LG, Bincoletto C, Han SW, et al. Nicotinic acid adenine dinucleotide phosphate (NAADP) regulates autophagy in cultured astrocytes. J Biol Chem. 2011;286(32):27875–81.

[76] Liu T-J, Koul D, LaFortune T, Tiao N, Shen RJ, Maira S-M, et al. NVP-BEZ235, a novel dual phosphatidylinositol 3-kinase/mammalian target of rapamycin inhibitor, elicits multifaceted antitumor activities in human gliomas. Mol Cancer Ther. 2009;8(8):2204–10.

[77] Serra V, Markman B, Scaltriti M, Eichhorn PJ a, Valero V, Guzman M, et al. NVP-BEZ235, a dual PI3K/mTOR inhibitor, prevents PI3K signaling and inhibits the growth of cancer cells with activating PI3K mutations. Cancer Res. 2008;68(16):8022–30.

[78] East DA, Fagiani F, Crosby J, Georgakopoulos ND, Bertrand H, Schaap M, et al. PMI: A $\Delta\Psi m$ independent pharmacological regulator of mitophagy. Chem Biol. 2014;21(11):1585–96.

[79] Apsel B, Blair JA, Gonzalez B, Nazif TM, Feldman ME, Aizenstein B, et al. Targeted polypharmacology: discovery of dual inhibitors of tyrosine and phosphoinositide kinases. Nat Chem Biol. 2008 Nov 12;4(11):691–9.

[80] Hsieh AC, Costa M, Zollo O, Davis C, Feldman ME, Testa JR, et al. Genetic Dissection of the Oncogenic mTOR Pathway Reveals Druggable Addiction to Translational Control via 4EBP-eIF4E. Cancer Cell. 2010 Mar;17(3):249–61.

[81] Feldman ME, Apsel B, Uotila A, Loewith R, Knight ZA, Ruggero D, et al. Active-site inhibitors of mTOR target rapamycin-resistant outputs of mTORC1 and mTORC2. PLoS Biol. 2009;7(2):e38.

[82] Brown EJ, Albers MW, Shin TB, Ichikawa K, Keith CT, Lane WS, et al. A mammalian protein targeted by G1-arresting rapamycin-receptor complex. Nature. 1994 Jun 30;369(6483):756–8.

[83] Pirola L, Fröjdö S. Resveratrol: one molecule, many targets. IUBMB Life. 2008 May;60(5):323–32.

[84] Puissant A, Auberger P. AMPK- and p62/SQSTM1-dependent autophagy mediate resveratrol-induced cell death in chronic myelogenous leukemia. Autophagy. 2010 Jul;6(5):655–7.

[85] Vingtdeux V, Giliberto L, Zhao H, Chandakkar P, Wu Q, Simon JE, et al. AMP-activated protein kinase signaling activation by resveratrol modulates amyloid- peptide metabolism. J Biol Chem. 2010;285(12):9100–13.

[86] Vingtdeux V, Chandakkar P, Zhao H, D'Abramo C, Davies P, Marambaud P. Novel synthetic small-molecule activators of AMPK as enhancers of autophagy and amyloid —peptide degradation. FASEB J. 2011;25(1):219–31.

[87] Wong VK, Li T, Law BY, Ma ED, Yip NC, Michelangeli F, et al. Saikosaponin-d, a novel SERCA inhibitor, induces autophagic cell death in apoptosis-defective cells. Cell Death Dis. 2013 Jul;4(7):e720.

[88] Shoji-Kawata S, Sumpter R, Leveno M, Campbell GR, Zou Z, Kinch L, et al. Identification of a candidate therapeutic autophagy-inducing peptide. Nature. 2013;494(7436): 201–6.

[89] Liu Y, Shoji-Kawata S, Sumpter RM, Wei Y, Ginet V, Zhang L, et al. Autosis is a Na +,K + –ATPase-regulated form of cell death triggered by autophagy-inducing peptides, starvation, and hypoxia-ischemia. Proc Natl Acad Sci U S A. 2013 Dec 17;110(51):20364–71.

[90] Frederick C, Ando K, Leroy K, Héraud C, Suain V, Buée L, et al. Rapamycin ester analog CCI-779/Temsirolimus alleviates tau pathology and improves motor deficit in mutant tau transgenic mice. J Alzheimers Dis. 2015;44(4):1145–56.

[91] Jiang T, Yu J-T, Zhu X-C, Zhang Q-Q, Cao L, Wang H-F, et al. Temsirolimus attenuates tauopathy in vitro and in vivo by targeting tau hyperphosphorylation and autophagic clearance. Neuropharmacology. 2014 Oct;85:121–30.

[92] Jiang T, Yu J-T, Zhu X-C, Tan M-S, Wang H-F, Cao L, et al. Temsirolimus promotes autophagic clearance of amyloid-β and provides protective effects in cellular and animal models of Alzheimer's disease. Pharmacol Res. 2014 Mar;81:54–63.

[93] Zhang L, Dai F, Cui L, Jing H, Fan P, Tan X, et al. Novel role for TRPC4 in regulation of macroautophagy by a small molecule in vascular endothelial cells. Biochim Biophys Acta. 2015 Feb;1853(2):377–87.

[94] Thoreen CC, Kang SA, Chang JW, Liu Q, Zhang J, Gao Y, et al. An ATP-competitive mammalian target of rapamycin inhibitor reveals rapamycin-resistant functions of mTORC1. J Biol Chem. 2009;284(12):8023–32.

[95] Sarkar S, Davies JE, Huang Z, Tunnacliffe A, Rubinsztein DC. Trehalose, a novel mTOR-independent autophagy enhancer, accelerates the clearance of mutant huntingtin and◎?-synuclein. J Biol Chem. 2007;282(8):5641–52.

[96] Zhang X, Chen S, Song L, Tang Y, Shen Y, Jia L, et al. MTOR-independent, autophagic enhancer trehalose prolongs motor neuron survival and ameliorates the autophagic flux defect in a mouse model of amyotrophic lateral sclerosis. Autophagy. 2014;10(4): 588–602.

[97] Carpenter JE, Jackson W, Benetti L, Grose C. Autophagosome formation during varicella-zoster virus infection following endoplasmic reticulum stress and the unfolded protein response. J Virol. 2011;85(18):9414–24.

[98] Yu K, Shi C, Toral-Barza L, Lucas J, Shor B, Kim JE, et al. Beyond rapalog therapy: preclinical pharmacology and antitumor activity of WYE-125132, an ATP-competitive and specific inhibitor of mTORC1 and mTORC2. Cancer Res. 2010;70(2):621–31.

[99] Ferlay J, Soerjomataram I, Dikshit R, Eser S, Mathers C, Rebelo M, et al. Cancer incidence and mortality worldwide: Sources, methods and major patterns in GLOBOCAN 2012. Int J Cancer. 2015 Mar 1;136(5):E359–86.

[100] Boyle P LB. World Cancer Report 2008. IARC Press. :510p.

[101] National Institutes of Health. Clinical Trial. Gov: A service of the U.S. National Institutes of Health. 2016.

[102] Amaravadi RK, Lippincott-Schwartz J, Yin XM, Weiss W a, Takebe N, Timmer W, et al. Principles and current strategies for targeting autophagy for cancer treatment. Vol. 17, Clinical Cancer Research. 2011. p. 654–66.

[103] Vakifahmetoglu-norberg H, Xia H, Yuan J. Pharmacologic agents targeting autophagy. J Clin Invest. 2015;125(1):5–13.

[104] Liu J, Lin M, Yu J, Liu B, Bao J. Targeting apoptotic and autophagic pathways for cancer therapeutics. Cancer Lett. 2011 Jan 28;300(2):105–14.

[105] Dalby KN, Tekedereli I, Lopez-Berestein G, Ozpolat B. Targeting the prodeath and prosurvival functions of autophagy as novel therapeutic strategies in cancer. Vol. 6, Autophagy. 2010. p. 322–9.

[106] Potze L, Mullauer FB, Colak S, Kessler JH, Medema JP. Betulinic acid-induced mitochondria-dependent cell death is counterbalanced by an autophagic salvage response. Cell Death Dis. Nature Publishing Group; 2014 Jan;5(4):e1169.

[107] Debnath J, Baehrecke EH, Kroemer G. Does autophagy contribute to cell death? Autophagy. 2005 Jul;1(2):66–74.

[108] White E, DiPaola RS. The double-edged sword of autophagy modulation in cancer. Clin Cancer Res. 2009;15(17):5308–16.

[109] Levy JMM, Thorburn A. Targeting autophagy during cancer therapy to improve clinical outcomes. Pharmacol Ther. Elsevier B.V.; 2011;131(1):130–41.

[110] Takeuchi H, Kondo Y, Fujiwara K, Kanzawa T, Aoki H, Mills GB, et al. Synergistic augmentation of rapamycin-induced autophagy in malignant glioma cells by phosphatidylinositol 3-kinase/protein kinase B inhibitors. Cancer Res. 2005 May 15;65(8): 3336–46.

[111] Yazbeck VY, Buglio D, Georgakis G V, Li Y, Iwado E, Romaguera JE, et al. Temsirolimus downregulates p21 without altering cyclin D1 expression and induces autophagy and synergizes with vorinostat in mantle cell lymphoma. Exp Hematol. 2008 Apr;36(4):443–50.

[112] Zhang H, Berel D, Wang Y, Li P, Bhowmick NA, Figlin RA, et al. A comparison of Ku0063794, a dual mTORC1 and mTORC2 inhibitor, and temsirolimus in preclinical renal cell carcinoma models. PLoS One. 2013;8(1):e54918.

[113] Crazzolara R, Cisterne A, Thien M, Hewson J, Baraz R, Bradstock KF, et al. Potentiating effects of RAD001 (Everolimus) on vincristine therapy in childhood acute lymphoblastic leukemia. Blood. 2009;113(14):3297–306.

[114] Rangwala R, Chang YC, Hu J, Algazy KM, Evans TL, Fecher LA, et al. Combined MTOR and autophagy inhibition: phase I trial of hydroxychloroquine and temsirolimus in patients with advanced solid tumors and melanoma. Autophagy. 2014;10(8):1391–402.

[115] Milano V, Piao Y, LaFortune T, de Groot J. Dasatinib-induced autophagy is enhanced in combination with temozolomide in glioma. Mol Cancer Ther. 2009;8(2):394–406.

[116] Shao Y, Gao Z, Marks PA, Jiang X. Apoptotic and autophagic cell death induced by histone deacetylase inhibitors. Proc Natl Acad Sci U S A. 2004;101(52):18030–5.

[117] Liu Y-L, Yang P-M, Shun C-T, Wu M-S, Weng J-R, Chen C-C. Autophagy potentiates the anti-cancer effects of the histone deacetylase inhibitors in hepatocellular carcinoma. Autophagy. 2010 Nov;6(8):1057–65.

[118] Yamamoto S, Tanaka K, Sakimura R, Okada T, Nakamura T, Li Y, et al. Suberoylanilide hydroxamic acid (SAHA) induces apoptosis or autophagy-associated cell death in chondrosarcoma cell lines. Anticancer Res. 2008;28 (3 A):1585–91.

[119] Min A, Im S-A, Kim DK, Song S-H, Kim H-J, Lee K-H, et al. Histone deacetylase inhibitor, suberoylanilide hydroxamic acid (SAHA), enhances anti-tumor effects of the poly (ADP-ribose) polymerase (PARP) inhibitor olaparib in triple-negative breast cancer cells. Breast Cancer Res. ???; 2015;17(1):33.

[120] Park JH, Ahn MY, Kim TH, Yoon S, Kang KW, Lee J, et al. A new synthetic HDAC inhibitor, MHY218, induces apoptosis or autophagy-related cell death in tamoxifen-resistant MCF-7 breast cancer cells. Invest New Drugs. 2012 Oct;30(5):1887–98.

[121] Kim DE, Kim Y, Cho D-H, Jeong S-Y, Kim S-B, Suh N, et al. Raloxifene induces autophagy-dependent cell death in breast cancer cells via the activation of AMP-activated protein kinase. Mol Cells. 2015;38(2):138–44.

[122] de Medina P, Silvente-Poirot S, Poirot M. Tamoxifen and AEBS ligands induced apoptosis and autophagy in breast cancer cells through the stimulation of sterol accumulation. Autophagy. 2009 Oct;5(7):1066–7.

[123] Leng S, Hao Y, Du D, Xie S, Hong L, Gu H, et al. Ursolic acid promotes cancer cell death by inducing Atg5-dependent autophagy. Int J Cancer. 2013;133(12):2781–90.

[124] Xavier CPR, Lima CF, Pedro DFN, Wilson JM, Kristiansen K, Pereira-Wilson C. Ursolic acid induces cell death and modulates autophagy through JNK pathway in apoptosis-resistant colorectal cancer cells. J Nutr Biochem. Elsevier Inc.; 2013;24(4):706–12.

[125] Pinto Ribeiro Xavier C, Lima C, Rohde M, Pereira-Wilson C. 253 Autophagy triggered by ursolic acid synergistically enhances 5-fluorouracil induced cell death in HCT15 (MSI p53 mutant) colorectal cancer cells. Eur J Cancer Suppl. Elsevier Ltd; 2010 Jun; 8(5):66–7.

[126] Ellington AA, Berhow M, Singletary KW. Induction of macroautophagy in human colon cancer cells by soybean B-group triterpenoid saponins. Carcinogenesis. 2005 Jan;26(1): 159–67.

[127] Carew JS, Medina EC, Esquivel JA, Mahalingam D, Swords R, Kelly K, et al. Autophagy inhibition enhances vorinostat-induced apoptosis via ubiquitinated protein accumulation. J Cell Mol Med. 2010;14(10):2448–59.

[128] Li J, Hou N, Faried A, Tsutsumi S, Kuwano H. Inhibition of autophagy augments 5-fluorouracil chemotherapy in human colon cancer in vitro and in vivo model. Eur J Cancer. 2010 Jul;46(10):1900–9.

[129] Verschooten L, Barrette K, van Kelst S, Rubio Romero N, Proby C, de Vos R, et al. Autophagy inhibitor chloroquine enhanced the cell death inducing effect of the flavonoid luteolin in metastatic squamous cell carcinoma cells. PLoS One. 2012;7(10): 1–11.

[130] Poklepovic A, Gewirtz DA. Outcome of early clinical trials of the combination of hydroxychloroquine with chemotherapy in cancer. Autophagy. 2014;10(8):1478–80.

[131] Liang X, Tang J, Liang Y, Jin R, Cai X. Suppression of autophagy by chloroquine sensitizes 5-fluorouracil-mediated cell death in gallbladder carcinoma cells. Cell Biosci. Cell & Bioscience; 2014;4(1):10.

[132] Koh Y, Lim HY, Ahn JH, Lee J-L, Rha SY, Kim YJ, et al. Phase II trial of everolimus for the treatment of nonclear-cell renal cell carcinoma. Ann Oncol. 2013 Apr;24(4):1026–31.

[133] Seiler M, Ray-Coquard I, Melichar B, Yardley DA, Wang RX, Dodion PF, et al. Oral ridaforolimus plus trastuzumab for patients with HER2+ trastuzumab-refractory metastatic breast cancer. Clin Breast Cancer. 2015 Feb;15(1):60–5.

[134] Bendell JC, Jones SF, Hart L, Spigel DR, Lane CM, Earwood C, et al. A phase Ib study of linsitinib (OSI-906), a dual inhibitor of IGF-1R and IR tyrosine kinase, in combination with everolimus as treatment for patients with refractory metastatic colorectal cancer. Invest New Drugs. 2015 Feb;33(1):187–93.

[135] Piha-Paul SA, Munster PN, Hollebecque A, Argilés G, Dajani O, Cheng JD, et al. Results of a phase 1 trial combining ridaforolimus and MK-0752 in patients with advanced solid tumours. Eur J Cancer. 2015 Sep;51(14):1865–73.

[136] Zibelman M, Wong Y-N, Devarajan K, Malizzia L, Corrigan A, Olszanski AJ, et al. Phase I study of the mTOR inhibitor ridaforolimus and the HDAC inhibitor vorinostat in advanced renal cell carcinoma and other solid tumors. Invest New Drugs. 2015 Oct; 33(5):1040–7.

[137] Nemunaitis J, Hochster HS, Lustgarten S, Rhodes R, Ebbinghaus S, Turner CD, et al. A phase I trial of oral ridaforolimus (AP23573; MK-8669) in combination with bevacizumab for patients with advanced cancers. Clin Oncol (R Coll Radiol). 2013 Jun;25(6):336–42.

[138] Oza AM, Pignata S, Poveda A, McCormack M, Clamp A, Schwartz B, et al. Randomized phase II trial of ridaforolimus in advanced endometrial carcinoma. J Clin Oncol. 2015 Nov 1;33(31):3576–82.

[139] Mita MM, Poplin E, Britten CD, Tap WD, Rubin EH, Scott BB, et al. Phase I/IIa trial of the mammalian target of rapamycin inhibitor ridaforolimus (AP23573; MK-8669) administered orally in patients with refractory or advanced malignancies and sarcoma. Ann Oncol. 2013;24(4):1104–11.

[140] Hanahan D, Weinberg RA. Hallmarks of cancer: the next generation. Cell. Elsevier Inc.; 2011 Mar 4;144(5):646–74.

[141] Kim MJ, Woo SJ, Yoon CH, Lee JS, An S, Choi YH, et al. Involvement of autophagy in oncogenic K-Ras-induced malignant cell transformation. J Biol Chem. 2011;286(15): 12924–32.

[142] Yang S, Wang X, Contino G, Liesa M, Sahin E, Ying H, et al. Pancreatic cancers require autophagy for tumor growth. Genes Dev. 2011;25(7):717–29.

[143] Eng CH, Wang Z, Tkach D, Toral-Barza L, Ugwonali S, Liu S, et al. Macroautophagy is dispensable for growth of KRAS mutant tumors and chloroquine efficacy. Proc Natl Acad Sci U S A. 2016 Jan 5;113(1):182–7.

[144] Armstrong JL, Corazzari M, Martin S, Pagliarini V, Falasca L, Hill DS, et al. Oncogenic B-RAF signaling in melanoma impairs the therapeutic advantage of autophagy inhibition. Clin Cancer Res. 2011 Apr 15;17(8): 2216–26.

[145] Li J, Hou N, Faried A, Tsutsumi S, Takeuchi T, Kuwano H. Inhibition of autophagy by 3-MA enhances the effect of 5-FU-induced apoptosis in colon cancer cells. Ann Surg Oncol. 2009;16(3):761–71.

[146] Giampietri C, Petrungaro S, Padula F, D'Alessio A, Marini ES, Facchiano A, et al. Autophagy modulators sensitize prostate epithelial cancer cell lines to TNF-alpha-dependent apoptosis. Apoptosis. 2012 Nov;17(11):1210–22.

[147] Zhu K, Dunner K, McConkey DJ. Proteasome inhibitors activate autophagy as a cytoprotective response in human prostate cancer cells. Oncogene. 2010 Jan 21;29(3): 451–62.

[148] Wu Z, Chang P-C, Yang JC, Chu C-Y, Wang L-Y, Chen N-T, et al. Autophagy blockade sensitizes prostate cancer cells towards Src family kinase inhibitors. Genes Cancer. 2010;1(1):40–9.

[149] Amaravadi RK, Yu D, Lum JJ, Bui T, Christophorou MA, Evan GI, et al. Autophagy inhibition enhances therapy-induced apoptosis in a Myc-induced model of lymphoma. J Clin Invest. 2007;117(2):326–36.

[150] Ding W-X, Ni H-M, Gao W, Chen X, Kang JH, Stolz DB, et al. Oncogenic transformation confers a selective susceptibility to the combined suppression of the proteasome and autophagy. Mol Cancer Ther. 2009;8(7):2036–45.

[151] Dikic I, Johansen T, Kirkin V. Selective autophagy in cancer development and therapy. Cancer Res. 2010;70(9):3431–4.

[152] Fischer S, Ronellenfitsch MW, Thiepold AL, Harter PN, Reichert S, Kögel D, et al. Hypoxia enhances the antiglioma cytotoxicity of b10, a glycosylated derivative of betulinic acid. PLoS One. 2014;9 (4).

[153] Vazquez-Martin A, López-Bonetc E, Cufí S, Oliveras-Ferraros C, Del Barco S, Martin-Castillo B, et al. Repositioning chloroquine and metformin to eliminate cancer stem cell traits in pre-malignant lesions. Drug Resist Updat. 2011;14(4–5):212–23.

[154] Le X-F, Mao W, Lu Z, Carter BZ, Bast RC. Dasatinib induces autophagic cell death in human ovarian cancer. Cancer. 2010 Nov 1;116(21):4980–90.

[155] Wilson EN, Bristol ML, Di X, Maltese WA, Koterba K, Beckman MJ, et al. A switch between cytoprotective and cytotoxic autophagy in the radiosensitization of breast tumor cells by chloroquine and vitamin D. Horm Cancer. 2011 Oct 2;2(5):272–85.

[156] Rossi M, Rotblat B, Ansell K, Amelio I, Caraglia M, Misso G, et al. High throughput screening for inhibitors of the HECT ubiquitin E3 ligase ITCH identifies antidepressant drugs as regulators of autophagy. Cell Death Dis. Nature Publishing Group; 2014;5(5):e1203.

[157] Hu T, Li P, Luo Z, Chen X, Zhang J, Wang C, et al. Chloroquine inhibits hepatocellular carcinoma cell growth in vitro and in vivo. Oncol Rep. 2015 Nov 2;43–9.

[158] Fan Q-W, Cheng C, Hackett C, Feldman M, Houseman BT, Nicolaides T, et al. Akt and Autophagy Cooperate to Promote Survival of Drug-Resistant Glioma. Sci Signal. 2010 Nov 9;3(147):ra81–ra81.

[159] Solomon VR, Lee H. Chloroquine and its analogs: A new promise of an old drug for effective and safe cancer therapies. Eur J Pharmacol. Elsevier B.V.; 2009;625(1–3):220–33.

[160] Sui X, Chen R, Wang Z, Huang Z, Kong N, Zhang M, et al. Autophagy and chemotherapy resistance: a promising therapeutic target for cancer treatment. Cell Death Dis. 2013;4:e838.

[161] Rosenfeld MR, Ye X, Supko JG, Desideri S, Grossman SA, Brem S, et al. A phase I/II trial of hydroxychloroquine in conjunction with radiation therapy and concurrent and adjuvant temozolomide in patients with newly diagnosed glioblastoma multiforme. Autophagy. 2014;10(8):1359–68.

[162] Sotelo J, Briceño E, López-González MA. Adding chloroquine to conventional treatment for glioblastoma multiforme: a randomized, double-blind, placebo-controlled trial. Ann Intern Med. 2006 Mar 7;144(5):337–43.

[163] Huang C, Liu W, Perry CN, Yitzhaki S, Lee Y, Yuan H, et al. Autophagy and protein kinase C are required for cardioprotection by sulfaphenazole. Am J Physiol Heart Circ Physiol. 2010;298(2):H570–9.

[164] Huang C, Andres AM, Ratliff EP, Hernandez G, Lee P, Gottlieb RA. Preconditioning involves selective mitophagy mediated by parkin and p62/SQSTM1. PLoS One. 2011;6(6).

[165] Martinet W, Knaapen MWM, Kockx MM, De Meyer GRY. Autophagy in cardiovascular disease. Trends Mol Med. 2007 Nov;13(11):482–91.

[166] Nemchenko A, Chiong M, Turer A, Lavandero S, Hill JA. Autophagy as a therapeutic target in cardiovascular disease. J Mol Cell Cardiol. 2011 Oct;51(4):584–93.

[167] Cao DJ, Wang Z V, Battiprolu PK, Jiang N, Morales CR, Kong Y, et al. Histone deacetylase (HDAC) inhibitors attenuate cardiac hypertrophy by suppressing autophagy. Proc Natl Acad Sci U S A. 2011;108(10):4123–8.

[168] Gottlieb RA, Mentzer RM. Cardioprotection through autophagy: Ready for clinical trial? Autophagy. 2011;7(4):434–5.

[169] Sala-Mercado JA, Wider J, Reddy Undyala VV, Jahania S, Yoo W, Mentzer RM, et al. Profound cardioprotection with chloramphenicol succinate in the swine model of myocardial ischemia-reperfusion injury. Circulation. 2010;122(11 SUPPL. 1):179–85.

[170] Przyklenk K, Undyala VVR, Wider J, Sala-Mercado JA, Gottlieb RA, Mentzer RM. Acute induction of autophagy as a novel strategy for cardioprotection: Getting to the heart of the matter. Autophagy. 2011;7(4):432–3.

[171] Shioi T, McMullen JR, Tarnavski O, Converso K, Sherwood MC, Manning WJ, et al. Rapamycin attenuates load-induced cardiac hypertrophy in mice. Circulation. 2003;107(12):1664–70.

[172] McMullen JR, Sherwood MC, Tarnavski O, Zhang L, Dorfman AL, Shioi T, et al. Inhibition of mTOR signaling with rapamycin regresses established cardiac hypertrophy induced by pressure overload. Circulation. 2004;109(24):3050–5.

[173] Martinet W, Verheye S, De Meyer GRY. Selective depletion of macrophages in atherosclerotic plaques via macrophage-specific initiation of cell death. Trends Cardiovasc Med. 2007 Feb;17(2):69–75.

[174] Verheye S, Martinet W, Kockx MM, Knaapen MWM, Salu K, Timmermans JP, et al. Selective clearance of macrophages in atherosclerotic plaques by autophagy. J Am Coll Cardiol. 2007;49(6):706–15.

[175] Martinet W, De Loof H, De Meyer GRY. mTOR inhibition: a promising strategy for stabilization of atherosclerotic plaques. Atherosclerosis. 2014 Apr;233(2):601–7.

[176] Buss SJ, Riffel JH, Katus HA, Hardt SE. Augmentation of autophagy by mTOR-inhibition in myocardial infarction: When size matters. Autophagy. 2010 Feb;6(2):304–6.

[177] Cherra SJ, Chu CT. Autophagy in neuroprotection and neurodegeneration: A question of balance. Future Neurol. 2008 May;3(3):309–23.

[178] Berger Z, Ravikumar B, Menzies FM, Oroz LG, Underwood BR, Pangalos MN, et al. Rapamycin alleviates toxicity of different aggregate-prone proteins. Hum Mol Genet. 2006;15(3):433–42.

[179] Webb JL, Ravikumar B, Atkins J, Skepper JN, Rubinsztein DC. -Synuclein is degraded by both autophagy and the proteasome. J Biol Chem. 2003 Jun 27;278(27):25009–13.

[180] Spencer B, Potkar R, Trejo M, Rockenstein E, Patrick C, Gindi R, et al. Beclin 1 gene transfer activates autophagy and ameliorates the neurodegenerative pathology in alpha-synuclein models of Parkinson's and Lewy body diseases. J Neurosci. 2009;29(43):13578–88.

[181] Ravikumar B, Duden R, Rubinsztein DC. Aggregate-prone proteins with polyglutamine and polyalanine expansions are degraded by autophagy. Hum Mol Genet. 2002;11(9):1107–17.

[182] Haung Yu W, Cuervo AM, Kumar A, Peterhoff CM, Schmidt SD, Lee JH, et al. Macroautophagy—A novel◎??-amyloid peptide-generating pathway activated in Alzheimer's disease. J Cell Biol. 2005;171(1):87–98.

[183] Rubinsztein DC, Bento CF, Deretic V. Therapeutic targeting of autophagy in neurodegenerative and infectious diseases. J Exp Med. 2015 Jun 29;212(7):979–90.

[184] Frake RA, Ricketts T, Menzies FM, Rubinsztein DC. Autophagy and neurodegeneration. J Clin Invest. 2015 Jan;125(1):65–74.

[185] Sarkar S, Krishna G, Imarisio S, Saiki S, O'Kane CJ, Rubinsztein DC. A rational mechanism for combination treatment of Huntington's disease using lithium and rapamycin. Hum Mol Genet. 2008 Jan 15;17(2):170–8.

[186] Ravikumar B, Vacher C, Berger Z, Davies JE, Luo S, Oroz LG, et al. Inhibition of mTOR induces autophagy and reduces toxicity of polyglutamine expansions in fly and mouse models of Huntington disease. Nat Genet. 2004 Jun;36(6):585–95.

[187] Rodrigues D, Viotto AC, Checchia RG, Gomide AB, Itri R, Severino D, et al. Mechanism of Aloe Vera extract protection against UVA: shelter of lysosomal membrane avoids photodamage. Photochem Photobiol Sci. 2016;

[188] Chiu H-C, Soni S, Kulp SK, Curry H, Wang D, Gunn JS, et al. Eradication of intracellular Francisella tularensis in THP-1 human macrophages with a novel autophagy inducing agent. J Biomed Sci. 2009;16:110.

[189] Chiu HC, Kulp SK, Soni S, Wang D, Gunn JS, Schlesinger LS, et al. Eradication of intracellular Salmonella enterica serovar typhimurium with a small-molecule, host cell-directed agent. Antimicrob Agents Chemother. 2009;53(12):5236–44.

[190] Kim JJ, Lee HM, Shin DM, Kim W, Yuk JM, Jin HS, et al. Host cell autophagy activated by antibiotics is required for their effective antimycobacterial drug action.©?? 2012 Elsevier Inc. Cell Host Microbe. Elsevier Inc.; 2012;11(5):457–68.

[191] Campbell GR, Spector SA. Vitamin D inhibits human immunodeficiency virus type 1 and Mycobacterium tuberculosis infection in macrophages through the induction of autophagy. PLoS Pathog. 2012;8(5).

[192] Gutierrez MG, Master SS, Singh SB, Taylor GA, Colombo MI, Deretic V. Autophagy is a defense mechanism inhibiting BCG and Mycobacterium tuberculosis survival in infected macrophages. Cell. 2004;119(6):753–66.

[193] Schiebler M, Brown K, Hegyi K, Newton SM, Renna M, Hepburn L, et al. Functional drug screening reveals anticonvulsants as enhancers of mTOR-independent autophagic killing of Mycobacterium tuberculosis through inositol depletion. EMBO Mol Med. 2015;7(2):127–39.

[194] Stanley SA, Barczak AK, Silvis MR, Luo SS, Sogi K, Vokes M, et al. Identification of host-targeted small molecules that restrict intracellular mycobacterium tuberculosis growth. PLoS Pathog. 2014;10 (2).

[195] Lam KKY, Zheng X, Forestieri R, Balgi AD, Nodwell M, Vollett S, et al. Nitazoxanide stimulates autophagy and inhibits mTORC1 signaling and intracellular proliferation of *Mycobacterium tuberculosis*. PLoS Pathog. 2012;8(5).

[196] Sundaramurthy V, Barsacchi R, Samusik N, Marsico G, Gilleron J, Kalaidzidis I, et al. Integration of chemical and RNAi multiparametric profiles identifies triggers of intracellular mycobacterial killing. Cell Host Microbe. Elsevier Inc.; 2013;13(2):129–42.

[197] Parihar SP, Guler R, Khutlang R, Lang DM, Hurdayal R, Mhlanga MM, et al. Statin therapy reduces the mycobacterium tuberculosis burden in human macrophages and in mice by enhancing autophagy and phagosome maturation. J Infect Dis. 2014 Mar 1;209(5):754–63.

[198] Tsujimoto Y, Shimizu S. Another way to die: autophagic programmed cell death. Cell Death Differ. 2005 Nov;12 Suppl 2:1528–34.

[199] Dzubak P, Hajduch M, Vydra D, Hustova A, Kvasnica M, Biedermann D, et al. Pharmacological activities of natural triterpenoids and their therapeutic implications. Nat Prod Rep. 2006 Jul;23(3):394–411.

[200] Fulda S. Activation of mitochondria and release of mitochondrial apoptogenic factors by betulinic acid. J Biol Chem. 1998 Dec 18;273(51):33942–8.

[201] Li Y, He K, Huang Y, Zheng D, Gao C, Cui L, et al. Betulin induces mitochondrial cytochrome c release associated apoptosis in human cancer cells. Mol Carcinog. 2010 Jul;49(7):630–40.

[202] Potze L, Di Franco S, Grandela C, Pras-Raves ML, Picavet DI, van Veen H a, et al. Betulinic acid induces a novel cell death pathway that depends on cardiolipin modification. Oncogene. Nature Publishing Group; 2015 Apr 20; (October 2014):1–11.

[203] Chen Y, Sun R, Wang B. Monolayer behavior of binary systems of betulinic acid and cardiolipin: thermodynamic analyses of Langmuir monolayers and AFM study of Langmuir-Blodgett monolayers. J Colloid Interface Sci. Elsevier Inc.; 2011 Jan 1;353(1): 294–300.

[204] Gao M, Lau PM, Kong SK. Mitochondrial toxin betulinic acid induces in vitro eryptosis in human red blood cells through membrane permeabilization. Arch Toxicol. 2014;88(3):755–68.

[205] Eder-Czembirek C, Erovic BM, Czembirek C, Brunner M, Selzer E, Pötter R, et al. Betulinic acid a radiosensitizer in head and neck squamous cell carcinoma cell lines. Strahlenther Onkol. 2010 Mar;186(3):143–8.

[206] Selzer E, Pimentel E, Wacheck V, Schlegel W, Pehamberger H, Jansen B, et al. Effects of betulinic acid alone and in combination with irradiation in human melanoma cells. J Invest Dermatol. 2000 May;114(5):935–40.

[207] Jeremias I, Steiner HH, Benner A, Debatin K-M, Herold-Mende C. Cell death induction by betulinic acid, ceramide and TRAIL in primary glioblastoma multiforme cells. Acta Neurochir (Wien). 2004 Jul;146(7):721–9.

[208] Sawada N, Kataoka K, Kondo K, Arimochi H, Fujino H, Takahashi Y, et al. Betulinic acid augments the inhibitory effects of vincristine on growth and lung metastasis of B16F10 melanoma cells in mice. Br J Cancer. 2004 May 19;90(8):1672–8.

[209] Schmidt ML, Kuzmanoff KL, Ling-Indeck L, Pezzuto JM. Betulinic acid induces apoptosis in human neuroblastoma cell lines. Eur J Cancer. 1997. pp. 2007–10.

[210] Fulda S, Susin SA, Kroemer G, Cells N. Molecular ordering of apoptosis induced by anticancer drugs in neuroblastoma cells molecular ordering of apoptosis induced by anticancer drugs in. 1998;4453–60.

[211] Kasperczyk H, La Ferla-Brühl K, Westhoff MA, Behrend L, Zwacka RM, Debatin K-M, et al. Betulinic acid as new activator of NF-kappaB: molecular mechanisms and implications for cancer therapy. Oncogene. 2005 Oct 20;24(46):6945–56.

[212] Pisha E, Chai H, Lee IS, Chagwedera TE, Farnsworth NR, Cordell GA, et al. Discovery of betulinic acid as a selective inhibitor of human melanoma that functions by induction of apoptosis. Nat Med. 1995 Oct;1(10):1046–51.

[213] Ehrhardt H, Fulda S, Führer M, Debatin KM, Jeremias I. Betulinic acid-induced apoptosis in leukemia cells. Leukemia. 2004 Aug;18(8):1406–12.

[214] Kessler JH, Mullauer FB, de Roo GM, Medema JP. Broad in vitro efficacy of plant-derived betulinic acid against cell lines derived from the most prevalent human cancer types. Cancer Lett. 2007 Jun 18;251(1):132–45.

[215] Mullauer FB, Kessler JH, Medema JP. Betulinic acid induces cytochrome c release and apoptosis in a Bax/Bak-independent, permeability transition pore dependent fashion. Apoptosis. 2009 Feb;14(2):191–202.

[216] Rzeski W, Stepulak A, Szymański M, Sifringer M, Kaczor J, Wejksza K, et al. Betulinic acid decreases expression of bcl-2 and cyclin D1, inhibits proliferation, migration and induces apoptosis in cancer cells. Naunyn Schmiedebergs Arch Pharmacol. 2006 Oct; 374(1):11–20.

[217] Yang L, Chen Y, Ma Q, Fang J, He J, Cheng Y, et al. Effect of betulinic acid on the regulation of Hiwi and cyclin B1 in human gastric adenocarcinoma AGS cells. Acta Pharmacol Sin. 2010;31(1):66–72.

[218] Raghuvar Gopal D V., Narkar Aa., Badrinath Y, Mishra KP, Joshi DS. Betulinic acid induces apoptosis in human chronic myelogenous leukemia (CML) cell line K-562 without altering the levels of Bcr-Abl. Toxicol Lett. 2005 Mar 15;155(3):343–51.

[219] Raghuvar Gopal D V., Narkar a. a., Badrinath Y, Mishra KP, Joshi DS. Protection of Ewing's sarcoma family tumor (ESFT) cell line SK-N-MC from betulinic acid induced apoptosis by alpha-DL-tocopherol. Toxicol Lett. 2004 Dec 2;153(2):201–12.

[220] Wick W, Grimmel C, Wagenknecht B, Dichgans J, Weller M. Betulinic acid-induced apoptosis in glioma cells: A sequential requirement for new protein synthesis, formation of reactive oxygen species, and caspase processing. J Pharmacol Exp Ther. 1999 Jun;289(3):1306–12.

[221] Chadalapaka G, Jutooru I, Burghardt R, Safe S. Drugs that target specificity proteins downregulate epidermal growth factor receptor in bladder cancer cells. Mol Cancer Res. 2010;8(5):739–50.

[222] Hsu T, Wang M, Chen S, Huang S, Yeh Y, Su W, et al. Betulinic acid decreases specificity protein 1 (Sp1) level via increasing the sumoylation of Sp1 to inhibit lung cancer growth. Mol Pharmacol. 2012;82(6):1115–28.

[223] Safe SH, Prather PL, Brents LK, Chadalapaka G, Jutooru I. Unifying mechanisms of action of the anticancer activities of triterpenoids and synthetic analogs. Anticancer Agents Med Chem. 2012;12(10):1211–20.

[224] Fulda S. Betulinic Acid for cancer treatment and prevention. Int J Mol Sci. 2008 Jun;9(6): 1096–107.

[225] Nakagawa-Goto K, Yamada K, Taniguchi M, Tokuda H, Lee K-HH. Cancer preventive agents 9. Betulinic acid derivatives as potent cancer chemopreventive agents. Bioorganic Med Chem Lett. Elsevier Ltd; 2009 Jul 1;19 (13):3378–81.

[226] Rajendran P, Jaggi M, Singh MK, Mukherjee R, Burman AC. Pharmacological evaluation of C-3 modified Betulinic acid derivatives with potent anticancer activity. Invest New Drugs. 2008;26(1):25–34.

[227] Yang S, Liang N, Li H, Xue W, Hu D, Jin L, et al. Design, synthesis and biological evaluation of novel betulinic acid derivatives. Chem Cent J. 2012;6(1):141.

[228] Li F, Goila-Gaur R, Salzwedel K, Kilgore NR, Reddick M, Matallana C, et al. PA-457: a potent HIV inhibitor that disrupts core condensation by targeting a late step in Gag processing. Proc Natl Acad Sci U S A. 2003;100(23):13555–60.

[229] Yu D, Morris-Natschke SL, Lee KH. New developments in natural products-based anti-AIDS research. Vol. 27, Med Res Rev. 2007. p. 108–32.

[230] Willmann M, Wacheck V, Buckley J, Nagy K, Thalhammer J, Paschke R, et al. Characterization of NVX-207, a novel betulinic acid-derived anti-cancer compound. Eur J Clin Invest. 2009;39(5):384–94.

[231] Bayer M, Proksch P, Felsner I, Brenden H, Kohne Z, Walli R, et al. Photoprotection against UVAR: Effective triterpenoids require a lipid raft stabilizing chemical structure. Exp Dermatol. 2011;20(11):955–8.

[232] Gonzalez P, Mader I, Tchoghandjian A, Enzenmüller S, Cristofanon S, Basit F, et al. Impairment of lysosomal integrity by B10, a glycosylated derivative of betulinic acid, leads to lysosomal cell death and converts autophagy into a detrimental process. Cell Death Differ. 2012 Aug;19(8):1337–46.

[233] Enzenmüller S, Gonzalez P, Karpel-Massler G, Debatin KM, Fulda S. GDC-0941 enhances the lysosomal compartment via TFEB and primes glioblastoma cells to lysosomal membrane permeabilization and cell death. Cancer Lett. 2013;329(1):27–36.

[234] Csuk R. Betulinic acid and its derivatives: a patent review (2008–2013). Expert Opin Ther Pat. 2014 Aug;24(8):913–23.

[235] Liby K, Honda T, Williams CR, Risingsong R, Royce DB, Suh N, et al. Novel semisynthetic analogues of betulinic acid with diverse cytoprotective, antiproliferative, and proapoptotic activities. Mol Cancer Ther. 2007 Jul;6(7):2113–9.

[236] Morselli E, Galluzzi L, Kepp O, Vicencio J-MM, Criollo A, Maiuri MC, et al. Anti- and pro-tumor functions of autophagy. Biochim Biophys Acta. 2009 Sep;1793(9):1524–32.

[237] Liu EY, Ryan KM. Autophagy and cancer—issues we need to digest. J Cell Sci. 2012 May 15;125(Pt 10):2349–58.

[238] Cufí S, Vazquez-Martin A, Oliveras-Ferraros C, Corominas-Faja B, Cuyàs E, López-Bonet E, et al. The anti-malarial chloroquine overcomes primary resistance and restores sensitivity to trastuzumab in HER2-positive breast cancer. Sci Rep. 2013;3:2469.

[239] Bacellar I, Tsubone T, Pavani C, Baptista M. Photodynamic efficiency: from molecular photochemistry to cell death. Int J Mol Sci. 2015;16(9):20523–59.

[240] Reiners JJ, Agostinis P, Berg K, Oleinick NL, Kessel D. Assessing autophagy in the context of photodynamic therapy. Autophagy. 2010 Jan;6(1):7–18.

[241] Oleinick NL, Morris RL, Belichenko I. The role of apoptosis in response to photodynamic therapy: what, where, why, and how. Photochem Photobiol Sci. 2002 Jan;1(1):1–21.

[242] Inguscio V, Panzarini E, Dini L. Autophagy Contributes to the Death/Survival Balance in Cancer PhotoDynamic Therapy. Vol. 1, Cells. 2012. p. 464–91.

Chapter 18

Choosing Lunch: The Role of Selective Autophagy Adaptor Proteins

Kahiry Leyva-Paredes ,
Nayeli Shantal Castrejón-Jiménez ,
Hugo Iván Arrieta-Oliva ,
Shantal Lizbeth Baltierra-Uribe and
Blanca Estela García-Pérez

Additional information is available at the end of the chapter

http://dx.doi.org/10.5772/63977

Abstract

Autophagy (macroautophagy) is a lysosome-dependent catabolic pathway that degrades damaged organelles, protein aggregates, microorganisms, and other cytoplasmic components. Autophagy was previously considered to be nonselective; however, studies have increasingly established that autophagy-mediated degradation is highly regulated. Selective autophagy regulates plenty of specific cellular components through specialized molecules termed autophagy receptors, which include p62, NBR1, NDP52, optineurin, and VCP among others. Autophagy receptors recognize ubiquitinated cargo and interact with the LC3/GABARAP/Gate16 protein on the membrane of nascent phagophore. In this review, we summarize the advances in the molecular mechanisms of selective autophagy adaptor proteins.

Keywords: selective autophagy, adaptor proteins, p62, optineurin, NDP52

1. Introduction

The various functions of eukaryotic organisms depend largely on the existence of highly efficient regulation mechanisms. Each physiologic activity involves the production of several molecules whose half-life must be controlled by degradation system to maintain homeostasis. Up to the present time, two systems of degradation molecules or organelles are known: (a) ubiquitin-proteasome system (UPS) and (b) autophagy, a lysosome-dependent degradation system. The precise mechanisms to lead the substrate to UPS or autophagy are not understood completely. However, it is known that ubiquitin is a key protein to regulate the substrate recognition through the conjugation of a single ubiquitin monomer (monoubiquitination) or sequential conjugation of several ubiquitin moieties (polyubiquitination). The conjugation of four ubiquitin monomers is sufficient signal to allow the ubiquitylated target protein to be recognized by UPS [1]. The specificity in the UPS is generated by the ability of ubiquitin to form eight different chain linkages on itself, through its seven lysine residues (K6, K11, K27, K29, K33, K48, and K63). K48 ubiquitin chain is the most well studied and was originally identified as the signal to target proteins to proteasomal degradation [2]. K11 and K63 ubiquitin chains are more related to signal for nonproteolytic functions as DNA repair and cell signaling, but a recent study shows that heterotypic K11/K48-polyUb chains bind to the proteasome and facilitate the degradation of cyclin B1 [3]. Whereas UPS is the major degradation pathway for short-lived and regulatory proteins, autophagy is more linked with the elimination of long-lived proteins and organelles. The selectivity of autophagy degradation is conferred by K63 ubiquitin chains [4, 5] (**Figure 1**). Autophagy was first described as a nonselective bulk degradation system, and now the accumulated evidence indicates that autophagy can be highly selective. Nonselective autophagy is triggered as a response to starvation and implies the random formation of the autophagosomes with the subsequent capture of any organelle or molecule near the autophagosome. In contrast, the selective autophagy is involved in the recruitment of different adaptor proteins that interact with Atg proteins and target organelles or molecules to be degraded [6, 7]. It is possible to distinguish

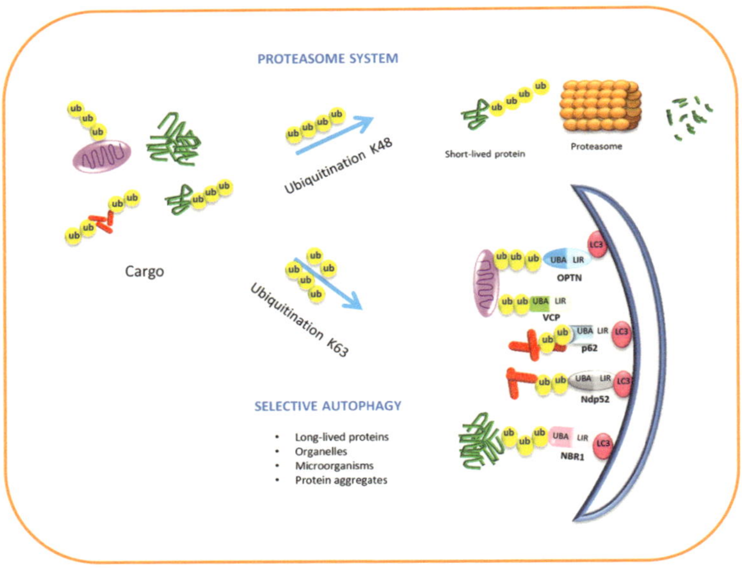

Figure 1. Schematic representation of selective autophagy. The two degradation pathways are shown. Mitochondria, misfolded proteins, and microorganisms are ubiquitinated and selected to proteasome system or autophagy. Selective autophagy involves the participation of adaptor proteins as p62, NDP52, optineurin, NBR1, and VCP, as bridges to cargo and nascent phagophore.

http://dx.doi.org/10.5772/63977

various types of selective autophagy, depending on the cargo that is captured and degraded: lipid droplets (lipophagy), mitochondria (mitophagy), ER (reticulophagy), pathogens (xenophagy), and aggregation-prone proteins (aggrephagy) [8], among others. Nevertheless, the precise mechanism of cargo recognition remains unclear; the molecular characterization of autophagy receptors, initially SQSTM1/p62 and NBR1, has revealed that ubiquitination is involved in substrate selectivity. Most autophagy receptors have a ubiquitin-binding domain (UBD) and LC3-interacting region (LIR). The UBD domains attach to target molecules or organelles, and LIR domain interacts with LC3/GABARAP/Gate16 protein in autophagy to facilitate autophagosome formation, transport, and/or maturation [8]. The core of LIR consists of D/E, D/E, D/E, F/W/Y, X, X, L/I/V, and D; the phosphorylation of this domain enhances the affinity with LC3/GABARAP/Gate16 protein [7]. In this chapter, we discuss the recent knowledge on autophagy receptors and their role in selective autophagy.

2. SQSTM1/p62

SQSTM1/p62 (referred to hereafter as p62) was initially identified as a phosphotyrosine-independent binding of a 62 kDa (gave p62 its name) to the Src homology 2 (SH2) domain [9]. Subsequently, the term sequestosome-1 (SQSTM1) was assigned for its capacity to sequester polyubiquitinated protein to cytoplasmic storage before its degradation by the proteasome, protecting the cytosol from the toxic effect of misfolded proteins [10]. Currently, p62 has been considered a protein with pleiotropic activities, derived from their multiple domains, which interact with several molecules involved in the cellular death, oxidative stress, inflammatory response, and recognition of molecules to be degraded by UPS or autophagy [10, 11]. Recent studies have reported that mTOR activation depends on p62 as a key regulator of the nutrient-sensing pathway [12].

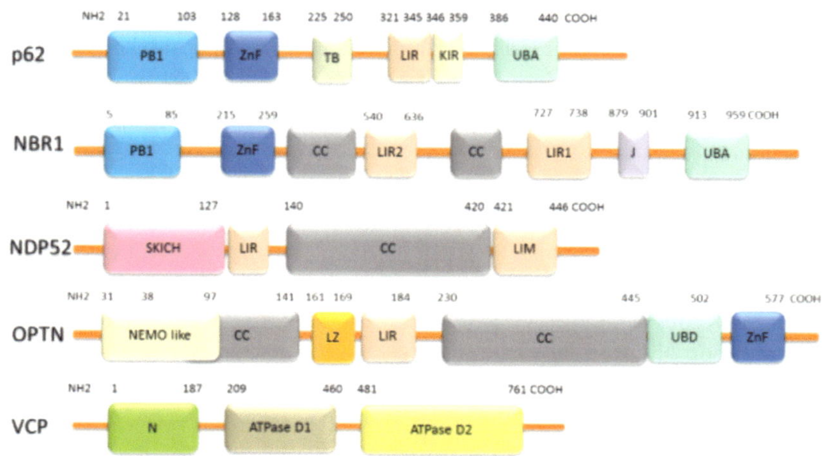

Figure 2. Structure of adaptor proteins involved in selective autophagy.

p62 is a multifunctional protein of 440-amino acids that is conserved in metazoans but not in plants and fungi [13]. Refractory to Sigma P (Ref(2)P) is a homologue protein of mammalian p62 in Drosophila melanogaster, that regulates protein aggregation in adult brain [14]. p62 has six functional domains: the N-terminal Phox/Bem1 domain (PB1, 21–103 aa), a ZZ-type zing finger domain (ZnF, 128–163 aa), TRAF6-binding domain (TB, 225–250 aa), a short LC3-interacting region (LIR, 321–345 aa), KEAP1-interacting region motifs (KIR, 346–359), and ubiquitin-associated domain (UBA, 386–440 aa) which is localized in C-terminal end (**Figure 2**) [11].

PB1 domains, ZZ zinc finger, KIR, and UBA can bind several proteins to participate in inflammatory responses and receptor-mediated signal transduction. PB1 has been associated with atypical protein kinase C (aPKC) to activate NF-kB signaling pathway. The ZZ zinc finger domain is responsible for binding the receptor interacting protein (RIP) to regulate the inflammatory response. The KIR region of this scaffold protein is a regulator for Nrf2, and its activation induces transcription of oxidative stress response genes [11, 15].

p62 provides a link between the degradation of ubiquitin cargo by UPS or autophagy through the UBA domain. The PB1, LIR, and UBA domains are implied in the degradation of ubiquitinated cargo by selective autophagy. UBA domain is responsible for noncovalent binding with polyubiquitinated cargos, through serine 403 phosphorylation by casein kinase 2 (CK2), which increases the affinity for ubiquitinated chains [16, 17]. PB1 domain is involved in self- and heteroligomerization with NBR1 (another receptor of selective autophagy) [18–20], and LIR region is important to LC3 interaction. It has been hypothesized that the activity of p62, in selective autophagy, requires a sequential interaction. Initially, there must be an interaction of p62 with ubiquitin proteins, and then aggregation of complex protein oligomerization to itself or with NBR1, and these aggregates are finally degraded by autolysosomes [7, 21, 22]. This order is altered in defects of autophagy, where firstly accumulations of p62 proteins are present and later are ubiquitinated [23, 24].

Furthermore, p62 is an autophagy receptor that can bind proteins to be degraded by selective autophagy as aggregates of misfolded proteins, damaged mitochondria, peroxisomes, and intracellular bacteria, which are ubiquitinated and targeted for clearance by autophagy [7, 25]. Several works have evidenced that p62 has a critical role in the normal functioning of mitochondria. p62 is localized to mitochondria under physiological conditions and plays an important role in mitochondrial morphology, genome integrity, and mitochondrial import of transcription factors. When p62 is deleted, it leads to mitochondrial fragmentation and mitochondrial dysfunction [26]. The role of p62 as adaptor receptor in mitophagy is currently debated. Geisler and colleagues reported that p62 with PINK1 y Parkin molecules has a key role in the sequential mitophagy process. p62 colocalized with Parkin on clustered mitochondria after the induction of mitophagy by carbonyl cyanide m-chlorophenylhydrazone (CCCP) treatment, and the silence of p62 by siRNA resulted in a significant loss of mitochondrial clearance [27, 28]. In contrast, Narendra and colleagues mentioned that p62 is not indispensable in the mitophagy; in HeLa cells with siRNA directed against p62, no difference was found in lacking mitochondria after the induction of mitophagy but is important in clustering of depolarized mitochondria [29, 30]. Similarly, in pexophagy, the role of p62 is not clear. p62 has been involved in the clustering of peroxisomes that were labeled with ubiquitin to selective degradation [31], but recent work suggests that p62 is responsible for clustering, and only NBR1 is essential for peroxisome degradation by autophagy [32].

Adaptor p62 protein has an important role in the xenophagy, which is responsible for restriction of the replication of several intracellular microorganisms. The role of p62 has been more explored in infections by intracellular bacteria. Bacteria such as *Shigella flexneri*, a nonmotile actA mutant of *Listeria monocytogenes* [33], *Salmonella enterica* serotype Typhimurium [34], and *Mycobacterium tuberculosis* [25, 35] are targeted selectively through p62 recruitment to deliver into nascent LC3-positive isolation membranes for autophagosomal degradation. Interestingly, p62 has an additional role in antibacterial effect against *M. tuberculosis*, through the delivery of cytosolic proteins to *M. tuberculosis* containing-autolysosomes, where they are processed to convert into new antimicrobial peptides [36]. In viral infections, it has been reported that p62 plays a role in the clearance of viral proteins. Orvedahl and colleagues demonstrate that Sindbis virus capsid protein can interact with p62 in an ubiquitin-independent pathway, suggesting that clearance of viral proteins by autophagy requires p62 and other molecules tag different from ubiquitin [37]. There is little evidence about the role of p62 in infections by parasites. In the infection of *Toxoplasma gondii*, it has been observed that p62 and ubiquitin were recruited to *T. gondii* parasitophorous vacuoles when infected cells were stimulated by INF-γ playing an important role in the antigen presentation to activate specific CD8+T cells [38].

The high relevance of p62 as a signaling hub implies their efficient regulation. When p62 is disregulated or dysfunctional, there are multiple consequences. Several studies have implicated p62 aggregates in cancer, inflammation, neurodegenerative disease, liver disease, and aging [10, 13, 20, 39].

3. NBR1

Neighbor of BRCA1 gene 1 (NBR1) is another cargo receptor that is selectively degraded by autophagy. NBR1 was originally cloned as a candidate gene for the ovarian cancer antigen CA125 [40]. Given the similarity and interaction with p62, NBR1 has been studied in cell signaling and differentiation [41]. In 2009, Kirkin and colleagues have shown that NBR1 was involved in autophagic degradation of ubiquitinated targets [42].

NBR1 was recognized as a direct binding partner of the autophagosome-specific ATG8/LC3/GABARAP modifiers both in vitro and in vivo. NBR1 has similar domain architecture as p62 and shares several key features with p62 though differs in sequence and size. Both proteins share a very similar overall domain architecture, consisting of an N-terminal PB1(residues 5–85) domain, a ZZ-like zinc finger domain (residues 215–259), a two-domain light-chain-3-interacting regions LRS1 (residues 540–636) and LRS2 (residues 727–738, the LRS2 does not have the core consensus motif W/YXXL/I, most likely representing a novel type of LC3-interacting sequence), and a C-terminal ubiquitin-associated (UBA) domain (**Figure 2**) [43].

NBR1 binds strongly to ubiquitin via its UBA domain with a bias toward the K63-linked polyUb chains [43]. NBR1 undergoes dimerization via the coiled-coil domain. NBR1 can directly bind to p62, and together they act as cargo receptors for autophagic degradation of polyubiquitylated aggregates and peroxisomes [43–45]. In the absence of p62, NBR1 interacts with misfolded and ubiquitinated proteins for degradation by autophagy [42]. Moreover, it is known that NBR1 promotes cell differentiation, may act as a tumor suppressor, and is also involved in bone remodeling [46, 47]. NBR1 also is involved in protein misfolding disorders such as body myositis sporadic inclusion, and autophagic degradation may have a role in the pathology [48].

4. NDP52

The nuclear dot protein 52 kD (NDP52) also named as calcium binding and coiled-coil domain 2 (CALCOCO2) is a 446-amino acid protein. Discovered in 1995, it was first erroneously found as part of the nuclear domains (ND10) called Kr bodies or PLM-containing oncogenic domains (POD), consisting of protein aggregates detected as dots (approximately 10) by autoimmune sera or monoclonal antibodies. By 1997, Sternsdorf and colleagues found by using polyclonal sera anti-NDP52 that the protein localization is restricted to the cytoplasm but not ND10 and confirmed an increase of NDP52 transcripts when cells were treated with IFN-γ [49].

NDP52 has a predicted molecular mass of 52 kD and exhibits an N-terminal skeletal muscle, and kidney-enriched inositol phosphatase carboxyl homology (SKICH) domain (aa 1–127), a central coil-coiled domain with zipper leucine motifs (aa 140–420) and the C-terminus presents homology with Lin11, Isl-1, and Mec-3 (LIM) domains containing two zinc finger arrangements involved in protein-protein interactions as ubiquitin (**Figure 2**) (aa 421–446) [50–52]. There are two paralogs existing, CoCoA (also known as Calcoco1) and Tax1BP1 (also known TXBP151). The CoCoA paralog comprises SKICH domains and LIM domains as NDP52, but it lacks an ubiquitin-binding domain. Therefore, it did not decorate bacteria when they escaped to the cytosol [53].

First studies linking NDP52 to a physiological process showed that after infection with *S. enterica*, the ubiquitin-coated bacteria in the cytosol are recognized by NDP52, which acts as a receptor. Then, NDP52 interacts with the adaptor proteins, Nap1 or Sintbad (also named TBKBP), and leads to the recruitment of TANK-binding kinase 1 (TBK1), which results in the control of bacterial growth [52]. The authors also found that NDP52 recruits and binds ATG8/LC3, an autophagosomal marker, and the knockdown of NDP52 impairs the autophagy of *Salmonella* [52, 54]. The same effect was observed with *Streptococcus pyogenes*-infected cells but not with *S. flexneri*. They conclude that NDP52 is a receptor that recognizes ubiquitin-coated bacteria and binds ATG8/LC3 leading to the control of bacterial growth by autophagy. Further studies demonstrated that NDP52 has a LC3-interacting region (LIR) domain and the sites of interaction with Nap1 or Sintbad are located at the SKICH domain (**Figure 2**) [55]. Later, Muhlinen and colleagues demonstrated that NDP52 binds all human ATG8/LC3 orthologs (LC3A, LC3B, LC3C, GABARAP, GABARAPL1, GABARAPL2), but only LC3C performs an antibacterial function when binds NDP52 through the LC3C-interacting region (CLIR) [56]. Other studies demonstrated that *S. flexneri* is also targeted by NDP52/P62 to autophagy pathway dependent upon septin and actin [57]. Additional to the ubiquitin-dependent pathway needed for the recruitment of NDP52, a carbohydrate-dependent galectin-8 pathway also mediates NDP52 recruitment to invading bacteria at early stages of infection, unlike the ubiquitin-dependent pathway which plays a major role at later points [58, 59].

As part of the autophagosome maturation, it was found that NDP52 interacts with myosin VI via RRL motif, and such interaction recruits myosin VI to deliver endosomal membranes to the nascent autophagosome [60, 61]. This function of NDP52 is independent of its function in xenophagy and involves a different binding domain [62].

Recently, it has been demonstrated that NDP52 also plays a role in regulation. The Toll-like receptor (TLR) signaling serves as, an example, the selective autophagic degradation of Toll/interleukin-1 receptor homology domain-containing adaptor-inducing interferon (TRIF), and TRAF6 is mediated by this receptor. The mechanism involves the polyubiquitination of NDP52 by TRAF6 to acquire the ability to form aggregates of polyubiquitinated TRAF6 [63, 64]. The regulation of the miRNA activity is another example; recently, it has been discovered that the miRNA-processing enzyme DICER and the miRNA effector AGO2 are a target by NDP52 for their degradation via autophagy [65]. Another target of NDP52 includes the RNA retrotransposon, and the degradation of this RNA via autophagy helps to maintain the stability of the genome [66]. More recently, Heo and colleagues found that the PINK1-PARKIN mitochondrial ubiquitination pathway promotes mitophagy by recruiting TBK1 kinase which binds to NDP52 and other autophagy receptors to induce autophagy of mitochondria [67, 68].

Some studies with viruses have related autophagy to anti- or pro-viral roles; NDP52 has been involved in promoting viral replication of chikungunya virus (CHIKV) when interacts with the nonstructural protein nsP2 in infected human cells [69].

Although autophagy has a protective role against some intracellular bacteria, some studies indicate that NDP52 has a relation with Crohn's disease, and Ellinghaus and colleagues found an association between the disease and a missense mutation in affected individuals [70, 71]. Additionally, NDP52 has been involved in Alzheimer's disease where it has a protective role in facilitating the clearance of phosphorylated tau [72, 73].

5. Optineurin

Optineurin was first described by Li and colleagues [74]. They were looking for proteins that interact with the early region 3 (E3) 14.7 K protein in the yeast two-hybrid system. E3 14.7 K is synthesized by E3 in group C of adenovirus and is an inhibitor of NF-kB cytolysis. They found a protein with the ability to interact with E3 14.7K and named as FIP-2 (for 14.7K interacting protein), and FIP-2 interacts with E3 14.7K in the cytoplasm and caused redistribution of the protein. Also, FIP-2 reversed the protective effect of E3 14.7K on cell death induced by TNF-α. After, Schwamborn and colleagues described that FIP-2 had a strong homology to NF-kB essential modulator (NEMO) and named as NEMO-related protein (NRP). They found that NRP was associated with Golgi apparatus and is de novo expressed by interferon and

TNF-α [75]. Rezaie and colleagues coined the name optineurin (optic neuropathy-inducing protein) after discovering that this molecule was associated with diseases such as normal tension glaucoma and a subtype of primary open-angle glaucoma [76].

Optineurin is a 67 KDa intracellular protein found in different tissues [77], and optineurin gene encodes an 884-amino acid protein and contains three noncoding exons and 13 exons that code for a 577-amino acid protein [78]. The mRNA can be found in 3 isoforms as a result of alternative splicing [74]. Optineurin has been described in different tissues such as spleen, kidney, skeletal muscle, brain, heart, lung, pancreas, and eyes of various species: human, mouse, and chicken [74, 78, 79]. The UPS is a very important pathway to recycle optineurin, but in situations where optineurin is upregulated, the UPS is compromised and autophagy is induced to control the optineurin levels [80].

Optineurin has different domains through which it can interact with different proteins. It contains putative domains such as C-terminal zinc finger, leucine zipper domain [74], a LIR domain [81], a NEMO-like domain [75], UBD domain [82], and various coiled-coil motifs [83]. In **Figure 2**, optineurin is shown and compared with other adaptor proteins involved in selective autophagy.

Optineurin can participate in different biological activities because it has multiple domains, which mediate the interaction with other proteins. For example, optineurin can interact with Rab8 [83–85], transferrin receptor [86], serine/threonine kinase receptor-interacting protein 1 (RIP) [87], ubiquitin [88], and Myosin VI [61, 84], among others. The interaction of optineurin with myosin VI [84, 89] is mediated by the UBD domain, and this interaction is important in the fusion of secretory vesicles with the plasma membrane [90]. Also, it has been described that macrophages from patients with Crohn's disease where optineurin was under-expressed fail to secrete pro-inflammatory cytokines [91], which suggest that optineurin can also be involved in the vesicular transport of the autophagosomes.

As it was mentioned, the importance of optineurin in selective autophagy relies on their UBD and LIR domains, through their interaction with specific cargo and the autophagy machinery respectively. However, it has been reported that optineurin can recognize a target like a superoxide dismutase 1 and huntingtin protein by an ubiquitin-independent pathway and degrade protein aggregates through autophagy [92], and this recognition was related to the C-terminal coiled-coil domain of optineurin.

LIR domain mediates the interaction of optineurin with autophagy machinery. The phosphorylation of serine 177 on LIR domain by TANK-binding kinase 1 (TBK1) increases the affinity of optineurin to LC3/GABARAP proteins [81, 93]. After the dominant phosphorylation at serine 177, optineurin forms a strand with the beta-strand 2 of LC3B and phenylalanine 178 and isoleucine 181 are inserted into a hydrophobic pocket on the LC3B [93]. On the other hand, the isomers of LC3 and the proteins of the GABARAP family interact with the machinery involved in autophagosome elongation, which is recruited to LC3/GABARAP-optineurin. Then, autophagy-related protein ATG4 is recruited to cleavage at the C-terminal of LC3/GABARAP and exhibits phenylalanine and glycine amino acids, which participate in the

http://dx.doi.org/10.5772/63977

conjugation of LC3/GABARAP with phosphatidylethanolamine (PE) [94] to complete autophagosome formation finally.

Mitochondria are important and dynamic organelles, depending on energetic requirements of the cells, mitochondria can undergo cycles of fusion and fission [95]. When mitochondria are damaged, they suffer an increase in the rate of fission, and this results in their fragmentation. Damaged and fragmented mitochondria are removed via mitophagy [28, 96]. The role of optineurin in mitophagy has been recently studied. Mitophagy requires the interaction with other proteins. First, damaged mitochondria are marked with ubiquitin by Parkin and PINK1. Both proteins act in the ubiquitination of Mitofusin 1 and Mitofusin 2 when mitochondrial depolarization was induced by carbonyl cyanide-m-chlorophenylhydrazone (CCCP) treatment [97]. When mitochondria become depolarized, PINK1 accumulates on the mitochondrial outer membrane and phosphorylates to Mitofusin 2, this allows their interaction with inactive Parkin, and the subsequent activation of Parkin by PINK1 activates the Parkin ubiquitin ligase activity [98]. This ubiquitination allows the interaction of optineurin with ubiquitinated mitochondria through its UBD domain, and this recognition is similar to p62 and NIX [28]. After this initial recognition, optineurin recruits autophagy machinery around damaged mitochondria to capture it into the autophagosome [96].

Xenophagy has been considered as an innate immune response against intracellular infections. Xenophagy guided by optineurin has been poorly described, so it represents an interesting and wide field of study. It has been described that xenophagy mediated by optineurin participates in the intracellular control of *S. enterica*. For this activity, optineurin requires UBD and LIR domains. It has been reported than when UBD domain was mutated, optineurin failed to colocalize with *S. enterica* and when LIR domain was mutated, optineurin colocalized with *S. enterica* but not with LC3. Also, TBK1 activity was necessary for the xenophagy mediated by optineurin [81]. Tumbarello and colleagues found that optineurin, TAX1BP1, and NDP52 are important in xenophagy response against *Salmonella typhimurium*. Also, they found that Myosin VI is necessary to restrict the replication of *S. typhimurium*, highlighting the role of Myosin VI in the vesicular transport of autophagosomes containing *S. typhimurium* to lysosomes [99].

Due to the role of optineurin in the capture of unnecessary or damaged organelles, the lack or the deficiency of optineurin has been associated with different pathologies such as amyotrophic lateral sclerosis [100], Paget's disease of bone [101], normal tension glaucoma, and primary open-angle glaucoma [102].

6. VCP/p97

Valosin-containing protein (VCP/97)—also called Cdc48p in yeast, p97 in Xenopus, CDC-48 in *Caenorhabditis elegans*, or TER94 in *Drosophila*—belongs to the hexameric AAA (ATPases associated with diverse cellular activities) family of proteins with two ATPase domains, D1 and D2. The structure of VCP/97 molecule includes N (1–187), D1 (209–460), D2 (481–761), and C (762–806) domains, with two linkers: N-D1 linker (188–208) and D1–D2 linker (461–480)

(**Figure 2**). VCP proteins form a barrel-like structure that comprises two ring-shaped layers made of D1 and D2 AAA modules [103]. The diversity of cellular functions and the activity of VCP/p97 are mediated by their interaction with a large number of protein cofactors. p97 forms core complexes with the major cofactors, which include the proteins ubiquitin-X (UBX) domain, the Ufd1 (ubiquitin fusion degradation 1)-Npl4 (nuclear protein localization homolog 4) heterodimer, and p47 [104]. Several works have established p97 as a principal element in emerging functions of the UPS as was described in a review by Meyer et al. [104]. Nevertheless, p97 lacks a LIR domain; recent reports now link to p97 with the autophagy. The first findings of the involvement of p97 with autophagy were reported in studies of the multisystem degenerative disorder characterized by inclusion body myopathy, frontotemporal dementia, and Paget's disease of bone (also known as IBMPFD) [105] which is associated with VCP mutations. This disorder is characterized by the extensive accumulation of ubiquitin conjugates in affected tissues, and in IBMPFD patients the VCP mutations cause no damage in ubiquitin-dependent degradation by the proteasome, but do impair maturation of ubiquitin-containing autophagosomes. Further, the myoblasts derived from IBMPFD patients showed accumulation of LAMP-1, LAMP-2, and LC3-II-positive vacuoles indicating that VCP/p97 is essential to autophagosome maturation [106]. Additional work has provided evidence about the role of Cdc48/p97 in the regulation of autophagosome biogenesis. In *Saccharomyces cerevisiae*, it has been demonstrated that the participation of Cdc48/p77 in autophagy is mediated by direct interaction of Shp1/Ubx1 cofactor with Atg8 PE-conjugated form [107]. Recently, it has been recognized the role of VCP/p97 in mitochondrial maintenance. On the one hand, VCP/p97 was accepted as part of outer mitochondrial membrane-associated degradation (OMMAD), functioning as retrotranslocase of ubiquitinated mitochondrial proteins for degradation by UPS [108]. On the other hand, Tanaka and colleges showed that p97 and proteasome activity are required to mitophagy mediated by Parkin protein [109]. The role of VCP/p97 in xenophagy remains unexplored, which opens a new window to investigate.

7. Conclusion

Currently, autophagy is an attractive area of investigation. It has been recognized the participation of autophagy in homeostatic cellular functions, such as clearance of damaged organelles, misfolded proteins, and microorganism, among others. Scientists have focused on describing the molecular mechanisms responsible for the autophagy. Now, we know that autophagy, far from beginning a random pathway, is a mechanism that elegantly regulates and is highly orchestrated by several proteins. Some proteins have been identified as bridges between the cargo and nascent phagophore, and recently studies are in process to know how these proteins are working and how to interact with the complex autophagy machinery. Several of these proteins share some structural characteristics, such as LIR domain, which allow the direct interaction with LC3 protein. However, recently, studies have identified new proteins that participate in selective autophagy but lack LIR domain, for example, VCP/p97 and Alfy. The studies to know the precise mechanisms of interaction of these proteins are in process. The understanding of the molecular mechanism that governs the autophagy represents an interesting field because many of these molecules could be manipulated to recover the cellular homeostasis in several pathologies, where autophagy is involved.

Acknowledgements

BEGP received fellowships from COFAA, EDI, and SNI. KLP, NSCJ, HIAO, and SLBU would like to acknowledge CONACYT and BEIFI for their fellowships. This work was supported by CONACYT (22001) and SIP/IPN 20160248.

Author details

Kahiry Leyva-Paredes[1], Nayeli Shantal Castrejón-Jiménez[1], Hugo Iván Arrieta-Oliva[1], Shantal Lizbeth Baltierra-Uribe[2] and Blanca Estela García-Pérez[1*]

*Address all correspondence to: abrilestela@hotmail.com

1 Microbiology Department, Escuela Nacional de Ciencias Biológicas. Instituto Politécnico Nacional, Mexico City, Mexico

2 Immunology Department, Escuela Nacional de Ciencias Biológicas. Instituto Politécnico Nacional, Mexico City, Mexico

References

[1] Thrower JS, Hoffman L, Rechsteiner M, Pickart CM. Recognition of the polyubiquitin proteolytic signal. EMBO J. 2000;19:94–102. doi:10.1093/emboj/19.1.94

[2] Hershko A, Ciechanover A. The ubiquitin system. Annu Rev Biochem. 1998;67:425–79. doi:10.1146/annurev.biochem.67.1.425

[3] Grice GL, Lobb IT, Weekes MP, Gygi SP, Antrobus R, Nathan JA. The proteasome distinguishes between heterotypic and homotypic lysine-11-Linked polyubiquitin chains. Cell Rep. 2015;12:545–53. doi:10.1016/j.celrep.2015.06.061

[4] Erpapazoglou Z, Walker O, Haguenauer-Tsapis R. Versatile roles of k63-linked ubiquitin chains in trafficking. Cells. 2014;3:1027–88. doi:10.3390/cells3041027

[5] Vasco Ferreira J, Rosa Soares A, Silva Ramalho J, Pereira P, Girao H. K63 linked ubiquitin chain formation is a signal for HIF1A degradation by Chaperone-Mediated Autophagy. Sci Rep. 2015;5:10210. doi:10.1038/srep10210

[6] He C, Klionsky DJ. Regulation mechanisms and signaling pathways of autophagy. Annu Rev Genet. 2009;43:67–93. doi:10.1146/annurev-genet-102808–114910

[7] Johansen T, Lamark T. Selective autophagy mediated by autophagic adapter proteins. Autophagy. 2011;7:279–96. doi:10.4161/auto.7.3.14487

[8] Isakson P, Holland P, Simonsen A. The role of ALFY in selective autophagy. Cell Death Differ. 2013;20:12–20. doi:10.1038/cdd.2012.66

[9] Park I, Chung J, Walsh CT, Yun Y, Strominger JL, Shin J. Phosphotyrosine-independent binding of a 62-kDa protein to the src homology 2 (SH2) domain of p56lck and its regulation by phosphorylation of Ser-59 in the lck unique N-terminal region. Proc Natl Acad Sci USA. 1995;92:12338–42. doi:10.1073/pnas.92.26.12338

[10] Komatsu M, Kageyama S, Ichimura Y. P62/SQSTM1/A170: Physiology and pathology. Pharmacol Res. 2012;66:457–62. doi:10.1016/j.phrs.2012.07.004

[11] Katsuragi Y, Ichimura Y, Komatsu M. p62/SQSTM1 functions as a signaling hub and an autophagy adaptor. FEBS J. 2015;282(24):4672–8. doi:10.1111/febs.13540

[12] Sahani MH, Itakura E, Mizushima N. Expression of the autophagy substrate SQSTM1/p62 is restored during prolonged starvation depending on transcriptional upregulation and autophagy-derived amino acids. Autophagy. 2014;10:431–41. doi:10.4161/auto.27344

[13] Bitto A, Lerner CA, Nacarelli T, Crowe E, Torres C, Sell C. p62/SQSTM1 at the interface of aging, autophagy, and disease. Age (Omaha). 2014;36:1123–37. doi:10.1007/s11357-014-9626-3

[14] Nezis IP, Simonsen A, Sagona AP, Finley K, Gaumer S, Contamine D, et al. Ref(2)P, the Drosophila melanogaster homologue of mammalian p62, is required for the formation of protein aggregates in adult brain. J Cell Biol. 2008;180:1065–71. doi:10.1083/jcb.200711108

[15] Geetha T, Wooten MW. Structure and functional properties of the ubiquitin binding protein p62. FEBS Lett. 2002;512:19–24. doi:10.1016/S0014-5793(02)02286-X

[16] Matsumoto G, Wada K, Okuno M, Kurosawa M, Nukina N. Serine 403 phosphorylation of p62/SQSTM1 regulates selective autophagic clearance of ubiquitinated proteins. Mol Cell. 2011;44:279–89. doi:10.1016/j.molcel.2011.07.039

[17] Vadlamudi RK, Joung I, Strominger JL, Shin J. p62, a phosphotyrosine-independent ligand of the SH2 domain of p56(lck), belongs to a new class of ubiquitin-binding proteins. J Biol Chem. 1996;271:20235–7. doi:10.1074/jbc.271.34.20235

[18] Wooten MW, Hu X, Babu JR, Seibenhener ML, Geetha T, Paine MG, et al. Signaling, polyubiquitination, trafficking, and inclusions: Sequestosome 1/p62's role in neurodegenerative disease. J Biomed Biotechnol. 2006;2006:62079. doi:10.1155/JBB/2006/62079

[19] Seibenhener ML, Geetha T, Wooten MW. Sequestosome 1/p62–more than just a scaffold. FEBS Lett. 2007;581:175–9. doi:10.1016/j.febslet.2006.12.027

[20] Salminen A, Kaarniranta K, Haapasalo A, Hiltunen M, Soininen H, Alafuzoff I. Emerging role of p62/sequestosome-1 in the pathogenesis of Alzheimer's disease. Prog Neurobiol. 2011;96:87–95. doi:10.1016/j.pneurobio.2011.11.005

[21] Ichimura Y, Komatsu M. Selective degradation of p62 by autophagy. Semin Immunopathol. 2010;32:431–6. doi:10.1007/s00281-010-0220-1

[22] Pankiv S, Clausen TH, Lamark T, Brech A, Bruun JA, Outzen H, et al. p62/SQSTM1 binds directly to Atg8/LC3 to facilitate degradation of ubiquitinated protein aggregates by autophagy. J Biol Chem. 2007;282:24131–45. doi:10.1074/jbc.M702824200

[23] Du Y, Wooten MC, Wooten MW. Oxidative damage to the promoter region of SQSTM1/p62 is common to neurodegenerative disease. Neurobiol Dis. 2009;35:302–10. doi:10.1016/j.nbd.2009.05.015

[24] Rusten TE, Stenmark H. P62, an autophagy hero or culprit? Nat Cell Biol. 2010;12:207–9. doi:10.1038/ncb0310-207

[25] Jo EK. Autophagy as an innate defense against mycobacteria. Pathog Dis. 2013;67:108–18. doi:10.1111/2049-632X.12023

[26] Seibenhener ML, Du Y, Diaz-Meco MT, Moscat J, Wooten MC, Wooten MW. A role for sequestosome 1/p62 in mitochondrial dynamics, Import and genome integrity. Biochim Biophys Acta—Mol Cell Res. 2013;1833:452–9. doi:10.1016/j.bbamcr.2012.11.004

[27] Geisler S, Holmström KM, Skujat D, Fiesel FC, Rothfuss OC, Kahle PJ, et al. PINK1/Parkin-mediated mitophagy is dependent on VDAC1 and p62/SQSTM1. Nat Cell Biol. 2010;12:119–31. doi:10.1038/ncb2012

[28] Ding WX, Ni HM, Li M, Liao Y, Chen X, Stolz DB, et al. Nix is critical to two distinct phases of mitophagy, reactive oxygen species-mediated autophagy induction and Parkin-ubiquitin-p62-mediated mitochondrial priming. J Biol Chem. 2010;285:27879–90. doi:10.1074/jbc.M110.119537

[29] Narendra DP, Kane LA, Hauser DN, Fearnley IM, Youle RJ. p62/SQSTM1 is required for Parkin-induced mitochondrial clustering but not mitophagy; VDAC1 is dispensable for both. Autophagy. 2010;6:1090–106. doi:10.4161/auto.6.8.13426

[30] Okatsu K, Saisho K, Shimanuki M, Nakada K, Shitara H, Sou YS, et al. P62/SQSTM1 cooperates with Parkin for perinuclear clustering of depolarized mitochondria. Gene Cell. 2010;15:887–900. doi:10.1111/j.1365-2443.2010.01426.x

[31] Kim PK, Hailey DW, Mullen RT, Lippincott-Schwartz J. Ubiquitin signals autophagic degradation of cytosolic proteins and peroxisomes. Proc Natl Acad Sci USA. 2008;105:20567–74. doi:10.1073/pnas.0810611105

[32] Yamashita S ichi, Abe K, Tatemichi Y, Fujiki Y. The membrane peroxin PEX3 induces peroxisome-ubiquitination-linked pexophagy. Autophagy. 2014;10:1549–64. doi: 10.4161/auto.29329

[33] Dupont N, Lacas-Gervais S, Bertout J, Paz I, Freche B, Van Nhieu GT, et al. Shigella Phagocytic vacuolar membrane remnants participate in the cellular response to pathogen invasion and are regulated by autophagy. Cell Host Microbe. 2009;6:137–49. doi:10.1016/j.chom.2009.07.005

[34] Zheng YT, Shahnazari S, Brech A, Lamark T, Johansen T, Brumell JH. The adaptor protein p62/SQSTM1 targets invading bacteria to the autophagy pathway. J Immunol. 2009;183:5909–16. doi:10.4049/jimmunol.0900441

[35] Seto S, Tsujimura K, Horii T, Koide Y. Autophagy adaptor protein p62/SQSTM1 and autophagy-related gene Atg5 mediate autophagosome formation in response to Mycobacterium tuberculosis infection in dendritic cells. PLoS One. 2013;8. doi:10.1371/journal.pone.0086017

[36] Ponpuak M, Davis AS, Roberts EA, Delgado MA, Dinkins C, Zhao Z, et al. Delivery of cytosolic components by autophagic adaptor protein p62 endows autophagosomes with unique antimicrobial properties. Immunity. 2010;32:329–41. doi:10.1016/j.immuni.2010.02.009

[37] Orvedahl A, MacPherson S, Sumpter R, Tallóczy Z, Zou Z, Levine B. Autophagy protects against Sindbis virus infection of the central nervous system. Cell Host Microbe. 2010;7:115–27. doi:10.1016/j.chom.2010.01.007

[38] Lee Y, Sasai M, Ma JS, Sakaguchi N, Ohshima J, Bando H, et al. P62 plays a specific role in interferon-γ-induced presentation of a toxoplasma vacuolar antigen. Cell Rep. 2015;13:223–33. doi:10.1016/j.celrep.2015.09.005

[39] Manley S, Williams JA, Ding W-X. Role of p62/SQSTM1 in liver physiology and pathogenesis. Exp Biol Med (Maywood). 2013;238:525–38. doi: 10.1177/1535370213489446

[40] Campbell IG, Nicolai HM, Foulkes WD, Senger G, Stamp GW, Allan G, et al. A novel gene encoding a B-Box protein within the BRCA1 region at 17q21.1. Hum Mol Genet. 1994;3:589–94. doi:10.1093/hmg/3.4.589

[41] Lange S, Xiang F, Yakovenko A, Vihola A, Hackman P, Rostkova E, et al. The kinase domain of titin controls muscle gene expression and protein turnover. Science. 2005;308:1599–603. doi:10.1126/science.1110463

[42] Kirkin V, Lamark T, Sou YS, Bjørkøy G, Nunn JL, Bruun JA, et al. A role for NBR1 in autophagosomal degradation of ubiquitinated substrates. Mol Cell. 2009;33:505–16. doi:10.1016/j.molcel.2009.01.020

[43] Kirkin V, McEwan DG, Novak I, Dikic I. A role for ubiquitin in selective autophagy. Mol Cell. 2009;34:259–69. doi:10.1016/j.molcel.2009.04.026

[44] Odagiri S, Tanji K, Mori F, Kakita A, Takahashi H, Wakabayashi K. Autophagic adapter protein NBR1 is localized in Lewy bodies and glial cytoplasmic inclusions and is involved in aggregate formation in α-synucleinopathy. Acta Neuropathol. 2012;124:173–86. doi:10.1007/s00401-012-0975-7

[45] Deosaran E, Larsen KB, Hua R, Sargent G, Wang Y, Kim S, et al. NBR1 acts as an autophagy receptor for peroxisomes. J Cell Sci. 2013;126:939–52. doi:10.1242/jcs.114819

[46] Kuo T-C, Chen C-T, Baron D, Onder TT, Loewer S, Almeida S, et al. Midbody accumulation through evasion of autophagy contributes to cellular reprogramming and tumorigenicity. Nat Cell Biol. 2011;13:1214–23. doi:10.1038/ncb2332

[47] Waters S, Marchbank K, Solomon E, Whitehouse CA. Autophagic receptors Nbr1 and p62 coregulate skeletal remodeling. Autophagy. 2010;6:981–3. doi:10.4161/auto.6.7.13155

[48] D´Agostino C, Nogalska A, Cacciottolo M, King Engel W, Askanas V. Abnormalities of NBR1, a novel autophagy-associated protein, in muscle fibers of sporadic inclusion-body myositis. Acta Neuropathol. 2011;122:627–36. doi:10.1007/s00401-011-0874-3

[49] Sternsdorf T, Jensen K, Züchner D, Will H. Cellular localization, expression, and structure of the nuclear dot protein 52. J Cell Biol. 1997;138:435–48. doi:10.1083/jcb.138.2.435

[50] Korioth F, Gieffers C, Maul GG, Frey J. Molecular characterization of NDP52, a novel protein of the nuclear domain 10, which is redistributed upon virus infection and interferon treatment. J Cell Biol. 1995;130:1–13. doi:10.1083/jcb.130.1.1

[51] Di Y, Li J, Zhang Y, He X, Lu H, Xu D, et al. HCC-associated protein HCAP1, a variant of GEMIN4, interacts with zinc-finger proteins. J Biochem. 2003;133:713–8. doi:10.1093/jb/mvg091

[52] Thurston TLM, Ryzhakov G, Bloor S, von Muhlinen N, Randow F. The TBK1 adaptor and autophagy receptor NDP52 restricts the proliferation of ubiquitin-coated bacteria. Nat Immunol. 2009;10:1215–21. doi:10.1038/ni.1800

[53] Ivanov S, Roy CR. NDP52: The missing link between ubiquitinated bacteria and autophagy. Nat Immunol. 2009;10:1137–9. doi:10.1038/ni1109-1137

[54] Von Muhlinen N, Thurston T, Ryzhakov G, Bloor S, Randow F. NDP52, a novel autophagy receptor for ubiquitin-decorated cytosolic bacteria. Autophagy. 2010;6:288–9. doi:10.4161/auto.6.2.11118

[55] Randow F. How cells deploy ubiquitin and autophagy to defend their cytosol from bacterial invasion. Autophagy. 2011;7:304–9. doi:10.4161/auto.7.3.14539

[56] von Muhlinen N, Akutsu M, Ravenhill BJ, Foeglein Á, Bloor S, Rutherford TJ, et al. LC3C, Bound selectively by a Noncanonical LIR motif in NDP52, is required for antibacterial autophagy. Mol Cell. 2012;48:329–42. doi:10.1016/j.molcel.2012.08.024

[57] Mostowy S, Sancho-Shimizu V, Hamon MA, Simeone R, Brosch R, Johansen T, et al. p62 and NDP52 proteins target intracytosolic Shigella and Listeria to different autophagy pathways. J Biol Chem. 2011;286:26987–95. doi:10.1074/jbc.M111.223610

[58] Thurston TLM, Wandel MP, von Muhlinen N, Foeglein Á, Randow F. Galectin 8 targets damaged vesicles for autophagy to defend cells against bacterial invasion. Nature. 2012;482:414–8. doi:10.1038/nature10744

[59] Li S, Wandel MP, Li F, Liu Z, He C, Wu J, et al. Sterical hindrance promotes selectivity of the autophagy cargo receptor NDP52 for the danger receptor galectin-8 in antibacterial autophagy. Sci Signal. 2013;6:ra9. doi:10.1126/scisignal.2003730

[60] Morriswood B, Ryzhakov G, Puri C, Arden SD, Roberts R, Dendrou C, et al. T6BP and NDP52 are myosin VI binding partners with potential roles in cytokine signalling and cell adhesion. J Cell Sci. 2007;120:2574–85. doi:10.1242/jcs.007005

[61] Tumbarello DA, Waxse BJ, Arden SD, Bright NA, Kendrick-Jones J, Buss F. Autophagy receptors link myosin VI to autophagosomes to mediate Tom1-dependent autophagosome maturation and fusion with the lysosome. Nat Cell Biol. 2012;14:1024–35. doi: 10.1038/ncb2589

[62] Verlhac P, Grégoire IP, Azocar O, Petkova DS, Baguet J, Viret C, et al. Autophagy receptor NDP52 regulates pathogen-containing autophagosome maturation. Cell Host Microbe. 2015;17:515–25. doi:10.1016/j.chom.2015.02.008

[63] Inomata M, Niida S, Shibata KI, Into T. Regulation of toll-like receptor signaling by NDP52-mediated selective autophagy is normally inactivated by A20. Cell Mol Life Sci. 2012;69:963–79. doi:10.1007/s00018-011-0819-y

[64] Into T, Inomata M, Takayama E, Takigawa T. Autophagy in regulation of toll-like receptor signaling. Cell Signal. 2012;24:1150–62. doi:10.1016/j.cellsig.2012.01.020

[65] Gibbings D, Mostowy S, Jay F, Schwab Y, Cossart P, Voinnet O. Selective autophagy degrades DICER and AGO2 and regulates miRNA activity. Nat Cell Biol. 2012;14:1314–21. doi:10.1038/ncb2611

[66] Guo H, Chitiprolu M, Gagnon D, Meng L, Perez-Iratxeta C, Lagace D, et al. Autophagy supports genomic stability by degrading retrotransposon RNA. Nat Commun. 2014;5:5276. doi:10.1038/ncomms6276

[67] Heo JM, Ordureau A, Paulo JA, Rinehart J, Harper JW. The PINK1-PARKIN mitochondrial ubiquitylation pathway drives a program of OPTN/NDP52 recruitment and TBK1 activation to promote mitophagy. Mol Cell. 2015;60:7–20. doi:10.1016/j.molcel.2015.08.016

[68] Lazarou M, Sliter DA, Kane LA, Sarraf SA, Wang C, Burman JL, et al. The ubiquitin kinase PINK1 recruits autophagy receptors to induce mitophagy. Nature. 2015;524:309–14. doi:10.1038/nature14893

[69] Judith D, Mostowy S, Bourai M, Gangneux N, Lelek M, Lucas-Hourani M, et al. Species-specific impact of the autophagy machinery on chikungunya virus infection. EMBO Rep. 2013;14:534–44. doi:10.1038/embor.2013.51

[70] Ellinghaus D, Zhang H, Zeissig S, Lipinski S, Till A, Jiang T, et al. Association between variants of PRDM1 and NDP52 and Crohn's disease, based on exome sequencing and functional studies. Gastroenterology. 2013;145:339–47. doi:10.1053/j.gastro.2013.04.040

[71] Till A, Lipinski S, Ellinghaus D, Mayr G, Subramani S, Rosenstiel P, et al. Autophagy receptor CALCOCO2/NDP52 takes center stage in Crohn's disease. Autophagy. 2013;9:1256–7. doi:10.4161/auto.25483

[72] Jo C, Gundemir S, Pritchard S, Jin YN, Rahman I, Johnson GVW. Nrf2 reduces levels of phosphorylated tau protein by inducing autophagy adaptor protein NDP52. Nat Commun. 2014;5:3496. doi:10.1038/ncomms4496

[73] Kim S, Lee D, Song JC, Cho SJ, Yun SM, Koh JH, et al. NDP52 associated with phosphorylated tau in brains of an Alzheimer disease mouse model. Biochem Biophys Res Commun. 2014; 454:196–201. doi:10.1016/j.bbrc.2014.10.066

[74] Li Y, Kang J, Horwitz MS. Interaction of an adenovirus E3 14.7-kilodalton protein with a novel tumor necrosis factor alpha-inducible cellular protein containing leucine zipper domains. Mol Cell Biol. 1998;18:1601–10. doi:10.1128/MCB.18.3.1601

[75] Schwamborn K, Weil R, Courtois G, Whiteside ST, Israël A. Phorbol esters and cytokines regulate the expression of the NEMO-related protein, a molecule involved in a NF-kB-independent pathway. J Biol Chem. 2000;275:22780–9. doi:10.1074/jbc.M001500200

[76] Rezaie T, Child A, Hitchings R, Brice G, Miller L, Coca-Prados M, et al. Adult-onset primary open-angle glaucoma caused by mutations in optineurin. Science. 2002;295:1077–9. doi:10.1126/science.1066901

[77] Ying H, Shen X, Park B, Yue BYJT. Posttranslational modifications, localization, and protein interactions of optineurin, the product of a glaucoma gene. PLoS One. 2010;5. doi:10.1371/journal.pone.0009168

[78] Rezaie T, Sarfarazi M. Molecular cloning, genomic structure, and protein characterization of mouse optineurin. Genomics. 2005;85:131–8. doi:10.1016/j.ygeno.2004.10.011

[79] Stroissnigg H, Repitz M, Miloloza A, Linhartova I, Beug H, Wiche G, et al. FIP-2, an IkappaB-kinase-gamma-related protein, is associated with the Golgi apparatus and translocates to the marginal band during chicken erythroblast differentiation. Exp Cell Res. 2002;278:133–45. doi:10.1006/excr.2002.5567

[80] Shen X, Ying H, Qiu Y, Park JS, Shyam R, Chi ZL, et al. Processing of optineurin in neuronal cells. J Biol Chem. 2011;286:3618–29. doi:10.1074/jbc.M110.175810

[81] Wild P, Farhan H, McEwan DG, Wagner S, Rogov VV, Brady NR, et al. Phosphorylation of the autophagy receptor optineurin restricts Salmonella growth. Science. 2011;333:228–33. doi:10.1126/science.1205405

[82] Shen WC, Li HY, Chen GC, Chern Y, Tu PH. Mutations in the ubiquitin-binding domain of OPTN/optineurin interfere with autophagy- mediated degradation of misfolded proteins by a dominant-negative mechanism. Autophagy. 2015;11:685–700. doi:10.4161/auto.36098

[83] Hattula K, Peränen J. FIP-2, a coiled-coil protein, links Huntingtin to Rab8 and modulates, cellular morphogenesis. Curr Biol. 2000;10:1603–6. doi:10.1016/S0960-9822(00)00864-2

[84] Sahlender DA, Roberts RC, Arden SD, Spudich G, Taylor MJ, Luzio JP, et al. Optineurin links myosin VI to the Golgi complex and is involved in Golgi organization and exocytosis. J Cell Biol. 2005;169:285–95. doi:10.1083/jcb.200501162

[85] Vaibhava V, Nagabhushana A, Chalasani MLS, Sudhakar C, Kumari A, Swarup G. Optineurin mediates a negative regulation of Rab8 by the GTPase-activating protein TBC1D17. J Cell Sci. 2012;125:5026–39. doi:10.1242/jcs.102327

[86] Park B, Ying H, Shen X, Park JS, Qiu Y, Shyam R, et al. Impairment of protein trafficking upon overexpression and mutation of optineurin. PLoS One. 2010;5. doi:10.1371/journal.pone.0011547

[87] Zhu G, Wu CJ, Zhao Y, Ashwell JD. Optineurin negatively regulates TNFα-induced NF-kB activation by competing with NEMO for ubiquitinated RIP. Curr Biol. 2007;17:1438–43. doi:10.1016/j.cub.2007.07.041

[88] Gleason CE, Ordureau A, Gourlay R, Arthur JSC, Cohen P. Polyubiquitin binding to optineurin is required for optimal activation of TANK-binding kinase 1 and production of interferon β. J Biol Chem. 2011;286:35663–74. doi:10.1074/jbc.M111.267567

[89] Chibalina MV, Poliakov A, Kendrick-Jones J, Buss F. Myosin VI and optineurin are required for polarized EGFR delivery and directed migration. Traffic. 2010;11:1290–303. doi:10.1111/j.1600-0854.2010.01101.x

[90] Bond LM, Peden AA, Kendrick-Jones J, Sellers JR, Buss F. Myosin VI and its binding partner optineurin are involved in secretory vesicle fusion at the plasma membrane. Mol Biol Cell. 2011;22:54–65. doi:10.1091/mbc.E10-06-0553

[91] Smith AM, Sewell GW, Levine AP, Chew TS, Dunne J, O'Shea NR, et al. Disruption of macrophage pro-inflammatory cytokine release in Crohn's disease is associated with reduced optineurin expression in a subset of patients. Immunology. 2015;144:45–55. doi:10.1111/imm.12338

[92] Korac J, Schaeffer V, Kovacevic I, Clement AM, Jungblut B, Behl C, et al. Ubiquitin-independent function of optineurin in autophagic clearance of protein aggregates. J Cell Sci. 2013;126:580–92. doi:10.1242/jcs.114926

[93] Rogov VV, Suzuki H, Fiskin E, Wild P, Kniss A, Rozenknop A, et al. Structural basis for phosphorylation-triggered autophagic clearance of Salmonella. Biochem J. 2013;454:459–66. doi:10.1042/BJ20121907

[94] Klionsky DJ, Schulman BA. Dynamic regulation of macroautophagy by distinctive ubiquitin-like proteins. Nat Struct Mol Biol. 2014;21:336–45. doi:10.1038/nsmb.2787

[95] Mishra P, Chan DC. Mitochondrial dynamics and inheritance during cell division, development and disease. Nat Rev Mol Cell Biol. 2014;15:634–46. doi:10.1038/nrm3877

[96] Wong YC, Holzbaur EL. Optineurin is an autophagy receptor for damaged mitochondria in parkin-mediated mitophagy that is disrupted by an ALS-linked mutation. Proc Natl Acad Sci USA. 2014;111:E4439–48. doi:10.1073/pnas.1405752111

[97] Gegg ME, Cooper JM, Chau KY, Rojo M, Schapira AH V, Taanman JW. Mitofusin 1 and mitofusin 2 are ubiquitinated in a PINK1/parkin-dependent manner upon induction of mitophagy. Hum Mol Genet. 2010;19:4861–70. doi:10.1093/hmg/ddq419

[98] Pallanck L. Mitophagy: Mitofusin recruits a mitochondrial killer. Curr Biol. 2013;23:R570–2. DOI: http://dx.doi.org/10.1016/j.cub.2013.05.032

[99] Tumbarello DA, Manna PT, Allen M, Bycroft M, Arden SD, Kendrick-Jones J, et al. The autophagy receptor TAX1BP1 and the molecular motor myosin VI are required for clearance of salmonella typhimurium by autophagy. PLoS Pathog. 2015;11:e1005174. doi:10.1371/journal.ppat.1005174

[100] Maruyama H, Morino H, Ito H, Izumi Y, Kato H, Watanabe Y, et al. Mutations of optineurin in amyotrophic lateral sclerosis. Nature. 2010;465:223–6. doi:10.1038/nature08971

[101] Ying H, Yue BYJT. Cellular and molecular biology of optineurin. Int Rev Cell Mol Biol. 2012;294:223–58. doi:10.1016/B978-0-12-394305-7.00005-7

[102] Ying H, Turturro S, Nguyen T, Shen X, Zelkha R, Johnson EC, et al. Induction of autophagy in rats upon overexpression of wild-type and mutant optineurin gene. BMC Cell Biol. 2015;16:1–13. doi:10.1186/s12860-015-0060-x

[103] Wang Q, Song C, Li CCH. Molecular perspectives on p97-VCP: Progress in understanding its structure and diverse biological functions. J Struct Biol. 2004;146:44–57. doi: 10.1016/j.jsb.2003.11.014

[104] Meyer H, Bug M, Bremer S. Emerging functions of the VCP/p97 AAA-ATPase in the ubiquitin system. Nat Cell Biol. 2012;14:117–23. doi:10.1038/ncb2407

[105] Kimonis VE, Mehta SG, Fulchiero EC, Thomasova D, Pasquali M, Boycott K, et al. Clinical studies in familial VCP myopathy associated with Paget's disease of bone and frontotemporal dementia. Am J Med Genet Part A. 2008;146:745–57. doi:10.1002/ajmg.a.31862

[106] Tresse E, Salomons FA, Vesa J, Bott LC, Kimonis V, Yao TP, et al. VCP/p97 is essential for maturation of ubiquitin-containing autophagosomes and this function is impaired

by mutations that cause IBMPFD. Autophagy. 2010;6:217–27. doi:10.4161/auto.6.2.11014

[107] Krick R, Bremer S, Welter E, Schlotterhose P, Muehe Y, Eskelinen EL, et al. Cdc48/p97 and Shp1/p47 regulate autophagosome biogenesis in concert with ubiquitin-like Atg8. J Cell Biol. 2010;190:965–73. doi:10.1083/jcb.201002075

[108] Fang L, Hemion C, Pinho Ferreira Bento AC, Bippes CC, Flammer J, Neutzner A. Mitochondrial function in neuronal cells depends on p97/VCP/Cdc48-mediated quality control. Front Cell Neurosci. 2015;9:16. doi:10.3389/fncel.2015.00016

[109] Tanaka A, Cleland MM, Xu S, Narendra DP, Suen DF, Karbowski M, et al. Proteasome and p97 mediate mitophagy and degradation of mitofusins induced by Parkin. J Cell Biol. 2010;191:1367–80. doi:10.1083/jcb.201007013

Chapter 19

Targeting Mitophagy in Combined Therapies of Haematological Malignancies

Laura I. Kornblihtt, María C. Carreras and
Guillermo A. Blanco

Additional information is available at the end of the chapter

http://dx.doi.org/10.5772/63306

Abstract

Mitophagy is a selective form of autophagy that eliminates mitochondria and is part of a larger network of mitochondrial quality control processes that respond to mitochondrial damage. Treatment of haematological malignancies often involves drugs that ultimately cause cell death by mitochondrial injury and initiation of apoptosis. Thus, mitophagy is a potential cause of resistance to anticancer drugs that target the mitochondria (mitocans). Since mitophagy is integrated to mitochondrial biogenesis, mitochondrial fission and fusion, the bioenergetics profile and metabolic reprogramming of tumour cells, the blockage of mitophagy may not be sufficient to overcome resistance. In addition, the mitochondrial unfolded protein response and the outer mitochondrial membrane-associated degradation have extensive crosstalk with mitophagy, and advanced forms of neoplasms will require targeting both systems. Proteasome inhibitors and vinca alkaloids target many of the critical steps involved in resistance to mitocans, while inducers of mitochondrial turnover (biogenesis and mitophagy) like valproic acid have a variable effect depending on metabolic reprograming and the activity of oxidative phosphorylation of tumour cells. Here we discuss the mechanisms of mitophagy and its associated mechanisms, and discuss its application to the rationale of targeted combined therapies of low- and high-grade B-cell neoplasms.

Keywords: mitocans, arsenic trioxide, BNIP3, Parkin, aggresome, Proteasome inhibitors, valproic acid, vincristine, mitochondrial dynamics, mitochondrial turnover, metabolic reprogramming, lymphoma, myeloma, chronic lymphocytic leukemia (CLL)

1. Introduction

Autophagy is a cell response that aims to recycle proteins, cytoplasmic components and even organelles particularly under starvation conditions [1]. Mitochondria are one of the many organelles and cytoplasmic components that can be identified as the cargo within autophagosomes. More recently, this kind of autophagy has been referred to as bulk or non-selective autophagy, to underscore that there is no particular selection of the cellular components that may enter the autophagy process. The main driver of nutrient depletion autophagy is a catabolic response to provide amino acids and support metabolic pathways such as gluconeogenesis and ketogenesis. By contrast, mitophagy is defined as the selective autophagy of mitochondria [2]. In fact, elimination of mitochondria is particularly avoided in nutrient depletion autophagy. Since mitochondria is the source of ATP in normoxia, supports lipid biosynthesis, gluconeogenesis, ketogenesis and many other metabolic functions that take place precisely under nutrient depletion, mitochondria are spare as much as possible under catabolic conditions [3]. One of the first metabolic contexts where mitochondria were found to be eliminated with selectivity was hypoxia. Even under normal nutrient conditions, cells exposed to hypoxia will undergo increased oxidative stress. The absence of O_2 causes abnormal function of the electron transport chain (ETC) and oxidative phosphorylation (OXPHOS), and this leads to increased superoxide anion leakage and reactive oxygen species (ROS) production [4]. The main driver of response to hypoxia is the increase in mitochondrial ROS, which stabilizes the hypoxia inducing factor 1α (HIF-1α) by preventing its proteasome degradation. HIF-1α is a transcription factor that orchestrates an array of changes in mitochondrial proteins to reduce OXPHOS and particularly promotes mitophagy resulting in the reduction of mitochondrial mass. Nevertheless, mitophagy may occur under many other conditions leading to abnormal function of mitochondria, particularly if it involves increased mitochondrial ROS (mtROS). Increased mtROS can lead to collapse of mitochondrial membrane potential (MMP), mitochondrial outer membrane permeabilization (MOMP), release of cytochrome c and initiation of intrinsic apoptosis. However, elimination of mitochondria by mitophagy prior to MOMP may prevent apoptosis particularly if the production of mtROS is not massive.

We review how mitophagy is integrated to mitochondrial quality control (QC), mitochondrial turnover and mitochondrial dynamics, and we discuss the interdependency with the bioenergetics profile of the cell. This knowledge will be further considered to discuss its implication in resistance to treatment of haematological malignancies, and how combined therapies can be selected to target mitophagy, mitochondrial biogenesis and compensating mechanisms of resistance that involve the ubiquitin proteasome system (UPS), particularly in the most aggressive forms of lymphomas. We discuss the use of drugs already in clinical use that may be potentially combined based on their recognized effects on mitophagy, autophagy and mitochondrial biogenesis. We finally discuss methods that can help determine the status of resistance mechanisms based on mitophagy in lymphoma cells obtained from patients.

2. Mitophagy and mitochondrial turnover

2.1. Mitophagy is coupled to biogenesis and both occur at perinuclear areas

Mitochondria are far from being static and frequently change their shape and size. Even in a single cell they are not equal to one another, a condition called heteroplasmy, and there is still a great variation depending on the cell type. Recently the term "mitochondrial behaviour" was proposed to embrace all these properties and states, underscoring the need to integrate the many dimensions of mitochondrial study [5]. Mitochondria spread around the cell by attaching to the microtubule network and move to the areas of high energy demands to provide OXPHOS-derived ATP. Even though mitophagy reduces the mitochondrial mass, this process is coupled to mitochondrial biogenesis through a homeostatic loop, in order to keep the mitochondrial mass in accordance with the bioenergetics demands of the cell [6]. Mitochondrial biogenesis requires duplication of mitochondrial DNA (mtDNA), the expression of new mtDNA-coded proteins, nuclear DNA-coded mitochondrial proteins and a huge work of mitochondrial protein import. This process is controlled by the peroxisome proliferator-activated receptor gamma coactivator 1-alpha (PGC-1α) transcription factor that regulates the expression of a large number of nuclear-encoded mitochondrial genes. Biogenesis involves elongation of existing mitochondria and further fission of the newly formed units (**Figure 1**). In fact, what is known as a mitochondrion may contain several units, each one with a single copy of mtDNA forming a nucleoid, and a completely assembled respiratory chain complex. Remarkably, the process of elongation and protein import associated with biogenesis occurs in perinuclear mitochondria [7]. The induction of biogenesis after mitophagy is often referred as the nuclear retrograde response, and is an important regulator of mitochondrial turnover because it leads to changes in the grade of heteroplasmy of the entire mitochondrial network. In cases of mutated mtDNA the grade of heteroplasmy may be critical for the severity of mitochondrial damage or malfunction.

2.2. Mitophagy is initiated after mitochondrial fission

Mitochondrial biogenesis involves elongation and fission, but fission also occurs apart from biogenesis [8]. In fact, mitochondria may be divided at positions dictated by a surrounding endoplasmic reticulum (ER) that corresponds to the joint between internal units. These sites are known as sites of ER-associated mitochondrial division (ERMD) [9]. At these sites dynamin relates protein 1 (DRP1) is recruited to constrict and further split mitochondria during fission (**Figure 1**). Fission allows the segregation of individual units with impaired or abnormal function that can no longer sustain a normal OXPHOS, the MMP, and the production of ATP [10]. Mitochondria with otherwise normal function can fuse again by the activation of Mfn1 and Mfn2, which accomplish fusion of the outer mitochondrial membrane (OMM), and OPA1 that completes fusion of internal mitochondrial membrane (IMM) [5]. In contrast, segregated mitochondria with abnormal function can no longer fuse again due to inactivation of Mfn1 and OPA1 particularly due to collapsed MMP. Thus, fission and segregation of mitochondria with collapsed MMP is the first step in the process of mitophagy (**Figure 1**) [10]. During mitophagy, the tip-ends of split mitochondria are kept sufficiently closed to prevent MOMP and cytochrome c release [11]. However, the particular phospholipid cardiolipin (CL), which is normally concentrated in the IMM, is partially exposed at the OMM. CL at the OMM acts as an "eat me" signal to initiate mitophagy, much in the same way as phophatidylserine acts as an "eat me" signal in the cell membrane during apoptosis [12]. The collapse of MMP acts to stabilize Pink1 and recruits the E3 ubiquitin ligase Parkin, while further attachment of p62 at the OMM initiates the formation of the mitophagosome [13]. In contrast to mitophagy, during intrinsic apoptosis the entire mitochondrial network undergoes fission, and at these tip-ends occurs a massive oxidation of CL that leads to MOMP, cytochrome c oxidation and further release to the cytosol.

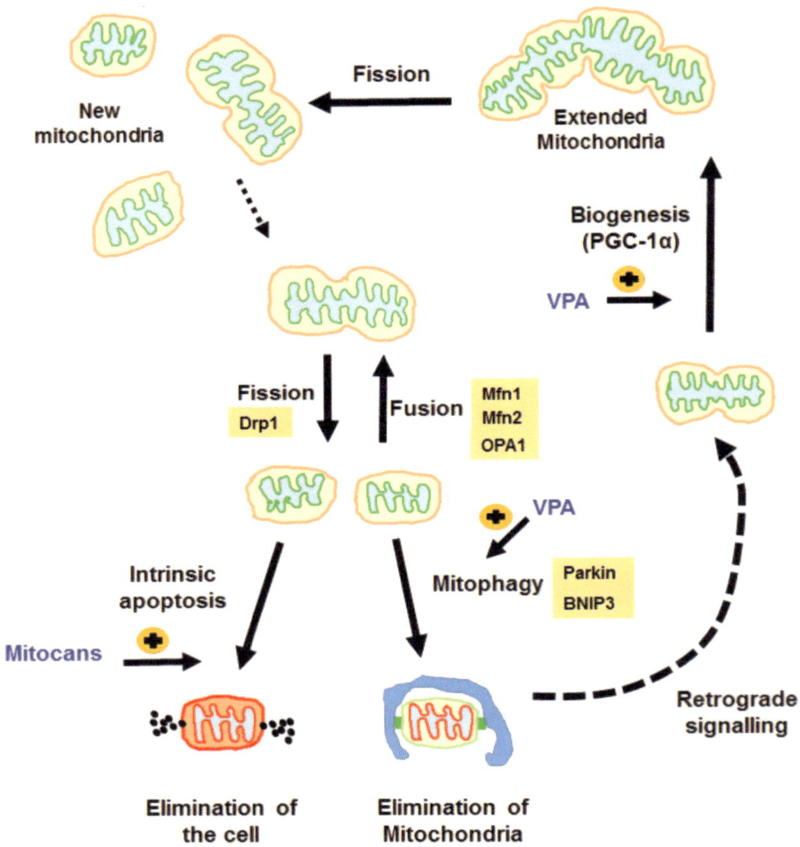

Figure 1. Transitional states of mitochondria.

Mitochondria undergo rounds of fusion and fission. Mitophagy is initiated after fission as part of mitochondrial quality control. Fission involves Drp1, while fusion requires Mfn1, Mfn2 and OPA1. Parkin and BNIP3 are the two most common mitophagy receptors. Mitophagy is coupled to biogenesis through the retrograde nuclear signalling. Mitochondrial biogenesis is initiated by mtDNA duplication and elongation, with assembly of a completely new nucleoid and respiratory unit. The elongated mitochondria start undergoing fission ad fusion. Mitocans are anticancer drugs that target the mitochondria and induce intrinsic apoptosis after inflicting different sorts of mitochondrial damage. Mitophagy can counterbalance this damage by removing compromised mitochondria; while mitochondrial biogenesis completes a turnover cycle by maintaining mitochondrial mass. Valproic acid (VPA) induces mitophagy by upregulation of BNIP3 and mitochondrial biogenesis by upregulating PGC-1α.

2.3. Bioenergetics implications of mitochondrial fusion and fission-metabolic reprogramming

Mitophagy and mitochondrial biogenesis do not provide a complete picture of the regulation of mitochondrial network, unless it is integrated to the metabolic changes associated with mitochondrial dynamics, which is a term that is used to refer collectively to mitochondrial fusion and fission. A hyperfused network in normoxia often denotes a high respiratory rate with high OXPHOS activity, O_2 consumption, ATP production, high MMP and increased mitochondrial mass. The cycles of fission and fusion contribute to exchange mitochondrial components, dilute any defects and improve efficiency of respiration. In contrast, a network with a predominance of fission is often indicative of low OXPHOS-derived ATP, low O_2 consumption, high glycolysis rate and decreased mitochondrial mass, with low exchange of mitochondrial components. Metabolic reprogramming is a characteristic of cancer cells that enhances their ability to survive under adverse conditions such as hypoxia and nutrient deprivation [14]. Cancer cells may use either glycolysis or OXPHOS as a source of ATP, and these two alternatives may be not mutually exclusive, and can vary during the progression of the disease depending on several factors. OXPHOS redirection towards lipid and protein synthesis, truncation of Krebs's cycle and glycolysis can all be regulated to support tumour growth, meeting particular demands that appear under harsh growing conditions, such as glucose and other nutrients deprivation, lack of oxygen and the expression of particular oncogenes [15]. The oncogene c-MYC, which is characteristically overexpressed in high-grade Burkitt´s lymphoma, participates in this regulation through the expression of genes required for either OXPHOS or glycolysis [15]. We will next consider a heterogeneous group of B-cell lymphomas as an example to discuss the role of mitophagy in disease progression and response to treatment with a particular emphasis on metabolic reprogramming.

2.4. Metabolic features of low- and high-grade lymphomas

Mature B-cell neoplasms are broadly classified as low-grade (indolent) and high-grade (aggressive) considering the severity of symptoms, rate of progression and response to treatment [16]. The most frequent low-grade B-cell neoplasm is chronic lymphocytic leukaemia (CLL) also denoted as small lymphocytic lymphoma. Among high-grade B-cell lymphomas, the most frequent ones are diffuse large B cell lymphoma (DLBCL) and Burkitt´s lymphoma. CLL is a small cell lymphoma with a large number of non-dividing mature B-cells circulating in the peripheral blood. These B-cells may enter the lymph nodes where they can survive under a hypoxic environment (hypoxic niche), receive growth signals and replicate in a location often protected from the effect of anticancer drugs. However, CLL cells circulating in the peripheral blood show activated HIF-1α, even though they are exposed to normoxic conditions. Progression of CLL often leads to a transformation into high-grade lymphomas such as DLBCL, a process designated as Ritcher's transformation [17]. Metabolic reprogramming parallels to some extent the grading and aggressiveness of these lymphomas [18]. In one extreme, CLL cells appear as small mature lymphocytes with metabolic features much similar to normal B-cells. This includes an efficient use of OXPHOS, low rates of glycolysis with a peripheral arrangement of predominantly fused mitochondria. By contrast, high-grade lymphoma cells appear as large immature cells with very high rates of glycolysis, low use of O_2 and OXPHOS, high production of lactic acid and an extensive use of glucose and glutamine. High-grade lymphomas may show high rates of mitophagy and autophagy with a network distribution that is often more perinuclear than peripheral [19]. When metabolically reprogrammed cells do not derive ATP from OXPHOS, they are less vulnerable to mtROS under hypoxia.

2.5. Mitophagy and metabolic reprogramming cause resistance to mitocans

Mitocans are a heterogeneous group of anticancer drugs that target the mitochondria and initiate apoptosis [20]. Mitocans may enhance OXPHOS (hexoquinase II blockers, sodium dichloroacetate, 2-bromopyruvate), block the ETC (tamoxifen, adaphostin), oxidize thiol groups and deplete mitochondrial glutathione (arsenic trioxide) or destabilize VDAC. Eventually, mitocans increase mtROS and initiate apoptosis by inducing MOMP. Mitocans like arsenic trioxide (ATO) are used to treat promyelocytic leukaemia but other haematological malignancies are often resistant [21]. ATO targets the ETC and oxidizes thiol groups by increasing mtROS. Metabolic reprogramming and mitophagy can have a great influence in resistance to the induction of apoptosis by ATO or other mitocans. In low-grade lymphomas mitocans like ATO that target the ETC and alter OXPHOS, induce massive mtROS and trigger apoptosis. However, the cytotoxicity of ATO is significantly reduced in high-grade lymphomas because of their low OXPHOS-dependency and high rates of mitophagy with increased mitochondrial turnover [19]. ATO as other mitocans can hardly inflict mitochondrial damage in these OXPHOS-independent cells, and in addition, damaged mitochondria are effectively replaced by a high mitochondrial turnover. A high rate of mitochondrial biogenesis is also required to assist rapid proliferation, which is a characteristic feature of these high-grade lymphoma cells.

2.6. Transport of mitophagosomes over the microtubule network

Mitophagy is initiated after fission and segregation of mitochondria with collapsed MMP. The first step involves the role of the E3 ubiquitin ligase Parkin [13]. Many proteins of OMM become ubiquitinated by Parkin. Among these Mfn1, Mfn2, Miro and Paris help to highlight the interaction of mitophagy with mitochondrial turnover and dynamics. Ubiquitination of Mfn1 precludes further fusion, ubiquitination of Miro immobilizes mitochondria and fixes it to the microtubule network, and ubiquitination of Paris causes an increase in the expression of PGC-1α, initiating the nuclear retrograde signalling that leads to biogenesis [22, 23]. Interestingly, Parkin is able to carry on K48 and K63 linked polyubiquitination, but recent studies support a preferential K63, K11 and K6 links for Parkin [24]. While K48 is a polyubiquitin modification known to target proteasome degradation, K63-linked polyubiquitin has been more specifically related to mitophagy, and even the mitochondrial deubiquitinase (DUB) USP30 was shown to block Parkin-mediated mitophagy by selectively eliminating this kind of polyubiquitins [25]. The ubiquitination of OMM proteins recruits p62, a linker molecule that has a ubiquitin binding domain (UBD) and a LC3 interaction region (LIR) domain, and recruits LC3 to initiate the formation of the mitophagosome by sequestering the ubiquitinated organelle (**Figure 2**) [26]. This initial phase has extensive opportunities for crosstalk between the UPS and mitophagy. A mitophagy receptor is defined as a molecule that is shuttled to the OMM, where it interacts with LC3 or gamma-aminobutyric acid receptor-associated protein (GABARAP) to initiate the mitophagosome formation [27]. BNIP3 and NIX are two mitophagy receptors that respond to increase in mtROS by inserting into the OMM [28]. These BH3-only molecules of the Bcl2 family have an LIR domain to initiate the formation of the LC3-decorated mitophagosome, even in the absence of p62 (**Figure 2**). BNIP3 can even cause the recruitment of Parkin to the OMM facilitating also the p62-dependent mitophagy pathway. Nix in particular is responsible for mitophagy during the normal maturation of erythrocytes [29]. The maturation of mitophagosomes ultimately leads to the fusion with lysosomes. This requires the transport of mitophagosomes along the microtubule network and is critical to complete the mitophagy process. The encounter of mitophagosomes and lysosomes often occurs at the perinuclear area at the minus end of the microtubule network. This is often quite evident in high-grade lymphoma cells as well as other aggressive tumour cells. In fact, the term mito-aggresome has been used to describe the occurrence of this end stage of mitophagy at the perinuclear area [30]. Mitochondrial elongation and biogenesis occur in mitochondria located at the perinuclear area. Thus, mitophagy and the nuclear retrograde response appear to occur at perinuclear location, having significance to interpret the occurrence of perinuclear mitochondrial clusters (PNMC) in high-grade B-cell neoplasms as suggestive of high basal mitochondrial turnover [19].

2.7. Mitophagy and the mitochondrial unfolded protein response (mtUPR)

The term aggresome is used to describe aggregates of ubiquitinated proteins located at perinuclear areas. These perinuclear areas close to the microtubule organizing centre (MTOC) are enriched with proteasomes and function as proteolytic centres. Aggresomes are formed when these proteolytic centres are no longer able to cope with an overload of misfolded ubiquitinated proteins [31, 32]. The transport of ubiquitinated proteins to the proteolytic centres depends on dynein motor-proteins that move along the microtubule network towards the minus end at the MTOC. Elimination of misfolded proteins and aggresome formation are part of the unfolded protein response (UPR) that occurs upon ER-stress [33]. Impaired proteolysis by inhibition of proteasomes increases the occurrence of aggresomes. Once aggresomes are formed, they cannot be cleared by proteasomes and in fact, persistence of aggresomes leads to proteasome inhibition and apoptosis. However, aggresomes may be cleared by autophagy. This involves a modification of K48-linked polyubiquitination to K63-linked polyubiquitination and involves the histone deacetylase HDAC6 [34]. Mitochondria have their own protein QC system that includes the proteases Lon and ClpX in the matrix, and i-AAA in the inter-membrane space. The mitochondrial unfolded protein response (mtUPR) is regulated by the transcription factor CHOP and involves upregulation of proteases and chaperones [35]. When this system is overloaded, misfolded mitochondrial proteins are exposed at the OMM where they become ubiquitinated and shuttled for proteasome degradation. Because of the similarity with ER-associated degradation (ERAD), this mitochondrial process has recently been designated as OMM-associated degradation (OMMAD) [2]. This stress pathway functions as a mitochondrial protein QC independently of mitophagy and may be similar to the ER-stress associated UPR, involving K48 ubiquitination of OMM misfolded proteins [35]. Mitochondrial UPR and mitophagy work in an integrated manner. If the amount of mitochondrial damage is low, mtUPR and OMMAD may be enough to achieve the required mitochondrial protein quality control. However, a more severe damage would require mitophagy as a higher level of mitochondrial quality control leading to recycling of the whole organelle.

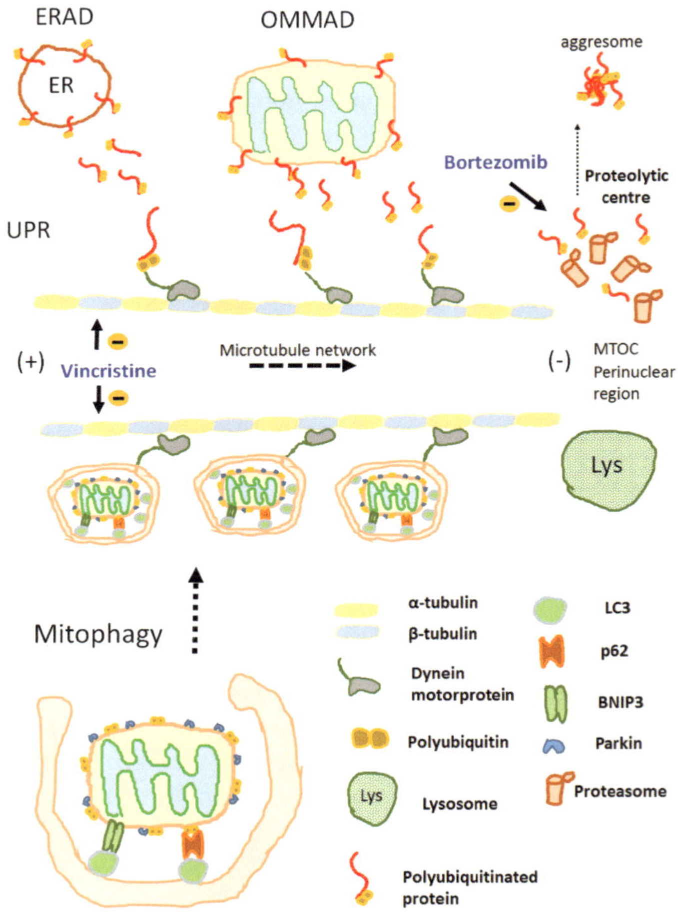

Figure 2. Mitophagy, OMMAD and ERAD require oriented transport through microtubules.

2.8. The transport to the perinuclear region in UPR and mitophagy

The transport of mitophagosomes and autophagosomes along the microtubule is required for fusion with lysosomes [36]. In fact, not only mitophagosomes and autophagosomes but also the entire mitochondrial network moves along microtubules. Mitochondrial biogenesis, fusion, fission and mitophagy also occur over the microtubule network, and drugs affecting the microtubule network have a profound impact on all of these processes. Microtubules metaphorically represent the railways over which cargo-carrying motor-proteins move along. Kinesins are motor-proteins that carry cargo towards the positive end of microtubules that is at the cell periphery. In contrast, dyneins carry cargo towards the minus end at the perinuclear area. In a pioneer study, Lee et al. used nocodazole and overexpression of dynamitin to cause inhibition of the motor-protein dynein, and prove that QC mitophagy involved the transport of mitophagosomes through motor-proteins towards the perinuclear area, to fuse with lysosomes at the PNMC (**Figure 2**) [30]. Even though mitophagy can be blocked very specifically by knocking down the motor-protein dynein, blocking the dynamics of the microtubule has a profound impact on all motor-proteins moving along the network. Vincristine, as other vinca alkaloids, binds to α tubulin and prevents further polymerization, leading to a complete destabilization of the network. These drugs compromise the entire dynamics of the microtubule network and have profound impact on all cargo-carrying motor-proteins (**Figure 2**). This explains why cells are arrested in metaphase during mitosis, but also explains why mitophagy and UPR are blocked, and why the PNMC and proteolytic centres are disrupted.

The UPR is a protein quality control mechanism that upregulates proteases and chaperones in response to misfolded and damaged proteins. The OMMAD functions in a similar way as the ERAD. In ERAD and OMMAD, ubiquitinated proteins are transported through microtubules to the proteolytic centres where proteasomes are clustered. Similarly, mitophagosomes are transported to the perinuclear region to fuse with lysosomes. Dynein motor-proteins transport cargo towards the minus end close to the microtubule-organizing centre (MTOC). Vincristine as other drugs affecting microtubule dynamics block transport and halt ERAD, OMMAD and mitophagy. Bortezomib may create an overload of toxic misfolded ubiquitinated proteins at the ER and mitochondria.

2.9. Cancer cells may be addicted to UPR and mitophagy

For several reasons, cancer cells may be adapted to an abnormally high load of misfolded proteins. One of the best examples is that of multiple myeloma cells [33]. These are malignant post-germinal centre B lymphocytes that actively produce and secrete immunoglobulins. Due to the huge amount of protein synthesis, myeloma cells are liable to ER-stress and therefore require activation of the UPR, induction of chaperones and autophagy for survival [37]. A significant number of genes involved in the UPR are frequently mutated in patients with multiple myeloma, and the UPR is highly active and increases in advanced disease stages [37]. Multiple myeloma cells may be addicted to the UPR for survival, and drugs that target protein homeostasis, such as proteasome inhibitors, shift the balance of the UPR from prosurvival to proapoptotic. However, myeloma cells further evolve alternative mechanisms to deal with ER-stress such as autophagy, and the disease remains incurable mainly due to therapy resistance.

Remarkably, the mitophagy receptor BNIP3 is also upregulated in advanced forms of multiple myeloma [38]. Most myeloma cells reside in a bone marrow hypoxic niche, and hypoxia increases ER-stress and upregulates UPR to deal with an overload of misfolded proteins at the ER and the mitochondria. Therefore, the UPR and mitophagy may be crucial to avoid the potential joint toxicity of undegraded misfolded proteins and damaged mitochondria.

2.10. Targeting mitophagy and the UPR with microtubule stabilizing drugs

Several mechanisms of resistance to proteasome inhibitors have been elucidated. The therapeutic targeting of UPS and UPR has similarities to the therapeutic targeting of mitophagy, and the potential resistance mechanisms may be comparable. Unfolded protein response and mitophagy are flows that may be potentially toxic after being halted. When proteasome is inhibited, the halted flow of proteolysis becomes lethal to the cells, due to accumulation of misfolded proteins. Similarly, when mitophagy is inhibited, the halted flow of elimination of damaged mitochondria becomes potentially lethal to the cell. However, as discussed before, mitochondria have an internal protein QC mechanism represented by the mtUPR that involves chaperones and proteases. In addition, it can target misfolded proteins to the OMM for further ubiquitination and shuttling to the proteasome (OMMAD). This means that blocking mitophagy does not warrant an accumulation of potentially lethal organelles that trigger apoptosis. Remarkably, by destabilizing the microtubule network, both the UPR and mitophagy become halted. Therefore, the microtubule network is a target where mitophagy, OMMAD and ERAD are convergent and lead to accumulation of ubiquitinated misfolded proteins and mitochondria (**Figure 2**). However, achieving a misfolded protein accumulation capable of causing mitochondrial toxicity and an increase in ER-stress, requires the concurrent inhibition of proteasomes. In a similar way, lethality of mitochondria accumulation depends much on the quality of mitochondria and factors such as mtROS level. The latter is much influenced by metabolic reprogramming and the rate of OXPHOS dependency. As commented above, aggressive neoplasm often have low OXPHOS activity, and damage inflicted to the ETC may have a low impact in increasing mtROS. This situation accounts for resistance to mitocans and tolerance to accumulation of damaged mitochondria. However, there are mitocan drugs that make tumour cells shift to OXPHOS, and thus they can become vulnerable to increased mtROS. This is the case of 2-deoxyglucose (2DG) and 3-bromopyruvate (3BP) that block mitochondrial-bound hexokinase and of dichloroacetate (DCA) that blocks pyruvate-dehydrogenase-kinase (PDK) [39]

2.11. The pro-death role of mitophagy receptor BNIP3

BNIP3 is a "BH3-only member" of the Bcl2 family that was originally described as a pro-death molecule that often triggers a caspase-independent mode of apoptosis. More recently, BNIP3 was also characterized as a mitophagy receptor, and created a controversy about whether it was a pro-death or pro-survival molecule [28]. Supporting its pro-death role, the knock- down of autophagy genes or the chemical inhibition of autophagy, enhanced apoptosis of cells where BNIP3 was overexpressed. However, BNIP3 upregulation enhances mitophagy and may help to eliminate potentially harmful damaged mitochondria [40]. The reconciliation of this dual role controversy requires the consideration of mitophagy as a flowing system. If accumulation of mitochondrial BNIP3 triggers cell death, then the rate of mitophagic flow must be sufficiently high to eliminate these BNIP3-bearing mitochondria. Otherwise, if mitophagy is blocked downstream, the result is that BNIP3 is accumulated at the mitochondria initiating cell death. In normal cells, BNIP3 is under transcriptional control of HIF-1α, and the most active inducer of BNIP3 is hypoxia and the increase of intracellular ROS [41]. BNIP3 targets to damaged mitochondria particularly under increase of intracellular ROS to initiate mitophagy. If mitophagic flow is not high enough to eliminate mitochondria, BNIP3 will induce cell death, otherwise mitophagy eliminates damaged mitochondria contributing to cell survival and tolerance to hypoxia. Thus, the pro-death role of BNIP3 is dependent on a balance between mitochondrial damage and mitophagic flow [19].

2.12. Valproic acid upregulates BNIP3 but also induces mitochondrial biogenesis

Downregulation of BNIP3 may result in failure of tumour cells to undergo cell death, and is associated with a chemo-resistant phenotype and decreased patient survival [42]. Samples from patients with multiple myeloma were found methylated at the BNIP3 promoter, and methylation was significantly correlated with poor patient survival rates [43]. The finding that many haematological and other tumour cells have epigenetic silencing of BNIP3, led to the hypothesis that epigenetic drugs that restore expression of BNIP3 could cause tumour cell death. In fact, this was confirmed in some kinds of lymphoma, leukaemia and epithelial tumours. Burkitt's lymphoma cells have epigenetic silencing of BNIP3 and VPA is a histone deacetylase (HDAC) inhibitor that upregulates expression of BNIP3. However, upregulation of VPA improves growth and resistance to death in these cells, and even antagonizes the effect of mitocans like ATO and improves tolerance to hypoxia [19]. Even though VPA increases the expression of BNIP3 in high-grade lymphoma, microarray data from studies conducted in low-grade lymphoma cells (CLL) showed that VPA upregulates BNIP3 and many other mitochondria-related genes, including PGC-1α that encodes a master regulator of mitochondrial biogenesis [44]. Upregulation of PGC1α by VPA was also confirmed in SH-SY5Y neuroblastoma cells, and even in fibroblasts from patients with mutation in mtDNA polymerase (POLG) as well as normal controls [45, 46]. Thus, VPA induces mitochondrial biogenesis through PGC-1 α and upregulates mitophagy through BNIP3, leading to an increase in mitochondrial mass and an enhanced mitochondrial turnover [44]. This results in an enhanced metabolic rate and increased OXPHOS. Patients with POLG mutation are extremely sensitive to VPA toxicity because, in contrast to normal patients, POLG-deficient cells cannot tolerate the increased function of mitochondrial respiratory chain [46]. This exemplifies the fact that defective mitochondria cannot tolerate increased OXPHOS derived from increased biogenesis. It also underscores the importance of metabolic reprogramming and the bioenergetics profile of cancer cells on the outcome of increased mitochondrial biogenesis. Metabolically reprogrammed cells of high-grade lymphomas that have a low rate of OXPHOS and a high rate of glycolysis can tolerate increased biogenesis. In contrast, low-grade lymphoma cells that are OXPHOS-dependent with low levels of glycolysis will increase OXPHOS upon induction of mitochondrial biogenesis by drugs like VPA, leading to increased mtROS and apoptosis. This was confirmed recently in studies conducted in peripheral blood circulating cells obtained from CLL patients, where VPA treatment is correlated with the upregulation of several genes involved in apoptosis [47]. In addition, VPA was synergic with fludarabine on the induction of apoptosis, and CLL patients treated with VPA plus fludarabine had a better outcome than patients treated with fludarabine alone. In contrast, the effect of VPA in high-grade B-cell neoplasms such as Burkitt's lymphoma is quite the opposite, providing survival advantage even under hypoxia and antagonizing the apoptotic effect of ATO that is a mitocan known to increase mtROS [19].

2.13. Mitophagy and bioenergetics profile in the cytotoxic response to mitocans

From the above discussion, it is clear that the pro-death effect of mitocans is dependent on several interacting factors such as the bioenergetics profile and the induction of increased mtROS, mitochondrial mass, mitochondrial dynamics and mitophagic flow. VPA influences many of these factors, but the bioenergetics profile appears critical in determining the sensitizing or resistance effect in combination with mitocans. The initiation of intrinsic apoptosis by mitocans demands achieving a critical threshold of damage across the mitochondrial network. If the mitochondrial damage inflicted by a mitocan is persistent, an increase in mitochondrial biogenesis may result in an increased amount of damaged mitochondria and an overload of mitophagic flow. However, mitophagic flow coupled to biogenesis may be high enough as to keep the cells below the apoptotic threshold. Thus, the efficacy of mitocans will increase if mitophagic flow is blocked, and conversely the anticancer efficacy of blocking mitophagy will be enhanced with the addition of mitocans. However, blocking mitophagy may still leave an overloaded mtUPR, and in the more aggressive forms, the addition of proteasome inhibitors will contribute to increase damage inflicted by mitocans and blockage of mitophagy [19]. Finally, mitocans such as 3BMP, 2DG and DCA may sensitize to mtROS production in OXPHOS-independent aggressive forms of neoplasms. The induction of mitochondrial biogenesis and blockage of mitophagy present a certain analogy with therapeutic strategies that induce ER-stress to create an overload of the UPS, while at the same time a proteasome inhibitor makes the overload even larger and more toxic. However, in the case of mitochondria, there is a further need to assure that treatment makes them a real source of toxicity to the cells, and this involves mitocans and evaluation of the bioenergetics profile of the target cells.

3. Assessing basal mitophagic flow in lymphoma cells

Lymphoma cells can be obtained from peripheral blood or lymph node biopsies of patients to be analysed for biological features involved in drug sensitivity, including mitophagic flow and mitochondrial biogenesis. Mitophagy and biogenesis can be considered in a steady state that determines the actual mitochondrial mass. As discussed above, this turnover can have a profound impact on the threshold of mitochondrial damage that initiates apoptosis. This is a reason why assessing the basal rate of mitophagy provides a first approach to predict sensitivity to mitocans, and provides a preliminary rationale for combined therapies. Since mitophagy has to be interpreted as a flow, inhibitors provide an indication of the magnitude of this flow. Accumulation of mitochondria occurs after halting mitophagy, while biogenesis is still active. Thus, the change in mitochondrial mass after halting mitophagy is an indicator of mitophagic flow [48]. Colocalization between autophagosomes and mitochondria can also provide an indicator of mitophagy. When mitophagosome formation is blocked upstream, such as by knocking down ATG genes or using chemical inhibitors of the initiation of autophagy, colocalization of mitochondria and autophagosomes decreases. By contrast, when mitophagy is blocked downstream, as for example with vincristine or other inhibitors of autophagosome-lysosome fusion, colocalization between autophagosomes and mitochondria will increase. The morphology and dynamics of the mitochondrial network also provide interesting information. Since fission is the first step in mitophagy and is required after elongation during biogenesis, a high rate of basal mitophagy will often show a fragmented network particularly at the perinuclear area. A fragmented PNMC is characteristic of high rates of basal mitophagy and biogenesis. The use of vincristine or other microtubule stabilizing drugs will show a disruption of the PNMC and a transition to a peripheral distribution of the fragmented network.

3.1. Methods to assess mitochondrial mass and mitophagosomes

Changes of mitochondrial mass can be traced by quantifying mtDNA using QPCR, by immunochemical methods such as western blot using antibodies directed against proteins of the OMM like TOM20, or at the single cell level using flow cytometry, with probes such as mitotracker or nonyl-acridine-orange (NAO) that binds to mitochondrial CL (**Figure 3**). Colocalization between mitochondria and autophagosomes requires a probe for each structure and microscopic analysis. One alternative is immunofluorescence with anti LC3 antibodies and the fixable series of mitotrackers [49]. A second alternative is transient transfection with a plasmid for expression of GFP-LC3 together with mitotracker. A third alternative is the use of mono-dancylcadaverine (with UV excitation and collecting a narrow band of blue fluorescent emission) and mitotracker [50, 51]. These are just some reference alternatives and many more can be found elsewhere [49, 50]. Some less frequently used, although more robust alternatives in quantitative aspects, are image flow cytometry and subcellular flow cytometry (**Figure 3**). The former relies on colocalization metrics derived from thousands of low magnification images obtained in flowing cells, while the latter refers to the analysis by flow cytometry of subcellular particles obtained from cells that have fluorescently labelled autophagosomes and mitochondria, in order to quantify those particles that have dual fluorescence as an indicator of colocalization [52, 53]. Although assessment of LC3-I and LC3-II by western blot can give a measure of bulk autophagy (which is often present in lymphoma cells), it does not provide the confidence that mitochondria are being part of that autophagic flow. Therefore, assessment of mitochondrial mass and colocalization is necessary. Another alternative to derive a measure of mitophagic flow using western blot, is the use of antibodies against mitophagy receptors such as BNIP3 or Parkin. The accumulation after the addition of flow inhibitors can also provide a measurement of mitophagic flow. This alternative is particularly interesting for clinical samples since no transfection is required, although caveats such as epigenetic silencing of BNIP3 should be considered in each case. The expression of tandem proteins combining monomeric red fluorescence protein and green fluorescence protein (mRFP-GFP) facilitates the monitoring of formation and maturation of autophagosomes, and the late stage after fusion with lysosomes. Because GFP but not mRFP fluorescence is quenched by low pH, the autophagosomes are "green-red" double fluorescent but autophagolysosomes are red-only fluorescent. This allows using the ratio single red versus double fluorescence as an indicator of autophagic flux without the need of inhibitors [54]. This measure should be complemented with colocalization between autophagosomes and mitochondria. A new alternative is a similar construction that includes the coding of mitochondria targeting domain, and allows tracking of mitophagosomes until fusion with lysosomes and a direct measurement of mitophagic flux [55]. Each of these methods has its strengths and limitations to measure mitophagic flux, and the use of more than one method has been widely recommended [50]. The measurement of turnover as a result of biogenesis and mitophagy has been recently facilitated with tandem probes that exploit the spectral change of the protein DsRed1-E5 over time (from green to red fluorescent). This probe is called mitotimer, includes a mitochondrial targeting sequence and its expression is under the control of doxycycline [56]

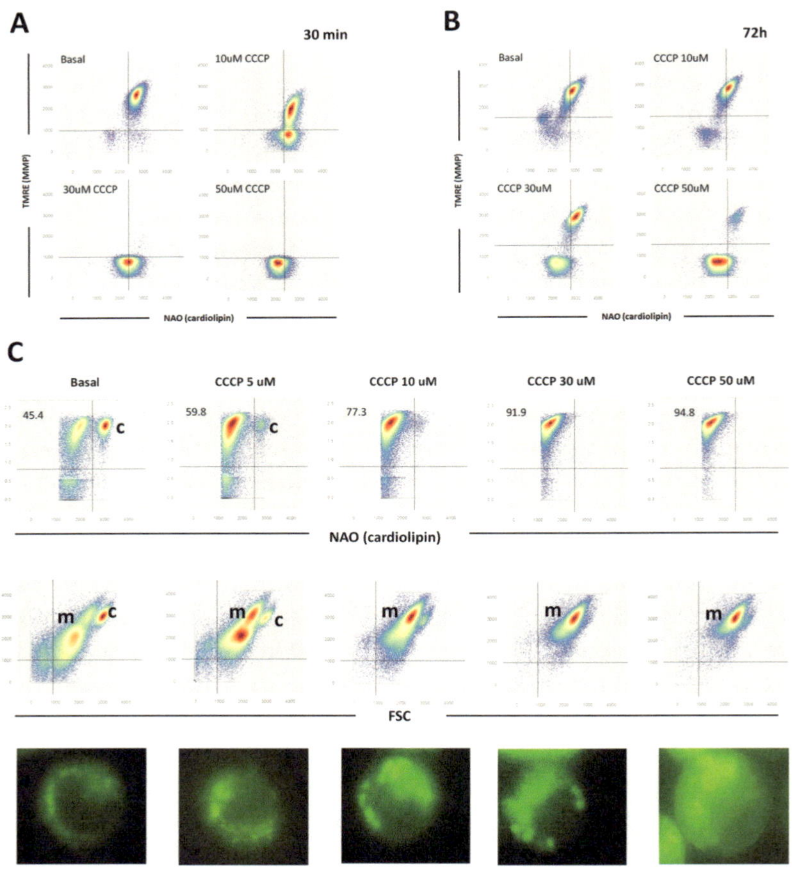

Figure 3. Fission is the first step of mitophagy.

Gamma-aminobutyric acid receptor-associated protein (CCCP) is an uncoupler of OXPHOS that causes massive depolarization and increases mtROS due to leakage of superoxide anion from the ETC. This generalized mitochondrial damage triggers mitophagy as a compensating response, and CCCP is thus commonly used as a positive control of bulk mitophagy. In panel A, Burkitt's lymphoma cells were exposed to increasing doses of CCCP and analysed by flow cytometry. CL content was assessed by the NAO probe (x-axis) and MMP was assessed by tetramethylrhodamine ethyl ester (TMRE) probe (y-axis). Although CCCP caused collapse of MMP as evaluated at 30 min, panel B shows that by 72 h MMP was completely recovered in cells treated with 10 and 30 μM CCCP. An evaluation of mitochondrial fission by subcellular flow cytometry is shown in panel C. Labelled cells were ruptured and immediately run in a flow cytometer, acquiring only particles having NAO fluorescence that corresponded to isolated mitochondria (m) or entire cells that were not ruptured (c). Numbers indicate percentage of mitochondria having a high SSC-A signal. The morphological changes, caused by massive fission of the mitochondrial network and initiation of mitophagy, correlate with an increase of the pulse area of the side scatter light signal (SSC-A). The fluorescence images of matching samples of non-ruptured NAO-labelled cells are shown at the bottom of panel C to illustrate the morphological counterpart of the SSC-A signal increase.

4. Conclusion

Mitophagy is a critical mechanism in the progression of B-cell neoplasms and other haematological malignancies. It is also critical in drug-resistance, particularly in advanced forms of myeloma and high-grade lymphomas. However, mitophagy is at the centre of an integrated system of resistance that involves mitochondrial biogenesis, dynamics, the UPS and metabolic reprogramming. Thus, even though mitophagy is critical, there is no single mechanism of resistance to drug treatment of advanced forms of B-cell neoplasms. Even though some particular mechanisms such as UPR and ERAD may prevail at some stages (as occurs in multiple myeloma), other mechanisms such as mitophagy, autophagy and metabolic reprogramming evolve with disease progression, and these neoplasms remain incurable. However, when these mechanisms are considered together as a system, and its occurrence is demonstrated in a particular case, a combined therapy can be designed to tackle one or all of them. For example, OXPHOS-dependent circulating CLL cells are quite sensitive to increased mitochondrial biogenesis and blockage of mitophagy. The more advanced forms become less OXPHOS-dependent, are more tolerant to increased mitochondrial biogenesis and have increased levels of basal mitophagy. These features make these neoplasm resistant to mitocans. In addition, a cross talk with the proteasome may compensate for blockage of mitophagy through OMMAD, and conversely mitophagy and autophagy may compensate for defective UPR under treatment with proteasome inhibitors. By characterizing this resistance system in particular patients at a particular disease stage, the combination of drugs that better tackle the biological behaviour can be defined.

Author details

Laura I. Kornblihtt[1], María C. Carreras[2] and Guillermo A. Blanco[3*]

*Address all correspondence to: gblanco@ffyb.uba.ar

1 Department of Hematology, Hospital de Clínicas San Martín, University of Buenos Aires, Argentina

2 Laboratory of Oxygen Metabolism, University of Buenos Aires, INIGEM-CONICET, Argentina

3 Laboratory of Immunotoxicology (LaITo), University of Buenos Aires, IDEHU-CONICET, Argentina

References

[1] Galluzzi L, Pietrocola F, Levine B, Kroemer G. Metabolic control of autophagy. Cell. 2014;159:1263–76. DOI: 10.1016/j.cell.2014.11.006

[2] Nunnari J, Suomalainen A. Mitochondria: in sickness and in health. Cell. 2012;148:1145–59. DOI: 10.1016/j.cell.2012.02.035

[3] Gomes LC, Di Benedetto G, Scorrano L. During autophagy mitochondria elongate, are spared from degradation and sustain cell viability. Nature Cell Biology. 2011;13:589–98. DOI: 10.1038/ncb2220

[4] Semenza GL. Oxygen-dependent regulation of mitochondrial respiration by hypoxia-inducible factor 1. The Biochemical Journal. 2007;405:1–9. DOI: 10.1042/BJ20070389

[5] Labbe K, Murley A, Nunnari J. Determinants and functions of mitochondrial behavior. Annual Review of Cell and Developmental Biology. 2014;30:357–91. DOI: 10.1146/annurev-cellbio-101011-155756

[6] Stotland A, Gottlieb RA. Mitochondrial quality control: easy come, easy go. Biochimica et Biophysica Acta. 2015;1853:2802–11. DOI: 10.1016/j.bbamcr.2014.12.041

[7] Neupert W, Herrmann JM. Translocation of proteins into mitochondria. Annual Review of Biochemistry. 2007;76:723–49. DOI: 10.1146/annurev.biochem.76.052705.163409

[8] Hoppins S, Lackner L, Nunnari J. The machines that divide and fuse mitochondria. Annual Review of Biochemistry. 2007;76:751–80. DOI: 10.1146/annurev.biochem.76.071905.090048

[9] Friedman JR, Lackner LL, West M, DiBenedetto JR, Nunnari J, Voeltz GK. ER tubules mark sites of mitochondrial division. Science. 2011;334:358–62. DOI: 10.1126/science.1207385

[10] Twig G, Elorza A, Molina AJ, Mohamed H, Wikstrom JD, Walzer G, et al. Fission and selective fusion govern mitochondrial segregation and elimination by autophagy. The EMBO Journal. 2008;27:433–46. DOI: 10.1038/sj.emboj.7601963

[11] Hamasaki M, Furuta N, Matsuda A, Nezu A, Yamamoto A, Fujita N, et al. Autophagosomes form at ER-mitochondria contact sites. Nature. 2013;495:389–93. DOI: 10.1038/nature11910

[12] Chu CT, Ji J, Dagda RK, Jiang JF, Tyurina YY, Kapralov AA, et al. Cardiolipin externalization to the outer mitochondrial membrane acts as an elimination signal for mitophagy in neuronal cells. Nature Cell Biology. 2013;15:1197–205. DOI: 10.1038/ncb2837

[13] Matsuda N, Sato S, Shiba K, Okatsu K, Saisho K, Gautier CA, et al. PINK1 stabilized by mitochondrial depolarization recruits Parkin to damaged mitochondria and activates latent Parkin for mitophagy. The Journal of Cell Biology. 2010;189:211–21. DOI: 10.1083/jcb.200910140

[14] Ward PS, Thompson CB. Metabolic reprogramming: a cancer hallmark even warburg did not anticipate. Cancer Cell. 2012;21:297–308. DOI: 10.1016/j.ccr.2012.02.014

[15] Wise DR, DeBerardinis RJ, Mancuso A, Sayed N, Zhang XY, Pfeiffer HK, et al. Myc regulates a transcriptional program that stimulates mitochondrial glutaminolysis and leads to glutamine addiction. Proceedings of the National Academy of Sciences of the United States of America. 2008;105:18782–7. DOI: 10.1073/pnas.0810199105

[16] Lu P. Staging and classification of lymphoma. Seminars in Nuclear Medicine. 2005;35:160–4. DOI: 10.1053/j.semnuclmed.2005.02.002

[17] Rossi D, Spina V, Deambrogi C, Rasi S, Laurenti L, Stamatopoulos K, et al. The genetics of Richter syndrome reveals disease heterogeneity and predicts survival after transformation. Blood. 2011;117:3391–401. DOI: 10.1182/blood-2010-09-302174

[18] Koczula KM, Ludwig C, Hayden R, Cronin L, Pratt G, Parry H, et al. Metabolic plasticity in CLL: adaptation to the hypoxic niche. Leukemia. 2016;30:65–73. DOI: 10.1038/leu.2015.187

[19] Cavaliere V, Lombardo T, Costantino SN, Kornblihtt L, Alvarez EM, Blanco GA. Synergism of arsenic trioxide and MG132 in Raji cells attained by targeting BNIP3, autophagy, and mitochondria with low doses of valproic acid and vincristine. European Journal of Cancer. 2014;50:3243–61. DOI: 10.1016/j.ejca.2014.09.012

[20] Neuzil J, Dong LF, Rohlena J, Truksa J, Ralph SJ. Classification of mitocans, anti-cancer drugs acting on mitochondria. Mitochondrion. 2013;13:199–208. DOI: 10.1016/j.mito.2012.07.112

[21] Lombardo T, Cavaliere V, Costantino SN, Kornblihtt L, Alvarez EM, Blanco GA. Synergism between arsenite and proteasome inhibitor MG132 over cell death in myeloid leukaemic cells U937 and the induction of low levels of intracellular superoxide anion. Toxicology and Applied Pharmacology. 2012;258:351–66.

[22] Gegg ME, Cooper JM, Chau KY, Rojo M, Schapira AH, Taanman JW. Mitofusin 1 and mitofusin 2 are ubiquitinated in a PINK1/parkin-dependent manner upon induction of mitophagy. Human Molecular Genetics. 2010;19:4861–70. DOI: 10.1093/hmg/ddq419

[23] Wang X, Winter D, Ashrafi G, Schlehe J, Wong YL, Selkoe D, et al. PINK1 and Parkin target Miro for phosphorylation and degradation to arrest mitochondrial motility. Cell. 2011;147:893–906. DOI: 10.1016/j.cell.2011.10.018

[24] Cunningham CN, Baughman JM, Phu L, Tea JS, Yu C, Coons M, et al. USP30 and parkin homeostatically regulate atypical ubiquitin chains on mitochondria. Nature Cell Biology. 2015;17:160–9. DOI: 10.1038/ncb3097

[25] Bingol B, Tea JS, Phu L, Reichelt M, Bakalarski CE, Song Q, et al. The mitochondrial deubiquitinase USP30 opposes parkin-mediated mitophagy. Nature. 2014;510:370–5. DOI: 10.1038/nature13418

[26] Geisler S, Holmstrom KM, Skujat D, Fiesel FC, Rothfuss OC, Kahle PJ, et al. PINK1/Parkin-mediated mitophagy is dependent on VDAC1 and p62/SQSTM1. Nature Cell Biology. 2010;12:119–31. DOI: 10.1038/ncb2012

[27] Hamacher-Brady A, Brady NR. Mitophagy programs: mechanisms and physiological implications of mitochondrial targeting by autophagy. Cellular and molecular life sciences: CMLS. 2016;73:775–95. DOI: 10.1007/s00018-015-2087-8

[28] Ney PA. Mitochondrial autophagy: origins, significance, and role of BNIP3 and NIX. Biochimica et Biophysica Acta. 2015;1853:2775–83. DOI: 10.1016/j.bbamcr.2015.02.022

[29] Sandoval H, Thiagarajan P, Dasgupta SK, Schumacher A, Prchal JT, Chen M, et al. Essential role for Nix in autophagic maturation of erythroid cells. Nature. 2008;454:232–5. DOI: 10.1038/nature07006

[30] Lee JY, Nagano Y, Taylor JP, Lim KL, Yao TP. Disease-causing mutations in parkin impair mitochondrial ubiquitination, aggregation, and HDAC6-dependent mitophagy. The Journal of Cell Biology. 2010;189:671–9. DOI: 10.1083/jcb.201001039

[31] Kawaguchi Y, Kovacs JJ, McLaurin A, Vance JM, Ito A, Yao TP. The deacetylase HDAC6 regulates aggresome formation and cell viability in response to misfolded protein stress. Cell. 2003;115:727–38

[32] Wojcik C, DeMartino GN. Intracellular localization of proteasomes. The International Journal of Biochemistry & Cell Biology. 2003;35:579–89

[33] Vincenz L, Jager R, O'Dwyer M, Samali A. Endoplasmic reticulum stress and the unfolded protein response: targeting the Achilles heel of multiple myeloma. Molecular Cancer Therapeutics. 2013;12:831–43. DOI: 10.1158/1535-7163.MCT-12-0782

[34] Hao R, Nanduri P, Rao Y, Panichelli RS, Ito A, Yoshida M, et al. Proteasomes activate aggresome disassembly and clearance by producing unanchored ubiquitin chains. Molecular Cell. 2013;51:819–28. DOI: 10.1016/j.molcel.2013.08.016

[35] Pellegrino MW, Nargund AM, Haynes CM. Signaling the mitochondrial unfolded protein response. Biochimica et Biophysica Acta. 2013;1833:410–6. DOI: 10.1016/j.bbamcr.2012.02.019

[36] Mackeh R, Perdiz D, Lorin S, Codogno P, Pous C. Autophagy and microtubules – new story, old players. Journal of Cell Science. 2013;126:1071–80. DOI: 10.1242/jcs.115626

[37] Nakamura M, Gotoh T, Okuno Y, Tatetsu H, Sonoki T, Uneda S, et al. Activation of the endoplasmic reticulum stress pathway is associated with survival of myeloma cells. Leukemia & Lymphoma. 2006;47:531–9. DOI: 10.1080/10428190500312196

[38] Ria R, Todoerti K, Berardi S, Coluccia AM, De Luisi A, Mattioli M, et al. Gene expression profiling of bone marrow endothelial cells in patients with multiple myeloma. Clinical Cancer Research: An Official Journal of the American Association for Cancer Research. 2009;15:5369–78. DOI: 10.1158/1078-0432.CCR-09-0040

[39] Shen YC, Ou DL, Hsu C, Lin KL, Chang CY, Lin CY, et al. Activating oxidative phosphorylation by a pyruvate dehydrogenase kinase inhibitor overcomes sorafenib resistance of hepatocellular carcinoma. British Journal of Cancer. 2013;108:72–81. DOI: 10.1038/bjc.2012.559

[40] Rikka S, Quinsay MN, Thomas RL, Kubli DA, Zhang X, Murphy AN, et al. Bnip3 impairs mitochondrial bioenergetics and stimulates mitochondrial turnover. Cell Death and Differentiation. 2011;18:721–31. DOI: 10.1038/cdd.2010.146

[41] Bellot G, Garcia-Medina R, Gounon P, Chiche J, Roux D, Pouyssegur J, et al. Hypoxia-induced autophagy is mediated through hypoxia-inducible factor induction of BNIP3 and BNIP3L via their BH3 domains. Molecular and Cellular Biology. 2009;29:2570–81. DOI: 10.1128/MCB.00166-09

[42] Murai M, Toyota M, Suzuki H, Satoh A, Sasaki Y, Akino K, et al. Aberrant methylation and silencing of the BNIP3 gene in colorectal and gastric cancer. Clinical Cancer Research: An Official Journal of the American Association for Cancer Research. 2005;11:1021–7

[43] Heller G, Schmidt WM, Ziegler B, Holzer S, Mullauer L, Bilban M, et al. Genome-wide transcriptional response to 5-aza-2'-deoxycytidine and trichostatin a in multiple myeloma cells. Cancer Research. 2008;68:44–54. DOI: 10.1158/0008-5472.CAN-07-2531

[44] Stamatopoulos B, Meuleman N, De Bruyn C, Mineur P, Martiat P, Bron D, et al. Antileukemic activity of valproic acid in chronic lymphocytic leukemia B cells defined by microarray analysis. Leukemia. 2009;23:2281–9. DOI: 10.1038/leu.2009.176

[45] Cowell RM, Talati P, Blake KR, Meador-Woodruff JH, Russell JW. Identification of novel targets for PGC-1alpha and histone deacetylase inhibitors in neuroblastoma cells. Biochemical and Biophysical Research Communications. 2009;379:578–82. DOI: 10.1016/j.bbrc.2008.12.109

[46] Sitarz KS, Elliott HR, Karaman BS, Relton C, Chinnery PF, Horvath R. Valproic acid triggers increased mitochondrial biogenesis in POLG-deficient fibroblasts. Molecular Genetics and Metabolism. 2014;112:57–63. DOI: 10.1016/j.ymgme.2014.03.006

[47] Yoon JY, Ishdorj G, Graham BA, Johnston JB, Gibson SB. Valproic acid enhances fludarabine-induced apoptosis mediated by ROS and involving decreased AKT and ATM activation in B-cell-lymphoid neoplastic cells. Apoptosis: An International Journal on Programmed Cell Death. 2014;19:191–200. DOI: 10.1007/s10495-013-0906-7

[48] Mauro-Lizcano M, Esteban-Martinez L, Seco E, Serrano-Puebla A, Garcia-Ledo L, Figueiredo-Pereira C, et al. New method to assess mitophagy flux by flow cytometry. Autophagy. 2015;11:833–43. DOI: 10.1080/15548627.2015.1034403

[49] Dolman NJ, Chambers KM, Mandavilli B, Batchelor RH, Janes MS. Tools and techniques to measure mitophagy using fluorescence microscopy. Autophagy. 2013;9:1653–62. DOI: 10.4161/auto.24001

[50] Klionsky DJ, Abdelmohsen K, Abe A, Abedin MJ, Abeliovich H, Acevedo Arozena A, et al. Guidelines for the use and interpretation of assays for monitoring autophagy (3rd edition). Autophagy. 2016;12:1–222. DOI: 10.1080/15548627.2015.1100356

[51] Iwai-Kanai E, Yuan H, Huang C, Sayen MR, Perry-Garza CN, Kim L, et al. A method to measure cardiac autophagic flux in vivo. Autophagy. 2008;4:322–9

[52] de la Calle C, Joubert PE, Law HK, Hasan M, Albert ML. Simultaneous assessment of autophagy and apoptosis using multispectral imaging cytometry. Autophagy. 2011;7:1045–51

[53] Degtyarev M, Reichelt M, Lin K. Novel quantitative autophagy analysis by organelle flow cytometry after cell sonication. PLoS One. 2014;9:e87707. DOI: 10.1371/journal.pone.0087707

[54] Kimura S, Noda T, Yoshimori T. Dissection of the autophagosome maturation process by a novel reporter protein, tandem fluorescent-tagged LC3. Autophagy. 2007;3:452–60

[55] Kim SJ, Khan M, Quan J, Till A, Subramani S, Siddiqui A. Hepatitis B virus disrupts mitochondrial dynamics: induces fission and mitophagy to attenuate apoptosis. PLoS Pathogens. 2013;9:e1003722. DOI: 10.1371/journal.ppat.1003722

[56] Gottlieb RA, Stotland A. MitoTimer: a novel protein for monitoring mitochondrial turnover in the heart. Journal of Molecular Medicine (Berl). 2015;93:271–8. DOI: 10.1007/s00109-014-1230-6

Section 6

Autophagy in metabolic homeostasis

Chapter 20

Autophagy in Multidrug-Resistant Cancers

Betty Law Yuen Kwan and Vincent Wong Kam Wai

Additional information is available at the end of the chapter

http://dx.doi.org/10.5772/64274

Abstract

Multidrug resistance (MDR) in cancers is the major challenge in cancer therapy, thus the development of sensitizing agents or small molecules with new mechanisms of action to kill the resistant cancers is highly desired. Autophagy is a cellular process responsible for the turnover of misfolded proteins or damaged organelles and recycling of nutrients to maintain cellular homeostasis. Recently, autophagy has been shown to regulate MDR in cancers. In this chapter, both intrinsic and acquired drug resistance affecting the efficiency of chemotherapy, and the MDR mechanisms including nonclassical MDR phenotype and classical transport-based MDR phenotype were discussed. In addition, the development of apoptosis-resistant cancer by the deregulation of apoptotic gene machinery, such as BCL-2, BAX, BAK, and TRAILR, was also covered. We then further discussed the controversial role of autophagy by illustrating how induction of autophagy could work as a tumor suppressor or promote tumor survival. The modulation of MDR in cancer by either induction or inhibition of autophagy was also discussed. We have further summarized the current compounds or drugs for modulating MDR cancers and how autophagy modulators could circumvent the MDR phenotypes in cancers. Finally, the new mechanisms participating in MDR phenotypes were proposed for future MDR drugs discovery.

Keywords: autophagy, multidrug-resistant cancers, apoptosis-resistant cancers, apoptosis, P-glycoprotein

1. Introduction

The efficiency of chemotherapy can be affected by both intrinsic and acquired drug resistance. Intrinsic resistance is caused by the existing resistance factors presented in the cancer cells before treatments, while acquired drug resistance is developed by mutations or adaptive responses arising during chemotherapy. Due to the high-molecular heterogeneity of cancer cells, drug resistance can therefore be acquired through a minor population of resistant cancer cells presented in the initially drug-sensitive tumor. Using the genomic, proteomic, bioinformatics and systems biology approaches, a wide range of molecular mechanisms, genotypes and therapeutic targets have been identified in developing drug-resistance cancers. For example, alterations in drug transport or metabolism, local tumor microenvironment and drug targets all contribute to chemoresistance. Recently, autophagy has also been identified as an important mechanism in regulating multidrug resistance (MDR) in cancers [1].

Autophagy is constitutively active in its basal level for maintaining normal homeostasis, quality control of protein and organelle and working with the ubiquitin proteasomal system to degrade the polyubiquitinated and aggregated proteins. This catabolic process is triggered by cellular stressful conditions such as nutrient or energy deprivation, pathogen infection or misfolded protein accumulation, for the recycle of energy and nutrients to sustain cellular metabolism. While most evidence supports the prosurvival role of autophagy, unrestrained autophagy can contribute to cell death resulted from excessive cellular consumption [2] (**Figure 1**). Therefore, targeting autophagy for modulating MDR cancers has become an attractive approach in anticancer therapy.

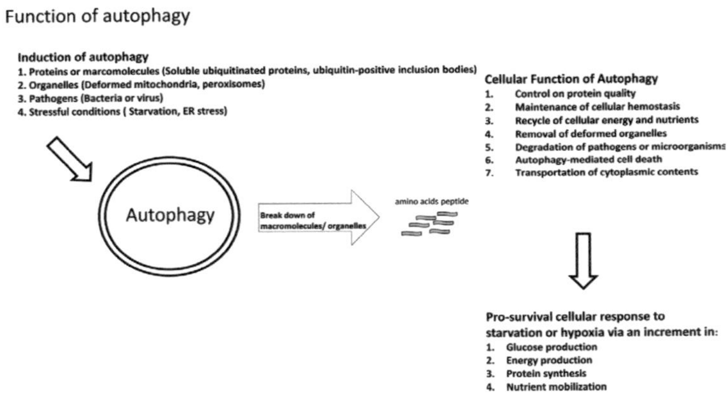

Figure 1. Cellular function and role of autophagy. Autophagy is a cellular-regulated degradation system for delivering unwanted cytoplasmic constituents to the lysosome for removal. Autophagy can be triggered by stressful conditions such as infection, nutrient deprivation or protein aggregates accumulation. Recent studies have depicted the physiological or pathological roles of autophagy in tumorgenesis, neurodegeneration, adaptation to starvation or development, clearance of misfolded proteins and organelles, antiaging, elimination of invading pathogens and regulation of cell death.

2. Autophagy and drug-/apoptosis-resistant cancers

2.1. Molecular mechanism of multidrug resistance in cancers

Drug-resistant mechanisms are classified into two major catergories: (1) nonclassical MDR phenotype and (2) classical transport-based MDR phenotype. Non-classical MDR refers to the decreased sensitivity of tumor to certain anticancer drugs due to the activity of specific enzymes, including topoisomerase or glutathione S-transferase. Defective apoptotic proteins may also contribute to the development of drug resistant in tumors. Classical type of transport-based MDR involves the ATP-binding cassette (ABC) family of transporters, which 49 types of human ABC transporters were identified. Among them, three ABC transporters were reported to cause human MDR, including (1) multidrug resistance protein 1 (MDR1/P-glycoprotein/ABCB1); (2) MDR-associated protein 1 (MRP1/ABCC1); and (3) breast cancer resistance protein (BCRP/ABCG2), which are responsible for the efflux of several hydrophobic chemotherapeutic agents such as antimetabolites, taxanes or topoisomerase inhibitors across the plasma membrane [3].

MDR1, a membrane-bound glycoprotein, which is most abundant on the surface of excretory epithelial cells of colon, small intestine, pancreatic or bile ductules, and kidney proximal tubules, was found overexpressed and associated with drug resistance in kidney, liver and colon cancers. Recently, overexpression of MRP1 has also been found in chemoresistance breast, prostate and lung cancers. BCRP is a MDR drug efflux pump, which was associated with drug resistance in leukemia and breast cancer. Chemotherapeutic agents, such as imatinib, erlotinib, sunitinib and nilotinib, are known substrates and modulators of both MDR1 and BCRP [1].

2.2. "Apoptosis-resistance cancers"

Deregulation of apoptosis can cause drug resistance of cancer cells, and therefore, targeting the overamplifications, mutations and chromosomal translocations of antiapoptotic proteins, such as BCL-2 family members, caspase 8 inhibitor FLIP and inhibitor of apoptosis proteins (IAPs), are important in cancer therapies. Extensive studies suggested the overexpression of BCL-2 confers resistance of leukemia cells and mouse thymocytes to chemotherapies. Apoptosis involving the permeabilization of outer mitochondrial membrane (MOMP) can be blocked by BCL-2. Antiapoptotic BCL-2 family members, including BCL-XL and MCL1, and proapoptotic family members, including BAX, BAD, BAK and BH3-only proteins (BIM) which antagonize the antiapoptotic BCL-2 family members, are key regulators of apoptosis through the induction of MOMP. Increased expression of BIM is responsible for apoptosis induced by chemotherapeutic agents, including imatinib, gefitinib and erlotinib, in various cancer models. Besides, responsiveness to inhibitors of EGFR, HER2 or PI3K is correlated with the level of BIM. Deletion in the gene-encoding BIM is associated with intrinsic resistance to tyrosine kinase inhibitors (TKIs) therapies in lung cancer models [1].

Overexpression of prosurvival protein BCL-2 was found in malignant cells with defective apoptosis; therefore, drugs mimicking its antagonists, BH3-only proteins, may be effective in chemotherapy. Preclinical data confirmed that both ABT-737 (a BH3 mimetic) and its orally bioavailable form ABT-263, antagonized the antiapoptotic function of BCL-2, BCL-XL and BCL-W and exhibited cytotoxicity through Bax/Bak-mediated apoptosis. The weak-binding affinity of ABT-737 to the antiapoptotic BCL-2 family member (MCL1) can affect the effectiveness of ABT-737 and lead to drug resistance in cells. Therefore, targeting the enzymatic degradation of antiapoptotic BCL-2 family members may be an alternative way to overcome the drug resistance in cancer cells [1].

Tumor necrosis factor (TNF)-related apoptosis-inducing ligand (TRAIL) and agonistic antibodies targeting the TRAILR death receptors demonstrated antitumor properties in both *in vitro* and xenograft models. Currently, combinational use of TRAIL receptor agonists with chemotherapeutics, including carboplatin, paclitaxel, BCL-2 antagonists, bevacizumab, or histone deacetylase (HDAC) inhibitors showed prominent effect in clinical evaluations. Several drugs that are able to improve sensitivity to TRAIL can also decrease the level of the caspase 8 inhibitor (FLIP) and induce apoptosis. Therefore, inhibition of FLIP may work as an attractive therapeutic strategy for apoptosis-resistance cancers [1, 4].

The small molecule obatoclax (GX15-070) has been reported to antagonize antiapoptotic BCL-2 proteins, such as BCL-2, BCL-XL, BCL-W and MCL-1, and induce both apoptosis and autophagic cell death. Although the molecular mechanisms underlying the effect of obatoclax remain unclear, both single and combinational therapeutic uses of GX15-070 are under clinical evaluation, suggesting the role of apoptosis and autophagy in cancer therapy [5].

2.3. Autophagy in tumor suppression

In fact, the role of autophagy in cancer is controversial as it can work as a tumor suppressor, which inhibits tumor initiation or facilitates the survival of cancer cells during metabolic stresses induced by anticancer agents. Although autophagy may play a prosurvival role in tumor cells, loss of function mutations in the autophagy pathways were associated with tumorgenesis. For example, defects in apoptosis and autophagy may lead to tumorgenesis through chronic wound-healing response triggered by necrosis and inflammation. Malfunction of autophagy lead to the mismanagement of metabolic stress, which contributes to damage of cellular proteins, organelles or DNAs; insufficient ATP levels that are essential for maintaining normal DNA replication and genomic integrity, all these finally lead to genomic damage and tumor progression in autophagy-deficient tumor cells. This conception is supported by evidence of increased DNA breaks or damage response and aneuploidy in autophagy-defective cancer cells [6] (**Figure 2**).

Beclin 1 is a haploinsufficient tumor suppressor gene responsible for the induction of autophagy. Overexpression of beclin 1 can inhibit tumorgenesis. Accumulation of the autophagic substrate, p62/SQSTM 1 protein aggregates, could lead to the damage of mitochondria, accumulation of misfolded proteins, and overproduction of reactive oxygen species (ROS) that lead to DNA damage and genomic instability, suggesting the tumor suppressive role of autophagy [6]. Genetic instability and mutation are causes for increased cancer cells evolution and resistance to chemotherapy. While autophagy is associated with the promotion of longevity, DNA damage facilitates both cancer and ageing. Therefore, it is proposed that autophagy may play a protective role in maintaining cellular and genome stabilities, which can eventually prevent cancer and extend lifespan [7].

Furthermore, the induction of autophagic cell death has been observed in cancer cells triggered by excessive activation of autophagy, which causes irreversible damage to cells through

http://dx.doi.org/10.5772/64274

redundant degradation of regulatory organelles, such as endoplasmic reticulum, mitochondria and golgi apparatus. Although evidence supporting the anticancer properties of autophagic death is limited and controversial, induction of autophagic cell death by novel small molecule activator of autophagy (e.g., STF-62247) was shown in renal cell carcinoma cells [6]. Proteins identified to be related to the autophagic cell death included steroid receptor coactivator, foxo1, histone deacetylases, ras or e4F1. Compound such as dasatinib, an inhibitor of Src/Abl family kinases, triggers autophagic cell death in ovarian cancer xenograft model; knockdown of

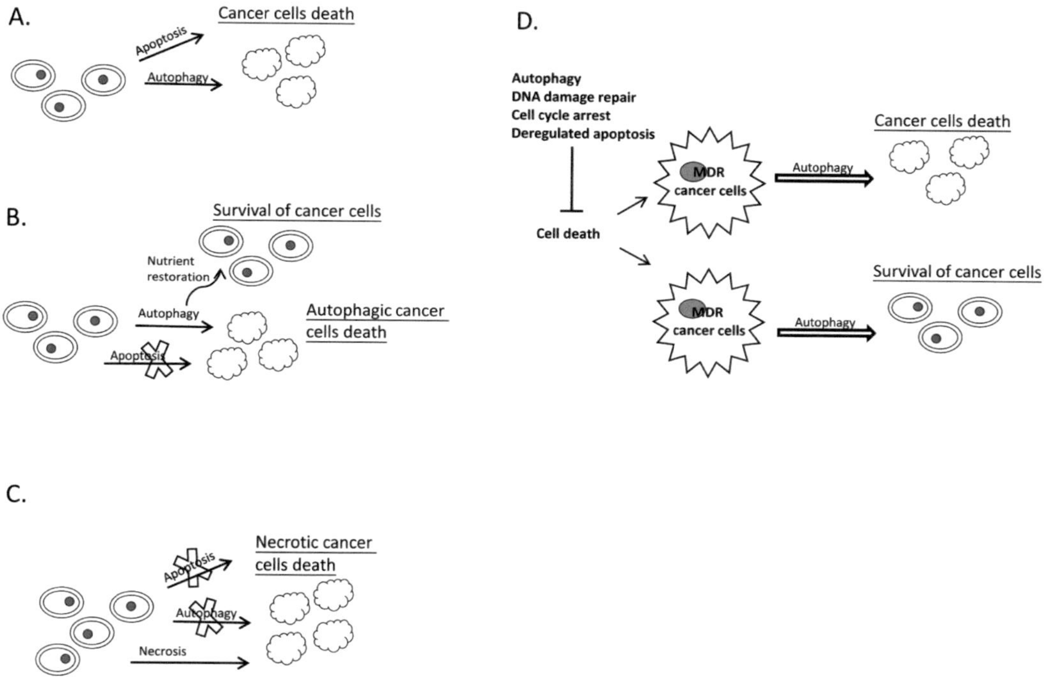

Figure 2. The interplay between autophagy, apoptosis and necrosis in the survival and death mechanisms of cancer cells. (A) Apoptosis and autophagy are programmed cell death mechanisms involved in various physiological and pathological conditions such as cancers. Failure in the induction of apoptosis or autophagy could result in malignant transformation of cells. Despite the basic physiological role of apoptosis and autophagy, targeting both apoptosis and autophagy pathways has been a popular strategy in chemotherapies. (B) Malfunction of autophagy is frequently found in human cancers. When apoptosis is inhibited, autophagy is promoted for inducing autophagic cancer cells death. However, evidence also indicated the induction of autophagy could confer stress tolerance to cancer cells and facilitates the survival of cancer cells under stressful condition, such as starvation, suggesting the controversial role of autophagy in cancer therapies. (C) The induction of necrotic cell death when both autophagy and apoptosis are inhibited. Necrosis is referred to the unprogrammed type of cell death characterized by an increase in cell volume and swelling, which finally lead to the rupture of the cell membrane and burst of cellular contents. (D) Inhibition of cell death is attributed to DNA damage induced by agents such as cisplatin, which can cause cell cycle arrest and allow the repairment of damaged cells, and finally lead to multidrug resistance (MDR). Besides, malfunction of apoptosis or induction of autophagy can also decrease drug-induced damage of DNA and cell death. All these mechanisms limit the effectiveness of chemotherapies via inhibition of specific enzymes or receptors in cells, leading to MDR in cancer cells.

beclin1 lead to reduction of autophagy and tumor growth. Interestingly, while oncogenic proteins, such as BCL-2, PI3K or AKT1, could inhibit autophagy, tumor suppressor proteins, such as beclin1, BIF-1, LKB1 or UVRAG, could induce autophagy. It was reported that autophagy may inhibit tumors growth by working with apoptosis and/or necrosis to cause cell death. For example, caffeine induces apoptosis through p70S6K-dependent activation of autophagy. A novel survivin suppressant, YM155, could induce autophagy-dependent apoptosis in prostate cancer cells. All these evidence suggested the potential role of autophagy in cancer therapy [3].

2.4. Inhibition of autophagy in cancer therapy

Increasing preclinical evidence supported the argument that pharmacological inhibition of autophagy or genetic knockdown of autophagy-related genes (ATGs) can increase sensitization of tumor cells to drug-induced cell death. Mechanistic pathways depicting the protective mechanisms of autophagy in cancers include the E3 ubiquitin ligase c-Cbl cascade that targets the active Src to promote survival of cancer cells via autophagy. Mutation of oncogene, *H-Ras* or *K-Ras*, upregulates basal autophagy for promoting the survival of cancer cells under nutrient deprivation conditions. Overexpression of inflammatory receptor, receptor for advanced glycation end products (RAGE), activated interleukin 6 (IL-6)-mediated mitochondrial pathway and transcription 3 (STAT3) signaling, induced autophagy and inhibited apoptosis to promote survival of pancreatic cancer cells. Lymphocyte-induced autophagy triggered tumorigenesis, suggesting a positive correlation between inflammation and cancer. An anticancer agent, suberoylanilide hydroxamic acid (SAHA), activated prosurvival autophagic pathway to attenuate both apoptotic and nonapoptotic cell death, suggesting the inhibition of autophagy may enhance the efficacy of SAHA [8].

With the prosurvival role of autophagy in cancer treatment, genetic or pharmacological inhibition on autophagy may therefore enhance the efficacy of cancer therapies. Pharmacological inhibition of autophagy can be induced by early phrase inhibitors such as 3-methyadenine (3-MA), wortmannin and LY294002. Late-phrase autophagy inhibitors include chloroquine (CQ), hydroxychloroquine (HCQ), bafilomycin A1 and monensin. CQ increased cyclophosphamide-induced cell death of tumor; reduced tumor size and enhanced apoptosis with the presence of anticancer drug vorinostat in colon cancer xenograft model. Besides, the treatment of saracatinib (Src inhibitor) with CQ in a prostate cancer mouse model demonstrated a higher percentage of tumor growth inhibition when compared with saracatinib alone [6].

Another autophagy inhibitor, 3-MA, enhanced 5-fluorouracil (5-FU)-induced apoptosis with increased regression of tumor in colon cancer models, suggesting the anticancer properties of autophagy inhibition in chemotherapy. While combinational use of bortezomib with CQ suppressed growth of tumor to a greater extent than bortezomib alone in colon cancer models, evaluation of phase I/II clinical trials is ongoing in patients with myeloma, suggesting the potential therapeutic use of autophagic inhibitor in cancer therapy [6].

2.5. Role of autophagy in MDR

2.5.1. Inhibition of autophagy in MDR

Overexpression of p-glycoproteins can efflux anticancer drugs and resulted in drug resistance and failure of anticancer therapies. It has been reported that the induction of autophagy may contribute to the survival of MDR cells during chemotherapy; defective autophagy contributes to MDR cells growth inhibition. In contrast, clinical trials have been performed to evaluate the beneficial role of autophagy enhancer in anticancer therapy, suggesting the double-edged sword of autophagy which its effect may highly dependent on the stage of tumor. The potent anticancer and apoptotic properties of gossypol, a BH3-mimetic small molecule isolated from cottonseeds, have been demonstrated in Ras-NIH 3T3-Mdr cells with overexpression of p-glycoproteins. However, results indicated that defective autophagy in the Ras-NIH 3T3-Mdr cells may enhance necrotic and apoptotic cell death induced by gossypol, suggesting the prosurvival role of autophagy in the MDR cells [9].

Further studies suggested that autophagy may contribute to resistance to chemotherapy drugs, such as tamoxifen, herceptin, paclitaxel (PTX) and epirubicin (EPI), in breast cancer. It was demonstrated that PTX- and vinorelbine (NVB)-induced autophagy facilitated the development of resistance to PTX and NVB in both MCF-7er and SK-BR-3er cells with overexpression of p-glycoproteins. The finding suggested that the use of autophagy inhibitors may be an attractive way for overcoming MDR in cancer therapy [10]. Further evidence suggested that the combinational use of autophagy and Src tyrosine kinase inhibitor was able to sensitize MDR cells to anticancer therapy. To increase the therapeutic efficacy of 5-fluorouracil (5-FU) to MDR colon cancer cells, many reports have confirmed the use of autophagy inhibitors, such as 3-MA or chloroquine, can increase the sensitivity of cancer cells toward treatment of 5-FU.

2.5.2. Activation of autophagy in MDR

Although mitigating MDR through inhibition of ABC transporter-mediated drug efflux or autophagy may be an effective approach in anticancer therapy, the induction of autophagy may also attenuate MDR. For example, vocamine (alkaloid) isolated from *Peschiera fuchsiaefolia* can mitigate drug resistance of osteosarcoma cells by inhibiting p-glycoproteins and inducing autophagic cell death when used with chemotherapeutic agent doxorubicin. Autophagy increases the efficacy of radiation therapy in both apoptosis-deficient lung cancer cells and xenograft model. Evidence demonstrating the potential anticancer efficacy of mTOR inhibitors was abundant. For example, combinational treatment by beclin1 expression and autophagic induction inhibited the growth of MDR (Ras-NIH 3T3) cells. Interestingly, constitutive expression of p-glycoproteins in hepatocellular cancer cells can lead to overexpression of BCL2 and mTOR, which contributed to resistance of cells to both apoptosis and autophagy [3].

In fact, autophagic cell death was observed in a p-glycoproteins overexpressing breast cancer cells that are resistant to paclitaxel. Further report demonstrated that long-term exposure to cisplatin leads to MDR and finally inhibition of both cisplatin-mediated apoptotic cell death and autophagy. Consistently, the presence of autophagic inducer can mitigate the resistance of cancer cells toward cisplatin-mediated apoptotic cell death, suggesting the anticancer property of autophagy induction. p53, a well-known tumor suppressor gene, can lead to cancer cell death. Mutations in p53 were highly correlated with failure of chemotherapy or radiotherapy. However, while p53 mutant utilizes autophagy to kill the ovarian MDR cancer cells, wild-type p53 reverses the MDR phenotype by autophagy inhibition. The observation has further suggested the dual role of autophagy in MDR cancer therapy, and special cautions are required when applying autophagy modulators in cancer therapies [3].

3. Current approaches, challenges and compounds in targeting MDR

3.1. Current compounds/drugs in modulating MDR cancers

Malfunction of the apoptosis contributed to the survival of cancer cells under oxidative stress and hypoxia conditions. These lead to the accumulation of genetic defects in cells, which resulted in the deregulation of cell proliferation or promotion of angiogenesis during tumorgenesis. Therapeutic strategies targeting apoptosis through inhibition of antiapoptotic proteins or stimulation of expression of proapoptotic proteins have been used in anti-MDR cancer treatments. In fact, mutation or alteration in the level of drug targets may also confer resistance to therapeutic agents in cancer cells. For example, fluoropyrimidine 5-fluorouracil (5-FU), a common chemotherapeutic agent for breast and colorectal cancers, triggers cancer cell death via inhibition of thymidylate synthase (TS) [11]. Consistently, many evidences suggest that high expression of TS contributed to the increase resistance of cancer cells to 5-FU. Another type of chemotherapeutic agents, including paclitaxel, docetaxel, vinblastine and vincristine (vinca alkaloids), inhibited tumor growth via suppressing the polymerization of microtubules, and eventually lead to apoptotic cell death. However, cancer cells can acquire resistance to paclitaxel or vinca alkaloids by alternating the levels of tubulin isotypes, since a reduction in tubulin levels was found in paclitaxel-resistant cells. Platinum-induced DNA damage by cisplatin can trigger apoptosis. While an increase in the BAX to BCL-2 ratio by cisplatin could induce apoptosis, overexpression of BCL-2 was reported in tumors that were resistance to cisplatin. Therefore, functional defects in the apoptotic pathway or inactivation of apoptosis may contribute to the development of drug resistance in cancers and revealed the current challenges of using chemotherapeutic agents in MDR cancer therapies [12].

New therapeutics against specific antiapoptotic targets has been applied to enhance apoptosis and sensitize MDR cancer cells to anticancer agents. For example, antiestrogens, such as tamoxifen, can inhibit estrogen-dependent BCL-2 expression and increase the sensitivity of cancer cells to anticancer agent doxorubicin. Expression of BCL-2 or BCL-XL can be downregulated by small molecule compounds regulating retinoic acid receptors (RAR), PPAR or retinoid receptors. For example, RAR and RXR ligands are used to treat leukemia and lymphoma in clinical trials; PPAR agonist troglitazone, decreased serum prostate-specific antigen (PSA) in patients with prostate cancer; inhibitors of histone deactylases (HDACs) can suppress the expression of antiapoptotic BCL-2-family genes in cancers and enhance sensitivity of cancer cells to conventional cytotoxic agents in xenograft models or clinical trials. Moreover, antisense oligonucleotides targeting the BCL-2 mRNA are in phase III clinical trials.

BH3-only proteins, endogenous antagonists targeting BCL-2 and MCL-1, have been revealed for their therapeutic potency to induce apoptosis. The anticancer effect of synthetic BH3 peptides occupying the BH3-binding site of BCL-2 or BCL-XL, and BH3-mimicking compounds have already been validated for their anticancer properties. Small molecule BCL-2 inhibitors (HA14-1), BH3 peptidomimetics and BCL-2 antagonists (ABT-737 and ABT-263) have been shown in the preclinical studies to enhance the sensitivity of cancer cells to ionizing radiation or chemotherapeutic agents, providing us an overview on current approach of targeting apoptosis in MDR cancer therapies [12].

Overcoming intrinsic apoptosis failure during anticancer therapies could be achieved by triggering extrinsic apoptosis. For example, TNF treatment by cytokines FasL and TRAIL can activate caspases without the proinflammatory activation of the NF-KappaB pathways. TRAIL and its agonistic antibodies that bind TRAIL receptors have been proved to possess potent antitumor effect in mouse xenograft models. Moreover, clinical trials evaluating the efficacy of agonistic antibodies were completed, suggesting the potential of applying these biological agents to circumvent the defective intrinsic mitochondrial apoptotic pathway. CDDO and CDDOm are synthetic triterpenoids that sensitize cells to TRAIL-induced apoptosis in chemoresistant leukemia cells. Byrostatin, a protein kinase C (PKC) modulator, triggers the production of TNF and apoptosis in myeloid leukemia cells. All-trans-retinoic acid (ATRA) induces the production of cytokine TRAIL and apoptosis in acute leukemia cells, confirming the beneficial role of targeting TRAIL to kill cancer cells that are resistance to apoptosis. Heat-shock protein 90 (Hsp90) is a molecular chaperone ubiquitously expressed for the maturation of a set of substrate proteins called clients. Hsp90 promotes invasion, angiogenesis and metastasis which all contribute to tumorgenesis. Hsp90 stabilizes Raf-1, Akt and ErbB2 proteins and lead to the resistance of cancer cells to radiation therapy. Geldanamycin, inhibitor of Hsp90, enhances the sensitivity of cancer cells to radiation therapy. However, due to the poor solubility, high toxicity and problems of drug efflux by p-glycoprotein, modification on synthetic Hsp90 inhibitors has been made. For example, pyrazole resorcinol compound NVP-AUY922 has higher binding affinity to Hsp90, and NVP-BEP800, a 2-aminothieno pyrimidine class Hsp90 inhibitor, possess high antiproliferative activity in cancer cells. These inhibitors increase the sensitivity of tumor cells to apoptosis via depletion and destabilization of Hsp90 proteins and finally lead to increased DNA damage and apoptosis [12].

3.2. Current compounds/drugs regulating MDR cancers through autophagy

Most, if not all, chemoresistant cancers have defects in apoptotic pathways. In addition, the mitochondrial/cytochrome c pathway of apoptosis is frequently perturbed in various types of human cancers [13]. For instance, gene deficiency of BAX or BAK is commonly found in many malignancies, and BAX-BAK double-knockout mouse embryonic fibroblasts (MEFs) are resistant to various apoptosis-inducing agents [14]. On the other hand, caspase-3 and caspase-7 are crucial mediators of mitochondrial-mediated apoptosis [15], whereas caspase-3, caspase-8 and caspase-9 contribute the signal transduction roles in apoptosis induced by anticancer agents. Deficiency of these caspases would ultimately develop apoptosis-resistance and drug resistance phenotypes [16]. Although autophagy is considered as a crucial player in drug resistance because cancer cells can circumvent cellular stress via autophagy induction, it has been shown that autophagy facilitates resistance of cancer cells to chemotherapeutic agents, and inhibition of autophagy could be therapeutically beneficial in some cases [17]. Nevertheless, significant number of studies demonstrated that the use of small molecules to induce autophagy-dependent cell death in apoptosis-defective or apoptosis-resistant cancer cells is an effective therapeutic approach [18]. These evidences are fuelling novel approaches to treat cancer and impede multidrug resistant through the induction of autophagy (**Table 1**).

Compounds or drugs	**Sources**	**Target pathways**	**Autophagy inducer or inhibitor**	**Types of MDR**	**Origin of cancers**	**References**
Saikosaponin-d (Ssd)	*Bupleurum falcatum L*	SERCA	Inducer	Caspases 3, 7 or 8 deficient Bax-Bak DKO	MEFs	[19]
Liensinine, isoliensinine, dauricine, cepharanthine	*Nelumbo nucifera, Asiatic moonseed rhizome, Stephania cepharantha*	AMPK-mTOR	Inducer	Caspases 3, 7 or 8 deficient Bax-Bak DKO	MEFs	[20]
Hernandezine	Thalictrum glandulosissimum	AMPK	Inducer	Caspases 3, 7 or 8 deficient Bax-Bak DKO	MEFs, colon	[21]
Ursolic acid	Apple peels	JNK signaling	Inducer	p53 mutation	Colorectal cancer	[22]
Rottlerin	Kamala powder	unknown	Inducer	Caspase 3 deficient	Breast cancer	[23]
Coibamide A	Marine cyanobacterium	mTOR-independent	Inducer	Apaf-1-null	MEFs, glioblas toma	[24]
Chalcone-24	Synthetic	c-FLIPL and c-IAPs degradation	Inducer	TRAIL resistant	Lung cancer	[25]
PG545	Synthetic	Heparanase	Inhibitor	Heparanase-overexpressing cancer	Cancer cells	[27]
Nelfinavir	FDA drug	Unfolded protein response (UPR) and endoplasmic reticulum	Inducer	Doxorubicin resistant	Breast cancer	[28]

Compounds or drugs	Sources	Target pathways	Autophagy inducer or inhibitor	Types of MDR	Origin of cancers	References
		(ER) stress				
Rapamycin	FDA drug	mTOR, mdr1 expression	Inducer	Doxorubicin resistant	Colon cancer	[29]
Lansoprazole	Synthetic	Proton pump	Inhibitor	Doxorubicin resistant	Solid tumors	[30]
Oxaliplatin derivatives with axial DCA ligands	Synthetic	Mitochondria and glucose metabolism	Inducer	Cisplatin resistant	Colorectal cancer	[31]
LY294002 and rapamycin	Synthetic and FDA drug	Akt/mTOR signaling, miR-222	Inducer	Cisplatin resistant	Bladder cancer	[32]
Rapamycin	FDA drug	mTOR	Inducer	Cisplatin resistant	Cervical cancer	[33]
Ursodeoxycholic acid	Synthetic	CD95/Fas	Inducer	Cisplatin resistant	Gastric cancer	[34]
Monanchocidin A	Marine sponge Monanchora pulchra	Lysosomal membrane permeabilization	Inducer	Cisplatin resistant	Germ cell tumor, prostate and bladder cancer	[35]
Resveratrol derivative, TMS	Synthetic	SERCA	Inducer	Gefitinb resistant	NSCLC cells	[36]
Chloroquine	FDA drug	Autophagy	Inhibitor	Gefitinb resistant	Liver cancer	[37]
Leu-Leu-O-methyl	Synthetic	Lysosome membranes	Inhibitor	Sunitinib resistant	Renal carcinoma	[39]
Metformin	FDA drug	GRP78	Inhibitor	Bortezomib resistant	Multiple myeloma	[40]

Table 1. Current compounds or drugs targeting MDR through modulation of autophagy.

In fact, the active components isolated from natural products have been found effective in inducing autophagic cell death or autophagy-dependent cell death in apoptosis resistance cells or cancers. For example, saikosaponin-d (Ssd), extracted from Chinese medicinal herb *Bupleurum falcatum L.*, is capable of inducing autophagic cell death in a panel of apoptosis-resistant cells [19]. Ssd increases cytosolic calcium via direct suppression of sarcoplasmic/endoplasmic reticulum $Ca^{2}+$ ATPase Pump (SERCA), leading to autophagy induction through the activation of the Ca^{2+}/calmodulin-dependent kinase kinase (CaMKK)—AMP-activated protein kinase (AMPK)—mammalian target of rapamycin (mTOR) pathway. Importantly, Ssd-disrupted calcium homeostasis stimulates endoplasmic reticulum (ER) stress as well as the unfolded protein responses (UPR) and eventually contributes to autophagic cell death in apoptosis-defective or apoptosis-resistant mouse embryonic fibroblasts (MEFs), which either lack caspases 3, 7 or 8 or had the BAX-BAK double knockout [19]. Concomitantly, a group of natural alkaloids, including liensinine (*Nelumbo nucifera*), isoliensinine (*Nelumbo nucifera*), dauricine (*Asiatic Moonseed Rhizome*) and cepharanthine (*Stephania cepharantha*), were identified to stimulate AMPK-mTOR-dependent induction of autophagy and autophagic cell death in the same panel of apoptosis-resistant cell [20]. These alkaloids were later confirmed as novel and direct AMPK activators [21]. Furthermore, hernandezine, an alkaloid isolated from Chinese medicinal herb *Thalictrum glandulosissimum,* sharing structural similarity with the above isoquinoline alkaloids, exhibits specific cytotoxicity and induces autophagy in a panel of cancer cells. Hernandezine is a new class of AMPK activator, which induces autophagy and autophagic cell death in a panel of caspases-deficient apoptosis-resistant MEF. Those studies further indicated that the MEFs with genes deficiency of BAX and BAK demonstrated drug-resistant phenotypes in various chemotherapeutic agents, such as cisplatin, adriamycin, taxol, etoposide and staurosporine, but not in hernandezine. And hernandezine also shows similar cytotoxicity toward both wild type and double knockout of BAX/BAK in human DLD-1 colon cancer cells, suggesting the cross sensitivity of hernandezine toward this apoptosis-deficient cancer [21].

Apart from AMPK-mTOR signaling, mutations of tumor suppressor p53 have been shown to confer cellular resistant to various chemotherapeutic agents. For instance, colorectal carcinomas (CRCs) with p53 mutations have developed resistance to widely used chemotherapeutic agent, 5-fluorouracil (5-FU) [22]. A natural triterpenoid, ursolic acid, was found to sensitize p53 mutant apoptosis-resistant colorectal cancer cells to 5-FU effects via activation of c-jun N-terminal kinase (JNK) [22]. In addition, rottlerin, a natural polyphenol purified from the kamala powder, induces autophagic death in caspase-$3^{-/-}$ apoptosis-resistant MCF-7 breast cancer cells [23]. Coibamide A, an N-methyl-stabilized depsipeptide isolated from a marine cyanobacterium, induces autophagy and cell death in apoptosis-resistant MEFs and glioblastoma cells [24]. Other synthetic compounds and derivatives can overcome chemoresistance by modulation of autophagy. For instance, a novel chalcone derivative, chalcone-24, potentiates the anticancer activity of TNF-related apoptosis-inducing ligand (TRAIL) through autophagy-mediated degradation of cellular FADD-like IL-1β-converting enzyme-inhibitory protein large (c-FLIPL) and cellular inhibitor of apoptosis proteins (c-IAPs), which could be an effective approach in alleviating drug resistance [25]. Besides, sepantronium bromide (YM155) is a selective survivin inhibitor that exhibits potent antitumor activities in head neck squamous cell carcinoma (HNSCC) *in vitro* and *in vivo* by inducing apoptosis and autophagic cell death [26]. On the other hand, heparanase is a mammalian enzyme capable of cleaving heparan sulfate, whose enzymatic activity contributed to tumor inflammation, angiogenesis and metastasis. Recent studies indicated that heparanase-overexpressing cancers were more resistant to chemotherapy in a manner associated with increased autophagy, and therefore, the suppression of heparanase and its mediated autophagy by heparanase inhibitor (PG545) is a promising strategy in treatment of these resistant cancers [27].

Several clinically approved drugs were found to be effective in treating doxorubicin-resistant cancer cells through modulation of autophagy. For example, nelfinavir, a clinically approved anti-HIV drug targets multidrug-resistant mechanism to enhance the efficacy of doxorubicin in doxorubicin-resistant MCF-7 breast cancer cells [28]. Rapamycin, a mTOR inhibitor, reverses drug resistance to doxorubicin in colon cancer through induction of autophagy and apoptosis, and suppression of multidrug resistance gene 1 (mdr1) expression [29]. Proton pump inhibitor lansoprazole potentiates the therapeutic effects of doxorubicin by improving its distribution and activity in solid tumors [30].

In cisplatin-resistant cancer, study reported that oxaliplatin derivatives with axial dichloroacetate (DCA) ligands induce autophagy and potentiate toxicity in cancer cells through modulation of mitochondria and glucose metabolism. These derivatives can also overcome inherent and acquired resistance to cisplatin and oxaliplatin [31]. Inhibition of Akt/mTOR signaling by LY294002 and rapamycin prevents miR-222-induced proliferation and restores the sensitivity of resistant bladder cancer cells to cisplatin. These findings indicated that miR-222 activates the protein phosphatase 2A subunit B/Akt/mTOR axis and thus plays a critical role in regulating proliferation and chemotherapeutic drug resistance [32]. Consistently, cisplatin cytotoxicity could be greatly enhanced in resistant cancer cells when mTOR had been inhibited prior to cisplatin treatment that was likely due to increased autophagy level [33]. Ursodeoxycholic acid effectively kills drug-resistant gastric cancer cell through induction of autophagic death [34]. Besides, a novel alkaloid, monanchocidin A isolated from the marine sponge *Monanchora pulchra*, overcomes drug resistance by induction of autophagy and lysosomal membrane permeabilization [35].

In other drugs-resistant cancers, a new analogue of resveratrol, (Z)3,4,5,4′-trans-tetramethoxystilbene (TMS), inhibits gefitinib-resistant non-small cell lung cancer (NSCLC) through inhibition of SERCA and induction of autophagy [36]. In contrast, co-delivery of gefitinib and chloroquine by chitosan nanoparticles overcomes gefitinb-resistant liver cancer by the inhibition of gefitinib-mediated autophagy [37]. Similarly, inhibition of lapatinib-mediated protective autophagy sensitizes HER2-positive breast resistance cancer cells to lapatinib, suggesting that autophagy is a promising target for circumventing lapatinib resistance of HER2-positive breast cancer cell [38]. Inhibition of sunitinib-mediated autophagy overcomes sunitinib resistance in metastatic renal cell carcinomas [39]. Similarly, metformin suppresses glucose-regulated protein 78 (GRP78), a key driver of bortezomib-induced autophagy and enhances the antimyeloma effect of bortezomib [40].

Collectively, MDR cancer cells defective in apoptosis signaling pathway are more sensitive to various types of autophagy inducers, which eventually induce autophagy-associated cell death or autophagic cell death. It is suggested that autophagy and its associated cell death provide an alternative promising approach for inducing cell death in MDR cancers. However, both autophagy inducers and inhibitors are commonly involved in the suppression of many other drugs-specific MDR cancer cells. Understanding the role of autophagy in particular chemotherapeutic treatment is crucial for establishing an effective treatment of MDR via modulation of autophagy.

4. New mechanisms participating in MDR phenotypes for future drug discovery

Heme oxygenase-1 (HO-1) contributes to imatinib resistance by promoting autophagy in chronic myeloid leukemia through disrupting the mTOR-signaling pathway [41], whereas Src/STAT3-dependent HO-1 induction contributes to doxorubicin resistance in breast cancer cells by promoting autophagy [42]. Thus, HO-1 may be a novel target for improving MDR phenotypes in leukemia and breast cancer therapy. Galectin-1 is a beta-galactoside-binding lectin, and its mediated autophagy facilitates cisplatin resistance of hepatocellular carcinoma (HCC). Thus, galectin-1 may be a potential target to improve the efficacy of cisplatin in the treatment of patients with HCC [43]. Other studies showed that activation of autophagy in tumor associated macrophages (TAMs) inhibits proliferation and induces apoptosis in colon cancer cells and alters the expression of radiosensitivity associated proteins. Therefore, stimulating TAM autophagy may increase the radiosensitivity of colorectal cancer cells [44]. BRAF is an oncogenic protein, which promotes protective autophagy in colorectal tumor cells. BRAFV600E is the mutant protein of BRAF, which contributes to the drug-resistant phenotype of colorectal tumors. Study showed that pretreatment of autophagy inhibitor 3-MA followed by combinational treatment with drug PLX4720 targeting BRAFV600E can synergistically sensitize resistant colorectal tumors and provide novel efficient approaches for the treatment of resistant colorectal tumors bearing BRAFV600E [45]. The CD44 isoform-containing variant exon v6 (CD44v6) exhibits a crucial role in the progression, metastasis and prognosis of colorectal cancer. Overexpression of CD44v6 contributes to acquired chemoresistance via upregulation of autophagy in colon cancer SW480 cells, indicating that CD44v6 may be the new therapeutic target for resistant colorectal cancer [46].

Early growth response gene-1 (Egr-1) enhances hypoxia-induced autophagy to contribute chemoresistance of HCC. Dominant negative Egr-1 inhibits autophagy and thus enhances the sensitivity of HCC cells to chemotherapeutic agents, indicating that hypoxia/Egr-1/autophagy axis might be a novel therapeutic target for improving drug resistance in HCC [47]. Urothelial carcinoma is characterized by therapeutic resistance and frequent tumor relapse. Study indicated that the synergistic cytotoxic effect of gemcitabine/mitomycin with autophagy inhibitor (chloroquine) or with glycolytic inhibitor (2-deoxyglucose) may be of help in improving the treatment outcome in patients with urothelial carcinoma [48]. Blockage of prosurvival autophagy by TGF-β inhibition in bone marrow fibroblasts circumvents bortezomib resistance in patients with multiple myeloma. Therefore, a combinational treatment of bortezomib and TGF-β inhibitor may provide the basis for a novel targeted therapeutic approach [49]. Metabolic reprogramming by activation of glutaminolysis induces resistance to anti-NOTCH1 therapies in T-cell acute lymphoblastic leukemia (T-ALL). Suppression of both glutaminolysis and autophagy synergistically potentiate the antileukemic effects of anti-

NOTCH1 therapy in mice harboring T-ALL, suggesting glutaminolysis as a novel therapeutic target for the treatment of T-ALL [50].

5. Conclusion

This chapter, with the main focus on the role of autophagy in multidrug-resistant cancers, has demonstrated the controversial role of autophagy in cellular mechanisms of defense against cancers and immunochemical homeostasis. Autophagy, a cellular process responsible for the turnover of misfolded proteins or damaged organelles, is important key mechanism for recycling of nutrients to maintain normal cellular homeostasis, DNA replication and genomic integrity. This prevents the overproduction of metabolic stress that contributes to damage of cellular proteins, organelles or DNAs, which finally lead to genomic damage and tumor progression in cells. This conception is supported by our listed evidence of increasing number of novel autophagic small molecules that could circumvent the MDR phenotypes in cancers via cellular regulation of autophagy. This chapter has therefore given a comprehensive summary on the role of autophagy or autophagic enhancers/ inhibitors in the cellular mechanisms of defence and homeostasis in cancers.

Author details

Betty Law Yuen Kwan and Vincent Wong Kam Wai*

*Address all correspondence to: kawwong@must.edu.mo

State Key Laboratory of Quality Research in Chinese Medicine, Macau University of Science and Technology, Macau, China

References

[1] Holohan C, Van Schaeybroeck S, Longley DB, Johnston PG. Cancer drug resistance: an evolving paradigm. Nat Rev Cancer. 2013;13:714–726. DOI: 10.1038/nrc3599

[2] Mathew R, Karantza-Wadsworth V, White E. Role of autophagy in cancer. Nat Rev Cancer. 2007; 7(12):961–7. DOI: 10.1038/nrc2254

[3] Kumar P, Zhang DM, Degenhardt K, Chen ZS. Autophagy and transporter-based multi-drug resistance.Cells. 2012; 1(3):558–75. DOI: 10.3390/cells1030558

[4] Van Delft MF, Wei AH, Mason KD, Vandenberg CJ, Chen L, Czabotar PE, Willis SN, Scott CL, Day CL, Cory S, Adams JM, Roberts AW, Huang DC. The BH3 mimetic

ABT-737 targets selective Bcl-2 proteins and efficiently induces apoptosis via Bak/Bax if Mcl-1 is neutralized. Cancer Cell. 2006; 10(5):389–99. DOI: 10.1016/j.ccr.2006.08.027

[5] Basit F, Cristofanon S, Fulda S. Obatoclax (GX15-070) triggers necroptosis by promoting the assembly of the necrosome on autophagosomal membranes. Cell Death Differ. 2013; 20(9):1161–73. DOI: 10.1038/cdd.2013.45. Epub 2013 Jun 7

[6] Yang ZJ, Chee CE, Huang S, Sinicrope FA. The role of autophagy in cancer: therapeutic implications. Mol Cancer Ther. 2011; 10(9):1533–41. DOI: 10.1158/1535-7163.MCT-11-0047

[7] Mathew R. Role of autophagy in cancer. Nat Rev Cancer. 2007; 7(12):961–7. DOI: 10.1038/nrc2254.

[8] Gammoh N, Lam D, Puente C, Ganley I, Marks PA, Jiang X. Role of autophagy in histone deacetylase inhibitor-induced apoptotic and nonapoptotic cell death. Proc Natl Acad Sci U S A. 2012; 109(17):6561–5. DOI: 10.1073/pnas.1204429109

[9] Ahn JH, Jang GH, Lee M. Defective autophagy in multidrug resistant cells may lead to growth inhibition by BH3-mimetic gossypol. J Cell Physiol. 2013; 228(7):1496–505. DOI: 10.1002/jcp.24305

[10] Sun WL, Lan D, Gan TQ, Cai ZW. Autophagy facilitates multidrug resistance development through inhibition of apoptosis in breast cancer cells. Neoplasma. 2015; 62(2): 199–208. DOI: 10.4149/neo_2015_025

[11] Rustum YM. Thymidylate synthase: a critical target in cancer. Front Biosci. 2004; 9:2467–73.

[12] Hassan M, Watari H, AbuAlmaaty A, Ohba Y, Sakuragi N. Apoptosis and molecular targeting therapy in cancer. BioMed Res Int. 2014; 2014(150845):23. DOI: 10.1155/2014/150845

[13] Kondo Y, Kanzawa T, Sawaya R, Kondo S. The role of autophagy in cancer development and response to therapy. Nat Rev Cancer. 2005;5(9):726–34. DOI: 10.1038/nrc1692

[14] Maiuri MC, Zalckvar E, Kimchi A, Kroemer G. Self-eating and self-killing: crosstalk between autophagy and apoptosis. Nat Rev Mol Cell Biol. 2007;8(9):741–52. DOI: 10.1038/nrm2239

[15] Lakhani SA, Masud A, Kuida K, Porter GA Jr, Booth CJ, Mehal WZ, Inayat I, Flavell RA. Caspases 3 and 7: key mediators of mitochondrial events of apoptosis. Science. 2006;311(5762):847–51. DOI: 10.1126/science.1115035

[16] Friesen C, Uhl M, Pannicke U, Schwarz K, Miltner E, Debatin KM. DNA-ligase IV and DNA-protein kinase play a critical role in deficient caspases activation in apoptosis-resistant cancer cells by using doxorubicin. Mol Biol Cell. 2008;19(8):3283–9. DOI: 10.1091/mbc.E08-03-0306

[17] Zhao Huang, Li Zhou, Zhibin Chen, Edouard C. Nice,Canhua Huang. Stress management by autophagy: Implications for chemoresistance. 2016;139:23-32. DOI: 10.1002/ijc.29990

[18] Alva AS, Gultekin SH, Baehrecke EH. Autophagy in human tumors: cell survival or death? Cell Death Differ. 2004; 11(9):1046–8. DOI: 10.1038/sj.cdd.4401445

[19] Wong VK, Li T, Law BY, Ma ED, Yip NC, Michelangeli F, Law CK, Zhang MM, Lam KY, Chan PL, Liu L. Saikosaponin-d, a novel SERCA inhibitor, induces autophagic cell death in apoptosis-defective cells. Cell Death Dis. 2013; 4:e720. DOI: 10.1038/cddis.2013.217

[20] Betty Yuen Kwan Law, Wai Kit Chan, Su Wei Xu, Jing Rong Wang, Li Ping Bai, Liang Liu, Vincent Kam Wai Wong. Natural small-molecule enhancers of autophagy induce autophagic cell death in apoptosis-defective cells. Scientific Reports. 2014; 4:5510. DOI: 10.1038/srep05510

[21] Law BY, Mok SW, Chan WK, Xu SW, Wu AG, Yao XJ, Wang JR, Liu L, Wong VK. Hernandezine, a novel AMPK activator induces autophagic cell death in drug-resistant cancers. 2016; 7(7):8090-8104. DOI: 10.18632/oncotarget.6980

[22] Xavier CP, Lima CF, Pedro DF, Wilson JM, Kristiansen K, Pereira-Wilson C. Ursolic acid induces cell death and modulates autophagy through JNK pathway in apoptosis-resistant colorectal cancer cells. J Nutr Biochem. 2013; 24(4):706–12. DOI: 10.1016/j.jnutbio.2012.04.004

[23] Torricelli C, Salvadori S, Valacchi G, Souček K, Slabáková E, Muscettola M, Volpi N, Maioli E. Alternative Pathways of Cancer Cell Death by Rottlerin: Apoptosis versus Autophagy. 2012;2012:1-11.980658. DOI: 10.1155/2012/980658

[24] Hau AM, Greenwood JA, Löhr CV, Serrill JD, Proteau PJ, Ganley IG, McPhail KL, Ishmael JE. Coibamide A induces mTOR-independent autophagy and cell death in human glioblastoma cells. PLoS One. 2013; 8(6):e65250. DOI: 10.1371/journal.pone.0065250

[25] Xu J, Xu X, Shi S, Wang Q, Saxton B, He W, Gou X, Jang JH, Nyunoya T, Wang X, Xing C, Zhang L, Lin Y. Autophagy-mediated degradation of IAPs and c-FLIPL potentiates apoptosis induced by combination of TRAIL and Chal-24. J Cell Biochem. 2016; 117(5):1136–44. DOI: 10.1002/jcb.25397

[26] Zhang L, Zhang W, Wang YF, Liu B, Zhang WF, Zhao YF, Kulkarni AB, Sun ZJ. Dual induction of apoptotic and autophagic cell death by targeting survivin in head neck squamous cell carcinoma. Cell Death Dis. 2015; 6:e1771. DOI: 10.1038/cddis.2015.139

[27] Shteingauz A, Boyango I, Naroditsky I, Hammond E, Gruber M, Doweck I, Ilan N, Vlodavsky I. Heparanase enhances tumor growth and chemoresistance by promoting autophagy. Cancer Res. 2015; 75(18):3946–57. DOI: 10.1158/0008-5472.CAN-15-0037

[28] Chakravarty G, Mathur A, Mallade P, Gerlach S, Willis J, Datta A, Srivastav S, Abdel-Mageed AB, Mondal D. Nelfinavir targets multiple drug resistance mechanisms to

increase the efficacy of doxorubicin in MCF-7/Dox breast cancer cells. 2016 ;124:53-64. DOI: 10.1016/j.biochi.2016.01.014

[29] Ma Q, Chang Z, Wang W, Wang B. Rapamycin-mediated mTOR inhibition reverses drug resistance to adriamycin in colon cancer cells. Hepatogastroenterology. 2015; 62(140):880–6. DOI: PMID:26902021

[30] Yu M, Lee C, Wang M, Tannock IF. Influence of the proton pump inhibitor lansoprazole on distribution and activity of doxorubicin in solid tumors. Cancer Sci. 2015; 106(10): 1438–47. DOI: 10.1111/cas.12756

[31] Zajac J, Kostrhunova H, Novohradsky V, Vrana O, Raveendran R, Gibson D, Kasparkova J, Brabec V. Potentiation of mitochondrial dysfunction in tumor cells by conjugates of metabolic modulator dichloroacetate with a Pt(IV) derivative of oxaliplatin. J Inorg Biochem. 2016; 156:89–97. DOI: 10.1016/j.jinorgbio.2015.12.003

[32] Zeng LP, Hu ZM, Li K, Xia K. miR-222 attenuates cisplatin-induced cell death by targeting the PPP2R2A/Akt/mTOR Axis in bladder cancer cells. J Cell Mol Med. 2016; 20(3):559–67. DOI: 10.1111/jcmm.12760

[33] Leisching GR, Loos B, Botha MH, Engelbrecht AM.The role of mTOR during cisplatin treatment in an in vitro and ex vivo model of cervical cancer. Toxicology. 2015; 335:72–8. DOI: 10.1016/j.tox.2015.07.010

[34] Lim SC, Han SI. Ursodeoxycholic acid effectively kills drug-resistant gastric cancer cells through induction of autophagic death. Oncol Rep. 2015; 34(3):1261–8. DOI: 10.3892/or.2015.4076

[35] Dyshlovoy SA, Hauschild J, Amann K, Tabakmakher KM, Venz S, Walther R, Guzii AG, Makarieva TN, Shubina LK, Fedorov SN, Stonik VA, Bokemeyer C, Balabanov S, Honecker F, von Amsberg G. Marine alkaloid Monanchocidin a overcomes drug resistance by induction of autophagy and lysosomal membrane permeabilization. Oncotarget. 2015; 6(19):17328–41. DOI: 10.18632/oncotarget.4175

[36] Fan XX, Yao XJ, Xu SW, Wong VK, He JX, Ding J, Xue WW, Mujtaba T, Michelangeli F, Huang M, Huang J, Xiao DK, Jiang ZB, Zhou YL, Kam RK, Liu L, Leung EL. (Z)3,4,5,4′-trans-tetramethoxystilbene, a new analogue of resveratrol, inhibits gefitinb-resistant non-small cell lung cancer via selectively elevating intracellular calcium level. Sci Rep. 2015; 5:16348. DOI: 10.1038/srep16348

[37] Zhao L, Yang G, Shi Y, Su C, Chang J. Co-delivery of Gefitinib and chloroquine by chitosan nanoparticles for overcoming the drug acquired resistance. J Nanobiotechnology. 2015;13:57. DOI: 10.1186/s12951-015-0121-5

[38] Chen S, Zhu X, Qiao H, Ye M, Lai X, Yu S, Ding L, Wen A, Zhang J. Protective autophagy promotes the resistance of HER2-positive breast cancer cells to lapatinib. 2016;37(2): 2321-31. DOI: doi: 10.1007/s13277-015-3800-9

[39] Giuliano S, Cormerais Y, Dufies M, Grépin R, Colosetti P, Belaid A, Parola J, Martin A, Lacas-Gervais S, Mazure NM, Benhida R, Auberger P, Mograbi B, Pagès G. Resistance

to sunitinib in renal clear cell carcinoma results from sequestration in lysosomes and inhibition of the autophagic flux. Autophagy. 2015; 11(10):1891–904. DOI: 10.1080/15548627.2015.1085742

[40] Jagannathan S, Abdel-Malek MA, Malek E, Vad N, Latif T, Anderson KC, Driscoll JJ. Pharmacologic screens reveal metformin that suppresses GRP78-dependent autophagy to enhance the anti-myeloma effect of bortezomib. Leukemia. 2015; 29(11):2184–91. DOI: 10.1038/leu.2015.157

[41] Cao L, Wang J, Ma D, Wang P, Zhang Y, Fang Q. Heme oxygenase-1 contributes to imatinib resistance by promoting autophagy in chronic myeloid leukemia through disrupting the mTOR signaling pathway. Biomed Pharmacother. 2016; 78:30–8. DOI: 10.1016/j.biopha.2015.12.029

[42] Tan Q, Wang H, Hu Y, Hu M, Li X, Aodengqimuge, Ma Y, Wei C, Song L. Src/STAT3-dependent heme oxygenase-1 induction mediates chemoresistance of breast cancer cells to doxorubicin by promoting autophagy. Cancer Sci. 2015; 106(8):1023–32. DOI: 10.1111/cas.12712

[43] Su YC, Davuluri GV, Chen CH, Shiau DC, Chen CC, Chen CL, Lin YS, Chang CP. Galectin-1-induced autophagy facilitates cisplatin resistance of hepatocellular carcinoma. PLoS One. 2016; 11(2):e0148408. DOI: 10.1371/journal.pone.0148408. eCollection 2016

[44] Shao LN, Zhu BS, Xing CG, Yang XD, Young W, Cao JP. Effects of autophagy regulation of tumor-associated macrophages on radiosensitivity of colorectal cancer cells. 2016;13(3):2661-70. DOI: 10.3892/mmr.2016.4820

[45] Goulielmaki M, Koustas E, Moysidou E, Vlassi M, Sasazuki T, Shirasawa S, Zografos G, Oikonomou E, Pintzas A. BRAF associated autophagy exploitation: BRAF and autophagy inhibitors synergise to efficiently overcome resistance of BRAF mutant colorectal cancer cells. 2016;7(8):9188-221. DOI: 10.18632/oncotarget.6942

[46] Lv L, Liu HG, Dong SY, Yang F, Wang QX, Guo GL, Pan YF, Zhang XH. Upregulation of CD44v6 contributes to acquired chemoresistance via the modulation of autophagy in colon cancer SW480 cells. 2016; Jan 9 :1-14. DOI: 10.1007/s13277-015-4755-6

[47] Peng WX, Xiong EM, Ge L, Wan YY, Zhang CL, Du FY, Xu M, Bhat RA, Jin J, Gong AH. Egr-1 promotes hypoxia-induced autophagy to enhance chemo-resistance of hepatocellular carcinoma cells. Exp Cell Res. 2016; 340(1):62–70. DOI: 10.1016/j.yexcr.2015.12.006

[48] Ojha R, Jha V, Singh SK. Gemcitabine and mitomycin induced autophagy regulates cancer stem cell pool in urothelial carcinoma cells. Biochim Biophys Acta. 2016; 1863(2):347–59. DOI: 10.1016/j.bbamcr.2015.12.002

[49] Frassanito MA, De Veirman K, Desantis V, Marzo LD, Vergara D, Ruggieri S, Annese T, Nico B, Menu E, Catacchio I, Ria R, Racanelli V, Maffia M, Angelucci E, Derudas D, Fumarulo R, Dammacco F, Ribatti D, Vanderkerken K, Vacca A. Halting pro-survival

autophagy by TGFβ inhibition in bone marrow fibroblasts overcomes bortezomib resistance in multiple myeloma patients. Leukemia. 2016; 30(3):640–8. DOI: 10.1038/leu.2015.289

[50] Herranz D, Ambesi-Impiombato A, Sudderth J, Sánchez-Martín M, Belver L, Tosello V, Xu L, Wendorff AA, Castillo M, Haydu JE, Márquez J, Matés JM, Kung AL, Rayport S, Cordon-Cardo C, DeBerardinis RJ, Ferrando AA. Metabolic reprogramming induces resistance to anti-NOTCH1 therapies in T cell acute lymphoblastic leukemia. Nat Med. 2015; 21(10):1182–9. DOI: 10.1038/nm.3955

Chapter 21

Autophagy in Non-Alcoholic Fatty Liver Disease (NAFLD)

Wilhelmus J. Kwanten , Wim Martinet and
Sven M. Francque

Additional information is available at the end of the chapter

http://dx.doi.org/10.5772/64534

Abstract

Autophagy is a mechanism involved in cellular homeostasis under basal and stressed conditions delivering cytoplasmic content to the lysosomes for degradation to macronutrients. The potential role of autophagy in disease is increasingly recognised and investigated. To date, a key role of autophagy in hepatic lipid metabolism is recognised and dysfunctional autophagy might be an underlying cause of non-alcoholic fatty liver disease (NAFLD). Nevertheless, the exact role of autophagy in lipid metabolism remains controversial, with both a lipolytic function of autophagy and lipogenic function reported. This chapter aims to review the current knowledge on autophagy in NAFLD, with a special focus on its role in hepatic lipid metabolism, hepatic glucose metabolism and insulin resistance, steatohepatitis, hepatocellular injury and hepatic fibrogenesis. Finally, interaction with another cellular homeostatic process, the unfolded protein response (UPR), will be briefly discussed.

Keywords: autophagy, non-alcoholic fatty liver disease (NAFLD), non-alcohlic steatohepatitis (NASH), lipid metabolism, glucose metabolism, insulin resistance, fibrogenesis, hepatocellular carcinoma

1. Introduction

The term autophagy has been introduced by de Duve *et al.* over 40 years ago [1] to define a process of vacuolisation for the transport of intracellular material to lysosomes for degradation. Because the importance of autophagy in (patho)physiology became more and more recognised, the knowledge and number of autophagy-related publications increased exponentially in the last decade. Indeed, autophagy is progressively acknowledged as an important regulator of intracellular homeostasis. Dysfunction of this process has been linked with cardiovascular, respiratory, neurodegenerative and metabolic diseases and with cancer [2, 3].

A growing body of evidence indicates that autophagy and lipid metabolism are correlated. Dysfunctional autophagy may therefore contribute to the pathogenesis of non-alcoholic fatty liver disease (NAFLD). However, controversies still exist and the exact role of autophagy in hepatic lipid metabolism is not entirely elucidated yet. This chapter aims to give a brief introduction about NAFLD and autophagy and subsequently reviews the current knowledge on autophagy in NAFLD.

2. Non-alcoholic fatty liver disease (NAFLD)

Even though histological features of NAFLD were recognised for decades, the first formal definition was introduced in 1980 to describe a small cohort with striking fatty changes in the liver with lobular hepatitis and focal necrosis, termed non-alcoholic steatohepatitis (NASH) [4]. Since then the concept has evolved to the definition of NAFLD, which covers a spectrum of fatty liver without evidence for any secondary cause of hepatic fatty liver accumulation, such as alcohol consumption or inherited disorders [5]. NAFLD is epidemiologically associated with the metabolic syndrome that encompasses obesity, diabetes mellitus, arterial hypertension and dyslipidaemia [6].

The hallmark of NAFLD is macrovesicular fat accumulation in more than 5% of the hepatocytes, ranging from scarce to panacinar steatosis and usually starting in the acinar zone 3 [7]. It is important to discriminate non-alcoholic fatty liver (NAFL, also known as simple steatosis) from non-alcoholic steatohepatitis (NASH). In the latter, not only macrovesicular steatosis is present, but also hepatocellular ballooning and lobular inflammation [5, 7] (**Figure 1**).

Simple steatosis is currently still considered as relatively innocent, as it has a slow evolution to advanced disease (though a subgroup of fast progressors was identified [8]) and in the absence of fibrosis mortality does not seem to be increased [9]. Potential consequences might therefore mainly be confined to the operative setting [10].

Once NASH has been established, patients are subjected to an increased risk of hepatic and non-hepatic comorbidities and mortality (**Figure 1**) [6, 11, 12]. NAFLD is an independent risk factor for cardiovascular disease, and cardiovascular events are the most important cause of death [12, 13]. Of all the features of advanced disease, fibrosis appears to be the most predictive for NAFLD-related outcomes [9, 14].

Since liver biopsy is the current gold standard for diagnosis but cannot be used routinely for population-based studies, the current prevalence of NAFLD can only be estimated based on a range of non-invasive diagnostic methods and/or in highly selected patient groups. The overall prevalence in the Western societies is estimated 20–30% for NAFLD and 3–5% for NASH. The incidence has substantially increased in the last decades and is expected to rise further [6, 16]. By 2025, NAFLD-related liver disease is anticipated to become the most important indication for liver transplantation in the United States of America [17].

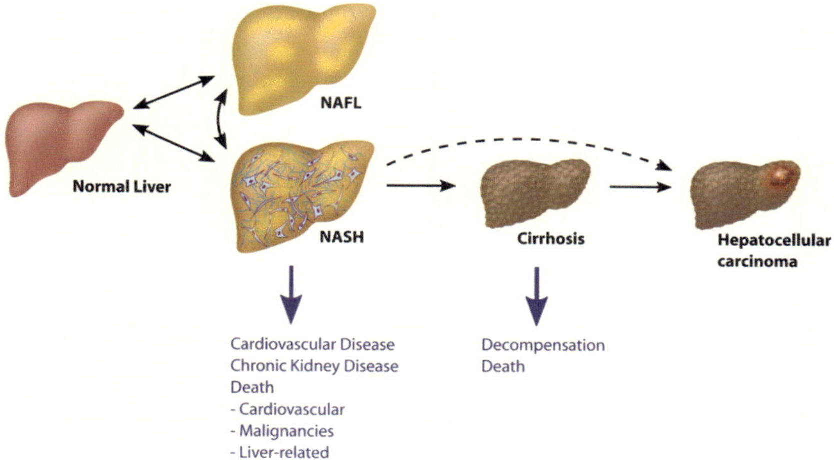

Figure 1. Non-alcoholic liver disease (NAFLD) spectrum. NAFLD encompasses a spectrum of fatty liver disease ranging from non-alcoholic fatty liver (NAFL, also simple steatosis) to non-alcoholic steatohepatitis (NASH). The latter might further evolve to advanced fibrosis and eventually cirrhosis that can be complicated by hepatocellular carcinoma (HCC). HCC can also develop outside the setting of cirrhosis. NASH is also associated with increased hepatic and non-hepatic comorbidities and mortality (adapted from Cohen *et al.* [15] and Torres *et al.* [11]).

With the growing importance of NAFLD, research on this topic further increased, which led to the formation of major research consortia in the United States and Europe [18]. Nevertheless, there are still major knowledge gaps concerning the exact pathophysiology behind the development and progression of NAFLD. It is generally believed that NAFLD is a very dynamic and multifactorial disease, in which different 'hits' contribute simultaneously and/or sequentially to the pathogenesis of NAFLD [19, 20]. Autophagy gained interest recently as one of those potential hits.

3. Autophagy

The term autophagy is derived from the Greek language and literally means "self-eating". Damaged or dysfunctional cellular contents are continuously removed via basally active autophagy to conserve cellular homeostasis and to supply the cell with substrates for energy production, though autophagy can be further stimulated via oxidative or metabolic stress [21, 22].

In mammalian cells, three types of autophagy are described: macroautophagy, chaperone-mediated autophagy (CMA) and microautophagy [2, 23, 24]. Microautophagy describes the direct engulfment of a small portion of cytoplasm by the lysosome. When proteins containing a special targeting motif are recognised by heat-shock cognate protein 70 (HSC70) and its cochaperones, they are selectively delivered to the lysosome via CMA. In macroautophagy, cytoplasmic material is either non-selectively or selectively (e.g. 'mitophagy', selective autophagy of mitochondria) [2, 3, 24, 25] sequestrated in a double membrane structure, the autophagosome. This process starts with the formation of an isolation membrane (also known as phagophore), which will lengthen to create an autophagosome. Autophagosomes fuse with a lysosome, after which the sequestrated content will be degraded.

Macroautophagy (henceforth autophagy) is generally considered to play the most important role in the (patho)physiology and is extensively studied in the last decades. The process of autophagy is dynamic and strictly regulated, with control mechanisms at the transcriptional and post-transcriptional level [26]. It is regulated at the molecular level by autophagy-related (*Atg*) genes and their products, which form the core machinery of autophagy [27]. Paramount in the regulation of autophagy is the mammalian target of rapamycin (mTOR) [28]. The initiation of autophagosome formation by phosphorylating UNC51-like kinase 1 (ULK1) is inhibited by mTOR. In response to growth factors (e.g. insulin), mTOR gets stimulated by the class I phosphatidylinositol 3-kinase (PI3K)/AKT pathway. However, in case of starvation, the AMP/ATP ratio increases and leads to adenosine 5′-monophosphate-activated protein kinase (AMPK) activation and consecutive mTOR inhibition and thus activation of autophagy [27, 28]. The nucleation of the phagophore is mediated by a beclin-1/VSP34 (a class III PI3K)-interacting complex [27, 28]. The elongation of the phagophore to form an autophagosome is performed by two ubiquitin-like conjugated complexes: the ATG5-ATG12-ATG16L1 complex and light chain 3 (LC3). The ATG7 protein (an E1-like protein) is needed to mediate the conjugation of both complexes and is an interesting target for the study of autophagy [29]. The active conjugated form of LC3, LC3-II, is frequently used as a marker for autophagy [30]. For further information on autophagy regulation, we refer to previously published reviews [24, 26, 27].

4. Autophagy in lipid metabolism

Autophagy was convincingly correlated to lipid metabolism for the first time by Singh and his colleagues [31] and considered as a novel selective pathway in lipid breakdown known as 'lipophagy'. Others claimed in the same year that autophagy was indispensible for the genesis of lipid droplets (LDs) rather than for the breakdown of LDs [32]. Ever since, supporting evidence for both lipid breakdown and lipogenesis has been published. After describing some common findings, both the opposing theories and contextual variations of autophagy in lipid metabolism will be discussed.

4.1. Common findings in autophagy and lipid metabolism

In spite of the opposing views present in current literature, some common findings supporting the relationship between autophagy and lipid metabolism in the liver deserve to be mentioned.

First, a close association between LDs and LC3, as well as between LDs and lysosomes, has been demonstrated. As demonstrated by immunofluorescence microscopy, LC3-positive structures and markers of LDs colocalise in liver tissue [32] and in cell lines [33–35]. Increased colocalisation of LDs with lysosomal markers such as lysosomal-associated membrane protein 1 (LAMP1) [31] or lysotracker [35] in fat-loaded cells was also demonstrated by immunofluorescence microscopy. Similarly, immunohistochemical LC3B-positive dots were localised on the surface of LDs [36]. Immunogold staining of LC3 on transmission electron microscopy (TEM) slides confirmed the colocalisation of LC3 with LDs and implies a LD-regulating function of autophagy [31–33].

The colocalisation of LC3 with LDs was not influenced by inhibition of autophagosome–lysosome fusion or knockout of autophagy. However, colocalisation of LDs with lysosomes decreased after inhibition of autophagosome formation or by knockdown of autophagy [31]. These findings suggest that processing of LC3 into LC3-II (the active form) not only occurs on autophagosomes but also on the surface of LDs [31].

Secondly, in parallel with the histological pattern of NAFLD [7], immunohistochemical staining of LC3 is more localised in acinar zone 3 (i.e. around the central veins) [36, 37]. Findings in glutamine metabolism postulate a theoretic zonal distribution of autophagy [38]. In this view, low rates of autophagy in the periportal areas and constitutively high levels in the pericentrally areas are assumed in case of well-nourished conditions. Accordingly, it might be a potential explanation for the pattern found in NAFLD.

4.2. Autophagy as a lipolytic mechanism

The liver is capable of mobilising free fatty acids (FFAs) rapidly when needed. Autophagy as contributing factor to lipolysis is hence an attractive theory, as it helps explaining this capability while hepatocytes have relatively low concentrations of cytosolic lipases [39].

When hepatocytes were cultured in the presence of lipid stimuli, hepatocyte triglyceride (TG) levels increase and LDs accumulate. The pharmacological inhibition or knockdown of autophagy (targeting ATG5) enhanced these findings [31]. It was shown that impaired lipolysis (fuelling β-oxidation) and not increased TG synthesis were responsible for these findings. When autophagy was pharmacologically induced, the opposite happened with decreased lipid stores in hepatocytes. Hepatocellular-specific, autophagy-deficient mice (targeting *Atg7*) confirmed these *in vitro* results. Indeed, compared with wild-type littermates, liver TG and cholesterol content increased [31]. Fasting-induced steatosis, which can be observed after 24 h starvation, was less pronounced in wild-type mice compared with autophagy-deficient littermates. Moreover, lysosomes and lipid-containing autophagosomes increased after fasting in the autophagy-competent mice, supporting lipolysis [31].

The oxidation of FFA and the production of very low-density lipoprotein (VLDL) appeared to be dependent on autophagy. Inhibition of autophagy decreases both FFA oxidation and VLDL production, while stimulation induces the opposite [40, 41]. The distribution of lysosomal lipases (LAL) changed towards the autophagosome fraction after starvation in rat liver and supports an increase in autophagy-mediated lysosomal lipolysis [40].

It has been shown that dietary-induced obesity induces decreased autophagy flux [42, 43]. In mice with dietary-induced obesity as well as in genetically induced obesity, ATG7 protein levels were reduced (although the mRNA expression was comparable) [42]. Autophagy induction via liver-specific overexpression of ATG7 in ob/ob mice [42] or via calcium channel blockers [43] restored autophagy flux, improved the metabolic state and reduced steatosis significantly. These findings further support a lipolytic function of autophagy.

Trafficking of autophagosomes and lysosomes, as well as their interaction, is just modestly understood. The Rab guanosine triphosphatases (GTPases) serve as master regulators of intracellular membrane traffic [44] and might be involved in regulation of lipophagy as well. Indeed, Rab7 is a fundamental component of both LDs and endolysosomal membranes and a central regulator for LD breakdown by autophagy [45]. Dynamin 2, another GTPase, is also involved in maintenance of lysosomal homeostasis by recycling of autophagosomes. Ablation of Dynamin 2 compromised the autolysosomal compartment, with subsequent depletion of lysosomes, and inhibited lipophagy [46].

Steatosis and dyslipidaemia are linked to defects in forkhead box class O (FOXO) [47], and liver-specific triple knockout of FOXO1/3/4 (LTKO) causes steatosis and hypertriglyceridemia [48]. Regulation of the key autophagy genes mediated by FOXO1 was demonstrated [49]. The autophagy gene *Atg14* is regulated by FOXO1 and 3. Hepatic and serum TG increase after knockdown of hepatic ATG14, whereas overexpression decreases steatosis in HFD fed animals. Overexpression of *Atg14* in LTKO mice could counteract the observed lipid disturbances including steatosis [48]. Surprisingly, an increase instead of a decrease in FOXO1 levels was described in a small cohort of NASH patients [50].

The longer-term transcriptional regulation of autophagy becomes further unravelled [51] and seems to be in favour of lipophagy. The transcription factor EB (TFEB) appears to be a master regulator of autophagy [52] and is involved in lipid metabolism as well. Steatosis is induced when TFEB is suppressed, while steatosis is inhibited when TFEB is overexpressed [53]. Both autophagy and the stimulation of the peroxisome proliferator-activated receptor γ coactivator 1 α–peroxisome proliferator-activated receptor α (pgc-1α-PPARα) pathway mediated the observed effects. The dependency of TFEB function on autophagy mechanisms was demonstrated by the inability to counteract steatosis caused by disruption of autophagy [53]. The cyclic adenosine monophosphate (cAMP) response element-binding protein (CREB) is able to promote lipophagy in the fasted state via activation of TFEB and transcription of autophagy genes, while the farnesoid X receptor (FXR) suppresses many autophagy genes in fed conditions [54]. Likewise, starvation-induced activation of PPARα impedes the inhibitory effects of FXR on autophagy [55].

It is well known that steatosis can be the consequence of some HIV antiretroviral drugs. Thymidine analogues were able to inhibit the autophagic flux of hepatocytes *in vitro* in a dose-dependent manner, with subsequent induction of lipid accumulation and mitochondrial dysfunction [56]. Even though currently only used as clinical treatment in trial [57], glucagon like peptide-1 (GLP-1) analogues were able to reduce fat accumulation *in vitro* and *in vivo* by the activation of autophagy [58–60], and by the reduction in endoplasmic reticulum (ER) stress [58]. Moreover, Roux-en-Y gastric bypass in obese diabetic rats was able to improve metabolic parameters and to restore hepatic autophagy and was correlated with increased plasma GLP-1 levels [61]. Carbamazepine and rapamycin induce autophagy and were effective in reducing steatosis in models of alcoholic and non-alcoholic fatty liver disease [62]. Also caffeine, of which epidemiological data suggest a protective effect on NAFLD, was shown to induce autophagy dose dependently with increased lipid clearance [35, 63]. Finally, the antioxidants resveratrol [64, 65] and tert-butylhydroquinone [66], but also metformin [67], were able to attenuate hepatic steatosis by inducing autophagy via activation of AMPK/Sirtuin-1. Other possible mechanisms parallel to changes in autophagy that might explain the observed effects are formally not excluded, as illustrated by the reduced expression of genes related to ER stress and inflammation and increased expression of genes involved in lipid oxidation in zebrafish exposed to caffeine [63]. However, given the alike effects of different compounds, involvement of autophagy in lipolysis is at least partially feasible.

Very recently, the role of autophagy appeared to be even more complex, since CMA emerged to control lipid homeostasis. Blockade of CMA induces severe hepatosteatosis, partially explained by defective breakdown of key enzymes involved in lipid binding, transport and synthesis [68]. Moreover, CMA is required for breakdown of LDs. LDs are covered with perilipins (PLINs), which regulate the accessibility of the LD for lipases. PLIN2 and PLIN3 are substrates of CMA, and upon fasting, they are removed of the LD surface in a CMA-mediated fashion. Afterwards, the lipid droplets are accessible for both cytosolic neutral lipases and for autophagy-mediated lipolysis. When CMA is blocked, lipid oxidation decreases and LDs accumulate [69].

Even though the exact relevance of microautophagy in normal cell physiology remains largely unknown at present, microautophagy might have a role in lipid breakdown. In a recent study, yeast cells were capable of translocating LDs in lysosomes by a process morphologically resembling microautophagy and independent of core autophagy proteins [70]. Whether this also holds true for mammalian cells remains to be proven.

Experimental restrictions, for example the impossibility to use specific drugs or to perform consecutive biopsies, limit the availability and interpretation of human data. Data of liver biopsy represent a snapshot of a very dynamic process and cannot accurately discern between increased autophagy and decreased degradation of autophagosomes [30, 36]. Furthermore, some markers need to be overexpressed for accurate identification of autophagic structures [36]. Nonetheless, a small post-mortem study demonstrated decreased LC3 and increased p62 staining in relation with the degree of steatosis, suggesting decreased autophagy in more severe steatosis [37]. Likewise, in proven NAFLD p62 accumulation, increased numbers of autophagic vesicles were demonstrated [71]. mRNA and protein analysis of liver biopsies were also indicative of an impaired autophagic flux in both NAFL and NASH patients [72]. A more in-depth analysis with gene set enrichment analysis of liver biopsies demonstrated that NASH has distinct patterns, compared with normal livers or NAFL. In these livers, the gene categories for apoptosis and autophagy were enriched for upregulated genes, while the gene categories for ER stress and lipogenesis were enriched for downregulated genes. In NAFL, no genes were significantly enriched except for an enrichment for upregulated genes related to autophagy [73]. This correlation can either indicate upregulation in an attempt to attenuate lipotoxicity and to increase lipolysis or indicate the direct involvement of autophagy in fat accumulation in the liver.

Finally, two clinical observations need to be mentioned. Hypothyroidism is more prevalent in patients with NAFLD [74–76] as well is hypovitaminosis D [77]. The thyroid hormone (T_3) is a known regulator of the basal metabolism and recently shown to be a powerful inducer of autophagy *in vitro* and *in vivo*. Autophagy plays a crucial role in T_3-stimulated β-oxidation [34]. Vitamin D acts also as a potent inducer of autophagy [28]. As a result, these associations might be explained via autophagy and are in line with lipophagy. A direct effect via autophagy, however, has hitherto not been investigated.

4.3. Autophagy as a lipogenic mechanism

Caenorhabditis elegans is a nematode increasingly used in metabolic research. Its intestine fulfils the role of a multifunctional organ reflecting the roles of the liver and adipose tissue [78]. Inhibition of autophagy via knockdown of several different genes involved in the autophagy process results in a strong reduction in lipid content. Importantly, there were no arguments for altered food uptake or defecation, nor for influenced differentiation of the tissue in case of autophagy deficiency [79].

The body possesses an adaptive mechanism to maintain homeostasis in case of fasting. The declined insulin levels no longer inhibit lipolysis in the adipose tissue (AT) and release FFA to the serum. The liver captures FFA, either for the formation of ketone bodies, or for temporary storage as TG in LDs [32, 80]. The latter can be seen with ^{1}H-magnetic resonance spectroscopy after 36-h fasting [81]. A substantial accumulation of TG in the liver, consequent to this mechanism, can be seen in rodents and is known as fasting-induced steatosis. The C57Bl/6 mouse strain showed to be very prone to develop fasting-induced steatosis [80]. However, compared with wild-type littermates, hepatocyte-specific autophagy-deficient mice lack fasting-induced steatosis. The total TG content in their livers is lower, and the remaining LDs are decreased in size and numbers [32]. This observation was not only the case in very young mice (22 days old) but was also demonstrated in 8- to 12-week-old mice [32, 36, 82]. As a consequence, autophagy seems to be implicated in the formation and growth of LDs. The colocalisation of LC3 (necessary for autophagosome formation) with LDs in starved wild-type mice further supports these findings [32]. Consistent with the overall nutrient shortage by fasting, mice exposed to a one-week dietary protein deficiency develop hepatic steatosis, accompanied by autophagy and ER stress. Leucine supplement, a known autophagy inhibitor, lowered autophagy, ER stress and liver TG content [83]

In different cell lines, amongst which hepatocytes, the indispensability of autophagy for LD formation was confirmed [33]. The knockdown of LC3 in these cells leads to reduced formation of LDs and reduced TG content compared with their controls. Since FFA uptake, TG synthesis or TG breakdown are unaltered after knockdown of LC3, an impaired ability to preserve synthesised TG within these cells is suggested [33].

Hepatocyte- or skeletal muscle-specific autophagy-deficient mice exhibit an improved metabolic profile [84]. When hepatocellular-specific autophagy-deficient mice are fed a control diet [82, 84], the aforementioned fasting-induced steatosis did not occur. Moreover, when these mice were fed a HFD, lipid accumulation was absent [84] or did not increase [82]. The expression of genes involved in fatty acid and TG synthesis, but also of those involved in β-oxidation and TG secretion were reduced in comparison with autophagy-competent littermates [82, 84]. Therefore, it is not clear whether these findings are epiphenomena or directly involved in the prevention of steatosis. Kim and colleagues held the 'mitokine' fibroblast growth factor 21 (FGF21), which is induced by mitochondrial stress, responsible as a central mediator of the metabolic alterations [84].

Studies with dietary and genetic models of obesity report decreased levels of autophagy, while the overexpression of ATG7 had beneficial metabolic effects [42] as discussed before. Nevertheless, in the same paper the suppression of ATG7 in lean mice failed to alter lipid accumulation in the liver (as well as TG or FFA in serum), while hepatic glycogen content did show an increase [42]. This study may therefore be considered as non-conclusive about a lipolytic or lipogenic function of autophagy.

4.4. Contextual variability of autophagy in lipid metabolism

Besides the opposing views regarding autophagy as a lipolytic or lipogenic process, autophagy is also subjected to context-dependent alterations. In most of these cases, these differences were described by those who support autophagy as a lipolytic mechanism.

Lipid metabolism seems to be more dependent on basal autophagy than on induced autophagy, since (sudden) lipid stimuli did not reveal signs of induced autophagy or autophagic flux in cultured hepatocytes [31]. *In vivo* there is also impaired adjustment of autophagy, as external lipid load by prolonged HFD decreases autophagy efficiency [31, 41, 42, 58]. Intriguingly, a detailed follow-up of autophagy reveals fluctuating levels of autophagy over time. Increased autophagic flux was observed after 2 weeks [41] or 4 weeks [85] of HFD, and a decrease was observed after, respectively, 10 or 16 weeks of HFD. Other data (only published in abstract form) suggest that autophagy decreases after short-term HFD (3 days) and normalises after long-term HFD (10 weeks) [86]. Instead of a decrease, 8 weeks of a diet high in fat load generates an increase in autophagy [87]. Recently, autophagy was shown to behave dynamically with an oscillating damping pattern under HFD, probably the consequence of a feedback loop mechanism between mTORC1 and TFEB (X.M. Yin, personal communication at the AASLD 2015). Overall, autophagic flux seems to be very dynamic in case of overnutrition.

The detrimental effects of fatty acids on the cellular integrity are called lipotoxicity [88, 89]. Thus, it is not surprising that lipids by itself may have impact on autophagy. Autophagy can be induced by short chain fatty acids [90] and ω3-fatty acids (mTOR independently) *in vitro* [91]. Unsaturated fatty acids (e.g. oleic acid) stimulate autophagy and protect against apoptosis, while saturated fatty acids (e.g. palmitic acid) inhibit autophagy and promote apoptosis [58, 92, 93]. Tu *et al.* observed the opposite effects, with inhibition of autophagy by oleic acid and induction by palmitic acid [94]. These conflicting results might reflect differences in cell type, concentration and duration of FFA application, but more importantly emphasise the contextual variability of autophagy.

The impairment of autophagy in case of saturated fatty acids is considered to be due to a diminished fusion capacity of autophagosomes with lysosomes [43, 95]. Long exposure to high lipid concentrations alters the lipid composition of membranes or vesicular compartments and in this way impairs their fusion [95]. Another explanation includes an inhibitory effect of increased cytoplasmic calcium concentrations via inhibitory effects of SFA on the sarco-ER calcium pump (SERCA) [43]. This may explain the altered autophagy after prolonged fatty diets. Attenuation of CMA was also observed after lipid challenge [23]. However, some authors did not observe an attenuated fusion capacity. Instead, they report a decrease in clearance of autophagosomes due to a disturbed acidification of lysosomal compartments [96, 97] and/or downregulated cathepsin expression [71, 96].

Variation is not only the case within the liver, but is also dependent on tissue type. Adipogenesis and transdifferentiation towards white AT, for example, depend on autophagy [98, 99], thus arguing against a potential lipolytic function of autophagy in AT. This is opposite to the observations in liver tissue, which mainly claim a lipolytic function (as discussed above). Additionally, in the AT of patients with metabolic syndrome or type 2 diabetes mellitus, autophagy increased [100–102].

4.5. Discrepancies and hypotheses

The role of autophagy in the liver seems to be more complicated than expected. At present, a clear-cut explanation for the discrepancies of autophagy with respect to lipid metabolism as stated above is missing [103]. However, several hypotheses have been put forward.

Firstly, some concerns were related to the age of the laboratory animals [104], since autophagy declines with age [105]. Older mice might be less dependent on autophagy than juvenile mice. However, experiments with both younger and older mice provide comparable results [32, 36, 82, 84], implying that only age cannot explain the observed differences.

Secondly, *in vivo* experiments can be subject to small variances in the mouse strains used [84]. However, this issue is not likely to offer an explanation. Most of the experiments were performed on a C57Bl/6 background, which is an inbred strain. Moreover, manual backtracking the cited resources of the hepatocyte-specific *Atg7* knockout mice leads to the same origin of the mice. *Atg7* flox mice were created by Komatsu *and colleagues* [29], while the albumin-Cre mice were created by the group of Magnuson [106]. In addition, conflicting results were also seen in hepatocyte cell lines as described above.

Thirdly, lipid accumulation and autophagy can be examined by many different methods. Liver steatosis can be induced by fasting, and by genetic and/or dietary interventions. However, these genetic modifications or pharmacological approaches can also alter autophagy by themselves. Genetic modification can target different autophagy-related genes as well, which might be reflected in the results. Whereas papers supporting autophagy as lipophagy use a wide range of methods (see above), articles claiming the contrary chiefly applied *in vivo* knockout and knockdown models and fasting-induced steatosis (see above). One also has to be aware of the potential influential consequences of knockout models on the developmental stages, for example as seen by the transdifferentiation of white adipocytes [98, 99]. As a result, the observed differences in autophagic lipid handling might be consequent to altered hepatocellular maturation as well as to the experimental method applied.

Fourthly, basal autophagy and stimulated autophagy have to be distinguished. The former is supposed to be the most important in the pathogenesis of NAFLD [31, 107]. Discernment between basal and induced autophagy is hampered by the fact that mostly total blockage of autophagy is used. Selective blockage of stimulus-induced autophagy (i.e. by exercise or starvation) can be achieved in a recently described *Bcl-2* knock-in model [108]. These mice have an exercise-related impairment of glucose metabolism and exhibit increased serum lipid levels when fed a HFD. Importantly, liver and pancreas morphology did not alter after HFD, supporting the importance of basal autophagy in lipid metabolism. These differences consequent to basal versus stimulated autophagy deserve further study to elucidate each particular role and potential influence on the divergent findings in lipid metabolism.

Whether the observed effects are a secondary/adaptive process or primary caused by autophagy is sometimes difficult to differentiate [99, 109], as well as the potential modified effects due to (compensatory upregulation of) other forms of autophagy (e.g. CMA) [107]. Furthermore, other organelles and cellular processes are impacted by dysfunctional autophagy and could in part explain observed differences in liver metabolism. Since autophagy can degrade apoB, a necessary protein for the VLDL formation, impaired VLDL production might be involved as well [109]. The involved ATG proteins might even exert non-autophagic (and autophagy independent) functions [39, 42, 110] in lipid metabolism.

While focussing on the role of autophagy in lipid metabolism, the role of cytosolic lipases may not become overlooked, while they still account for a substantial part of the lipolysis [88]. Total blockage of lipolysis by diethylumbelliferyl phosphate (DEUP) causes a greater increase in the cellular TG content than blocking autophagy alone [31]. In addition, if LD formation is autophagy dependent, small LD-like bodies are still observed on TEM in autophagy-deficient cells, suggesting that LDs formed out of the ER are unaffected [32].

Finally, the microscopical techniques currently available do not allow visualisation of the smallest LDs in living cells [111]. Hence, it is possible that the observed effects only reflect autophagy-related modulation once LDs are formed. In that case, autophagy can be considered as a dynamically active process that controls LD size and the amount of lipotoxic FFA in the cytoplasm. The effects of autophagy will be rather context dependent [79, 88]. In this perspective, lipolysis and lipogenesis are no longer mutually exclusive and in fact coexist [112].

5. Autophagy in glucose metabolism and insulin resistance

The liver has a key role in glucose metabolism and autophagy substantially contributes to maintain glucose homeostasis. In case of a conditional whole-body knockout of *Atg7* in adult mice, liver glycogen stores were totally depleted and serum amino acids and glycaemia dropped severely when fasted and led to death. The underlying cause was a lack of sufficient substrates because of deficient autophagy, since liver ketogenesis and gluconeogenesis in the liver remained intact [113].

Similar to whole-body autophagy, hepatic autophagy is necessary to deliver sufficient substrates to maintain blood glucose levels in the fasting state [114, 115], which is under strict control of insulin [114]. Additionally, long-term maintenance of blood glucose levels is dependent on growth hormone-stimulated autophagy [115]. Adenoviral overexpression of

TFEB, a master regulator of autophagy, improves the metabolic syndrome in HFD fed and ob/ob mice, amongst which improved glucose metabolism [53].

CMA is also involved in glucose metabolism. Mice with knockout of LAMP2A, necessary for internalisation of CMA-dedicated proteins, demonstrated increased levels of glycolysis. As a consequence, they exhibit lower blood glucose levels after fasting, have decreased glucose tolerance testing and recover less from insulin tolerance testing [68].

Impaired insulin signalling is an important feature of NAFLD [11]. The exact interactions between the action of insulin and autophagy, explaining the observed alterations in glucose metabolism, are not entirely clarified yet. **Figure 2** summarises the current knowledge of autophagy and insulin resistance.

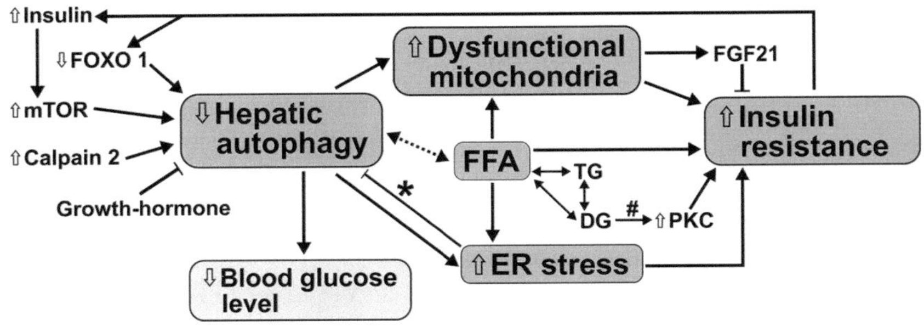

Figure 2. Autophagy and insulin resistance. Reciprocal influences are described between autophagy and insulin resistance (IR). Reduced autophagy causes a decrease in blood glucose level and is subject to different influences. Growth hormone regulates blood glucose on long term through the stimulation of hepatic autophagy. Increased levels of calpain 2 can induce IR through the effects of decreased autophagy on mitochondria and the endoplasmic reticulum (ER) stress levels. On the other hand, when normal insulin sensitivity remains present, secondary hyperinsulinism due to IR can decrease autophagy through the effects on mammalian target of rapamycin (mTOR) and through reduced forkhead box class O 1 (FOXO1). IR is also influenced directly through modulation of ER stress and the mitochondrial function by free fatty acids (FFA). Controversy exists on how autophagy might influence the level of lipids and thus FFA. Likewise, the formation of bioactive stereoisomers of diacylglycerol (DG), which might induce protein kinase C (PKC)-dependent IR, is also dependent on the effects of autophagy on FFA. The fibroblast growth factor 21 (FGF21) is produced in response to mitochondrial dysfunction and capable of reducing IR. Arrows indicate a consequence of a certain alteration, bar-headed arrows denote an inhibition. Double-headed arrows present a reciprocal influence. The dashed arrow denotes the uncertain relation between FFA and autophagy. *ER stress actually increases autophagy; #Only certain bioactive stereoisomers induce PKC (adapted from [112]).

Insulin inhibits autophagy via the stimulation of mTOR [2, 27]. When the liver presents a normal insulin sensitivity, reduced hepatocellular autophagy might be explained by insulin-dependent stimulation of mTOR in hyperinsulinaemic states. In case of insulin resistance (IR), an alternative inhibitory pathway of autophagy was described as well [49]. In mice fed HFD, diminished autophagy was the consequence of reduced FOXO1-mediated expression of key autophagy genes. The IR was believed to be due to the reduced clearance of dysfunctional mitochondria, as oxidative stress and altered mitochondrial integrity (and mass) are related to IR [49].

In contrast to the abovementioned study, IR was the result of decreased autophagy rather than the cause of reduced autophagy [42]. Insulin sensitivity and glucose tolerance improved, while the hepatic glucose production and steatosis decreased when obese mice overexpressed *Atg7*. Knockdown of *Atg7* in lean mice induced severe IR. Decreased protein levels of ATG7, with the following increase in cellular stress, with emphasis on ER stress, might be the cause of the observed IR. For this, increased levels of the calcium-dependent protease calpain 2, capable of cleaving several autophagy-related proteins, were held responsible [42].

In line with the potential underlying mechanism of ER stress-mediated increase in the hepatic IR, developing secondary to decreased autophagy, are the findings that intracellular saturated fatty acids can contribute to IR by an increase in ER stress [116]. However, ER stress-independent mechanism was described as well [116]. Moreover, ER stress can stimulate autophagy [27, 117]. In this viewpoint, autophagy might potentially prevent cell injury and IR particular by serving as an escape mechanism in an attempt to reduce ER stress.

Autophagy not only plays a role on cellular level, but appears to exert endocrine and metabolic functions as well [84]. Defective clearance of mitochondria, due to dysfunctional autophagy, induces cellular stress and subsequent stress responses. One of those is activation of transcription factor 4 (ATF4), which promotes the expression of FGF21. FGF21 in turn has several beneficial metabolic effects including improvement in insulin sensitivity and glucose tolerance [84, 118]. Paradoxically, in patients with NAFLD a positive association was observed between plasma FGF21 levels, IR and steatosis [119]. This phenomenon might be explained by FGF21 resistance, since less IR is actually expected with increasing levels of FGF21, or can either be explained by an adaptive increase in FGF21 once IR has been established. Taking into account these positive metabolic consequences, mitochondrial dysfunction no longer has to be seen as just detrimental, but also as beneficial by improving glucose metabolism and reducing fasting-induced steatosis.

Finally, insulin resistance is linked to protein kinase C (PKC) [120]. PKC in an important cellular effector enzyme involved in several signal transduction cascades and several isoforms exist. Many of them can be activated by diacylglycerol (DG). PKC showed to be able to inhibit and stimulate autophagy [121, 122]. DG is a product of lipolysis, but also a TG intermediate, and might be an additional crosslink between IR and autophagy. Importantly, not all DG are able to activate PKC, only specific stereoisomers can. Of these, no bioactive DG can be produced by lipolysis, and therefore, potential crosstalk is less likely [88, 120]. On the contrary, bioactive stereoisomers of DG (i.e. 1,2-diacyl-glycerol) can still be generated in lipid synthesis and interfere with insulin signalling.

6. Autophagy and hepatocellular injury and hepatocellular carcinoma

As mentioned in Section 1, NAFL can evolve to NASH and is prone to subsequent development of advanced fibrosis or even development of hepatocellular carcinoma (HCC) in a subset of patients [6]. Because autophagy plays a central role in cellular homeostasis, dysfunction likely results in cellular injury. Indeed, stimulation of autophagy could reduce liver injury in animal models of ethanol-induced steatohepatitis [62, 123] and NAFLD [62].

Mitochondrial damage is often observed in autophagy deficiency [29, 36, 56, 84, 92, 124]. Dysfunctional mitochondria lead to the production of reactive oxygen species (ROS), which are involved in the pathogenesis of NASH [125]. Increased ROS production was indeed observed in autophagy deficiency [56, 126]. Autophagy-deficient hepatocytes are more susceptible to cell death via menadione-induced oxidative stress [107]. Compensatory upregulation of CMA, capable of protecting against menadione-induced cell death via different mechanisms, is unable to overcome the induced oxidative stress [107]. TNF-induced hepatic injury causes similar alterations as menadione, with increased cell death, JNK/c-JUN overactivation and activation of the mitochondrial death pathway in hepatocyte-specific autophagy-deficient mice. However, this seems to be independent of oxidative stress or impaired cellular energy homeostasis secondary to mitochondrial dysfunction [127].

In autophagy-deficient cells, SQSTM1/p62-positive proteinaceous aggregates accumulate as they are no longer degraded. p62 seems to contribute considerably to the hepatocellular injury seen in autophagy deficiency, as double knockouts (DKO) of autophagy (*Atg7*) and p62 have less hepatocyte injury compared with autophagy knockout (*Atg7*) alone [128]. However, overexpression of p62 is not cytotoxic [129]. The effects of p62 might be executed by NF-E2-related factor 2 (NRF2), of which p62 is an endogenous protein inducer. Indeed, DKO of *Atg7* and *Nrf2* was able to prevent hepatic injury, similar to the DKO of *Atg7* and p62 [129, 130]. This is a paradoxical finding, as NRF2-dependent gene products are known to be cytoprotective, but is believed to be the consequence of increased cellular stress due to imbalance between increased protein synthesis (NRF2-driven) and reduced breakdown (by autophagy) [129]. Of note, despite the alleviation of cellular injury by these DKOs, the observed phenotypes cannot be attributed solely to NRF2 activation. Since turnover of damaged organelles is still not corrected in these DKOs, complete prevention of the cellular injury comparable to control levels cannot be achieved either.

In adipocytes, autophagy is linked to inflammatory cytokines and inflammation. Knockout of p62 in adipocytes induces infiltration of macrophages and the production of pro-inflammatory cytokines in AT [131]. Furthermore, obesity and glucose intolerance are observed, while this is not the case in p62 knockout in hepatocytes. Whether these inflammatory changes also occur in liver is not investigated. Human and mice adipocytes also increase their pro-inflammatory cytokine production when autophagy is inhibited [102, 132].

Autophagy exerts a dual role in tumorigenesis depending on the stage of tumour development. In normal tissue, autophagy acts as a tumour suppressor and thus prevents the development of malignant neoplasia. However, when a tumour already has developed, autophagy aids the survival of tumour cells by supplying nutrients [133].

Hepatocellular autophagy deficiency, either by *Atg5* or *Atg7* knockout, causes the development of multiple spontaneous liver tumours [124, 126, 130, 134]. In line with the observed effects of the contributing role of p62-NRF2 to cellular injury, DKO of *Atg7* and p62 was able to reduce tumour size [124], and DKO of *Atg7* and *Nrf2* totally prevented tumour formation [130]. Intriguingly, all these tumours were not hepatocellular carcinomas (HCCs) but hepatocellular adenoma, which are benign tumours. Even stimulation with diethylnitrosamine, an establish-ed chemical inducer of HCC, was not able to induce HCC in autophagy-deficient livers compared with wild-type livers [126]. The induction of several tumour suppressors (e.g. p53) could explain the prevention of carcinogenesis in case of autophagy deficiency [126]. Once HCC has developed in autophagy-competent rat livers, differences in autophagy pattern correlate with aggressiveness of the tumours, as determined by the marker cytokeratin-19 [135]. For further extensive discussion of the role(s) of autophagy in liver tumour biology, we refer to other published reviews [21, 136].

7. Autophagy in liver fibrosis

As stated before, liver fibrosis is the main predictor for long-term outcomes in patients with NAFLD [9, 14]. Interestingly, hepatocellular autophagy affects liver fibrosis too. In mice with hepatocellular autophagy deficiency, the degree of fibrosis was significantly increased [82, 130, 137]. This could be only an indirect effect considering the protective role of hepatocellular autophagy on liver injury and inflammation, which are the driving forces of fibrogenesis [82, 137].

Available evidence suggests also a direct elementary role for autophagy in different fibrogenic cells [138]. Autophagy seems to provide nutrients to fuel the processes involved in the activation of these cells. The hepatic stellate cells (HSCs) are considered major fibrogenic cells in the liver. When these cells transdifferentiate from a quiescent state to active myofibroblasts, their lipid stores (in HSC mainly vitamin A) become depleted [139].

During HSC activation, the autophagic flux increases and pharmacological or genetic inhibition could prevent the activation of HSCs [140, 141]. In these cells, autophagy interferes with LD metabolism, as shown by colocalisation of LC3B with LDs. Interestingly, only larger LDs seem to be affected and colocalisation disappeared once HSCs were activated [140]. More specifically, autophagy plays a role as energy supplier through delivery of FFA out of LDs for β-oxidation, necessary for the transdifferentiation of HSCs [141]. Even though there is no effect on fibrogenesis in autophagy-competent cells, oleic acid could partly restore HSC activation in autophagy-deficient cells [141].

HSC activation through autophagy activation was very recently ratified in an *in vitro* model using rat HSCs [142]. When hypoxic stress was applied, HSCs increased autophagic flux and got activated. Pharmacological intervention and knockdown of autophagy demonstrated that HSC activation was autophagy dependent and mediated by the activation of Ca^{2+}-AMPK-mTOR and PKCθ signalling pathways.

8. Autophagy and ER stress

Next to autophagy, cells possess another homeostatic mechanism to protect cells by alleviating cellular stress or by inducing cell death under extreme conditions: the unfolded protein response (UPR). The UPR is activated in response to the accumulation of unfolded proteins in the endoplasmic reticulum (ER) (ER stress) [143, 144]. ER stress results from perturbation of the normal protein folding capacity of the ER and induces inflammation and oxidative stress [145]. The UPR encompasses three major adaptive mechanisms to restore protein homeostasis, named after the respective ER stress sensor: activating transcription factor 6 (ATF6), protein kinase RNA-like ER kinase (PERK) and inositol-requiring enzyme-1α (IRE1α) [143].

The UPR and autophagy can function independently, but are dynamically interconnected. The classical view is that ER stress induces autophagy in order to restore cellular integrity, though ER stress can both induce and inhibit autophagy, even in a selective way [144]. Reciprocal feedback also exists, where autophagy influences the turnover of ER and the removal of misfolded proteins and hence regulates ER stress [146].

Interestingly, impaired autophagy is associated with increased levels of ER stress (mRNA and protein levels) in patients with NASH [72]. Gene analysis showed a general enrichment of downregulated genes related to the UPR in patients with NASH [73]. However, microarray data of the different UPR branches showed a more scattered pattern with both up- and downregulated gene expression. Finally, there were increased levels of IRE1α-regulated spliced X-box binding protein 1 (XBP1s) at protein level, with congruent increased nuclear staining, and equal levels of other UPR chaperones (though with large variability) [73].

In a methionine-choline-deficient diet (MCDD) and HFD model of NAFLD, an analogous association between impaired autophagy and ER stress was observed [72]. The same authors demonstrated alleviation of ER stress *in vitro* after induction of autophagy in palmitic acid induced fat accumulation. High fructose feeding, a model for diabetes, induces ER stress and reduces autophagy after two weeks. Autophagy and ER stress occurred prior to lipid accumulation, wherein autophagy preceded ER stress. Vice versa, induction of autophagy could alleviate ER stress, restored insulin signalling and reduced liver fat content [147].

In line with these results, knockdown of *Atg7* increased ER stress levels in lean mice, while overexpression of *Atg7* in obese ob/ob mice showed the opposite [42]. Furthermore, HFD induced obesity or *in vitro* addition of SFA was able to impair autophagy by inhibiting the fusion of autophagosomes with lysosomes. This impairment induced increased ER stress. Intriguingly, ER stress *in se* was not able to inhibit autophagy, but the underlying inhibition of SERCA pumps with subsequent rise in cytosolic calcium levels was. Calcium channel blockers were able to restore autophagy, ER stress and the metabolic consequences of HFD or SFA [43].

Finally, C1q/TNF-related protein 9 (CTRP9) is the closest known paralog of adiponectin and also thought to serve as an adipokine. CTRP9 has shown *in vitro* and *in vivo* to induce hepatocellular autophagy, reduce ER stress and subsequently alleviate TG accumulation and apoptosis. The reduction in ER stress was independent of direct effects on UPR chaperones and proven to be dependent on its actions on autophagy [148].

These studies subscribe the reciprocal effects of autophagy on ER stress and their role (albeit possibly indirectly) on lipid metabolism.

9. Conclusion

The current literature clearly emphasises the importance of autophagy in the liver. Nevertheless, when focussing on its role in liver lipid metabolism, controversy still exists regarding lipolytic or lipogenic features of autophagy. Moreover, autophagy is a highly dynamic process and appears to act in a context- and tissue-specific way.

Autophagy is not only involved in lipid metabolism but also in glucose metabolism, liver fibrogenesis and cellular injury. At the cellular level, there is a close interaction between ER stress and the corresponding UPR, another cellular homeostatic defence mechanism.

Unravelling the exact function of autophagy in the complex pathophysiology of metabolic disturbances and NAFLD could make autophagy an interesting target for treatment of the metabolic syndrome or for NAFLD.

Author details

Wilhelmus J. Kwanten[1], Wim Martinet[2] and Sven M. Francque[1,3*]

*Address all correspondence to: sven.francque@uza.be

1 Laboratory of Experimental Medicine and Paediatrics (LEMP), University of Antwerp, Antwerp, Belgium

2 Laboratory of Physiopharmacology, University of Antwerp, Antwerp, Belgium

3 Department of Gastroenterology Hepatology, Antwerp University Hospital, Edegem, Belgium

References

[1] De Duve C, Wattiaux R. Functions of lysosomes. Annu Rev Physiol 1966;28:435–92. doi: 10.1146/annurev.ph.28.030166.002251.

[2] Martinet W, Agostinis P, Vanhoecke B, Dewaele M, De Meyer GRY. Autophagy in disease: a double-edged sword with therapeutic potential. Clin Sci (Lond) 2009;116:697–712. doi:10.1042/CS20080508.

[3] Choi AMK, Ryter SW, Levine B. Autophagy in human health and disease. N Engl J Med 2013;368:651–62. doi:10.1056/NEJMra1205406.

[4] Ludwig J, Viggiano TR, McGill DB, Oh BJ. Nonalcoholic steatohepatitis: Mayo Clinic experiences with a hitherto unnamed disease. Mayo Clin Proc 1980;55:434–8.

[5] Chalasani N, Younossi Z, Lavine JE, Diehl AM, Brunt EM, Cusi K, et al. The diagnosis and management of non-alcoholic fatty liver disease: practice guideline by the American Association for the Study of Liver Diseases, American College of Gastroenterology, and the American Gastroenterological Association. Hepatology 2012;55:2005–23. doi: 10.1002/hep.25762.

[6] Satapathy SK, Sanyal AJ. Epidemiology and natural history of nonalcoholic fatty liver disease. Semin Liver Dis 2015;35:221–35. doi:10.1055/s-0035-1562943.

[7] Kleiner DE, Brunt EM. Nonalcoholic fatty liver disease: pathologic patterns and biopsy evaluation in clinical research. Semin Liver Dis 2012;32:3–13. doi:10.1055/s-0032-1306421.

[8] Singh S, Allen AM, Wang Z, Prokop LJ, Murad MH, Loomba R. Fibrosis progression in nonalcoholic fatty liver vs nonalcoholic steatohepatitis: a systematic review and meta-analysis of paired-biopsy studies. Clin Gastroenterol Hepatol 2015;13:643–54.e1–9; quiz e39–40. doi: 10.1016/j.cgh.2014.04.014.

[9] Angulo P, Kleiner DE, Dam-Larsen S, Adams LA, Bjornsson ES, Charatcharoenwitthaya P, et al. Liver fibrosis, but no other histologic features, is associated with long-term outcomes of patients with nonalcoholic fatty liver disease. Gastroenterology 2015;149:389–97.e10. doi: 10.1053/j.gastro.2015.04.043.

[10] de Meijer VE, Kalish BT, Puder M, Ijzermans JNM. Systematic review and meta-analysis of steatosis as a risk factor in major hepatic resection. Br J Surg 2010;97:1331–9. doi: 10.1002/bjs.7194.

[11] Torres DM, Williams CD, Harrison SA. Features, diagnosis, and treatment of nonalcoholic fatty liver disease. Clin Gastroenterol Hepatol 2012;10:837–58. doi: 10.1016/j.cgh.2012.03.011.

[12] Byrne CD, Targher G. NAFLD: a multisystem disease. J Hepatol 2015;62:S47–64. doi: 10.1016/j.jhep.2014.12.012.

[13] Francque SM, van der Graaff D, Kwanten WJ. Non-alcoholic fatty liver disease and cardiovascular risk: pathophysiological mechanisms and implications. J Hepatol 2016. doi: 10.1016/j.jhep.2016.04.005.

[14] Ekstedt M, Hagström H, Nasr P, Fredrikson M, Stål P, Kechagias S, et al. Fibrosis stage is the strongest predictor for disease-specific mortality in NAFLD after up to 33 years of follow-up. Hepatology 2015;61:1547–54. doi: 10.1002/hep.27368.

[15] Cohen JC, Horton JD, Hobbs HH. Human fatty liver disease: old questions and new insights. Science 2011;332:1519–23. doi:10.1126/science.1204265.

[16] Vernon G, Baranova A, Younossi ZM. Systematic review: the epidemiology and natural history of non-alcoholic fatty liver disease and non-alcoholic steatohepatitis in adults. Aliment Pharmacol Ther 2011;34:274–85. doi:10.1111/j.1365-2036.2011.04724.x.

[17] Charlton MR, Burns JM, Pedersen RA, Watt KD, Heimbach JK, Dierkhising RA. Frequency and outcomes of liver transplantation for nonalcoholic steatohepatitis in the United States. Gastroenterology 2011;141:1249–53. doi:10.1053/j.gastro.2011.06.061.

[18] Day CP, Anstee QM. Foreword. Nonalcoholic fatty liver disease. Semin Liver Dis 2015;35:203–6. doi:10.1055/s-0035-1562941.

[19] Tilg H, Moschen AR. Evolution of inflammation in nonalcoholic fatty liver disease: the multiple parallel hits hypothesis. Hepatology 2010;52:1836–46. doi:10.1002/hep.24001.

[20] Brunt EM, Wong VW-S, Nobili V, Day CP, Sookoian S, Maher JJ, et al. Nonalcoholic fatty liver disease. Nat Rev Dis Prim 2015:15080. doi:10.1056/NEJMra011775.

[21] Czaja MJ, Ding W-X, Donohue TM, Friedman SL, Kim J-S, Komatsu M, et al. Functions of autophagy in normal and diseased liver. Autophagy 2013;9:1131–58. doi:10.4161/auto.25063.

[22] Galluzzi L, Pietrocola F, Levine B, Kroemer G. Metabolic control of autophagy. Cell 2014;159:1263–76. doi:10.1016/j.cell.2014.11.006.

[23] Rodriguez-Navarro JA, Kaushik S, Koga H, Dall'Armi C, Shui G, Wenk MR, et al. Inhibitory effect of dietary lipids on chaperone-mediated autophagy. Proc Natl Acad Sci U S A 2012;109:E705–14. doi:10.1073/pnas.1113036109.

[24] Mizushima N, Komatsu M. Autophagy: renovation of cells and tissues. Cell 2011;147:728–41. doi:10.1016/j.cell.2011.10.026.

[25] Ding W-X, Yin X-M. Mitophagy: mechanisms, pathophysiological roles, and analysis. Biol Chem 2012;393:547–64. doi:10.1515/hsz-2012-0119.

[26] Feng Y, Yao Z, Klionsky DJ. How to control self-digestion: transcriptional, post-transcriptional, and post-translational regulation of autophagy. Trends Cell Biol 2015;25:354–63. doi:10.1016/j.tcb.2015.02.002.

[27] Ravikumar B, Sarkar S, Davies JE, Futter M, Garcia-Arencibia M, Green-Thompson ZW, et al. Regulation of mammalian autophagy in physiology and pathophysiology. Physiol Rev 2010;90:1383–435. doi:10.1152/physrev.00030.2009.

[28] Rubinsztein DC, Codogno P, Levine B. Autophagy modulation as a potential therapeutic target for diverse diseases. Nat Rev Drug Discov 2012;11:709–30. doi:10.1038/nrd3802.

[29] Komatsu M, Waguri S, Ueno T, Iwata J, Murata S, Tanida I, et al. Impairment of starvation-induced and constitutive autophagy in Atg7-deficient mice. J Cell Biol 2005;169:425–34. doi:10.1083/jcb.200412022.

[30] Klionsky DJ, Abdalla FC, Abeliovich H, Abraham RT, Acevedo-Arozena A, Adeli K, et al. Guidelines for the use and interpretation of assays for monitoring autophagy. Autophagy 2012;8:445–544.

[31] Singh R, Kaushik S, Wang Y, Xiang Y, Novak I, Komatsu M, et al. Autophagy regulates lipid metabolism. Nature 2009;458:1131–5. doi:10.1038/nature07976.

[32] Shibata M, Yoshimura K, Furuya N, Koike M, Ueno T, Komatsu M, et al. The MAP1-LC3 conjugation system is involved in lipid droplet formation. Biochem Biophys Res Commun 2009;382:419–23. doi:10.1016/j.bbrc.2009.03.039.

[33] Shibata M, Yoshimura K, Tamura H, Ueno T, Nishimura T, Inoue T, et al. LC3, a microtubule-associated protein 1A/B light chain3, is involved in cytoplasmic lipid droplet formation. Biochem Biophys Res Commun 2010;393:274–9. doi:10.1016/j.bbrc.2010.01.121.

[34] Sinha RA, You S-H, Zhou J, Siddique MM, Bay B-H, Zhu X, et al. Thyroid hormone stimulates hepatic lipid catabolism via activation of autophagy. J Clin Invest 2012;122:2428–38. doi:10.1172/JCI60580.

[35] Sinha RA, Farah BL, Singh BK, Siddique MM, Li Y, Wu Y, et al. Caffeine stimulates hepatic lipid metabolism by the autophagy-lysosomal pathway in mice. Hepatology 2014;59:1366–80. Doi: 10.1002/hep.26667.

[36] Martinet W, Schrijvers DM, Timmermans J-P, Bult H, De Meyer GRY. Immunohistochemical analysis of macroautophagy: recommendations and limitations. Autophagy 2013;9:386–402. doi:10.4161/auto.22968.

[37] Kashima J, Shintani-Ishida K, Nakajima M, Maeda H, Unuma K, Uchiyama Y, et al. Immunohistochemical study of the autophagy marker microtubule-associated protein 1 light chain 3 in normal and steatotic human livers. Hepatol Res 2013:1–9. doi:10.1111/hepr.12183.

[38] Gebhardt R, Coffer PJ. Hepatic autophagy is differentially regulated in periportal and pericentral zones – a general mechanism relevant for other tissues? Cell Commun Signal 2013;11:21. doi:10.1186/1478-811X-11-21.

[39] Zechner R, Madeo F. Cell biology: another way to get rid of fat. Nature 2009;458:1118–9. doi:10.1038/4581118a.

[40] Skop V, Cahová M, Papáčková Z, Páleníčková E, Daňková H, Baranowski M, et al. Autophagy-lysosomal pathway is involved in lipid degradation in rat liver. Physiol Res 2012;61:287–97.

[41] Papáčková Z, Daňková H, Páleníčková E, Kazdová L, Cahová M. Effect of short- and long-term high-fat feeding on autophagy flux and lysosomal activity in rat liver. Physiol Res 2012;61(Suppl 2):S67–76.

[42] Yang L, Li P, Fu S, Calay ES, Hotamisligil GS. Defective hepatic autophagy in obesity promotes ER stress and causes insulin resistance. Cell Metab 2010;11:467–78. doi: 10.1016/j.cmet.2010.04.005.

[43] Park H-W, Park H, Semple IA, Jang I, Ro S-H, Kim M, et al. Pharmacological correction of obesity-induced autophagy arrest using calcium channel blockers. Nat Commun 2014;5:4834. doi:10.1038/ncomms5834.

[44] Hutagalung AH, Novick PJ. Role of Rab GTPases in membrane traffic and cell physiology. Physiol Rev 2011;91:119–49. doi:10.1152/physrev.00059.2009.

[45] Schroeder B, Schulze RJ, Weller SG, Sletten AC, Casey CA, McNiven MA. The small GTPase Rab7 as a central regulator of hepatocellular lipophagy. Hepatology 2015;61:1896–907. doi:10.1002/hep.27667.

[46] Schulze RJ, Weller SG, Schroeder B, Krueger EW, Chi S, Casey CA, et al. Lipid droplet breakdown requires dynamin 2 for vesiculation of autolysosomal tubules in hepatocytes. J Cell Biol 2013;203:315–26. doi:10.1083/jcb.201306140.

[47] Tikhanovich I, Cox J, Weinman S a. Forkhead box class O transcription factors in liver function and disease. J Gastroenterol Hepatol 2013;28 Suppl 1:125–31. doi:10.1111/jgh.12021.

[48] Xiong X, Tao R, DePinho RA, Dong XC. The autophagy-related gene 14 (Atg14) is regulated by forkhead box O transcription factors and circadian rhythms and plays a critical role in hepatic autophagy and lipid metabolism. J Biol Chem 2012;287:39107–14. doi:10.1074/jbc.M112.412569.

[49] Liu H-Y, Han J, Cao SY, Hong T, Zhuo D, Shi J, et al. Hepatic autophagy is suppressed in the presence of insulin resistance and hyperinsulinemia: inhibition of FoxO1-dependent expression of key autophagy genes by insulin. J Biol Chem 2009;284:31484–92. doi:10.1074/jbc.M109.033936.

[50] Valenti L, Rametta R, Dongiovanni P, Maggioni M, Fracanzani AL, Zappa M, et al. Increased expression and activity of the transcription factor FOXO1 in nonalcoholic steatohepatitis. Diabetes 2008;57:1355–62. doi:10.2337/db07-0714.

[51] Zhang Z, Guo M, Zhao S, Xu W, Shao J, Zhang F, et al. The update on transcriptional regulation of autophagy in normal and pathologic cells: A novel therapeutic target. Biomed Pharmacother 2015;74:17–29. doi:10.1016/j.biopha.2015.06.003.

[52] Settembre C, Fraldi A, Medina DL, Ballabio A. Signals from the lysosome: a control centre for cellular clearance and energy metabolism. Nat Rev Mol Cell Biol 2013;14:283–96. doi:10.1038/nrm3565.

[53] Settembre C, De Cegli R, Mansueto G, Saha PK, Vetrini F, Visvikis O, et al. TFEB controls cellular lipid metabolism through a starvation-induced autoregulatory loop. Nat Cell Biol 2013;15:647–58. doi:10.1038/ncb2718.

[54] Seok S, Fu T, Choi S-E, Li Y, Zhu R, Kumar S, et al. Transcriptional regulation of autophagy by an FXR–CREB axis. Nature 2014;516:108–11. doi:10.1038/nature13949.

[55] Lee JM, Wagner M, Xiao R, Kim KH, Feng D, Lazar MA, et al. Nutrient-sensing nuclear receptors coordinate autophagy. Nature 2014;516:112–5. doi:10.1038/nature13961.

[56] Stankov M V, Panayotova-Dimitrova D, Leverkus M, Vondran FWR, Bauerfeind R, Binz A, et al. Autophagy inhibition due to thymidine analogues as novel mechanism leading to hepatocyte dysfunction and lipid accumulation. AIDS 2012;26:1995–2006. doi: 10.1097/QAD.0b013e32835804f9.

[57] Armstrong MJ, Gaunt P, Aithal GP, Barton D, Hull D, Parker R, et al. Liraglutide safety and efficacy in patients with non-alcoholic steatohepatitis (LEAN): a multicentre, double-blind, randomised, placebo-controlled phase 2 study. Lancet (London, England) 2016;387:679–90. doi:10.1016/S0140-6736(15)00803-X.

[58] Sharma S, Mells JE, Fu PP, Saxena NK, Anania FA. GLP-1 analogs reduce hepatocyte steatosis and improve survival by enhancing the unfolded protein response and promoting macroautophagy. PLoS One 2011;6:e25269. doi:10.1371/journal.pone. 0025269.

[59] Zhou S-W, Zhang M, Zhu M. Liraglutide reduces lipid accumulation in steatotic L-02 cells by enhancing autophagy. Mol Med Rep 2014;10:2351–7. doi:10.3892/mmr. 2014.2569.

[60] Tong W, Ju L, Qiu M, Xie Q, Cheng Y, Shen W, et al. Liraglutide ameliorate non-alcohol fatty liver disease by enhancing mitochondrial architecture and promoting autophagy through SIRT1/SIRT3-FOXO3a pathway. Hepatol Res 2015:n/a – n/a. doi:10.1111/hepr. 12634. DOI [Epub ahead of print].

[61] He B, Liu L, Yu C, Wang Y, Han P. Roux-en-Y Gastric bypass reduces lipid overaccumulation in liver by upregulating hepatic autophagy in obese diabetic rats. Obes Surg 2015;25:109–18. doi:10.1007/s11695-014-1342-7.

[62] Lin C-W, Zhang H, Li M, Xiong X, Chen X, Dong XC, et al. Pharmacological promotion of autophagy alleviates steatosis and injury in alcoholic and non-alcoholic fatty liver conditions in mice. J Hepatol 2013;58:993–9. doi:10.1016/j.jhep.2013.01.011.

[63] Zheng X, Dai W, Chen X, Wang K, Zhang W, Liu L, et al. Caffeine reduces hepatic lipid accumulation through regulation of lipogenesis and ER stress in zebrafish larvae. J Biomed Sci 2015;22:105. doi:10.1186/s12929-015-0206-3.

[64] Li L, Hai J, Li Z, Zhang Y, Peng H, Li K, et al. Resveratrol modulates autophagy and NF-kB activity in a murine model for treating non-alcoholic fatty liver disease. Food Chem Toxicol 2014;63:166–73. doi:10.1016/j.fct.2013.08.036.

[65] Zhang Y, Chen M, Zhou Y, Yi L, Gao Y, Ran L, et al. Resveratrol improves hepatic steatosis by inducing autophagy through the cAMP signaling pathway. Mol Nutr Food Res 2015;59:1443–57. doi:10.1002/mnfr.201500016.

[66] Li S, Li J, Shen C, Zhang X, Sun S, Cho M, et al. Tert-Butylhydroquinone (tBHQ) protects hepatocytes against lipotoxicity via inducing autophagy independently of Nrf2

activation. Biochim Biophys Acta Mol Cell Biol Lipids 2014;1841:22–33. doi:10.1016/j.bbalip.2013.09.004.

[67] Song YM, Lee Y, Kim J-W, Ham D-S, Kang E-S, Cha BS, et al. Metformin alleviates hepatosteatosis by restoring SIRT1-mediated autophagy induction via an AMP-activated protein kinase-independent pathway. Autophagy 2015;11:46–59. doi:10.4161/15548627.2014.984271.

[68] Schneider JL, Suh Y, Cuervo AM. Deficient chaperone-mediated autophagy in liver leads to metabolic dysregulation. Cell Metab 2014;20:417–32. doi:10.1016/j.cmet.2014.06.009.

[69] Kaushik S, Cuervo AM. Degradation of lipid droplet-associated proteins by chaperone-mediated autophagy facilitates lipolysis. Nat Cell Biol 2015;17:759–70. doi:10.1038/ncb3166.

[70] van Zutphen T, Todde V, de Boer R, Kreim M, Hofbauer HF, Wolinski H, et al. Lipid droplet autophagy in the yeast Saccharomyces cerevisiae. Mol Biol Cell 2014;25:290–301. doi:10.1091/mbc.E13-08-0448.

[71] Fukuo Y, Yamashina S, Sonoue H, Arakawa A, Nakadera E, Aoyama T, et al. Abnormality of autophagic function and cathepsin expression in the liver from patients with non-alcoholic fatty liver disease. Hepatol Res 2013;44:1–11. doi:10.1111/hepr.12282.

[72] González-Rodríguez A, Mayoral R, Agra N, Valdecantos MP, Pardo V, Miquilena-Colina ME, et al. Impaired autophagic flux is associated with increased endoplasmic reticulum stress during the development of NAFLD. Cell Death Dis 2014;5:e1179. doi:10.1038/cddis.2014.162.

[73] Lake AD, Novak P, Hardwick RN, Flores-Keown B, Zhao F, Klimecki WT, et al. The adaptive endoplasmic reticulum stress response to lipotoxicity in progressive human nonalcoholic fatty liver disease. Toxicol Sci 2014;137:26–35. doi:10.1093/toxsci/kft230.

[74] Liangpunsakul S, Chalasani N. Is hypothyroidism a risk factor for non-alcoholic steatohepatitis? J Clin Gastroenterol 2003;37:340–3.

[75] Chung GE, Kim D, Kim W, Yim JY, Park MJ, Kim YJ, et al. Non-alcoholic fatty liver disease across the spectrum of hypothyroidism. J Hepatol 2012;57:150–6. doi:10.1016/j.jhep.2012.02.027.

[76] Pagadala MR, Zein CO, Dasarathy S, Yerian LM, Lopez R, McCullough AJ. Prevalence of hypothyroidism in nonalcoholic fatty liver disease. Dig Dis Sci 2012;57:528–34. doi:10.1007/s10620-011-2006-2.

[77] Eliades M, Spyrou E, Agrawal N, Lazo M, Brancati FL, Potter JJ, et al. Meta-analysis: vitamin D and non-alcoholic fatty liver disease. Aliment Pharmacol Ther 2013;38:246–54. doi:10.1111/apt.12377.

[78] Mak HY. Lipid droplets as fat storage organelles in *Caenorhabditis elegans*: thematic review series: lipid droplet synthesis and metabolism: from yeast to man. J Lipid Res 2012;53:28–33. doi:10.1194/jlr.R021006.

[79] Lapierre LR, Silvestrini MJ, Nuñez L, Ames K, Wong S, Le TT, et al. Autophagy genes are required for normal lipid levels in C. elegans. Autophagy 2013;9:278–86. doi: 10.4161/auto.22930.

[80] Guan H-P, Goldstein JL, Brown MS, Liang G. Accelerated fatty acid oxidation in muscle averts fasting-induced hepatic steatosis in SJL/J mice. J Biol Chem 2009;284:24644–52. doi:10.1074/jbc.M109.034397.

[81] Gan SK, Watts GF. Is adipose tissue lipolysis always an adaptive response to starvation? Implications for non-alcoholic fatty liver disease. Clin Sci (Lond) 2008;114:543–5. doi: 10.1042/CS20070461.

[82] Ma D, Molusky MM, Song J, Hu C-R, Fang F, Rui C, et al. Autophagy deficiency by hepatic FIP200 deletion uncouples steatosis from liver injury in NAFLD. Mol Endocrinol 2013;27:1643–54. doi:10.1210/me.2013-1153.

[83] Yokota S-I, Ando M, Aoyama S, Nakamura K, Shibata S. Leucine restores murine hepatic triglyceride accumulation induced by a low-protein diet by suppressing autophagy and excessive endoplasmic reticulum stress. Amino Acids. Springer Vienna; 2016 Apr 26;48(4):1013–21.

[84] Kim KH, Jeong YT, Oh H, Kim SH, Cho JM, Kim Y-N, et al. Autophagy deficiency leads to protection from obesity and insulin resistance by inducing Fgf21 as a mitokine. Nat Med 2013;19:83–92. doi:10.1038/nm.3014.

[85] Hsu H-C, Liu C-H, Tsai Y-C, Li S-J, Chen C-Y, Chu C-H, et al. Time-dependent cellular response in the liver and heart in a dietary-induced obese mouse model: the potential role of ER stress and autophagy. Eur J Nutr 2015. doi:10.1007/s00394-015-1017-8.

[86] Zhang H, Chen X, Yin X-M. The dynamics of liver autophagy status during steatosis development (1440). Hepatology 2012;56:869A. doi: 10.1002/hep.26040.

[87] Xiao J, Guo R, Fung M-L, Liong EC, Chuen R, Chang C, et al. Garlic-derived S-allylmercaptocysteine ameliorates nonalcoholic fatty liver disease in a rat model through inhibition of apoptosis and enhancing autophagy. Evid Based Complement Alternat Med 2013;2013:642920. doi:10.1155/2013/642920.

[88] Zechner R, Zimmermann R, Eichmann TO, Kohlwein SD, Haemmerle G, Lass A, et al. FAT SIGNALS – lipases and lipolysis in lipid metabolism and signaling. Cell Metab 2012;15:279–91. doi:10.1016/j.cmet.2011.12.018.

[89] Fuchs M, Sanyal AJ. Lipotoxicity in NASH. J Hepatol 2012;56:291–3. doi:10.1016/j.jhep.2011.05.019.

[90] Tang Y, Chen Y, Jiang H, Nie D. Short-chain fatty acids induced autophagy serves as an adaptive strategy for retarding mitochondria-mediated apoptotic cell death. Cell Death Differ 2011;18:602–18. doi:10.1038/cdd.2010.117.

[91] Chen Y, Xu C, Yan T, Yu C, Li Y. ω-3 Fatty acids reverse lipotoxity through induction of autophagy in nonalcoholic fatty liver disease. Nutrition 2015;31:1423–9.e2. doi:10.1016/j.nut.2015.05.022.

[92] Mei S, Ni H, Manley S, Bockus A, Kassel KM, Luyendyk JP, et al. Differential roles of unsaturated and saturated fatty acids on autophagy and apoptosis in hepatocytes. J Pharmacol Exp Ther 2011;339:487–98. doi:10.1124/jpet.111.184341.

[93] Tanaka S, Hikita H, Aono S, Kawaguchi T, Mukai K, Shimizu S, et al. Differential roles of palmitic acid and oleic acid in regulation of apoptosis and autophagy in non-alcoholic fatty liver disease (1399). Hepatology 2012;56:850A–851A. doi:10.1002/hep.26040.

[94] Tu Q, Zheng R, Li J, Hu L, Chang Y, Li L, et al. Palmitic acid induces autophagy in hepatocytes via JNK2 activation. Acta Pharmacol Sin 2014;35:504–12. doi:10.1038/aps.2013.170.

[95] Koga H, Kaushik S, Cuervo AM. Altered lipid content inhibits autophagic vesicular fusion. FASEB J 2010;24:3052–65. doi:10.1096/fj.09-144519.

[96] Inami Y, Yamashina S, Izumi K, Ueno T, Tanida I, Ikejima K, et al. Hepatic steatosis inhibits autophagic proteolysis via impairment of autophagosomal acidification and cathepsin expression. Biochem Biophys Res Commun 2011;412:618–25. doi:10.1016/j.bbrc.2011.08.012.

[97] Nakadera E, Yamashina S, Izumi K, Inami Y, Sato T, Fukushima H, et al. Inhibition of mTOR improves the impairment of acidification in autophagic vesicles caused by hepatic steatosis. Biochem Biophys Res Commun 2016;469:1104–10. doi:10.1016/j.bbrc.2015.12.010.

[98] Zhang Y, Zeng X, Jin S. Autophagy in adipose tissue biology. Pharmacol Res 2012;66:505–12. doi:10.1016/j.phrs.2012.09.004.

[99] Maixner N, Kovsan J, Harman-Boehm I, Blüher M, Bashan N, Rudich A. Autophagy in adipose tissue. Obes Facts 2012;5:710–21. doi:10.1159/000343983.

[100] Kovsan J, Blüher M, Tarnovscki T, Klöting N, Kirshtein B, Madar L, et al. Altered autophagy in human adipose tissues in obesity. J Clin Endocrinol Metab 2011;96:E268–77. doi:10.1210/jc.2010-1681.

[101] Ost A, Svensson K, Ruishalme I, Brännmark C, Franck N, Krook H, et al. Attenuated mTOR signaling and enhanced autophagy in adipocytes from obese patients with type 2 diabetes. Mol Med 2010;16:235–46. doi:10.2119/molmed.2010.00023.

[102] Jansen HJ, van Essen P, Koenen T, Joosten L a B, Netea MG, Tack CJ, et al. Autophagy activity is up-regulated in adipose tissue of obese individuals and modulates proin-

flammatory cytokine expression. Endocrinology 2012;153:5866–74. doi:10.1210/en.2012-1625.

[103] Kim KH, Lee M-S. Autophagy – a key player in cellular and body metabolism. Nat Rev Endocrinol 2014;10:322–37. doi:10.1038/nrendo.2014.35.

[104] Singh R, Cuervo AM. Lipophagy: connecting autophagy and lipid metabolism. Int J Cell Biol 2012;2012:282041. doi:10.1155/2012/282041.

[105] Cuervo AM, Bergamini E, Brunk UT, Dröge W, Ffrench M, Terman A. Autophagy and aging: the importance of maintaining "clean" cells. Autophagy 2005;1:131–40.

[106] Postic C, Shiota M, Niswender KD, Jetton TL, Chen Y, Moates JM, et al. Dual roles for glucokinase in glucose homeostasis as determined by liver and pancreatic beta cell-specific gene knock-outs using Cre recombinase. J Biol Chem 1999;274:305–15.

[107] Wang Y, Singh R, Xiang Y, Czaja MJ. Macroautophagy and chaperone-mediated autophagy are required for hepatocyte resistance to oxidant stress. Hepatology 2010;52:266–77. doi:10.1002/hep.23645.

[108] He C, Bassik MC, Moresi V, Sun K, Wei Y, Zou Z, et al. Exercise-induced BCL2-regulated autophagy is required for muscle glucose homeostasis. Nature 2012;481:511–5. doi:10.1038/nature10758.

[109] Christian P, Sacco J, Adeli K. Autophagy: Emerging roles in lipid homeostasis and metabolic control. Biochim Biophys Acta 2013;1831:819–24. doi:10.1016/j.bbalip.2012.12.009.

[110] Subramani S, Malhotra V. Non-autophagic roles of autophagy-related proteins. EMBO Rep 2013;14:143–51. doi:10.1038/embor.2012.220.

[111] Fujimoto T, Ohsaki Y, Cheng J, Suzuki M, Shinohara Y. Lipid droplets: a classic organelle with new outfits. Histochem Cell Biol 2008;130:263–79. doi:10.1007/s00418-008-0449-0.

[112] Kwanten WJ, Martinet W, Michielsen PP, Francque SM. Role of autophagy in the pathophysiology of nonalcoholic fatty liver disease: a controversial issue. World J Gastroenterol 2014;20:7325–38. doi:10.3748/wjg.v20.i23.7325.

[113] Karsli-Uzunbas G, Guo JY, Price S, Teng X, Laddha S V., Khor S, et al. Autophagy is required for glucose homeostasis and lung tumor maintenance. Cancer Discov 2014;4:915–27. doi:10.1158/2159-8290.CD-14-0363.

[114] Ezaki J, Matsumoto N, Takeda-Ezaki M, Komatsu M, Takahashi K, Hiraoka Y, et al. Liver autophagy contributes to the maintenance of blood glucose and amino acid levels. Autophagy 2011;7:727–36. doi:10.4161/auto.7.7.15371.

[115] Zhang Y, Fang F, Goldstein JL, Brown MS, Zhao T-J. Reduced autophagy in livers of fasted, fat-depleted, ghrelin-deficient mice: reversal by growth hormone. Proc Natl Acad Sci 2015;112:1226–31. doi:10.1073/pnas.1423643112.

[116] Flamment M, Hajduch E, Ferré P, Foufelle F. New insights into ER stress-induced insulin resistance. Trends Endocrinol Metab 2012;23:381–90. doi:10.1016/j.tem.2012.06.003.

[117] Knævelsrud H, Simonsen A. Lipids in autophagy: constituents, signaling molecules and cargo with relevance to disease. Biochim Biophys Acta 2012;1821:1133–45. doi:10.1016/j.bbalip.2012.01.001.

[118] Camporez JPG, Jornayvaz FR, Petersen MC, Pesta D, Guigni B a, Serr J, et al. Cellular mechanisms by which FGF21 improves insulin sensitivity in male mice. Endocrinology 2013;154:3099–109. doi:10.1210/en.2013-1191.

[119] Yan H, Xia M, Chang X, Xu Q, Bian H, Zeng M, et al. Circulating fibroblast growth factor 21 levels are closely associated with hepatic fat content: a cross-sectional study. PLoS One 2011;6:e24895. doi:10.1371/journal.pone.0024895.

[120] Schmitz-Peiffer C. The tail wagging the dog – regulation of lipid metabolism by protein kinase C. FEBS J 2013;280:5371–83. doi:10.1111/febs.12285.

[121] Jiang H, Cheng D, Liu W, Peng J, Feng J. Protein kinase C inhibits autophagy and phosphorylates LC3. Biochem Biophys Res Commun 2010;395:471–6. doi:10.1016/j.bbrc.2010.04.030.

[122] Patergnani S, Marchi S, Rimessi A, Bonora M, Giorgi C, Mehta KD, et al. PRKCB/protein kinase C, beta and the mitochondrial axis as key regulators of autophagy. Autophagy 2013;9:1367–85. doi:10.4161/auto.25239.

[123] Ding W-X, Li M, Chen X, Ni H-M, Lin C-W, Gao W, et al. Autophagy reduces acute ethanol-induced hepatotoxicity and steatosis in mice. Gastroenterology 2010;139:1740–52. doi:10.1053/j.gastro.2010.07.041.

[124] Takamura A, Komatsu M, Hara T, Sakamoto A, Kishi C, Waguri S, et al. Autophagy-deficient mice develop multiple liver tumors. Genes Dev 2011;25:795–800. doi:10.1101/gad.2016211.

[125] Farrell GC, van Rooyen D, Gan L, Chitturi S. NASH is an inflammatory disorder: pathogenic, prognostic and therapeutic implications. Gut Liver 2012;6:149–71. doi:10.5009/gnl.2012.6.2.149.

[126] Tian Y, Kuo C-F, Sir D, Wang L, Govindarajan S, Petrovic LM, et al. Autophagy inhibits oxidative stress and tumor suppressors to exert its dual effect on hepatocarcinogenesis. Cell Death Differ 2015;22:1025–34. doi:10.1038/cdd.2014.201.

[127] Amir M, Zhao E, Fontana L, Rosenberg H, Tanaka K, Gao G, et al. Inhibition of hepatocyte autophagy increases tumor necrosis factor-dependent liver injury by

promoting caspase-8 activation. Cell Death Differ 2013;20:878–87. doi:10.1038/cdd. 2013.21.

[128] Komatsu M, Waguri S, Koike M, Sou Y-S, Ueno T, Hara T, et al. Homeostatic levels of p62 control cytoplasmic inclusion body formation in autophagy-deficient mice. Cell 2007;131:1149–63. doi:10.1016/j.cell.2007.10.035.

[129] Komatsu M, Kurokawa H, Waguri S, Taguchi K, Kobayashi A, Ichimura Y, et al. The selective autophagy substrate p62 activates the stress responsive transcription factor Nrf2 through inactivation of Keap1. Nat Cell Biol 2010;12:213–23. doi:10.1038/ncb2021.

[130] Ni H-M, Woolbright BL, Williams J, Copple B, Cui W, Luyendyk JP, et al. Nrf2 promotes the development of fibrosis and tumorigenesis in mice with defective hepatic autophagy. J Hepatol 2014;61:617–25. doi:10.1016/j.jhep.2014.04.043.

[131] Müller TD, Lee SJ, Jastroch M, Kabra D, Stemmer K, Aichler M, et al. P62 Links B-adrenergic input to mitochondrial function and thermogenesis. J Clin Invest 2013;123:469–78. doi:10.1172/JCI64209.

[132] Harris J. Autophagy and cytokines. Cytokine 2011;56:140–4. doi:10.1016/j.cyto. 2011.08.022.

[133] Galluzzi L, Pietrocola F, Pedro JMB, Amaravadi RK, Baehrecke EH, Cecconi F, et al. Autophagy in malignant transformation and cancer progression. EMBO J 2015;34:e201490784. doi:10.15252/embj.201490784.

[134] Inami Y, Waguri S, Sakamoto A, Kouno T, Nakada K, Hino O, et al. Persistent activation of Nrf2 through p62 in hepatocellular carcinoma cells. J Cell Biol 2011;193:275–84. doi: 10.1083/jcb.201102031.

[135] Kowalik MA, Perra A, Ledda-Columbano GM, Ippolito G, Piacentini M, Columbano A, et al. Induction of autophagy promotes the growth of early preneoplastic rat liver nodules. Oncotarget 2015. doi:10.18632/oncotarget.6810.

[136] Lee YJ, Jang BK. The role of autophagy in hepatocellular carcinoma. Int J Mol Sci 2015;16:26629–43. doi:10.3390/ijms161125984.

[137] Chen R, Wang Q, Song S, Liu F, He B, Gao X. Protective role of autophagy in methionine-choline deficient diet-induced advanced nonalcoholic steatohepatitis in mice. Eur J Pharmacol 2016;770:126–33. doi:10.1016/j.ejphar.2015.11.012.

[138] Hilscher M, Hernandez-Gea V, Friedman SL. Autophagy and mesenchymal cell fibrogenesis. Biochim Biophys Acta 2012;1831:972–8. doi:10.1016/j.bbadis.2012.11.001.

[139] Friedman SLS. Hepatic stellate cells: protean, multifunctional, and enigmatic cells of the liver. Physiol Rev 2008;88:125–72. doi:10.1152/physrev.00013.2007.

[140] Thoen LFR, Guimarães ELM, Dollé L, Mannaerts I, Najimi M, Sokal E, et al. A role for autophagy during hepatic stellate cell activation. J Hepatol 2011;55:1353–60. doi: 10.1016/j.jhep.2011.07.010.

[141] Hernández-Gea V, Ghiassi-Nejad Z, Rozenfeld R, Gordon R, Fiel MI, Yue Z, et al. Autophagy releases lipid that promotes fibrogenesis by activated hepatic stellate cells in mice and in human tissues. Gastroenterology 2012;142:938–46. doi:10.1053/j.gastro.2011.12.044.

[142] Jin Y, Bai Y, Ni H, Qiang L, Ye L, Shan Y, et al. Activation of autophagy through calcium-dependent AMPK/mTOR and PKCθ pathway causes activation of rat hepatic stellate cells under hypoxic stress. FEBS Lett. 2016 Mar 5;590(5):672–82.

[143] Hetz C. The unfolded protein response: controlling cell fate decisions under ER stress and beyond. Nat Rev Mol Cell Biol 2012;13:89–102. doi:10.1038/nrm3270.

[144] Rashid H-O, Yadav RK, Kim H-R, Chae H-J. ER stress: autophagy induction, inhibition and selection. Autophagy 2015;11:1956–77. doi:10.1080/15548627.2015.1091141.

[145] Dara L, Ji C, Kaplowitz N. The contribution of endoplasmic reticulum stress to liver diseases. Hepatology 2011;53:1752–63. doi:10.1002/hep.24279.

[146] Senft D, Ronai ZA. UPR, autophagy, and mitochondria crosstalk underlies the ER stress response. Trends Biochem Sci 2015. doi:10.1016/j.tibs.2015.01.002.

[147] Wang H, Sun RQ, Zeng XY, Zhou X, Li S, Jo E, et al. Restoration of autophagy alleviates hepatic ER stress and impaired insulin signalling transduction in high fructose-fed male mice. Endocrinology 2014;156:en20141454. doi:10.1210/en.2014-1454.

[148] Jung TW, Hong HC, Hwang H-J, Yoo HJ, Baik SH, Choi KM. C1q/TNF-Related Protein 9 (CTRP9) attenuates hepatic steatosis via the autophagy-mediated inhibition of endoplasmic reticulum stress. Mol Cell Endocrinol 2015;417:131–40. doi:10.1016/j.mce.2015.09.027.

Chapter 22

Autophagy and Coagulation in Liver Cancer and Disorders

Chih-Che Lin, Chih-Wen Lin, Ming-Chao Tsai,
Kuang-Tzu Huang, Chao-Long Chen and
Kuang-Den Chen

Additional information is available at the end of the chapter

http://dx.doi.org/10.5772/64685

Abstract

The physiological role of autophagy in metabolism of the body involves both protein synthesis and degradation. The autophagy-lysosome and the ubiquitin-proteasome systems are the two major intracellular proteolytic mechanisms. Autophagy in hepatocytes is known to be quite active and contribute to its normal functions and the pathogenesis of liver diseases. The role of autophagy in liver diseases has been widely studied, and growing evidence has now shown that autophagy is involved in the pathogenesis of cirrhosis and hepatocellular carcinoma (HCC). However, the role of autophagy in the progression of liver fibrosis and prognosis of human HCC is not well known. Recent studies have demonstrated that tissue factor (TF) combined with coagulation factor VII (FVII) has a pathological role by activating a protease-activated receptor 2 (PAR2) for tumor growth. Autophagy-related LC3A/B-II formation induced by the inhibition of TF/FVII/PAR2 coagulation axis, particularly by FVII knockdown, was selectively mediated by the Atg7 induction. These results are consistent with clinical observations that indicate the important role of FVII activation in regulating autophagy in HCC. In this chapter, we discuss our findings in which FVII promotes growth and progression in HCC through ERK-TSC/mTOR signaling to repress autophagy and may play a pivotal role in conferring cirrhosis and other liver diseases.

Keywords: autophagy, coagulation, hepatocellular carcinoma, cirrhosis, tissue factor, factor VII, protease-activated receptor 2, mammalian target of rapamycin

1. Introduction

Autophagy comprises some of the most fundamental reactions in which the cell sequesters part of its own material for degradation and converts proteins and lipids into life-preserving fuel through times of energy deprivation. In addition, the cell also uses autophagy to remove misfolded protein aggregates and dysfunctional organelles under certain stress conditions [1, 2]. During autophagy, multiple signaling pathways converge on the autophagy-related (Atg) proteins, mediating the formation of a double-membraned structure known as the autophagosome. Recruitment of the Atg12-Atg5 complex and microtubule-associated protein 1 light chain 3 (LC3) is essential for this process and correlates with the level of autophagy. These autophagosomes then fuse with lysosomes to form autolysosomes that lead to the degradation and recycling of their content [3]. Despite the general acceptance that autophagy is a protective mechanism toward cell survival, recent studies have shown an active role of autophagy in cell death [1]. Autophagic cell death is known as the type II programmed cell death in response to several anti-tumor therapies in various types of cancer [3]. Recent studies have revealed the involvement of autophagy in major fields of liver physiology and pathology, including acute/chronic liver injury, lipid accumulation, viral infection, and hepatocellular carcinoma (HCC) [4]. The autophagy pathway may be used by liver cells to generate energy during periods of starvation or exploited as a tumor suppressor mechanism depending on different biological contexts [5]. Studies assessing autophagy in HCC have demonstrated an anti-tumor role in various cellular, animal and clinical models. However, the mechanisms underlying low incidence of autophagy in HCC are not fully elucidated.

The mammalian target of rapamycin (mTOR) is a serine/threonine kinase that controls cell growth and survival and is regulated by various stimuli [6]. The mTOR pathway is abnormally activated in a proportion of HCC patients. As the name implies, mTOR is the intracellular target of rapamycin, a naturally occurring small molecule inhibitor that is currently used clinically as an immunosuppressant and in some cases, to inhibit tumor growth and metastasis [7, 8]. In HCC animal models, inhibition of the mTOR/ribosomal protein p70 S6 kinase (p70S6K) pathway demonstrates anti-proliferative and anti-angiogenic effects [9, 10]. Studies using histone deacetylase (HDAC) inhibitors OSU-HDAC42 and SAHA to induce autophagy in HCC cells further reveal that their anti-cancer properties are mainly through blockade of Akt/mTOR signaling [11].

Our recent publications tied together autophagy and the coagulation cascade in HCC and highlighted the important role of factor VII (FVII)/protease-activated receptor 2 (PAR2) signaling in regulation of autophagy, which is dependent upon mTOR activity. More importantly, an inverse correlation between FVII/PAR2 and LC3 expression in HCC tissues and their contiguous normal regions suggests that components of this particular pathway may serve as potential therapeutic targets in HCC and other cancers which exhibit aberrant FVII/PAR2/LC3 signaling.

2. mTOR interconnects coagulation cascade and autophagy in HCC and related disorders

An established relation between cancer and thrombosis can be dated back to more than a century ago, and the association was further refined and supported among patients who showed venous thromboembolism (VTE) and were subsequently diagnosed with clinically overt cancer [12, 13]. Thrombosis is considered as one of the common complications related to the cancer itself as well as interventions for treating the disease. Approximately 15–20% of VTE events are associated with malignancy [14, 15]. In addition, patients with cancer have a fourfold higher risk for VTE compared with non-cancer patients. This risk is even higher in patients who undergo chemotherapy [16]. Recurrent VTE is also twice as likely to occur in cancer patients, even in those who experience oral anti-coagulant therapy [17, 18]. Notably, patients with cancer who have a thrombotic event have reduced survival compared with those who do not [19].

The transmembrane tissue factor (TF) initiates blood coagulation cascade that complexes with activated FVII and transmits signals through direct activation of protease-activated receptors (PARs) [20–22]. In fact, dysregulated cancer cells may have an important contribution to the elevated levels of circulating TF, which in turn activates the coagulation system [23]. Cancer cells release plasma membrane vesicles with pro-coagulant activity. Such activity has been shown to behave as TF, requiring FVII for activation [24]. The role of direct TF signaling in cancer is not fully understood. The best known pathway of TF-dependent signaling is through the activation of PAR2 [25, 26] and in part through the thrombin-PAR1 pathway [27, 28]. It has also been reported a constitutive association of TF with β1 integrins in cancer [25]. Inhibition using specific antibodies or peptide inhibitors concludes that blockade of the FVII/TF/PAR2 signaling independent of the coagulation response can suppress cancer progression [25, 29]. TF expression varies among different types of malignancies; some may be more pro-thrombotic than others. TF levels can also be influenced by tumor staging as well as associated therapeutic interventions [30, 31].

Plasma TF levels are closely related to occurrence of chronic liver diseases [32, 33]. The diagnostic/prognostic value of TF in tumor pathology has also been demonstrated [34]. Angiogenesis is recognized as an important factor in the development, progression, and recurrence of HCC, and targeting vascularization has been vigorously studied for potential therapeutic strategies [35–37]. A recent study to evaluate the correlation between TF expression with tumor angiogenesis and invasiveness in HCC suggests that tissue TF levels have a significant association with microvascular density, venous invasion, microsatellite nodules, tumor staging, and survival [38]. Zhou et al. also demonstrates that TF is overexpressed in both plasma and liver tissue of HCC patients, and it is closely related to many invasive and metastasis indexes [39]. Furthermore, hepatocytes occupy more than half of the total liver volume and carry out critical functions in coagulation factor synthesis (TF, FVII, etc.) in the liver. It has also been discussed that extremely low levels of hepatocyte TF compared with other organs/tissues [40, 41], however, may be the potential source for the activation of coagulation in liver diseases [42].

TF is an essential cofactor of FVII, which accelerates the conversion of an inactive FVII into an active FVIIa, propagating a series of serine protease activation events. In addition to pathogenic mechanisms in which TF triggers a pro-coagulant state and aberrant signaling, when TF-expressing tumor cells are in contact with the blood or when TF-positive membrane particles are shed into the circulation, constitutive expression of FVII also participates in the process of tumor invasion and metastasis [43]. Studies that target the TF/FVIIa signaling pathway with specific inhibitor or RNA interference provide a logical path for the development of potential therapeutics [44, 45]. In our clinical observations in HCC, we were the first to demonstrate that the expressions of FVII and PAR2 were inversely correlated with the amount of autophagic effector proteins LC3A/B-II [46]. We have also demonstrated that the treatment with recombinant TF (rTF), rFVII, or PAR2 agonist decreased expression of LC3A/B-II protein in cultured Hep3B cells suggesting a crucial impact of the TF/FVII/PAR2 coagulation pathway on tumor malignancy under certain circumstances. The dependence of mTOR activation on pathological thrombosis has been seen in various studies, in which the risk of thrombotic events while patients were receiving organ transplants [47–52] and coronary stents [53] was associated with TF expression. In this study, treatment with rTF, rFVII, and the PAR2 peptide agonist, rather than thrombin and PAR1 agonist, induced activation and expression of mTOR whereas silencing of TF, FVII, or PAR2 by siRNAs repressed its phosphorylation and expression. Additionally, rTF, rFVII, or PAR2 activation drastically inhibited the LC3A/B-II expression levels which were fully rescued by mTOR knockdown or treatment of mTOR inhibitors *in vitro*. The results illustrated that repression of autophagy by TF/FVII/PAR2 relies upon mTOR activity.

Our study indicated that recombinant FVII, TF, and a PAR2 agonist increased expression of mTOR whereas thrombin or PAR1 agonists did not. Gene silencing of FVII, TF, or PAR2 decreased mTOR. The decrease of LC3A/B-II initiated by recombinant FVII/TF/PAR2 agonist was rescued by mTOR knockdown *in vitro*. These results indicated an mTOR-dependent repression of autophagy by the FVII/TF/PAR2 signaling.

Although our observations suggest that the inhibition of autophagy by the FVII/TF/PAR2 signaling involves mTOR activity, targeting mTOR itself may result in differential outcomes. The relationship between autophagy and mTOR signaling has been comprehensively studied in terms of tumor suppression. However, administration of metformin, which negatively regulates mTOR by activating adenosine monophosphate kinase (AMPK), shows a contradictory effect, suggesting that blockade of mTOR activity does not always lead to autophagy [54]. Other signaling companions may be responsible. Another example is sirolimus (rapamycin). Our results showed that an increase of LC3A/B-II by sirolimus was partially inhibited by recombinant TF and completely blocked in the presence of recombinant FVII or a PAR2 agonist.

Taken together, our results highlight an essential role of the FVII/TF/PAR2 signaling in regulation of autophagy in HCC, which nicely correlates with our observations in clinical and animal research (**Figure 1**). In addition, a recent study demonstrated differential effects of valproic acid-induced autophagy among various human prostate cancer cell lines. The differences were likely due to the existence of alternatively spliced, inactive forms of Atg5 that dampens the formation of Atg12-Atg5 conjugates [55]. Therefore, careful interpretation is

http://dx.doi.org/10.5772/64685

necessary especially in cancer research when aberrant signaling in the autophagic pathway may result from gene variants that do not normally exist in cells.

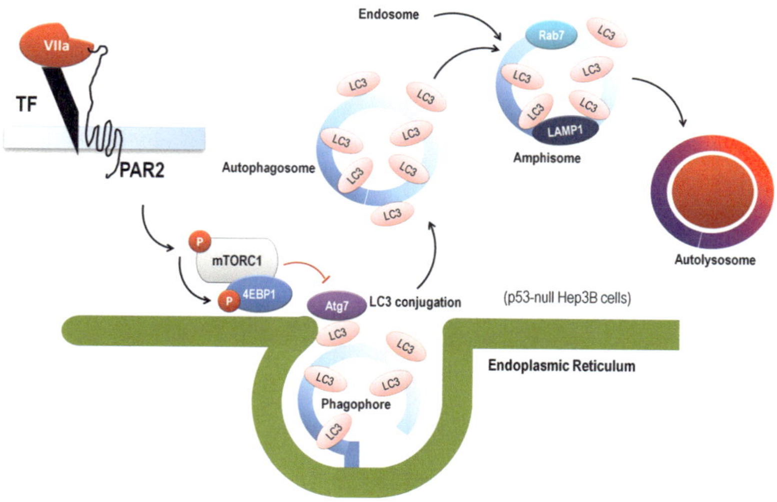

Figure 1. Schematic representation shows that PAR2 transduces the TF/FVII coagulation signaling through mTOR-dependent inhibition in autophagy flux, which may facilitate development and progression of HCC.

Our recent study has demonstrated that the activation of the FVII/TF/PAR2 axis correlates with increased migratory and invasive properties in HCC cells [56]. This finding can be consistently recapitulated in tumor tissues of HCC patients, in which we found that increased levels of FVII and PAR2 were significantly correlated with clinical staging, increased invasion and worse disease-free survival. Notably, the signals that drive FVII/PAR2 stimulation toward cell migration are mainly through ERK-TSC, independent of other coagulation effectors such as thrombin/PAR1. Interestingly, in our cellular model, only FVII, but not soluble TF, activates ERK1/2 through PAR2 signaling. Gene knockdown of FVII abrogates migration and invasion of HCC cells more effectively than knocking down TF. Our speculation is that TF is constitutively expressed in HCC tissue, and therefore, the cells are less sensitive to changes in TF levels. Therefore, the amount of FVIIa determines the ratio that is engaged to regulate PAR2 signaling. Although TF upregulation is well linked to aggressiveness in several cancers, our results are more consistent with the findings from Rullier et al., in which no association between TF levels and clinicopathological characteristics of HCC was found [57]. Several studies also show limited contribution of TF to tumor growth [58–60]. Therefore, the role of TF in cancer progression may be essential in some but not all cancers, and TF may not be a reliable prognostic marker at least for HCC progression. In addition, the animal model using mouse xenografts can well reflect our clinical findings in which we found that FVIIa administration only positively affected vascular density but not size and number of the inoculated tumor. These results were consistent with the expression of FVII in liver tissues of HCC patients which was associated with vascular invasion and capsulations but not the number and size of tumor.

It has been generally accepted that tumor cell motility is necessary for cancer metastasis [61]. The molecular basis to acquire ability to colonize other organs by invading tumor cells has been long studied, but it remains an unmet challenge in therapeutic control on disseminated tumors [62, 63]. Especially in China and other East Asian countries, survival of HCC patients has improved due to advanced surgical skills and technologies such as orthotopic liver transplantation and perioperative medication, prognosis and long-term survival after surgical resection remains low owing to risk of invasive recurrence [64, 65]. Thus, there is an urgent need to identify new targets responsible for impaired metastatic mechanisms and develop of novel therapeutic strategies as well as preoperative biomarkers that supplement current treatment protocols. Although potential risk of bleeding using specific FVII antagonists exists, a recent clinical study has claimed that PCI-27483, a selective inhibitor of FVIIa, is well tolerated in advanced pancreatic cancer patients [44]. Our previous studies also found that two metastatic suppressors, NME/NM23 nucleoside diphosphate kinase 1 (NME1) and the basic helix-loop-helix family member e41 (BHLHE41), were highly induced in FVII/PAR2 knocked down HCC cells. These findings support the idea of targeting the FVII/PAR2 pathway and provide mechanistic insights that specific members involved in autophagic flux could be potential targets for treatment of metastatic HCC.

Emerging cohort studies indicate that HCC is currently the major cause of death in patients with compensated hepatic cirrhosis. The mortality rate of HCC associated with cirrhosis is increasing, whereas mortality rate from non-HCC complications with cirrhosis is reducing and stable. Viral-related cirrhosis especially with hepatitis C virus infection is associated with the highest HCC incidence in cirrhotic cases, occurring with almost two times in East Asia than in the West [66]. Liver cirrhosis is a slowly progressive disease that enhances extracellular matrix (ECM) accumulation after chronic injury, in which healthy liver tissue is replaced with scar tissue and poor function is seen at the terminal stages.

It is generally accepted that the vast majority of chronic liver disease patients with cirrhosis have a dysregulated coagulation system [67, 68]. A growing number of studies represented a thrombotic risk in patients with chronic liver disease [69–72]. Thus, the reevaluation of homeostasis in patients with thrombotic tendency in cirrhosis challenges the dogma that considered this coagulopathy an acquired bleeding disorder and featured in most hematology textbooks [67]. However, the mechanistic basis for this hypercoagulable state in cirrhotic patients of chronic liver disease is not yet understood. A recent study have suggested that hepatocytes are the source of increased TF microparticles, and hepatocyte TF may contribute to the activation of coagulation in patients with chronic liver disease [42]. We recently demonstrated that hepatic steatosis and liver injury by alcohol (AFLD) were exacerbated by chloroquine (an autophagy inhibitor), but alleviated by carbamazepine (an autophagy promoter) or rapamycin (an mTOR inhibitor) [73]. The protective effects of carbamazepine and rapamycin in reducing steatosis were also represented in high fat diet-induced non-alcoholic fatty liver conditions (NAFLD). Furthermore, we also found that a second autophagy promoter amiodarone significantly reduced liver injury and improved liver regeneration and survival after 90% partial hepatectomy in a mouse model [74]. Our data suggests that pharmacological modulation of autophagy in the liver can be an effective strategy for alleviating liver injury, improve proliferative recovery, and may also ameliorate progression of liver cirrhosis.

Taken together, TF/FVII signaling is the main initiator of the extrinsic coagulation cascade, which is also the major contributor in modulating the systemic balance of homeostasis in healthy persons as well as in response to the pathogenesis in patients with chronic liver disease. Amiodarone is now a potential drug to treat HCC through the modulation of autophagy to decrease oncogenic miR-224 expression [75]. Thus, increasing evidences including our studies support a close relationship between TF/FVII coagulation and liver disease in association with reduced autophagy, suggesting pharmacological modulation of autophagy for AFLD, NAFLD, and/or HCC could be a potential strategy for clinical uses.

Conflicts

I confirm there are no conflicts of interest.

Author details

Chih-Che Lin[1], Chih-Wen Lin[2], Ming-Chao Tsai[3], Kuang-Tzu Huang[1], Chao-Long Chen[1] and Kuang-Den Chen[1*]

*Address all correspondence to: dennis8857@gmail.com

1 Institute for Translational Research in Biomedicine, Liver Transplantation Program and Department of Surgery, Kaohsiung Chang Gung Memorial Hospital and Chang Gung University College of Medicine, Niao-Sung District, Kaohsiung, Taiwan

2 Graduate Institute of Medicine, College of Medicine, Kaohsiung Medical University, Kaohsiung, Taiwan

3 Division of Hepato-Gastroenterology, Department of Internal Medicine, Kaohsiung Chang Gung Memorial Hospital, Kaohsiung, Taiwan

References

[1] Baehrecke EH. Autophagy: dual roles in life and death? Nat Rev Mol Cell Biol. 2005;6:505–510. doi:10.1038/nrm1666

[2] Komatsu M, Waguri S, Ueno T, Iwata J, Murata S, et al. Impairment of starvation-induced and constitutive autophagy in Atg7-deficient mice. J Cell Biol. 2005;169:425–434. doi:10.1083/jcb.200412022

[3] Kondo Y, Kanzawa T, Sawaya R, Kondo S. The role of autophagy in cancer development and response to therapy. Nat Rev Cancer. 2005;5:726–734. doi:10.1038/nrc1692

[4] Rautou PE, Mansouri A, Lebrec D, Durand F, Valla D, et al. Autophagy in liver diseases. J Hepatol. 2010;53:1123–1134. doi:10.1016/j.jhep.2010.07.006

[5] Carew JS, Kelly KR, Nawrocki ST. Autophagy as a target for cancer therapy: new developments. Cancer Manag Res. 2012;4:357–365. doi:10.2147/CMAR.S26133

[6] Guertin DA, Sabatini DM. Defining the role of mTOR in cancer. Cancer Cell. 2007;12:9–22. doi:10.1016/j.ccr.2007.05.008

[7] Wang Z, Zhou J, Fan J, Tan CJ, Qiu SJ, et al. Sirolimus inhibits the growth and metastatic progression of hepatocellular carcinoma. J Cancer Res Clin Oncol. 2009;135:715–722. doi:10.1007/s00432-008-0506-z

[8] Wang Z, Zhou J, Fan J, Qiu SJ, Yu Y, et al. Effect of rapamycin alone and in combination with sorafenib in an orthotopic model of human hepatocellular carcinoma. Clin Cancer Res. 2008;14:5124–5130. doi:10.1158/1078-0432.CCR-07-4774

[9] Semela D, Piguet AC, Kolev M, Schmitter K, Hlushchuk R, et al. Vascular remodeling and antitumoral effects of mTOR inhibition in a rat model of hepatocellular carcinoma. J Hepatol. 2007;46:840–848. doi:10.1016/j.jhep.2006.11.021

[10] Bu X, Jia F, Wang W, Guo X, Wu M, et al. Coupled down-regulation of mTOR and telomerase activity during fluorouracil-induced apoptosis of hepatocarcinoma cells. BMC Cancer. 2007;7:208. doi:10.1186/1471-2407-7-208

[11] Liu YL, Yang PM, Shun CT, Wu MS, Weng JR, et al. Autophagy potentiates the anti-cancer effects of the histone deacetylase inhibitors in hepatocellular carcinoma. Autophagy. 2010;6:1057–1065. doi:10.4161/auto.6.8.13365

[12] Sack GH, Jr., Levin J, Bell WR. Trousseau's syndrome and other manifestations of chronic disseminated coagulopathy in patients with neoplasms: clinical, pathophysiologic, and therapeutic features. Medicine (Baltimore). 1977;56:1–37.

[13] Prandoni P, Lensing AW, Buller HR, Cogo A, Prins MH, et al. Deep-vein thrombosis and the incidence of subsequent symptomatic cancer. N Engl J Med. 1992;327:1128–1133. doi:10.1056/NEJM199210153271604

[14] Rickles FR, Edwards RL. Activation of blood coagulation in cancer: Trousseau's syndrome revisited. Blood. 1983;62:14–31.

[15] Donati MB. Cancer and thrombosis: from Phlegmasia alba dolens to transgenic mice. Thromb Haemost. 1995;74:278–281.

[16] Heit JA. Cancer and venous thromboembolism: scope of the problem. Cancer Control. 2005;12(Suppl 1):5–10.

[17] Levitan N, Dowlati A, Remick SC, Tahsildar HI, Sivinski LD, et al. Rates of initial and recurrent thromboembolic disease among patients with malignancy versus those without malignancy. Risk analysis using Medicare claims data. Medicine (Baltimore). 1999;78:285–291.

[18] Bona RD, Hickey AD, Wallace DM. Warfarin is safe as secondary prophylaxis in patients with cancer and a previous episode of venous thrombosis. Am J Clin Oncol. 2000;23:71–73.

[19] Sorensen HT, Mellemkjaer L, Olsen JH, Baron JA. Prognosis of cancers associated with venous thromboembolism. N Engl J Med. 2000;343:1846–1850. doi:10.1056/NEJM200012213432504

[20] Mackman N. Role of tissue factor in hemostasis, thrombosis, and vascular development. Arterioscler Thromb Vasc Biol. 2004;24:1015–1022. doi:10.1161/01.ATV.0000130465.23430.74

[21] Camerer E, Huang W, Coughlin SR. Tissue factor- and factor X-dependent activation of protease-activated receptor 2 by factor VIIa. Proc Natl Acad Sci U S A. 2000;97:5255–5260.

[22] Mann KG. Biochemistry and physiology of blood coagulation. Thromb Haemost. 1999;82:165–174.

[23] Rickles FR, Hair GA, Zeff RA, Lee E, Bona RD. Tissue factor expression in human leukocytes and tumor cells. Thromb Haemost. 1995;74:391–395.

[24] Dvorak HF, Van DeWater L, Bitzer AM, Dvorak AM, Anderson D, et al. Procoagulant activity associated with plasma membrane vesicles shed by cultured tumor cells. Cancer Res. 1983;43:4434–4442.

[25] Versteeg HH, Schaffner F, Kerver M, Petersen HH, Ahamed J, et al. Inhibition of tissue factor signaling suppresses tumor growth. Blood. 2008;111:190–199. doi:10.1182/blood-2007-07-101048

[26] Ruf W, Disse J, Carneiro-Lobo TC, Yokota N, Schaffner F. Tissue factor and cell signalling in cancer progression and thrombosis. J Thromb Haemost. 2011;9(Suppl 1):306–315. doi:10.1111/j.1538-7836.2011.04318.x

[27] Ruf W, Mueller BM. Thrombin generation and the pathogenesis of cancer. Semin Thromb Hemost. 2006;32(Suppl 1):61–68. doi:10.1055/s-2006-939555

[28] Nierodzik ML, Karpatkin S. Thrombin induces tumor growth, metastasis, and angiogenesis: Evidence for a thrombin-regulated dormant tumor phenotype. Cancer Cell. 2006;10:355–362. doi:10.1016/j.ccr.2006.10.002

[29] Carneiro-Lobo TC, Schaffner F, Disse J, Ostergaard H, Francischetti IM, et al. The tick-derived inhibitor Ixolaris prevents tissue factor signaling on tumor cells. J Thromb Haemost. 2012;10:1849–1858. doi:10.1111/j.1538-7836.2012.04864.x

[30] Kakkar AK, Lemoine NR, Scully MF, Tebbutt S, Williamson RC. Tissue factor expression correlates with histological grade in human pancreatic cancer. Br J Surg. 1995;82:1101–1104.

[31] White RH, Chew H, Wun T. Targeting patients for anticoagulant prophylaxis trials in patients with cancer: who is at highest risk? Thromb Res. 2007;120(Suppl 2):S29–S40. doi:10.1016/S0049-3848(07)70128-7

[32] Panasiuk A, Zak J, Panasiuk B, Prokopowicz D. Increase in expression of monocytic tissue factor (CD142) with monocytes and blood platelet activation in liver cirrhosis. Blood Coagul Fibrinolysis. 2007;18:739–744. doi:10.1097/MBC.0b013e3282ef99f6

[33] Tacke F, Schoffski P, Trautwein C, Manns MP, Ganser A, et al. Tissue factor and thrombomodulin levels are correlated with stage of cirrhosis in patients with liver disease. Blood Coagul Fibrinolysis. 2001;12:539–545.

[34] Lwaleed BA, Francis JL, Chisholm M. Tissue factor assays as diagnostic tools for cancer? Correlation between urinary and monocyte tissue factor activity. J Hematother Stem Cell Res. 1999;8:659–668.

[35] Coulon S, Heindryckx F, Geerts A, Van Steenkiste C, Colle I, et al. Angiogenesis in chronic liver disease and its complications. Liver Int. 2011;31:146–162. doi:10.1111/j.1478-3231.2010.02369.x

[36] Sugimachi K, Tanaka S, Terashi T, Taguchi K, Rikimaru T. The mechanisms of angiogenesis in hepatocellular carcinoma: angiogenic switch during tumor progression. Surgery. 2002;131:S135–S141.

[37] Wu XZ. New strategy of antiangiogenic therapy for hepatocellular carcinoma. Neoplasma. 2008;55:472–481.

[38] Poon RT, Lau CP, Ho JW, Yu WC, Fan ST, et al. Tissue factor expression correlates with tumor angiogenesis and invasiveness in human hepatocellular carcinoma. Clin Cancer Res. 2003;9:5339–5345.

[39] Zhou Q, Huang T, Wang YF, Zhou XB, Liang LJ, et al. Role of tissue factor in hepatocellular carcinoma genesis, invasion and metastasis. Chin Med J (Engl). 2011;124:3746–3751.

[40] Mackman N, Sawdey MS, Keeton MR, Loskutoff DJ. Murine tissue factor gene expression in vivo. Tissue and cell specificity and regulation by lipopolysaccharide. Am J Pathol. 1993;143:76–84.

[41] Parry GC, Erlich JH, Carmeliet P, Luther T, Mackman N. Low levels of tissue factor are compatible with development and hemostasis in mice. J Clin Invest. 1998;101:560–569. doi:10.1172/JCI814

[42] Rautou PE, Tatsumi K, Antoniak S, Owens AP, 3rd, Sparkenbaugh E, et al. Hepatocyte tissue factor contributes to the hypercoagulable state in a mouse

model of chronic liver injury. J Hepatol. 2016;64:53–59. doi:10.1016/j.jhep.2015.08.017

[43] Koizume S, Jin MS, Miyagi E, Hirahara F, Nakamura Y, et al. Activation of cancer cell migration and invasion by ectopic synthesis of coagulation factor VII. Cancer Res. 2006;66:9453–9460. doi:10.1158/0008-5472.CAN-06-1803

[44] Liu Y, Jiang P, Capkova K, Xue D, Ye L, et al. Tissue factor-activated coagulation cascade in the tumor microenvironment is critical for tumor progression and an effective target for therapy. Cancer Res. 2011;71:6492–6502. doi:10.1158/0008-5472.CAN-11-1145

[45] Yu JL, May L, Lhotak V, Shahrzad S, Shirasawa S, et al. Oncogenic events regulate tissue factor expression in colorectal cancer cells: implications for tumor progression and angiogenesis. Blood. 2005;105:1734–1741. doi:10.1182/blood-2004-05-2042

[46] Chen KD, Wang CC, Tsai MC, Wu CH, Yang HJ, et al. Interconnections between autophagy and the coagulation cascade in hepatocellular carcinoma. Cell Death Dis. 2014;5:e1244. doi:10.1038/cddis.2014.212

[47] Guba M, Yezhelyev M, Eichhorn ME, Schmid G, Ischenko I, et al. Rapamycin induces tumor-specific thrombosis via tissue factor in the presence of VEGF. Blood. 2005;105:4463–4469. doi:10.1182/blood-2004-09-3540

[48] Trotter JF. Sirolimus in liver transplantation. Transplant Proc. 2003;35:193S–200S. doi:10.1016/S0041-1345(03)00234-3

[49] Fortin MC, Raymond MA, Madore F, Fugere JA, Paquet M, et al. Increased risk of thrombotic microangiopathy in patients receiving a cyclosporin-sirolimus combination. Am J Transplant. 2004;4:946–952. doi:10.1111/j.1600-6143.2004.00428.x

[50] Paramesh AS, Grosskreutz C, Florman SS, Gondolesi GE, Sharma S, et al. Thrombotic microangiopathy associated with combined sirolimus and tacrolimus immunosuppression after intestinal transplantation. Transplantation. 2004;77:129–131. doi:10.1097/01.TP.0000092522.36410.D0

[51] Barone GW, Gurley BJ, Abul-Ezz SR, Gokden N. Sirolimus-induced thrombotic microangiopathy in a renal transplant recipient. Am J Kidney Dis. 2003;42:202–206. doi:10.1016/S0272-6386(03)00424-4

[52] Robson M, Cote I, Abbs I, Koffman G, Goldsmith D. Thrombotic micro-angiopathy with sirolimus-based immunosuppression: potentiation of calcineurin-inhibitor-induced endothelial damage? Am J Transplant. 2003;3:324–327. doi:10.1034/j.1600-6143.2003.00051.x

[53] Choi SB. CYPHER coronary stents and risk of thrombosis. CMAJ. 2003;169:218.

[54] Ben Sahra I, Laurent K, Giuliano S, Larbret F, Ponzio G, et al. Targeting cancer cell metabolism: the combination of metformin and 2-deoxyglucose induces p53-depend-

ent apoptosis in prostate cancer cells. Cancer Res. 2010;70:2465–2475. doi: 10.1158/0008-5472.CAN-09-2782

[55] Ouyang DY, Xu LH, He XH, Zhang YT, Zeng LH, et al. Autophagy is differentially induced in prostate cancer LNCaP, DU145 and PC-3 cells via distinct splicing profiles of ATG5. Autophagy. 2013;9:20–32. doi:10.4161/auto.22397

[56] Chen KD, Huang KT, Tsai MC, Wu CH, Kuo IY, et al. Coagulation factor VII and malignant progression of hepatocellular carcinoma. Cell Death Dis. 2016;7:e2110. doi: 10.1038/cddis.2015.395

[57] Rullier A, Senant N, Kisiel W, Bioulac-Sage P, Balabaud C, et al. Expression of protease-activated receptors and tissue factor in human liver. Virchows Arch. 2006;448:46–51. doi:10.1007/s00428-005-0078-0

[58] Toomey JR, Kratzer KE, Lasky NM, Broze GJ, Jr. Effect of tissue factor deficiency on mouse and tumor development. Proc Natl Acad Sci U S A. 1997;94:6922–6926.

[59] Bromberg ME, Sundaram R, Homer RJ, Garen A, Konigsberg WH. Role of tissue factor in metastasis: functions of the cytoplasmic and extracellular domains of the molecule. Thromb Haemost. 1999;82:88–92.

[60] Palumbo JS, Talmage KE, Massari JV, La Jeunesse CM, Flick MJ, et al. Tumor cell-associated tissue factor and circulating hemostatic factors cooperate to increase metastatic potential through natural killer cell-dependent and-independent mechanisms. Blood. 2007;110:133–141. doi:10.1182/blood-2007-01-065995

[61] Liotta LA. Tumor invasion and metastases—role of the extracellular matrix: Rhoads Memorial Award lecture. Cancer Res. 1986;46:1–7.

[62] Iaccarino I, Martins LM. Therapeutic targets in cancer cell metabolism and death. Cell Death Differ. 2011;18:565–570. doi:10.1038/cdd.2010.174

[63] Steeg PS. Perspective: the right trials. Nature. 2012;485:S58–S59. doi:10.1038/485S58a

[64] Block TM, Mehta AS, Fimmel CJ, Jordan R. Molecular viral oncology of hepatocellular carcinoma. Oncogene. 2003;22:5093–5107. doi:10.1038/sj.onc.1206557

[65] Bosch FX, Ribes J, Diaz M, Cleries R. Primary liver cancer: worldwide incidence and trends. Gastroenterology. 2004;127:S5–S16. doi:10.1053/j.gastro.2004.09.011

[66] Fattovich G, Stroffolini T, Zagni I, Donato F. Hepatocellular carcinoma in cirrhosis: incidence and risk factors. Gastroenterology. 2004;127:S35–S50. doi:10.1053/j.gastro.2004.09.014

[67] Tripodi A, Mannucci PM. The coagulopathy of chronic liver disease. N Engl J Med. 2011;365:147–156. doi:10.1056/NEJMra1011170

[68] Lisman T, Caldwell SH, Burroughs AK, Northup PG, Senzolo M, et al. Hemostasis and thrombosis in patients with liver disease: the ups and downs. J Hepatol. 2010;53:362–371. doi:10.1016/j.jhep.2010.01.042

[69] Northup PG, McMahon MM, Ruhl AP, Altschuler SE, Volk-Bednarz A, et al. Coagulopathy does not fully protect hospitalized cirrhosis patients from peripheral venous thromboembolism. Am J Gastroenterol. 2006;101:1524–1528; quiz 1680. doi:10.1111/j.1572-0241.2006.00588.x

[70] Sogaard KK, Horvath-Puho E, Gronbaek H, Jepsen P, Vilstrup H, et al. Risk of venous thromboembolism in patients with liver disease: a nationwide population-based case-control study. Am J Gastroenterol. 2009;104:96–101. doi:10.1038/ajg.2008.34

[71] Gulley D, Teal E, Suvannasankha A, Chalasani N, Liangpunsakul S. Deep vein thrombosis and pulmonary embolism in cirrhosis patients. Dig Dis Sci. 2008;53:3012–3017. doi:10.1007/s10620-008-0265-3

[72] Tripodi A, Anstee QM, Sogaard KK, Primignani M, Valla DC. Hypercoagulability in cirrhosis: causes and consequences. J Thromb Haemost. 2011;9:1713–1723. doi:10.1111/j.1538-7836.2011.04429.x

[73] Lin CW, Zhang H, Li M, Xiong X, Chen X, et al. Pharmacological promotion of autophagy alleviates steatosis and injury in alcoholic and non-alcoholic fatty liver conditions in mice. J Hepatol. 2013;58:993–999. doi:10.1016/j.jhep.2013.01.011

[74] Lin CW, Chen YS, Lin CC, Chen YJ, Lo GH, et al. Amiodarone as an autophagy promoter reduces liver injury and enhances liver regeneration and survival in mice after partial hepatectomy. Sci Rep. 2015;5:15807. doi:10.1038/srep15807

[75] Lan SH, Wu SY, Zuchini R, Lin XZ, Su IJ, et al. Autophagy suppresses tumorigenesis of hepatitis B virus-associated hepatocellular carcinoma through degradation of microRNA-224. Hepatology. 2014;59:505–517. doi:10.1002/hep.26659

Chapter 23

Role of Autophagy in Cancer Metabolism

Heesun Cheong

Additional information is available at the end of the chapter

http://dx.doi.org/10.5772/64025

Abstract

Cancer cells undergo a wide range of metabolic reprogramming to take advantage for supporting rapid growth and survival. Autophagy plays a critical role in directly regulating cellular metabolism as a main catabolic process mediated by lysosomal degradation in response to the metabolic stress. During cancer development, autophagy plays opposite functions in suppressing or promoting tumors dependent of distinct stage. Autophagy maintains cellular homeostasis by degrading unnecessary cellular molecules and oncogenic products, thereby suppressing tumorigenesis. By contrast, autophagy enables to promote cancer growth in advanced tumor by supplying nutrients and relieving metabolic stress.

In this book chapter, recent progress indicates how autophagy is integrated with cellular metabolic alteration during cancer development, particularly focusing on distinct metabolic substrates including glucose or glutamine. Multiple mechanisms would be suggested to explain the functions of autophagy at distinct stage of tumor progression. Cancer metabolic alterations associated with autophagy can be determined by certain oncogenic activators and/or tumor suppressors. Understanding the molecular mechanism of autophagy and metabolic alteration during cancer development may suggest potential targets for therapeutic intervention.

Keywords: autophagy, glucose metabolism, glutamine metabolism, macropinocytosis

1. Introduction

Autophagy is a lysosome-mediated self-degradation process in which cytosolic components and organelles are sequestered into membrane-bound vesicles called autophagosomes and are delivered to lysosomes. Autophagic cargo contents are ultimately degraded and recycled back to the cytoplasm for supporting cell metabolic processes. Most Atg genes have been identified and characterized by genetic screening in yeast. The mutants harboring autophagy genes showed severe growth defect upon nitrogen starvation, although some genes were identified by the other-distinct nutrient starvation conditions. Autophagy process is evolutionally conserved from yeast to mammals, and more than 30 Atg proteins have also been found in mammals. Accordingly, autophagy process is utilized to maintain constant nutrient balance, which is important for certain stages of cellular development and physiology. Autophagy process can be categorized at distinct steps from phagophore induction, vesicle nucleation, expansion, fusion to the lysosome, and degradation of autophagic cargoes [1–3].

A key nutrient-signaling molecule, mammalian target of rapamycin (mTOR), has been identified as a major regulator of autophagy activity. mTOR activated by nutrients and growth factors usually suppresses autophagy through the direct phosphorylation of ULK1/2 and Atg13 [4, 5]. A well-known energy-sensing factor, adenosine monophosphate-activated protein kinase (AMPK), also positively regulates autophagy depending on the ratio of intracellular AMP/ATP levels. Activated AMPK by low-energy levels phosphorylates a series of autophagy proteins including ULK1 and Beclin1/VPS34complex distinctly, thereby enhancing autophagy activity. AMPK-mediated autophagy regulation occurs either through mTOR inactivation or through direct phosphorylation of ULK1 [6–8].

Cancer cells generally enhanced metabolic demands for supporting rapid proliferation, thereby ultimately altering cell metabolism toward anabolic-addicted condition. Metabolic stress often occurs in fast-growing tumor cells and the tumor microenvironment, which is caused by the lack of sufficient nutrient and oxygen. To overcome this metabolic hurdle, tumor cells engage in metabolic reprogramming and autophagy to increase intracellular nutrient supplies to support cell growth and survival [9, 10]. A typical catabolic process, autophagy, might support anabolic pathways such as macromolecule synthesis by supplying intracellular metabolites to the cell through degradation of cellular constitutes in lysosome-mediated manner. Substantial evidence for the integration of autophagy and metabolic alterations is reported, even though it is still not completely understood how these two processes are mechanistically balanced to promote cancer development.

2. Tumor-suppressing role of autophagy

Autophagy is considered to have both tumor-suppressing and tumor-promoting roles during cancer progression. This functional duality can be determined by the oncogenic feature of the primary tumor including oncogene types or the levels of tumor suppressors. In addition, the cellular context of the tumor such as tumor type and tumor stage might also be critical determinants for explaining the complex interactions of autophagy and tumor development.

The tumor-suppressing role of autophagy was characterized that monoallelic deletion of *BECN1* was observed in various human cancers in breast, ovarian, and prostate [11]. Mice with monoallelic loss of BECN1 spontaneously develop lymphoma, hepatocellular carcinoma, and lung adenocarcinomas, suggesting that Beclin1 is a haploin-sufficient tumor-suppressor protein [12]. Moreover, Beclin1-interacting autophagic proteins such as UVRAG and Bif-1 exhibited tumor-suppressor roles in different mouse models [13, 14]. Similarly, mice harboring monoallelic deletion for the BECN1 interactor autophagy/beclin-1 regulator 1 (AMBRA1) also exhibit increased spontaneous tumorigenesis, through that preventing AMBRA1-mediated deregulation of c-Myc [15]. Moreover, mice bearing a systemic mosaic deletion of Atg5 or a liver-specific knockout of Atg7 spontaneously develop benign hepatic neoplasms more frequently than their wild-type counterparts [16].

To address the role of autophagy as tumor suppressor, a series of reports have suggested that multiple oncogenes can be degraded by autophagy processes. An autophagy cargo receptor p62/SQSM1, an autophagic cargo receptor which is expected to be degraded by autophagy, plays an oncogenic role in promoting cancer progression. Overexpression of p62/SQSM1 in KRas-induced tumor cells exhibits to increase pro-inflammatory responses through the Nrf2 and nuclear factor-kappa B (NF–κB) activation, thereby leading to tumor progression. These results indicate that p62 accumulation due to autophagy defect is strongly correlated with tumor development [17–19].

The mechanism of how p62/SQSTM1 regulates Nrf2 activity elucidates the tumor-suppressing role of autophagy. The transcription factor Nrf2 is known to activate the expression of oncogenes involved in angiogenesis and cell survival. The p62/SQSTM1 is competing with Nrf2 for binding Keap1 in the E3 ubiquitin ligase complex. Keap1 usually enables Nrf2 to be ubiquitinated, thereby inducing its degradation under normal conditions. Under autophagy-defective conditions, accumulated p62/SQSTM1 directly competes with Nrf2 to interact with Keap1, thereby preventing Keap1-mediated Nrf2 degradation. Thus, Keap1 sequestration by p62/SQSTM1 prevents Nrf2 degradation, which facilitates Nrf2-mediated tumor survival and aggressive angiogenesis [20].

Moreover, p62/SQSTM1 is phosphorylated by mTORC1 at S351 and increases its affinity for Keap1, which eventually enhances Nrf2-associated tumor progression [21]. ULK1 also plays a role in phosphorylation of p62/SQSTM1 in response to proteotoxic stress including defective proteasome or protein aggregate insult. In this condition, the phosphorylation directs p62/SQSTM1 to be ubiquitinated, thereby leading to efficient degradation of p62/SQSTM1 [22]. Although mTOR and autophagy protein ULK1 inversely regulate autophagy activity, p62/SQSTM1 is a substrate of both protein kinases and p62/SQSTM1 can be phosphorylated at distinct sites and regulate autophagy activity distinctly. In addition, p62/SQSTM1 also acts as an important role for activating mTORC1 through interaction with TNF receptor-associated factor 6 (TRAF6), showing that TRAF6-p62 complex recruits mTORC1 to the lysosomal membrane to be activated under the amino acids-abundant conditions [23]. Furthermore, significant activation of Nrf2 through p62 accumulation was observed in multiple cancer types including hepatocellular carcinoma cells (HCCs) [24]. Taken together, the levels of p62/SQSM1, an autophagic cargo receptor can be regulated by multiple mechanisms, thereby influencing tumor progression.

Interestingly, the function of p62/SQSTM1 in tumor microenvironment is also critical for tumor progression. Tumor-associated stromal cells contain reduced p62/SQSTM1 levels compared to cancer cells, which eventually enhances malignant tumorigenesis of epithelial prostate tumor. Low levels of p62 in stromal cells inactivate mTOR and c-Myc pathway resulting in downregulation of glucose and glutamine metabolism. Associated metabolic defects in tumor microenvironment ultimately fail to maintain redox balance and increase interleukin-6 (IL-6) secretion, thereby leading to promote adjacent tumor progression [25]. Accordingly, the levels of p62/SQSTM1 enable tumor-associated stromal cells to work coordinately with adjacent tumor cells, which ultimately alters stromal metabolism and influences on tumor development

3. Tumor-promoting functions of autophagy

The importance of autophagy during tumor development can be elaborated as a feature of its survival mechanism. Autophagy supports cell survival and growth by supplying degraded and recycled nutrients, in response to various metabolic stresses, often facing rapidly proliferating or hypovascularizing well-developed tumors. Cancer cells can utilize autophagy to provide alternative bioenergetics and effective precursors for macromolecule biosynthesis, which is required for fulfilling metabolic alteration in malignant tumor.

As a direct example, when hematopoietic cells dependent of IL-3 are exposed to IL-3-deprived conditions, glucose utilization is decreased, instead autophagy process is upregulated, which provides energy and nutrients to prolong cell survival [26]. The tumor-promoting role of autophagy has been largely investigated in multiple oncogene-driven cancers in vivo and in vitro system, including oncogenic Ras expression. Genetic deletion of ATG genes in both oncogenic HRas-transformed MEFs and human breast carcinoma cells, harboring oncogenic KRas, leads to reduced tumorigenic transformation and proliferation as well as decreased glycolysis [27]. Similarly, a breast cancer mouse model driven by the polyoma middle T (PyMT) oncogene, when FIP200, an essential protein for autophagy initiation, was deleted, exhibited defective glycolysis in vitro and significantly blocked mammary tumor progression in vivo [28].

Rapidly proliferating tumor cells primarily depend on glycolysis as main glucose metabolism, which is mediated by the activation of oncogenes or the inactivation of tumor suppressors. This metabolic alteration of glycolysis is proposed to provide a major portion of metabolic intermediates for newly activated biosynthetic pathways [29]. Established tumors exhibited Increasing anabolic reactions as the main cellular metabolism, which is supplied with metabolic precursors that are generated by autophagic degradation. Specific oncogenic transformation such as oncogenic Ras promotes autophagic catabolic pathways, although most oncogenic pathways are clearly associated with anabolic processes such as cell growth and proliferation.

Multiple in vivo studies using genetically engineered mouse models (GEMMs) of cancer have provided additional support for cancer-promoting functions of autophagy. Genetic deletion of Atg5 or Atg7 showing early tumorigenesis, however, revealed to reduce advanced tumor development driven from certain oncogene activation.

Using KRas mutant-driven PDAC or lung mouse model, autophagy is an important pathway that exacerbates tumor development. In a pancreatic cancer mouse model harboring a pancreas-specific KRas mutant, when autophagy genes Atg5 or Atg7 were deleted, the progression of PDAC was significantly inhibited [30]. In a lung cancer model driven by oncogenic KRas or BRAF mutant, autophagy deficiency due to the deletion of Atg5 or Atg7 significantly decreased the tumor burden. These autophagy-deficient mice still harbored benign oncocytomas, which are different from adenocarcinoma generally induced by additional oncogenic insult [31, 32]. As a mechanism for generating oncocytomas, the importance of p53 was raised. The loss of p53 in the KRas-induced lung cancer model suppressed fatty acid oxidation and showed lipid-accumulated oncocytomas, which phenotype might be due to defective mitophagy caused from when autophagy genes were deleted [33].

Interestingly, the suppression of autophagy in oncogenic KRas-driven PDAC mouse models revealed conflicting results depending on the p53 status. Tumor-promoting effect of autophagy mostly was observed in p53-intact condition. In the background of p53 deletion, autophagy inhibition is not sufficient to block tumor progression in oncogenic Ras-mutant mice. Moreover, in the oncogenic KRas-mutant mice with p53 deletion, genetic or pharmacological inhibition of autophagy significantly increased PDAC development. As a survival mechanism, glycolysis especially pentose phosphate pathway(PPP) is activated in tumor cell lines derived from KRas G12D-mutant mice with both deletion of p53 and Atg7, which contribute to tumor progression in PDAC [30]. Since the pentose phosphate pathway can generate NADPH as reductive molecule to scavenge reactive oxygen species (ROS) and produce the metabolic intermediates supporting for biosynthesis efficiently, glycolysis and PPP activated in Ras-driven, p53-deficient tumors might play a role in supplying the metabolic precursors, which are reduced due to the lack of autophagy. Therefore, the p53 can determine cellular metabolic status in coordinating with autophagy and directs to undergo tumor progression.

Moreover, the loss of Atg5 with oncogenic KRas-driven p53-deficient lung tumors markedly increases tumor progression, due to the recruitment of regulatory T cells (T reg) on the tumors. These accumulated Treg cells in tumor lesion might prevent immune surveillance system against tumors and further promote lung cancer progression [34]. The distinct role of p53 in particularly autophagy-defective conditions might be associated with various aspects of tumor-favorable mechanisms including metabolic rewiring including increasing glycolysis, regulating redox balance in addition to controlling immune cell populations adjacent to tumor.

According to a recent report, dormant populations of tumor cells can be survived even after oncogene ablation which are derived from inducible KRas mutant in a heterozygous p53 mouse model. These surviving tumor cell exhibited substantial dependency of oxidative phosphorylation (OXPHOS) for generating energy and utilized autophagy for the survival of these cell populations. This result suggests that metabolic rewiring including autophagic catabolism widely occurs even in heterozygous p53 mouse model. Autophagy and its related mitochondria function are particularly crucial for the survival of tumor cells harboring features of cancer stem cells or tumor relapse [35].

Accordingly, autophagy defect in oncogenic Ras-driven tumor confers accumulated cellular stress including metabolic and redox imbalance leading to cell death, when a tumor suppressor, p53, might have limitation of massive metabolic reprogramming. Thus, loss of p53 in oncogenic Ras driven cells enables autophagy-defective cancer to avoid cell death through substantial metabolic rewiring to support cell proliferation.

Similar to the function of autophagy in normal cells, autophagy basically plays a role in the effective clearance of unnecessary intracellular products, thereby maintaining cell viability in malignant-transformed cancer cells. However, since cancer cells are frequently exposed to metabolic stress condition as well as high anabolic demand for proliferation, the requirement of autophagy might be more crucial for satisfying metabolic demand of malignant cancer cells. Additionally, autophagy can be activated by multiple anticancer therapies to sustain cancer survival against the treatment, implying the ability of autophagy for drug resistance.

4. Autophagy in glucose metabolism

Autophagy regulates aerobic glycolysis, which supports rapid growth and proliferation in cancer cells. In HRas- or KRas-mutant cells, the deletion of Atg5 or Atg7 leads to reduced glycolysis significantly and then suppresses anchorage-independent colony formation, indicating inhibitory effect on tumor progression [27]. However, additional deletion of p53 in tumor driven from oncogenic KRas mutant enables autophagy-defective mice to increase the levels of glycolysis and markedly facilitates pentose phosphate pathway, thereby promoting PDAC progression [30].

Accordingly, the molecular regulatory mechanism between autophagy and glucose metabolism during cancer development should be studied. Particularly, the function of p53 as a metabolic determinant in autophagy-defective conditions should be investigated more.

Recent study identified specific glycolytic enzymes including hexokinase II (HK II) and phosphofructokinase (PFK) that regulate autophagy [36–38]. Inhibition of hexokinase II (HK II), the enzyme involved in the first step of glycolysis, markedly decreases autophagy and facilitates cell death under the glucose-starvation conditions. Autophagy is induced by HK II upon glucose deprivation through HK II-mediated mTOR inactivation [37]. Moreover, hexokinase II (HK II) is phosphorylated by Akt, leading to increased mitochondrial binding and mitochondrial protection against ROS, where phenotype is abrogated by the addition of glucose-6-phosphate [39].

Another key glycolytic enzyme, 6-phosphofructo-2-kinase/fructose-2,6-bisphosphatases (PFKFBs), is also involved in regulating autophagy, which converts fructose-6-phosphate to and from fructose-2,6-bisphosphate. An isoform of PFKFBs, PFKFB3 acts as a positive regulator of autophagy in T-effector cells [36], but PFKFB3 in human cancers shows inverse phenotype that the inhibition of PFKFB3 significantly increases autophagy activity due to suppression of glucose uptake and utilizes this pathway as cancer-survival mechanism. Therefore, the concomitant inhibition of autophagy and PFKFB3, using chloroquine (CQ) and PFKFB3 inhibitor, 3-(3-pyridinyl)-1-(4-pyridinyl)-2-propen-1-one (3PO), might provide a reasonable combinatorial strategy as an anticancer therapeutics, which can be more effective to the cancers harboring highly expressed PFKFB [38]. PFKFB4 is also suggested as a putative autophagy regulator resulting from the shRNA screening, which activates pentose phosphate pathway. Inhibition of PFKFB4 increases autophagy activity due to lowering NADPH and enhancing cellular ROS levels, which was as a result of defective pentose phosphate pathway, but paradoxically this PFKFB4 knockdown revealed the accumulation of p62 [40].

In addition to the function of glycolytic enzymes on autophagy activity, conversely autophagy regulates specific steps of the glycolytic pathway. HK II is known as an oncogenic kinase to promote metabolic pathways that are important to overcome metabolic stress. HK II overexpression defends cell death after growth factor withdrawal through increasing glucose metabolism [41]. The mutation of a receptor tyrosine kinase FLT3 in nonacute myeloid leukemia cells activates autophagy to overcome metabolic stress. However, when both autophagy and FLT3 are blocked, chaperone-mediated autophagy (CMA) markedly promotes to degrade hexokinase II. Cellular degradation of HK II by CMA primarily inhibits glycolysis and increases metabolic stress in cancers, thereby facilitating cancer cell death [42].

CMA also directly degrades another glycolytic enzyme, pyruvate kinase, M isoform (PKM2) in lysosome-mediated manner, which supports tumor-promoting role during tumor progression [43]. Distinct from HK II, the degradation of PKM2 by CMA may enhance the accumulation of diverse glycolytic intermediates, which could be converted to biosynthetic precursors. Consequently, this process is beneficial to proliferation of cells during tumor progression [44].

Dimeric form of PKM2 harboring low activity is known to be abundant in cancer, which mainly converts from pyruvate to lactate [45].

Recently, acetyl-CoA is reported as an essential metabolite for regulating autophagy activity. The metabolic enzymes involved in the multiple nodes for generating acetyl-CoA have a role in the inhibition of autophagy activity. By contrast, the enzymes participate in reducing the levels of acetyl-CoA, which generally induces autophagy in vitro and in vivo system. Suppression of either glucose, branched chain amino acids (BCAAs) or fatty acid catabolism generates acetyl-CoA, which markedly induces autophagy regardless of intracellular ATP levels. Upregulated autophagy is eventually restored back to normal levels by exogenous treatment with acetyl-CoA [46].

5. Autophagy in glutamine metabolism

Glutamine exists in mammalian plasma with the highest levels among 20 amino acids, which and is utilized for diverse purpose depending on the cellular environment conditions. Glutamine has important functions as a nitrogen source for contributing biosynthesis of nucleotide, other nonessential amino acids (NEAAs), and hexosamine, and is also utilized as a key component of an antioxidant to maintain redox homeostasis.

In various cancers, glycolytic intermediates from the enhanced glycolysis largely support anabolic process, which is essential for rapid cell proliferation. Similarly, the tricarboxylic acid (TCA) cycle can provide metabolic intermediate for supporting biosynthetic pathways in addition to generating energy. Eventually, TCA cycle intermediates themselves can convert to nonessential amino acids and fatty acids, which are used as primary precursors for anabolic processes. As a crucial carbon source, glutamine is converted to glutamate and then turns to α-ketoglutarate (α-KG), which replenishes intermediates for the TCA cycle and preserves mitochondrial function [9, 47, 48].

The levels of glutamine are elevated in multiple cancers, which are indispensable for cancer growth and survival [49]. Particularly, in certain cancer cells, glutamine tends to replace glucose for playing a role in carbon source through glutaminolysis. Distinct from glutaminolysis to replenish TCA cycle, glutamine can also be utilized to generate oxaloacetate by nonconventional metabolic pathways, which ultimately increase NADPH to maintain redox homeostasis and support cancer cell growth in pancreatic ductal carcinoma (PDAC) [50].

In addition to metabolic reprogramming of glutamine, autophagy is markedly upregulated in response to glutamine deprivation although glutamine is one of NEAAs and is not absolutely essential for regulating autophagy and growth in normal cell conditions. However, autophagy is required for tumor transformation and growth in PDACs. As a similar concept, glutamine might be supplied from the autophagic degradation of cellular macromolecules, which can compensate or restore metabolic stress often shown in progressed malignant tumor.

Recently, multiple reports have suggested the mechanism of how glutamine controls autophagy activity, which is that glutamine and leucine act together to regulate mTORC1 activity and thereby regulating autophagy. Import of glutamine tends to be enhanced in cancer cells, which elevated levels of intracellular glutamine contribute to import leucine into cells with a bidirectional alpha-ketoglutarate transporting system. These bidirectional transport mechanisms of two amino acids control mTORC1 activity, thereby inversely regulating autophagy [51]. In addition, glutamine and leucine work coordinately to activate glutaminolysis and α-ketoglutarate (α-KG), production, which increases the activity and lysosomal localization of mTORC1, resulting in the inhibition of autophagy [52]. A key enzyme in the glutaminolysis, glutamate dehydrogenase (GLUD1) involved in converting glutamine to glutamate, shows a critical role in autophagy regulation, which works as a leucine sensor. Intracellular leucine levels and ROS levels activate mTORC1, respectively, to influence autophagy activity [53]. As a consequence of rewired glutamine metabolism, ammonia at physiological concentration is produced from the amino acid catabolism or glutaminolysis and eventually increases autophagy. This ammonia-mediated autophagy occurs independent of mTORC1 and ULK1/2. These results can suggest direct evidence for autophagy induction mediated by metabolite by-products [54, 55].

As aforementioned in tumor-promoting role of autophagy, Braf-driven lung cancer model harboring the Atg7 deletion showed significant reduction of tumor progression [31]. Dysfunctional mitochondria accumulation in Atg7-deficient cell lines generated from these tumors might be due to defective mitophagy. The addition of glutamine to the cells rescued from the mitochondrial functional defects and slow growth of autophagy-defective tumors. By contrast, the treatment with antioxidant reagents, N-acetyl-cysteine (NAC), is not completely restored starvation-mediated cellular growth defect. These results suggested that glutamine is one of metabolic intermediates derived from autophagy, which are critical for regulating metabolic homeostasis and sustaining mitochondrial intactness, rather than the function for redox homeostasis [31].

Moreover, the functional connection between glutamine-dependent metabolism and autophagy during metabolic stress conditions is recently reported. Using WT and *atg5*$^{-/-}$ MEFs, metabolomic profiling, oxygen consumption, and quantitative real-time polymerase chain reaction (RT-PCR) analyses about the metabolic enzymes are altered, upon glutamine deprivation, suggesting that novel regulatory pathways between autophagy and glutamine utilization. Autophagy deficiency shows significantly decreased levels of intracellular glutamine, indicating that glutamine can be supplied from activated autophagy, especially under the nutrient starvation conditions. Interestingly, autophagy-deficient cells increase the uptake of essential amino acids (EAAs) and branched chain amino acids catabolism upon glutamine deprivations, implying that the defect of glutamine generation caused from autophagy deficiency in ATG5 null cells might lead to activation of an alternative mechanism to compensate glutamine limitation.

Furthermore, mRNA levels encoding enzymes used for glutamine-dependent anaplerosis to the TCA cycle and encoding glutamine/EAA transporters are upregulated in *atg5*$^{-/-}$ MEFs, indicating that autophagy can function at transcriptional regulation to compensate glutamine-deprived conditions, although the exact mechanism is still not understood yet [56]. Taken together, glutamine supplied from autophagy plays critical roles in fueling mitochondrial function and regulating gene expression. Glutamine generated from autophagy is also important for cell growth and survival under the specific conditions including nutrient starvation.

Moreover, general amino acid control (GAAC) pathway, which usually maintains intracellular amino acid levels, also regulates autophagy activity in addition to the uptake of amino acids. Upon glutamine deprivation, the uptake of amino acids is enhanced by activated GAAC pathway, which eventually restores mTORC1 activity and suppresses autophagy. This feedback mechanism can explain how GAAC controls the degree of autophagy by the regulation of amino acid uptake and amino acid synthesis [57].

Accordingly, glutamine metabolism and autophagy are not only reciprocally regulated to compensate each other, along with metabolic and transcriptional alteration, but also more complex and diverse molecular networks act to regulate for cell growth and survival.

In glioblastoma (GBM), glutamine metabolic alteration provides drug resistance to the mTOR inhibition because targeting mTOR increases the expression levels of glutaminase (GLS), a key enzyme for glutaminolysis. Inhibition of GLS can be expected to be more effective in addition to the treatment with mTOR inhibitor [58]. Accordingly, it is speculated that autophagy can associate with glutamine metabolism and glutamine derived from autophagy might ameliorate metabolic stress due to glutamine deprivation, especially in mTOR-activated tumors. Therefore, concomitant-targeting glutamine metabolism and autophagy in diverse cancer including mTOR active cells would be considered as a promising anticancer strategy.

6. Autophagy in macromolecule catabolism

In addition to massive metabolic alteration in various cancers, oncogenic Ras-driven tumors rely on macromolecule degradation including autophagy. Multiple reports have suggested that oncogenic Ras-driven tumors stimulate a unique endocytosis process called macropinocytosis, which engulf and break down extracellular macromolecules. Ras-driven tumors utilize this process as an effective nutrient-supply strategy. Moreover, associated with macropinocytosis driven by oncogenic Ras expression, core autophagy machinery is largely required for macropinocytosis process.

Lysosomal degradation of the extracellular cargoes commonly occurs dependent on nutrient availability and growth signaling. However, under the oncogenic Ras-activating conditions, autophagy is significantly activated despite the fact that this traditional degradation pathway is suppressed by increasing anabolic-signaling pathway.

Clearance of specific cargoes through activated lysosomal degradation, tends to maintain intracellular homeostasis, and recycling of nonspecific cargoes more likely ameliorates cellular metabolic stress of rapidly growing cells by supplying nutrients effectively.

As a typical scavenging pathway, autophagy is upregulated to support cancer cell proliferation and survival in oncogenic Ras-driven tumors including PDAC. In addition, oncogenic Ras-driven tumorigenic cells show increased uptake of extracellular materials for utilizing them as metabolic fuels after lysosomal degradation.

Macropinocytosis is a unique type of endocytosis that engulfs random portion of extracellular fluid without the need for specific vesicle-coat proteins. Multiple growth factor signals positively regulate macropinocytosis, which facilitates the plasma membrane ruffles and engulfs extracellular fluid to be internalized into the cell forming as vesicles, called macropinosome. Ultimately, these vesicles can fuse either with the lysosome, thereby degrading its cargo contents, or might follow regular secretory pathway to release its cargo contents out of the cells [59, 60].

Although oncogenic Ras is known to induce macropinocytosis, the exact function of macropinocytosis on cancer development is largely unknown. Recently, oncogenic Ras-mediated macropinocytosis has revealed to contribute to cancer cell growth and survival through the supply of the essential nutrients to overcome metabolic stress conditions of the cancer cells [59, 60].

In a recent report, macropinocytosis promotes the uptake of extracellular albumin as a cargo molecule, which can be degraded by the lysosome. Thus, overall nutrients including amino acids were generated from this macromolecule degradation. A couple of specific amino acids including glutamine are essential nutrients for supporting high metabolic demand of cancers. ^{13}C-labeled whole protein treated in the media is utilized as a macropinocytosis cargo molecule and degraded to generate amino acids through the detection of ^{13}C-labeled amino acid form. Amino acids labeled with ^{13}C can be thought as degraded products from extracellular macromolecule in oncogenic KRas mutant, indicating that these amino acids were derived from the macropinocytosis-mediated degradation and mostly utilized for replenishing the TCA cycle [61].

Recent reports have demonstrated a novel regulatory mechanism for macropinocytosis, which is correlated with a representative anabolic signaling molecule, mTORC1 activity. mTORC1 acts as a key regulator to determine metabolic pathways depending on nutrient status. In nutrient-rich condition, active mTORC1 suppresses lysosomal catabolism including the degradation of extracellular proteins, whereas mTORC1 inhibition increases lysosomal degradation of proteins to supply nutrients and support cell growth under nutrient-deprivation conditions. Therefore, mTORC1 activity depending on environmental-nutrient conditions determines metabolic status in tumor either to addict to the anabolism or to rely on the degradation of extracellular macromolecules.

mTORC1 also shows its activity by the intracellular localization of this protein. mTORC1 is redistributed from cytoplasmic localization to the lysosomal membrane by nutrient abundance. Lysosomal localization of active mTORC1 is also exhibited by adding exogenous albumin to the nutrient-starved condition, similar to adding amino acids. This albumin-mediated mTORC1 re-localization is not restored by the blockade of macropinocytosis and lysosomal degradation [62, 63].

However, oncogenic KRas-expressing MEFs deleting Atg5 gene or PDAC cell line harboring Atg7 shRNA show the accumulation of extracellular proteins internalized by macropinocytosis, compared to complete degradation of the protein shown in WT control. Moreover, the lysosomal degradation of extracellular proteins leads to restore the decreased mTORC1 activity, which phenotype is not observed in $atg5^{-/-}$ MEFs. These results suggest that the generation nutrients from environmental extracellular proteins are mostly directed by the major autophagy machineries [63]. When the effect of mTORC1 inhibition was examined using a mouse model with pancreas-specific Kras mutations or xenograft experiment with KRas-mutant PDAC cell lines, PDAC-bearing mice showed rapid tumor growth along with the treatment with mTORC1 inhibitor rapamycin, compared to the nontreated group of mice. As results of histology analysis after rapamycin treatment, well-vascularized outer regions of the tumor revealed low number of Ki-67-positive, proliferating cells, whereas tumor cells in interior and hypovascularized regions are markedly increasing the number of proliferating cells, suggesting that poorly vascularized tumor microenvironment of PDAC is easily exposed to the nutrients and oxygen deprivation. This tumor microenvironment tends to alter tumor metabolism, which requires lysosomal degradation of extracellular macromolecule to generate an alternative nutrient source to support tumor growth.

In addition to the effect of mTORC1 on tumor progression in KRas- and p53-mutant mouse model (KPC), concomitant inhibition of mTORC1 and upregulated macropinocytosis-driven autophagy leads to inhibit tumor growth significantly compared to single inhibition of either mTORC1 or macropinocytosis/autophagy in mouse xenograft experiment using KRas-mutant PDAC cell lines. These results implied that macropinocytosis associated with autophagy can fulfill cellular metabolic requirements for promoting cell growth under the mTOR-compromised conditions. In other words, anabolic perturbation by mTORC1 inhibition results in more active access of nutrients from the degradation of extracellular macromolecules, which ultimately promotes cell proliferation and survival under nutrient-deprived conditions. Thus, mTORC1 plays an opposite regulating role in tumor growth depending on environmental nutrient availability, implying that the utilized catabolic pathways could be distinguished depending on the nutrient status of the tumor microenvironment.

Accordingly, it raises the possibility of potential novel anticancer strategies interrupting these metabolic balances during tumor progression can open promising avenues.

7. Concluding remarks

Cancer cells undergo metabolic change to support cell proliferation and survival during tumor development.

As a representative catabolic process, autophagy can be suggested a key regulator. Most cancers in advanced stage show "autophagy-addiction" phenotype and need autophagy as a type of metabolic reprogramming during cancer development. Despite the controversial role of autophagy in cancer development, metabolically dynamic cancer cells utilize autophagy to supply the bioenergetic fuels and biosynthetic precursors that support cancer cell growth and survival.

As a new functional mechanism, autophagy also contributes to the metabolism of tumor microenvironment including stroma and immune cells adjacent tumor, which integrated with cancer metabolic alterations. These functional interactions and metabolic re-modulation within heterogeneous tumor microenvironment allow to overcome metabolic stress often facing to the cancer and to sustain in the harsh tumor microenvironment.

Furthermore, autophagy is induced by oncogenic stress such as Ras activation, which is implicated in oncogene-mediated transformation and proliferation. Most cancers driven by oncogenic Ras require autophagy to recover from the metabolic stress. Understanding the molecular mechanism on how autophagy is integrated to major metabolic change including glucose or glutamine metabolic rewiring in cancer and how these pathways are mutually regulated to each other to support cancer development are important for the development of cancer therapeutics with novel strategy. Accordingly, accumulated knowledge of molecular interactions among growth-signaling pathways and the metabolic alteration including anabolic-addicted phenotypes and autophagy dependency during cancer development shed a light to identify the effective target combination for anticancer therapeutics.

Author details

Heesun Cheong

Address all correspondence to: heesunch@ncc.re.kr

Comparative Biomedicine Research Branch, Division of Cancer Biology, National Cancer Center, Ilsandong-gu, Goyang, Gyeonggi-do, Republic of Korea

References

[1] Nakatogawa, H., et al., Dynamics and diversity in autophagy mechanisms: lessons from yeast. Nat Rev Mol Cell Biol, 2009. 10(7): pp. 458–67.

[2] Feng, Y., et al., The machinery of macroautophagy. Cell Res, 2014. 24(1): pp. 24–41.

[3] Klionsky, D.J. and S.D. Emr, Autophagy as a regulated pathway of cellular degradation. Science, 2000. 290(5497): pp. 1717–21.

[4] Jung, C.H., et al., ULK-Atg13-FIP200 complexes mediate mTOR signaling to the autophagy machinery. Mol Biol Cell, 2009. 20(7): pp. 1992–2003.

[5] Ganley, I.G., et al., ULK1.ATG13.FIP200 complex mediates mTOR signaling and is essential for autophagy. J Biol Chem, 2009. 284(18): pp. 12297–305.

[6] Kim, J., et al., AMPK and mTOR regulate autophagy through direct phosphorylation of Ulk1. Nat Cell Biol, 2011. 13(2): pp. 132–41.

[7] Mihaylova, M.M. and R.J. Shaw, The AMPK signalling pathway coordinates cell growth, autophagy and metabolism. Nat Cell Biol, 2011. 13(9): pp. 1016–23.

[8] Kim, J., et al., Differential regulation of distinct Vps34 complexes by AMPK in nutrient stress and autophagy. Cell, 2013. 152(1–2): pp. 290–303.

[9] DeBerardinis, R.J., et al., The biology of cancer: metabolic reprogramming fuels cell growth and proliferation. Cell Metab, 2008. 7(1): pp. 11–20.

[10] Lozy, F. and V. Karantza, Autophagy and cancer cell metabolism. Semin Cell Dev Biol, 2012. 23(4): pp. 395–401.

[11] Liang, X.H., et al., Induction of autophagy and inhibition of tumorigenesis by beclin 1. Nature, 1999. 402(6762): pp. 672–6.

[12] Yue, Z., et al., Beclin 1, an autophagy gene essential for early embryonic development, is a haploinsufficient tumor suppressor. Proc Natl Acad Sci U S A, 2003. 100(25): pp. 15077–82.

[13] Kim, M.S., et al., Frameshift mutation of UVRAG, an autophagy-related gene, in gastric carcinomas with microsatellite instability. Hum Pathol, 2008. 39(7): pp. 1059–63.

[14] Takahashi, Y., et al., Bif-1 interacts with Beclin 1 through UVRAG and regulates autophagy and tumorigenesis. Nat Cell Biol, 2007. 9(10): pp. 1142–51.

[15] Cianfanelli, V., et al., AMBRA1 links autophagy to cell proliferation and tumorigenesis by promoting c-Myc dephosphorylation and degradation. Nat Cell Biol, 2015. 17(1): pp. 20–30.

[16] Takamura, A., et al., Autophagy-deficient mice develop multiple liver tumors. Genes Dev, 2011. 25(8): pp. 795–800.

[17] Duran, A., et al., The signaling adaptor p62 is an important NF-kappaB mediator in tumorigenesis. Cancer Cell, 2008. 13(4): pp. 343–54.

[18] Kim, T.H., et al., NRF2 blockade suppresses colon tumor angiogenesis by inhibiting hypoxia-induced activation of HIF-1alpha. Cancer Res, 2011. 71(6): pp. 2260–75.

[19] Mathew, R., et al., Autophagy suppresses tumorigenesis through elimination of p62. Cell, 2009. 137(6): pp. 1062–75.

[20] Komatsu, M., et al., The selective autophagy substrate p62 activates the stress responsive transcription factor Nrf2 through inactivation of Keap1. Nat Cell Biol, 2010. 12(3): pp. 213–23.

[21] Ichimura, Y., et al., Phosphorylation of p62 activates the Keap1-Nrf2 pathway during selective autophagy. Mol Cell, 2013. 51(5): pp. 618–31.

[22] Lim, J., et al., Proteotoxic stress induces phosphorylation of p62/SQSTM1 by ULK1 to regulate selective autophagic clearance of protein aggregates. PLoS Genet, 2015. 11(2): p. e1004987.

[23] Linares, J.F., et al., K63 polyubiquitination and activation of mTOR by the p62-TRAF6 complex in nutrient-activated cells. Mol Cell, 2013. 51(3): pp. 283–96.

[24] Sun, X., et al., Activation of the p62-Keap1-NRF2 pathway protects against ferroptosis in hepatocellular carcinoma cells. Hepatology, 2016. 63(1): pp. 173–84.

[25] Valencia, T., et al., Metabolic reprogramming of stromal fibroblasts through p62-mTORC1 signaling promotes inflammation and tumorigenesis. Cancer Cell, 2014. 26(1): pp. 121–35.

[26] Lum, J.J., et al., Growth factor regulation of autophagy and cell survival in the absence of apoptosis. Cell, 2005. 120(2): pp. 237–48.

[27] Lock, R., et al., Autophagy facilitates glycolysis during Ras-mediated oncogenic transformation. Mol Biol Cell, 2011. 22(2): pp. 165–78.

[28] Wei, H., et al., Suppression of autophagy by FIP200 deletion inhibits mammary tumorigenesis. Genes Dev, 2011. 25(14): pp. 1510–27.

[29] Ward, P.S. and C.B. Thompson, Metabolic reprogramming: a cancer hallmark even warburg did not anticipate. Cancer Cell, 2012. 21(3): pp. 297–308.

[30] Rosenfeldt, M.T., et al., p53 status determines the role of autophagy in pancreatic tumour development. Nature, 2013.504(7479):pp.296–300.

[31] Strohecker, A.M., et al., Autophagy sustains mitochondrial glutamine metabolism and growth of BrafV600E–driven lung tumors. Cancer Discov, 2013. 3(11): pp. 1272–85.

[32] Karsli-Uzunbas, G., et al., Autophagy is required for glucose homeostasis and lung tumor maintenance. Cancer Discov, 2014.4(8):pp.914–927.